# BLEACHING AGENTS AND TECHNIQUES

# BLEACHING AGENTS AND TECHNIQUES

Jules A. Szilard

NOYES DATA CORPORATION
Park Ridge, New Jersey London, England
1973

Library of Congress Catalog Card Number: 72-82918
ISBN: 0-8155-0490-X
Printed in the United States

Published in the United States of America by
Noyes Data Corporation
Noyes Building, Park Ridge, New Jersey 07656

# FOREWORD

The detailed, descriptive information in this book is based on U.S. patents since 1962 relating to bleaching agents and the techniques of using them.

This book serves a double purpose in that it supplies detailed chemical information and can be used as a guide to the U.S. patent literature in this field. By indicating all the information that is significant and eliminating legal jargon and juristic phraseology, this book presents an advanced, commercially oriented review of bleaching agents and their applications.

The U.S. patent literature is the largest and most comprehensive collection of technical information in the world. There is more practical, commercial, timely process information assembled here than is available from any other source. The technical information obtained from a patent is extremely reliable and comprehensive; sufficient information must be included to avoid rejection for "insufficient disclosure."

The patent literature covers a substantial amount of information not available in the journal literature. The patent literature is a prime source of basic commercially useful information. This information is overlooked by those who rely primarily on the periodical journal literature. It is realized that there is a lag between a patent application on a new process development and the granting of a patent, but it is felt that this may roughly parallel or even anticipate the lag in putting that development into commercial practice.

Many of these patents are being utilized commercially. Whether used or not, they offer opportunities for technological transfer. Also, a major purpose of this book is to describe the number of technical possibilities available, which may open up profitable areas of research and development. One should have to go no further than this condensed information to establish a sound background before launching into research in this field.

Advanced composition and production methods developed by Noyes Data are employed to bring these durably bound books to you in a minimum of time. Specialized techniques are used to close the gap between "manuscript" and "completed book". Industrial technology is progressing so rapidly that time-honored, conventional typesetting, printing, binding and shipping methods can render a technical or scientific book quite obsolete before the potential user gets to see it.

The Table of Contents is organized in such a way as to serve as a subject index. Other indexes by company, inventor and patent number help in providing easy access to the information contained in this book.

**15 Reasons Why the U.S. Patent Office Literature Is Important to You –**

(1) The U.S. patent literature is the largest and most comprehensive collection of technical information in the world. There is more practical commercial process information assembled here than is available from any other source.

(2) The technical information obtained from the patent literature is extremely comprehensive; sufficient information must be included to avoid rejection for "insufficient disclosure."

(3) The patent literature is a prime source of basic commercially utilizable information. This information is overlooked by those who rely primarily on the periodical journal literature.

(4) An important feature of the patent literature is that it can serve to avoid duplication of research and development.

(5) Patents, unlike periodical literature, are bound by definition to contain new information, data and ideas.

(6) It can serve as a source of new ideas in a different but related field, and may be outside the patent protection offered the original invention.

(7) Since claims are narrowly defined, much valuable information is included that may be outside the legal protection afforded by the claims.

(8) Patents discuss the difficulties associated with previous research, development or production techniques, and offer a specific method of overcoming problems. This gives clues to current process information that has not been published in periodicals or books.

(9) Can aid in process design by providing a selection of alternate techniques. A powerful research and engineering tool.

(10) Obtain licenses — many U.S. chemical patents have not been developed commercially.

(11) Patents provide an excellent starting point for the next investigator.

(12) Frequently, innovations derived from research are first disclosed in the patent literature, prior to coverage in the periodical literature.

(13) Patents offer a most valuable method of keeping abreast of latest technologies, serving an individual's own "current awareness" program.

(14) Copies of U.S. patents are easily obtained from the U.S. Patent Office at 50¢ a copy.

(15) It is a creative source of ideas for those with imagination.

# CONTENTS AND SUBJECT INDEX

# INTRODUCTION

This review on bleaching agents and methods covers processes from two hundred and three U.S. Patents, issued between 1962 and 1973. The processes describe both the manufacture and the use of the bleaching agents.

The purpose of bleaching is to make materials white or whiter. In the context of this review, the usage includes the bleaching of textile materials of all types, natural and synthetic, including specially treated textiles, such as wash and wear fabrics. The use also includes the bleaching of wood pulp, the removing of stains, the stripping of dyes, and the scouring of floor and wall coverings. Finally, it includes the use bleach in both household and commercial laundry processing. This last application is, next to textile mill processing, the largest utilization of bleaching agents.

The material is divided into ten chapters, based on the chemical nature of the bleaching agents. The large chapters are subdivided, either according to the indicated primary use, or according to the activators, stabilizers, detergents, enzymes or other additives used to support the bleaching process.

The four largest chapters are on hypochlorites, peroxides, persalts and the chlorine releasing organic compounds, the so-called chloramines. The largest single group is on persalts, with fifty-eight processes. In addition to the use of persalts in textile mill processing, there is an increasing use of the perborates in laundry products. Another area of extensive development is the use of activators, for both peroxides and persalts, in order to increase their bleaching efficiency at lower washing temperatures. There is also an increasing use of the chloramines for their combined bleaching and disinfecting effect.

Almost all the bleaching processes are designed for use in aqueous medium. There are a few exceptions, such as the use of peroxygens in dry cleaning fluids or the use of organic solvents to remove dyestuff stains. In a few of the processes environmental problems are considered. There are also some processes using fluidized bed techniques to coat and encapsulate bleaching agents or to prepare concentrated hypochlorite solutions.

Many of the bleaching agents are recommended for multiple use, such as for textile mills, or for commercial or household laundry. The title of the processes listed in the Contents and Subject Index gives a clue to their primary use.

# CHLORINE AND CHLORINE DIOXIDE

## BLEACHING CELLULOSIC TEXTILES WITH AQUEOUS CHLORINE DIOXIDE

H.L. Robson and J.F. Synan; U.S. Patent 3,291,559; December 13, 1966; assigned to Olin Mathieson Chemical Corporation describe a process to bleach cellulosic textiles with an aqueous chlorine dioxide solution in a closed container with continuous replenishment of chlorine dioxide. The process is carried out in conventional closed dyeing or bleaching machines, with liquor to cloth ratio between 5:1 and 10:1. The concentration of the chlorine dioxide solution used is 1%.

Burlington machines with perforated beams are used in the process. The beam wound with cloth is loaded in and the machine is filled with water or aqueous solution, which may contain wetting agents, buffers and/or other adjuncts and the machine is closed. The treating solution is pumped into the beam and through perforations therein into the cloth and collected in a closed tank. During passage through the cloth, the bleaching solution is depleted according to the amount of bleaching done by the solution. For example, pumping a solution carrying 40 ppm chlorine dioxide through 1,000 lbs. of cloth on the beam at 160°F. at a rate of 250 gpm for 120 minutes exposes the cloth to sufficient chlorine dioxide to effect excellent bleaching.

To maintain the chlorine dioxide content of the bleach solution pumped from the enclosing tank to the beam, a mixing T is placed downstream from the outlet of the regular pump which circulates liquor from the tank to the beam inlet. A strong aqueous solution of chlorine dioxide is pumped into the mixing T by a proportioning pump.

For example, 2 gpm of a 2,000 parts per million chlorine dioxide stock solution is added to 100 gpm of solution circulated from the tank through the beam inlet and thus through the cloth. The amount of bleaching liquor circulated through the cloth varies with the design of the machine and the cloth to be bleached. Up to 1,000 gpm is pumped to a beam 36 inches in diameter wound with cloth in an open or folded width to a depth of 10 to 12 inches. Other machines, for example, with a beam 18 inches in diameter wound with cloth to a depth of 2 feet require only 100 to

300 gpm of circulating liquor. Operating at 160°F. it is preferable to pump sufficient of the strong solution, containing 4,000 to 10,000 ppm of chlorine dioxide, to bring the solution to 40 ppm as it enters the beam. During passage through the cloth, the content of chlorine dioxide falls substantially and the concentration of chlorine dioxide in the solution in the closed tank may be as low as 10 ppm after the first hour.

When the content of chlorine dioxide in the solution in the tank reaches 20 ppm the supply of strong chlorine dioxide solution cloth varies with the design of the machine and the cloth is stopped while continuing to circulate the solution from the tank to the beam. If this recirculation results in a rapid drop in the chlorine dioxide content of the tank liquor, addition of strong solution is started again and continued until the tank liquor again reaches 20 ppm of chlorine dioxide.

When the solution in the tank shows only a slow drop in chlorine dioxide content after the supply of fresh solution has been cut off, the bleaching is deemed to be complete. Operating at higher temperatures, such as 190° to 210°F., the same procedure is followed, except that the solution in the tank is preferably allowed to rise only to 5 or most 10 ppm while strong solution is being added. These operations are eminently safe because the rate of circulation is high and the high concentration of chlorine dioxide in the incoming liquor is rapidly reduced until near the end of the bleaching process.

Example 1: A Burlington machine was charged with a beam wound with 1,000 yards of 80 x 120 greige cotton cloth which had previously been desized. The wound beam was placed in position in the Burlington machine which was then closed. The machine was filled by pumping in 750 gal. of water at the rate of 250 gpm. During a period of 3 minutes, while the complete contents were recycled, 7.5 gal. of a 1% solution of chlorine dioxide in water was proportioned into the discharge of the main pump thus forming the bleaching solution with an average content of 100 ppm of chlorine dioxide.

Maintaining the temperature at 160°F., the chlorine dioxide solution was circulated from the Burlington machine tank through the pump to the beam and through the cloth for a period of 120 minutes. During the 10 minutes from 100 to 110 minutes of the operation, a 1% chlorine dioxide solution was metered into the pump in sufficient quantities to maintain the concentration of chlorine dioxide in the tank liquor at 20 ppm.

At the end of 110 minutes of operation, the supply of concentrated chlorine dioxide solution was cut off and the chlorine dioxide content of the liquor in the tank dropped to only 18 ppm at 120 minutes, when the bleaching appeared to be complete. The liquor was pumped to the sewer and replaced by fresh water which was circulated through the cloth for 5 minutes and then discarded. The machine was opened and the beam and textile were removed. No odor of chlorine dioxide was noticed and the textile was homogeneously bleached to an excellent white.

Example 2: A percolator type closed kier was loaded with 3 tons of 80 x 100 thread count broadcloth (running 3.65 yards per pound) folded into the kier in 48 x 60 inch folds. The broadcloth filled 80% of the volume of the kier. The kier was closed and filled with an aqueous solution containing 0.5 gram per liter of monosodium-disodium phosphate buffer to maintain the pH at 6.5 and 0.3% of Polytergent B-350, a polyoxyethylated nonylphenol wetting agent. The solution was drawn from the top of the kier and returned to the bottom of the kier where it passed over closed steam coils to maintain the temperature at 160°F. The treating solution was then forced to flow upwardly through the folded layers of the broadcloth. A 1% aqueous chlorine dioxide solution was metered into the pump discharge to provide 120 ppm of chlorine dioxide in the solution delivered to the kier.

After 75 minutes, the chlorine dioxide content of the liquor effluent from the kier, which had been at 5 to 10 ppm, began to rise sharply and the introduction of the 1% solution was discontinued. After an additional 30 minutes of circulation, the chlorine dioxide content of the effluent had leveled off at 15 ppm of chlorine dioxide. The liquor was pumped out of the kier and replaced by circulating wash water. After 30 minutes of washing, the broadcloth was removed from the kier and dried. An excellent bleach was obtained.

## CHLORINE DIOXIDE STABILIZED BY SODIUM CHLORIDE

G. Gordon; U.S. Patent 3,585,147; June 15, 1971; assigned to International Dioxcide, Incorporated describes a process to add or form in situ alkali metal chlorides, such as sodium chloride in chlorine dioxide solutions, stabilized or converted to chlorite by alkaline peroxygen compounds. Solutions of the stabilized or complexed chlorine dioxide may be prepared by dissolving the alkali metal peroxygen compound or compounds in water to form an aqueous solution containing 5 to 15%, and preferably 8 to 13%, of the peroxygen compound, and then bubbling chlorine dioxide gas through the aqueous solution, preferably until at least 40,000 ppm, and preferably between 50,000 and 75,000 ppm, of chlorine dioxide is absorbed therein. For good stability, the solution should be maintained at a pH value between 7 and 13 or higher, and preferably between 8 and 12.

In general, the chlorine dioxide gas which is to be used in the stabilized or complex solution should preferably be purified to remove all traces of free chlorine. The alkali metal peroxygen stabilizing or complexing compounds are sodium carbonate peroxide or hydrogen peroxide with alkaline carbonates and bicarbonates, but alkali metal perborates may also be used if the toxicity imparted by the perborates is not objectionable. Moreover, the process is similarly applicable to aqueous alkali metal chlorite solutions prepared by methods other than the stabilization of chlorine dioxide gas in water by a peroxygen compound, e.g., by direct solution of the appropriate dry chlorite salt in water or in dilute caustic.

The chlorides which are included in the aforesaid solutions are chlorides of alkali metal or alkaline earth metals. Sodium chloride and potassium chloride are preferred, but lithium chloride, calcium chloride, barium chloride, magnesium chloride and the like are also useful. For maximum effect such chlorides are best added to the stabilized chlorine dioxide solutions while the latter are at a pH of at least 5, or preferably 6 or higher, i.e., either before acidification or substantially concurrently therewith. Alternatively, the chloride can first be added to the water which is to be used in preparing the stabilized chlorine dioxide or chlorite solution. As still another alternative, the chloride may be formed in situ by adding hydrochloric acid to a solution which contains an alkali metal hydroxide or carbonate or bicarbonate. Such chloride formation preferably is at least partially carried out while the pH of the stablized chlorine dioxide or chlorite solution is at a pH of 5 or more, e.g., between 8 and 12, but additional chloride may be formed when the solution is acidified for the purpose of releasing the chlorine dioxide oxidant.

Example: A stabilized chlorine dioxide solution is prepared by dissolving powdered sodium carbonate peroxide in water to form a 10% solution. The stabilizing compound, sodium carbonate peroxide, is an addition compound of sodium carbonate and hydrogen peroxide, corresponding to the formula $2Na_2CO_3 \cdot 3H_2O_2$. This compound as available commercially is a white powder containing 14% active oxygen and 29% hydrogen peroxide. Its solubility in water at 20°C. is 13.3%. In powdered form, the compound is relatively stable.

When the stabilized chlorine dioxide solution is intended to serve as a source of dry sodium chlorite upon evaporation, or when accidental evaporation to dryness in use is to be allowed for and the attendant explosive hazard is to be minimized, it is desirable to include between 0.1 and 20, preferably between 0.3 and 2, and most preferably between 0.5 and 1, mols sodium chloride or similar soluble hygroscopic chloride per mol of chlorine dioxide or chlorite present in the solution and to avoid exposing the dry residue to temperatures above 90°C. Solutions of sodium carbonate peroxide in water have characteristics similar to a solution prepared by separately dissolving hydrogen peroxide and sodium carbonate in water. The former solution, however, is more stable.

Chlorine dioxide gas which contains substantially no free chlorine is then bubbled through the solution of sodium carbonate peroxide. Approximately 568 mg. of gaseous chlorine dioxide is taken up per gram dry weight of sodium carbonate peroxide. A stabilized chlorine dioxide solution prepared in accordance with this method contains 50,000 ppm of chlorine dioxide at a pH value between 8 and 12.

This solution was then diluted 200-fold. In rapid succession sulfuric acid was added to obtain an acidity of 1.0 mol of hydrogen ions per liter of diluted solution and 6 grams of sodium chloride was then added per liter of diluted solution. It was then found that 98% of the oxidizing power of this solution appeared as chlorine dioxide and only 2% as sodium chlorate. On the other hand, for the same solution (200 times diluted), upon addition of the same amount of sulfuric acid as before but in the absence of any added sodium chloride, only 60% of the oxidizing power appeared as chlorine dioxide and 40% as sodium chlorate.

# SODIUM AND OTHER CHLORITES

## CHLORITES WITH ACTIVATORS AND STABILIZERS

### Hexamethylene Tetramine Activated Chlorite Bleach

A. Lehn; U.S. Patent 3,050,359; August 21, 1962; assigned to Deutsche Gold- und Silber-Scheideanstalt vormals Roessler, Germany found that chlorite bleach can be activated with hexamethylene tetramine. Chlorite bleaching solutions to which hexamethylene tetramine has been added do not develop chlorine dioxide when cold and they only cause bleaching at elevated temperatures, especially at temperatures of 50°C. and over, preferably between 50° and 100°C., in the presence of the goods to be bleached, especially cellulose containing goods. Preferably, the quantity of hexamethylene tetramine employed is 0.1 to 0.5 mol per mol of chlorite. The bleaching solutions contain preferably 0.5 to 50 g. of the chlorite per liter solution. The commonly used chlorite is the sodium chlorite. In some instances the chlorites of potassium, magnesium or calcium are used.

Example 1: A desized cotton nettle cloth was impregnated with an aqueous bleaching solution containing 12 g. $NaClO_2$, 1.2 g. hexamethylene tetramine and 1 g. of a wetting agent polyglycol ether of a fatty alcohol per liter and then held for three hours in a closed vessel at 78°C. The pH of the bleaching solution originally was 9.2 and was 4.6 after completion of the bleach. The grade of whiteness of the cloth increased from 60.3 to 85.6%. Upon completion of the bleach the cloth only contained 0.1 part by weight of unused chlorite per 100 parts by weight of cloth.

Example 2: A desized coarse cotton nettle cloth containing seed shells was impregnated with an aqueous solution containing 16 g. $NaClO_2$, 1 g. hexamethylene tetramine and 1 g. of a wetting agent polyglycol ether of a fatty alcohol per liter and then held for two hours at 93°C. in a closed vessel. The pH of the bleaching solution originally was 9.2 and was 4.6 after completion of the bleach. The grade of whiteness of the cloth increased from 55.5 to 84.6%. Upon completion of the bleach the cloth only contained 0.2 part by weight of unused chlorite per 100 parts by weight of cloth.

## Chlorite Bleach Stabilized with Hydrazide and Formaldehyde

G.M. Wagner; U.S. Patent 3,063,783; November 13, 1962; assigned to Olin Mathieson Chemical Corporation found that hydrazides of organic acids, in combination with formaldehyde, stabilize slightly acidic bleach solutions for cotton textiles as well as cotton-polyester or cotton-rayon fabrics. Suitable hydrazides have the formulas: $(CONHNH_2)_2$, $R(CONHNH_2)_x$, $R'(CONHNH_2)_x$ and $R''(CONHNH_2)_x$ wherein R is selected from the group consisting of hydrogen, saturated hydrocarbon chains and hydroxyl-substituted hydrocarbon chains, R' is a saturated hydrocarbon chain interrupted by an oxygen atom and R" is a saturated hydrocarbon chain interrupted by a sulfur atom. Each of the R groups should have less than about 9 carbon atoms. In the formulas, x is a whole number less than 4.

Examples of some suitable hydrazides include those of formic, acetic, glycolic, adipic, oxalic, diglycolic, tartaric and citric acids. The use of hydrazides of acids with a higher molecular weight than about that of suberic acid is hampered by water insolubility. However, any organic acid hydrazide which is soluble in water to the extent required can be used. According to this process textiles are bleached with a solution having the following composition:

| | Parts by Weight |
|---|---|
| Alkali or alkaline earth metal chlorite | 0.002 to 2.0 |
| Formaldehyde or paraformaldehyde | 0.002 to 0.05 |
| Hydrazide | 2 to 10 times $CH_2O$ |
| Water | 100 |

A weak acid such as acetic or cyanuric acid is added to the solution to make it slightly acid, but not below a pH of about 5. Cyanuric acid is especially good because of the excellent buffering action of the cyanurates which are formed in solution. They can hold the pH in the preferred range of 5.5 to 7.0 and have a beneficial effect on the degree of bleaching. Other substances can also be used as buffers as long as they are nonreactive with any of the bleach bath components. The use of more hydrazide than about 10 times the formaldehyde weight is uneconomical and has no further effect. A ratio of less than 2 parts of hydrazide to formaldehyde can, under certain conditions, cause precipitation of polymethylol hydrazide polymers.

Using such a bath the textile can be bleached in the bath in about one or two hours at 80° to 100°C. Alternatively, the textile can be passed through the bath at room temperature and the excess liquid expressed from the cloth by passing it between rollers at suitable pressure. The cloth is then heated at 80° to 100°C. for one or two hours to effect the bleaching. The heating should be done in a steam box or other suitable apparatus which provides a hot, humid atmosphere substantially saturated with water vapor to prevent the cloth from losing moisture during this bleaching period. A suitable solution content on the cloth to be heated is 80 to 150% by weight based on the weight of the cloth.

Example 1: Several portions of cotton broadcloth were passed through an aqueous solution containing 0.8% by weight of sodium chlorite, 0.5% of diglycolic hydrazide, 0.02% of formaldehyde, enough acetic acid to bring the pH to about 5.5 and 0.23% of sodium nitrate as corrosion inhibitor. The cloths were squeezed to contain a weight of solution equal to their own and then maintained in an atmosphere saturated with water vapor at 90° to 100°C. for one hour. They were then removed, rinsed and dried and had an average brightness of 87.7 units, compared to an original value of 63 units.

Example 2: An aqueous bleach bath was prepared containing the following components in percentages by weight based on the entire bath:

0.0025% formaldehyde
0.0075% diglycolic hydrazide
0.025% cyanuric acid
0.4% sodium chlorite
99.5% water

A cotton poplin test fabric was passed through the bath, squeezed to contain a weight of solution equal to its own weight and then placed in an atmosphere saturated with water vapor at 100°C. for one hour. After this time the cloth was cooled, and rinsed in water. The brightness, or reflectance, of the cloth before treatment was 81. After the bleaching the brightness was 88, which is near the maximum obtainable with cotton poplin.

Chlorite Solutions Containing Salts of Strong Bases and Weak Acids

W. Waibel; U.S. Patent 3,065,040; November 20, 1962; assigned to Farbwerke Hoechst AG, Germany describes a process of bleaching textiles by first padding with an alkaline chlorite solution, to which acetic acid is added to form sodium acetate followed by steaming. Instead of forming the salt in the solution, it may be added directly to the alkaline chlorite solution. The addition of salts of strong bases and weak acids prevents the textile fiber from being damaged. However, the complete elimination of seed husks requires an increased bleaching period since activation of the chlorite is delayed by the buffer action of the salt added. It is of special advantage to add weak acids, rather than weak acid salts, since the buffering salts which inhibit acid corrosion and are subsequently activated by heat are formed between the added weak acid and the alkali of the chlorite solution.

Using this method, the bleaching reaction is not delayed, which is of utmost importance since padding and steaming are usually performed as continuous processes and bleaching periods as short as possible give the highest yields. A similar effect is obtained by adding to the impregnation baths mixtures of salts of strong bases and weak acids together with weak acids different from those acids from which the salts are derived. When using such mixtures, bleaching is delayed only shortly due to the higher quantity of buffer substances present. It has proved advantageous to fix the pH of the impregnation baths of the process at a value lower than the pH (about 8 to 9) of alkaline chlorite baths consisting of sodium chlorite and salts of strong acids and weak bases. It is most favorable to operate at a pH between about 6 and 7.5, preferably between 6.5 and 7.5. The amount of weak acid or of weak acid salts depends on the pH desired.

Suitable salts of weak acids and strong bases are, for example: sodium formate or the corresponding acetate, propionate, citrate, lactate, tartrate, and benzoate. Naturally, the corresponding potassium salts can be used instead of sodium salts, as can the salts of the alkaline earth metals if sufficiently soluble. As weak acid, the organic acids are especially suitable, such as formic acid, acetic acid, propionic acid, lactic acid, tartaric acid, citric acid, benzoic acid and the like. The first dissociation constants of these acids all lie between about $1 \times 10^{-3}$ and about

$1.4 \times 10^{-5}$, which are the first dissociation constants of tartaric and propionic acids, respectively.

Example 1: Raw desized cotton fabric was impregnated with a solution (pH of 8.8) containing 15 grams/liter of 100% sodium chlorite, 8 grams/liter of ammonium sulfate and 5 grams/liter of a chlorite resistant wetting agent. Solution in excess of that giving a humidity of 100% was removed by squeezing the fabric, which was then heated to 95°C. with steam and then rolled up. The goods remained for one hour at this same temperature while being slowly rotated in the steaming chamber. The goods were then washed. The average degree of polymerization of the fabric after bleaching was 1,820, as compared with an average degree of polymerization in the raw material of 2,960. The bleached fabric had a degree of whiteness of 84.3%. Seed husks were completely destroyed.

Example 2: By proceeding as in Example 1, but adjusting the solution to a pH of 7 by the addition of acetic acid to the chlorite bath, a bleached fabric with a degree of whiteness of 84.2% was obtained. The average degree of polymerization in the bleached fibers was 2,760, as compared with 2,960 in the raw material. Thus, the fiber has not been damaged, although the seed husks are completely destroyed.

Hydrazine Salt to Activate and Stabilize Chlorite Bleach

R.L. Doerr; U.S. Patent 3,111,358; November 19, 1963; assigned to Olin Mathieson Chemical Corporation describes a process to bleach cotton and other cellulosic textiles, also mixed textiles, with sodium chlorite and a water-soluble hydrazine salt as stabilizer and activator. Chlorite is used in a concentration of 0.1 to 3.0% by weight at a pH between 2 and 7. The hydrazine salt is used in a concentration between 0.001 and 3%. Any water-soluble chlorite may be employed. Alkali metal and alkaline earth metal chlorites are normally used, with the alkali metal chlorites being preferred. Sodium chlorite is especially preferred because it is commercially available in the form of Textone, a product containing about 80% sodium chlorite. Examples of suitable chlorites are lithium chlorite, potassium chlorite, sodium chlorite, calcium chlorite, barium chlorite and strontium chlorite.

Any water-soluble hydrazine salt may be used in the process, including partially substituted, water-soluble hydrazine salts. Partially substituted hydrazine salts are those in which the hydrogen attached to the nitrogen of the hydrazine moiety is substituted by such groups as, for example, alkyl or aryl groups. Free hydrazine or partially substituted hydrazine may be added to the acidified solution, in which case the free hydrazine or the partially substituted hydrazine reacts with the acid to form the corresponding salt. The hydrazine salt may be either inorganic or organic.

Typical hydrazine salts include the following: mineral acid salts such as mono- or dihydrazine sulfate, mono- or dihydrazine phosphate; aliphatic organic acid salts such as alkane monobasic carboxylic acid salts, for example, hydrazine formate, hydrazine acetate and hydrazine propionate; saturated dibasic carboxylic acid salts, for example, mono- or dihydrazine oxalate, mono- or dihydrazine malonate; unsaturated dibasic carboxylic acid salts, for example, mono- or dihydrazine maleate; tribasic carboxylic acid salts, for example, mono-, di- or trihydrazine propane-1, 2,3-tricarboxylate; alkane sulfonic acid salts, for example, hydrazine methyl sulfonate; halogen or hydroxy substituted salts, for example, hydrazine mono-, di- or trichloroacetate; amino acid salts, for example, hydrazine aminoacetate; aromatic acid salts such as hydrazine benzene sulfonate, hydrazine benzoate, hydrazine ortho-, meta- or parachlorobenzoate; cycloaliphatic salts such as hydrazine hexahydrobenzoate; and partially substituted hydrazine salts such as mono-, di- or trihydroxyalkyl-substituted hydrazine salts.

In this process the textile is immersed in an aqueous solution containing the requisite quantity of chlorite and hydrazine salt and acidified to a pH of between 2 and 7. The acid used for the acidification is not critical, for example, typical acids include formic, acetic, hydrochloric, sulfuric, phosphoric, or nitric. The textile is then removed from the solution and excess liquid removed so that the textile has a pickup of from 50 to 120% based on the weight of the textile. The excess liquid may be removed by, for example, passing through preset squeeze rolls. After the textile has been wetted with the aqueous solution containing chlorite and hydrazine it is stored until bleaching is complete.

This process is greatly accelerated by heat; a humid atmosphere at about 50° to 100°C. for 0.5 to 3 hours is excellent, although a temperature range of from 20° to 130°C. for from 0.25 to 4 hours may be employed. The humidity can be supplied by applying the heat in the form of steam in a steam box. The humidity may also be supplied by evaporation of a minor amount of the water on the textile if the textile is in a container of small volume which is quickly saturated with water vapor. The solution on the textile should not lose a material proportion of the water, however, since bleaching is preferably effected by dilute chlorite solutions. After completion of the holding period the textile is rinsed in water and dried according to conventional procedures.

The bleaching solution may optionally contain various conventional additives which are stable to the components of the bleaching solution. Typical additives include the conventional buffers, such as sodium phosphate and sodium acetate, wetting agents, such as members of the Igepon series (salts of acylalkyl taurides), corrosion inhibitors, such as water-soluble nitrate salts, and chelating agents such as those amines having a plurality of hydrogen groups of the amino group replaced by fatty carboxylic acid groups.

Examples 1 through 16: In the following examples the textile was immersed in an acidified dilute solution containing 100 cc of a 2% solution of sodium chlorite and the indicated amount of a 2% solution of the hydrazine salt. The textile was squeezed through preset squeeze rolls to a pickup of about 85% based on the weight of the textile. The textile was then heated in a steam box at 110° to 120°C. for about one hour, rinsed thoroughly in water and dried. None of the textiles treated showed any signs of textile damage. The reflectance was measured with a Photovolt Brightness Meter equipped with a tristimulus blue filter. With the brightness scale used a difference of two units is noticeable to the unaided eye, while a difference of ten units represents a striking difference in whiteness.

In the brightness range above 80 it becomes increasingly difficult to obtain each additional point of brightness without damage to the textile; therefore, the improvements shown in the following examples are of particular commercial importance. The following examples utilized a greige cotton textile having an initial whiteness of 52. When this textile was treated in the above manner in a solution acidified to a pH of 3.5 but which did not contain a hydrazine salt the brightness was 85. In the table that follows, the abbreviations used have the meanings attached.

MHS = monohydrazine sulfate
DHS = dihydrazine sulfate
DHP = dihydrazine phosphate
MHN = monohydrazine nitrate
HA = hydrazine acetate

| Example | cc of a 2% Solution of Hydrazine Salt | cc of a 2% Solution of Sodium Chlorite | cc of Water | pH of Solution | Brightness |
|---|---|---|---|---|---|
| 1 | -- | 100 | - | 3.5 | 85.0 |
| 2 | 5 cc MHS | 100 | - | 6.3 | 88.4 |
| 3 | 6 cc MHS | 100 | - | 4.9 | 88.5 |
| 4 | 7 cc MHS | 100 | - | 4.1 | 88.9 |
| 5 | 6.34 cc DHS | 100 | 85 | 6.5 | 87.0 |
| 6 | 6.34 cc DHS | 100 | 85 | 5.5 | 87.6 |
| 7 | 6.34 cc DHS | 100 | 85 | 4.5 | 88.1 |
| 8 | 6.34 cc DHS | 100 | 85 | 3.5 | 88.5 |
| 9 | 6.34 cc DHP | 100 | 80 | 5.5 | 88.5 |
| 10 | 6.34 cc DHP | 100 | 80 | 4.5 | 88.9 |
| 11 | 8.24 cc MHN | 100 | 80 | 6.5 | 88.0 |
| 12 | 8.24 cc MHN | 100 | 80 | 5.5 | 88.6 |
| 13 | 8.24 cc MHN | 100 | 80 | 4.5 | 88.2 |
| 14 | 8.24 cc MHN | 100 | 80 | 3.5 | 89.0 |
| 15 | 7.2 cc HA | 100 | 80 | 5.5 | 87.8 |
| 16 | 7.2 cc HA | 100 | 80 | 4.5 | 89.0 |

## Sodium Chlorite with Amides, Carbodiimides, Amino Acids or Sulfamic Acid Activators

K. Hintzmann, K.-H. Lange and H. Bergs; U.S. Patent 3,173,749; March 16, 1965; assigned to Farbenfabriken Bayer AG, Germany found that the efficiency of sodium chlorite bleaching solutions is increased by the use of amides, carbodiimides, amino acids or sulfamic acid compounds. In the presence of these activators temperatures as low as 70° to 80°C. may be used and the length of the bleaching process is shortened. Examples of suitable amides are: formamide, acetamide, chloroacetamide, acetoacetic acid anilide, acetoacetic acid-2-chloroanilide, acetoacetic acid-2-anisidide, stearoyl amide, benzamide, benzosulfonic acid amide, 1-methyl benzene-2-sulfonic acid amide and cyclohexyl amino sulfonic amide. Di-isopropyl carbodiimide may be used as the carbodiimide.

Suitable representatives of the amino acids are, for example, amino acetic acid, methyl amino acetic acid, amino ethane sulfonic acid or methyl amino ethane sulfonic acid. As sulfamic acid compounds the compounds of the general formula $HO—SO_2—NXY$, may be used, wherein X and Y stand for hydrogen and/or aliphatic, aromatic or hydroaromatic hydrocarbon radicals which may be substituted, or wherein X and Y together with the nitrogen atom form a heterocyclic radical. As hydrocarbon radicals the methyl, ethyl, phenyl or cyclohexyl radicals and as heterocyclic radicals the ethylene imine, pyrrolidine, piperidine or morpholine radicals are examples. The chlorite containing bleaching bath is suitably rendered weakly alkaline or neutral before the commencement of bleaching.

Example 1: A desired calico containing husks is soaked with a solution which contains per liter 13 g. of 80% sodium chlorite, 4 g. of a surface active paraffin sulfonate, 3 g. of sodium pyrophosphate and 4 g. of monochloroacetimide. The fabric is then squeezed until it shows 100% added weight and is maintained in a steam atmosphere at 80°C. for 2 to 3 hours. Finally the fabric is rinsed as usual; it is then outstandingly bleached and the cotton husks are completely removed. If desired, the bleaching time can be reduced to 1 hour if the temperature is raised to about 95°C. Instead of the amide applied above, one of the other aforementioned amides or carbodiimides can be used in approximately equal quantitites.

Example 2: A calico containing husks is soaked with a solution which contains per liter 18 g. of 80% sodium chlorite, 4 g. of a surface active paraffin sulfonate and 0.18 g. sulfamic acid. The fabric is then squeezed to give 100% added weight and maintained in a steam atmosphere at 90°C. for 1 1/2 hours. Finally the fabric is rinsed as usual; it is then outstandingly bleached, the cotton husks being completely removed. The bleaching time can be reduced to 40 minutes, if the temperature is raised to 100°C.

Sodium Chlorite Activated with Sodium Bromite

J. Tourdot and J. Breiss; U.S. Patent 3,547,573; December 15, 1970; assigned to L'Air Liquide, SA and Ste. d'Etudes Chimiques pour l'Industrie et l'Agriculture, France describe a process to bleach and in the same operation desize cotton fabrics with sodium chlorite, in the presence of sodium bromite and a small amount of ammonium nitrate. The advantages of the process are the simultaneous desizing and bleaching, combined with improved whiteness without any degradation of the fabric. In addition to cotton fabrics, the process is suitable for cotton-polyester blends. The bromite is added to the chlorite bath in such a quantity that preferably the active bromine concentration in this bath, in the form of bromite, is between 0.8 and 3 g. per liter. The initial pH of the bleaching bath is of the order of 9 to 11 and the final pH is of the order of 6.

It has further been found that the efficiency of bleaching and desizing can be increased by the further addition of a small quantity of ammonium nitrate to the chlorite-bromite bath. When used, the ammonium nitrate is added in a quantity such that the concentration in the bath of ammonium nitrate is between 0.5 and 4 g./l., preferably 2 g./l., when the chlorite-bromite bath contains sufficient sodium bromite to deposit 0.8 to 3 g. of active bromine per kilogram of cloth. The initial pH of the bleaching bath is of the order of 9.5 to 10.

Example: The treated fabric is a raw poplin initially containing 9.1% of starch. The weight of this poplin is 140 g. per square meter.

(A) In this example, the raw poplin is treated after it has been initially subjected to an enzymatic desizing. The bleaching of this poplin is carried out by the impregnation and steaming method, this second operation being performed on a roll in a housing, of the "Pad Roll" type, for 1 hour, 30 minutes at a temperature of 95° to 100°C. The bleaching bath contains 80% sodium chlorite, a wetting agent and sodium bromite employed in the form of a commercial solution containing 180 g. per liter of active bromine.

A number of bleaching tests are applied to this desized fabric, the operating conditions, the bath composition and the pH of which are set out in Table 1 with the results. After treatment, the fabrics were only subjected to one wash in hot water.

TABLE 1

| Test No.: | Composition of the bath | Initial pH | Final pH | Degree of whiteness: Green filter | Degree of whiteness: Blue filter |
|---|---|---|---|---|---|
| 1 | Sodium chloride, 15 g./l. | 9.6 | 7 | 81 | 69 |
| 2 | Sodium chlorite, 15 g./l. plus 0.9 g. of active bromine in the form of $NaBrO_2$. | 9.9 | 7 | 83 | 73 |
| 3 | Sodium chlorite plus 1.8 g. of active bromine in the form of $NaBrO_2$ (15 g./l. sodium chlorite). | 10.3 | 6.4 | 89.5 | 84 |
| 4 | Sodium chlorite, 10 g./l. plus acid sodium pyrophosphate to obtain a pH of 6.4. | 6.4 | 6 | 84.5 | 73.5 |
| 5 | Sodium chlorite, 10 g./l. plus 1.8 g. of active bromine in the form of $NaBrO_2$. | 10.3 | 5.8 | 86.5 | 77.5 |

Comparison of the results shows the increase of the degree of whiteness in the case of tests 2 and 3 as compared with the degree obtained by the action of sodium chlorite alone in test 1. The increase is equally appreciable in the case of test 5 as compared with the degree obtained by the action of a bath containing sodium chlorite in the presence of acid sodium pyrophosphate.

(B) In this example, the same poplin is treated as in (A) but after having undergone boiling after desizing. The operation is carried out by impregnation followed by steaming for 1 hour, 30 minutes at 95°C., under the same conditions as before. The impregnation bath contains 80% sodium chlorite, a wetting agent and sodium bromite employed in the form of a commercial solution containing 180 g. per liter of active bromine. A series of bleaching tests are applied to this desized and boiled fabric, of which the various operating, bath composition and pH conditions are set out in the following Table 2 together with the results obtained. After treatment, the fabrics were subjected to a simple wash in boiling water and then in cold water.

TABLE 2

| Test No.: | Composition of the bath | Initial pH | Final pH | Degree of whiteness: Green filter | Degree of whiteness: Blue filter |
|---|---|---|---|---|---|
| 1 | Sodium chlorite, 15 g./l. | 9.6 | 7 | 87 | 81 |
| 2 | Sodium chlorite, 15 g./l. plus acid sodium pyrophosphate to obtain a pH of 6.4. | 6.4 | 6.4 | 90.5 | 86 |
| 3 | Sodium chlorite, 15 g./l. plus 0.9 g. of active bromine per liter in the form of $NaBrO_2$. | 9.9 | 6.4 | 90 | 85 |
| 4 | Sodium chlorite, 15 g./l. plus 1.8 g. of active bromine per liter in the form of $NaBrO_2$. | 10.3 | 6.7 | 90.5 | 85.5 |
| 5 | Sodium chlorite, 10 g./l. plus acid sodium pyrophosphate to obtain a pH of 6.4. | 6.4 | 6.4 | 89 | 82.5 |
| 6 | Sodium chlorite, 10 g./l. plus 1.8 g. active bromine per liter in the form of $NaBrO_2$. | 10.3 | 6.4 | 90 | 86 |

It will be seen on studying these results that there is an appreciable improvement in the degree of whiteness in tests 3 and 4 as compared with test 1 corresponding to a bath containing chlorite alone. It may also be observed on comparing results of tests 2 and 6 that the addition of bromite results in a reduction of the quantity of bleaching agent from 15 g. per liter to 10 g. per liter of sodium chlorite, while affording the same result as with a conventional activator in a bath containing 15 g. per liter of chlorite.

## Alkaline Chlorite Solutions Stabilized with Other Carboxylic Acids

C. Heid and K.-H. Keil; U.S. Patent 3,580,851; May 25, 1971; assigned to Cassella Farbwerke Mainkur AG, Germany found that aqueous mixtures of alkali chlorite are stabilized by ether carboxylic acids of ethoxylated alkyl phenols and alcohols or their alkali salts having the following general formula.

$$R—O—(CH_2—CH_2—O)_x—CH_2—CH_2—O—(CH_2)_y—\overset{R_1}{\overset{|}{C}}H—COOH$$

wherein R is $—C_nH_{2n+1}$ or

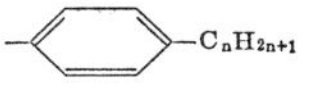

n denotes an integer of from 6 to 18, preferably from 7 to 12; x is an integer of from 6 to 30, preferably from 9 to 19; y is 0 or 1; and $R_1$ is hydrogen or methyl. Particularly preferred ether carboxylic acids are the reaction products of ethoxylated alkylphenols and ethoxylated alcohols with monochloroacetic acid and ether carboxylic acids obtained by the addition of acrylonitrile or methacrylonitrile to ethoxylated alkylphenols or alcohols and subsequent saponification of the formed nitriles. Preferably, the weight ratio of sodium chlorite to surfactant in the compositions ranges from 10:1 to 1:1, the most preferred compositions having a weight ratio ranging from 3:1 to 2:1. It is also preferred that the solid content of the aqueous mixtures comprises from 10 to 60% by weight of the total weight thereof.

As compared to the mixtures described in French Patent 1,453,380, those of this process distinguish themselves by a considerably higher stability and certain other advantages. They have, for instance, higher points of turbidity. Moreover, it was found that in the mixtures with alkali chlorites, unlike the esters described in the French patent, it is possible to replace up to 50% by weight of the ether carboxylic acids with their precursor ethoxylated alkylphenols or ethoxylated alcohols, whereby the stability of the mixtures is not affected. This is advantageous inasmuch as the last mentioned nonionogenic products are generally superior to the corresponding anionic ether carboxylic acids as to dispersing effect.

Although it is unnecessary to include a buffering agent in these compositions, the presence of such agent exerts an advantageous effect when utilizing the mixtures for bleaching and oxidizing purposes. The preferred buffering agents include sodium carbonate and sodium borate. Generally the amount ranges from 0.1 to 10% by weight based on the total weight of the aqueous composition.

French Patent 1,453,380 describes mixtures having the pH value thereof adjusted to above 10. They consist of alkali chlorites and surfactants. The latter are aliphatic or arylaliphatic polyglycol ethers that contain an oxethylene chain consisting of at least 8 mols ethylene oxide, and having a terminal hydroxyl group esterified with organic or inorganic acids which are stable to oxidation. A typical example is the potassium salt of the diester of phosphoric acid and the condensation product of nonylphenol and 10 mols ethylene oxide, which has the following formula:

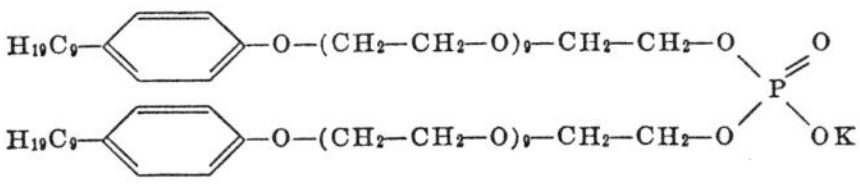

According to this French patent, a mixture of sodium chlorite and the potassium salt of the diester of phosphoric acid is homogeneous at temperatures ranging from 20° to 60°C. and stable, for a maximum of 200 hours, at 60°C., the pH value decreasing during this period only from 11.4 to 10.8. Furthermore, the patent discloses that storage of such a mixture at 20°C. for three months or at 40°C. for twenty-eight days changed neither its content of chlorite nor its pH value.

## CONTINUOUS TEXTILE MILL PROCESSING

### Chlorite-Permanganate Bleaching for Cotton and Rayon

R.R. Heinze and R.L. Ostrozynski; U.S. Patent 3,035,883; May 22, 1962; assigned to Olin Mathieson Chemical Corporation describe a process to bleach cotton, rayon and linen with aqueous chlorite solutions without significant chlorine dioxide evolution and at pH above 4 by the addition of permanganate. In the first step, the cloth or fiber is wetted with an aqueous solution containing between 0.3 and 3% by weight of a water-soluble chlorite. Examples of some suitable chlorites are those of lithium, potassium, sodium, calcium, barium and strontium. The wetting can be done by spraying the solution on the cloth or by putting the cloth in a chlorite bath. This bath should not be alkaline and must be made slightly acidic with almost any acid, for example, formic, acetic, hydrochloric or sulfuric.

The preferred pH range is about 4.5 to 7. There is no problem with chlorine dioxide evolution in this pH range. The substance being bleached should finally contain about 50 to 100% by weight, based on its own weight, of the chlorite solution. Certain exceptionally absorbent cloths such as huckaback toweling can hold up to twice their weight of liquid and can be impregnated with well over 100% by weight of the chlorite solution. If the cloth has been immersed in a chlorite bath, the excess liquid must be removed, for example by pressing or wringing. If this is not done it will drop when the permanganate solution is applied, thus losing the active agents and resulting in poorer bleaching. The range of chlorite applied is about 0.25 to 3% by weight based on the weight of the material being bleached.

Different types of cloth require different amounts of chlorite to attain a maximum whiteness, e.g., a heavy linen may require twice as much as a light cotton. The substance impregnated with the aqueous chlorite is next impregnated with a dilute solution of a water-soluble permanganate to the extent of 0.5 to 20% by weight based on the weight of the chlorite thereon. Although more than 20% can be used, it is uneconomical and not any more effective. This small amount of permanganate can also be applied by spraying or dipping.

Textiles and fibers can conveniently be passed between horizontal rolls wherein the bottom roller is partially immersed in the permanganate solution (dip roll method), by rapidly passing the textile or fiber through an aqueous permanganate bath and wringing it to contain the desired amount of permanganate. The latter method may tend to leach some of the chlorite solution out of the cloth. The dip roll method is preferred because of convenience. Only the bottom roller need be wet because a slight pressure between the rollers causes the permanganate solution to permeate the cloth quickly. The permanganate solution is conveniently applied in the form of a 0.005 to 0.5% by weight aqueous solution.

The amount applied is dependent upon the chlorite on the cloth; however, about 10 to 50% by weight of the solution, based on the weight of the cloth is usually sufficient if the permanganate concentration of the solution is within the above limits. Since most cellulosic textiles and fibers cannot retain more than about 150% by weight of liquid without dripping, it is preferable not to use permanganate and chlorite solutions which are both in the lower end of the concentration ranges disclosed above. It is preferred not to exceed about 120% by weight of liquid on the cloth or fiber. Permanganates, in contrast to manganese salts, cannot be added to the chlorite solution because they are reactive with chlorites and quickly cause them to lose their bleaching power. Thus, manganous chloride or sulfate can be added to a chlorite bath to effect some improvements in bleaching; however, a permanganate employed in this way causes the bath to become exhausted, by chlorine dioxide evolution, more quickly than it would have otherwise.

After the cloth is wet with the chlorite and permanganate solutions, it is stored until bleaching is complete. This process is greatly accelerated by heat; a humid atmosphere at about 50° to 100°C. for 1/2 to 3 hours is excellent. The humidity can be supplied by applying the heat in the form of steam in a steam box or it can arise by evaporation of a minor amount of the water on the cloth if it is in a container of small volume which is quickly saturated with water vapor. The solution on the cloth should not lose a material proportion of water, however, because the bleaching is preferably effected by dilute chlorite solutions.

When the holding period is over, the cloth is passed through a dilute solution of a reducing agent effective to convert the manganese compounds on the cloth to soluble salts of manganese. The latter are white while the former are water-insoluble brown stains. A solution of 0.3 to 1% of such agents as hydrogen peroxide, oxalic acid, alkali metal sulfite or hydrosulfite is adequate. This being done, the cloth is rinsed in water and conventionally dried. For any given textile or fiber, the process of this method produces an exceptional whiteness more quickly and with less chlorite than prior processes.

Examples 1 through 5: In the Examples 1 through 5 shown in the table below, the cloth indicated was treated according to this process. It was impregnated with dilute solution of chlorite and activators of the type and to the extent shown. After the heating period these cloths were passed through a 0.5% by weight oxalic acid solution, rinsed thoroughly in water and the reflectance was measured using a Photovolt Brightness Meter equipped with a tristimulus blue filter. With the brightness scale used, a difference of 2 units is noticeable to the unaided eye, while a difference of 10 units represents a striking difference in whiteness.

| | 1 | 2 | 3 | 4 | 5 |
|---|---|---|---|---|---|
| Activator | $KMnO_4$ | $Mg(MnO_4)_2$ | $NaMnO_4$ | $KMnO_4$ | $KMnO_4$ |
| Percent by wt. on cloth | 0.2 | 0.18 | 0.28 | 0.01 | 0.005 |
| Percent $NaClO_2$ on cloth | 0.95 | 1.8 | 2 | 1 | 1 |
| pH of Chlorite Bath | 6 | 5.6 | 5.2 | 6 | 6 |
| Heating time, Hrs | 1 | 3 | 2 | 1 | 1 |
| Heating temp., °C | 95 | 60 | 80 | 100 | 100 |
| Final Brightness | 88 | 83.6 | 86.4 | 81.9 | 81.1 |
| Initial Brightness of cloth | 59 | 63 | 61 | 57 | 57 |
| Type of cloth | Cotton Broadcloth. | Rayon | Rayon | Cotton-Poplin. | Cotton-Poplin. |

## Continuous Chlorite Bleaching with Ammonium Salt Activator

E. Ruedi; U.S. Patent 3,120,424; February 4, 1964; assigned to FMC Corporation describes a continuous bleaching process for cotton and synthetic fibers utilizing sodium chlorite with an ammonium salt activator. The process consists of impregnating the material to be bleached at room temperature with the bleaching liquor followed by squeezing to 60 to 100% and then steaming to at least 60°C. to effect bleaching of the material. The activator salts used are those which at temperatures up to 50°C. do not yield acid solutions, so as not to activate the chlorite bleaching liquor, but which at temperatures of at least 60°C., decompose and become strongly acid. Potentially acid salts used in the process may be exemplified by salts of a strong acid and a volatile base, e.g., the ammonium salts of strong acids such as ammonium chloride, ammonium nitrate, ammonium sulfate, ammonium phosphate, ammonium tartrate and ammonium oxalate, or the salts of strong acids with weak bases, such as magnesium chloride, calcium chloride, magnesium sulfate, calcium sulfate, magnesium phosphate, calcium phosphate, magnesium nitrate and calcium nitrate. Other potentially acid salts include the salts of strong acids with organic bases, such as triethanolamine hydrochloride, diethanolamine hydrochloride, triethylamine hydrochloride, diethylamine hydrochloride, ethylenediamine hydrochloride, etc.

The bleaching liquors contain also a wetting agent. Suitable wetting agents include alkylarylsulfonates, nonionic wetting agents, and fatty acid condensation products. While sodium chlorite is by far the most economic, and therefore the preferred chlorite used in the bleaching liquors of this process, potassium and ammonium chlorite, though far less advantageous, would also be suitable.

The neutral alkali chlorite solution with which the textile material to be bleached is impregnated, may be either cold or mildly heated to a temperature not substantially exceeding 50°C. Once the excess bleaching liquor is squeezed off, the material thus impregnated is heated to a temperature of at least 60°C. While considerable activation occurs at temperatures of 60°C. and slightly above, the activation is materially stronger at temperatures of 70° to 80°C. Heating to temperatures above 100°C. a very strong, and very fast activation effect is observed at these higher temperatures. Above 140°C., however, a strong development of chlorine dioxide occurs which militates against further increases of the temperature.

This process is applicable to a great variety of textile materials. It is recommended for use on cellulose fibers such as cotton staple fiber, viscose or cuprammonium rayon, linen, hemp and other bast fibers; on cellulose acetate fibers such as diacetate and triacetate fibers; on synthetic fibers such as polyamide, polyethyleneterephthalate and polyacrylonitrile fibers, to mention some representative examples. Fibers may be bleached by the method at any stage of their manufacture: in the form of flakes, roves, fleece, yarn, felt (nonwoven material), or else when woven into fabrics, etc.

Example: A bleaching solution is prepared containing the following essential ingredients:

| | Grams/Liter |
|---|---|
| Sodium chlorite (80%) | 37.5 |
| Leonil ART, an aryl-alkyl sulfate | 5.0 |
| Ammonium chloride | 10.0 |

The bleaching liquor thus prepared is used as follows. A cotton fabric is desized on a covered jigger, and then washed. Thereafter, a suction desiccator is employed to reduce the moisture content to 50%. The fabric is again put on the jigger and impregnated, in the cold, with a bleaching liquor of the composition noted above. The material is squeezed off lightly so the squeeze-off effect amounts to 100%. The excess liquor is pumped off, for use on a new section of material.

At this point, the jigger is closed and heated, by a direct supply of steam, to the reaction temperature of 90° to 95°C. Now the material is passed through four times in order to heat the impregnated fabric to the bleaching temperature of about 90°C. Subsequently, the material is turned slowly on a jigger roller for about one hour, the jigger temperature being maintained at 95°C. by means of saturated steam. The material is twice rinsed hot, then rinsed cold, with the result that a beautiful, perfectly white fabric of excellent absorptive capacity is obtained.

## Continuous Chlorite Bleaching of Cellulose and Cellulose-Synthetic Textiles

J.F. Synan and W.W. Northgraves; U.S. Patent 3,140,146; July 7, 1964; assigned to Olin Mathieson Chemical Corporation describe a process and equipment to continuously bleach at speeds of 5,000 to 15,000 yards per hour cellulosic textiles which have been given a previous alkaline wash and contain therefore residual alkali of up to 0.25% or more. Acidified sodium chlorite solution is used as the bleach. The process is suitable for blends of cotton with synthetics.

In Figure 2.1 the alkaline cloth is shown entering at the left into acid saturator (11) over roller (12) and passing through the aqueous acid (13) over immersed rollers (14). Acid is introduced via line (15). Excess of aqueous acid is removed by squeeze rollers (16) and the cloth is guided by rollers (17) and (18) and stacked into the longer arm of J-chute (19) to provide time for interaction of the aqueous solutions, usually at room temperature, and acidification of the cloth. Guided over rollers (20) and (21), the cloth is washed in a first washer (22) passing through squeeze rollers (23) via guide roller (24) into a second washer (25). Water is introduced via lines (26) and (27) into washer (22) and (25) respectively.

The cloth is removed from washer (25) through squeeze rolls (28) and over guide roller (29) into saturator (30) into which chlorite is introduced via line (31). The impregnated textile passes through squeeze rollers (32) and is guided by rollers (33) and (34) into the longer arm of J-chute (35). Here time is provided for diffusion of

FIGURE 2.1: CONTINUOUS CHLORITE BLEACHING PROCESS

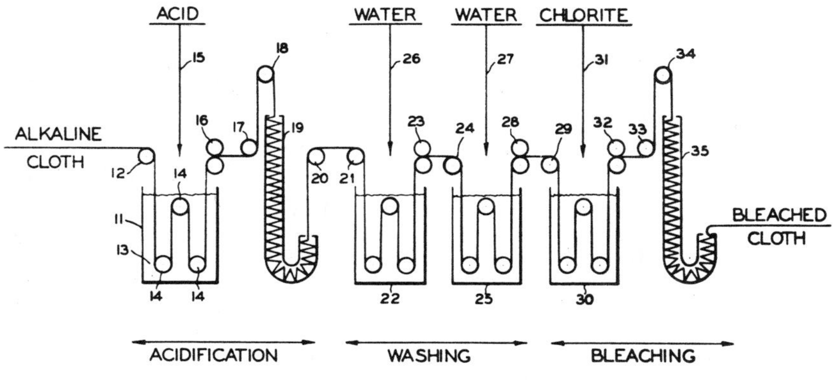


Source: J.F. Synan and W.W. Northgraves; U.S. Patent 3,140,146; July 7, 1964

the solutions on the cloth and for bleaching to occur. The J-chute (19) is usually at room temperature and J-chute (35) is usually heated by steam either by means of a jacket (not shown) or by direct steam lines (not shown) into the chute.

Example 1: About 60,000 yards of cotton cloth, varying in weight from light lawn to heavy oxford, were sewn end to end and passed serially through the several process steps. The goods were desized, mercerized, impregnated with dilute caustic soda (3 to 4%), steamed for 75 minutes, washed, impregnated with dilute sulfuric acid (1.0 to 1.5%) and stacked in a J-chute without heat for about 45 minutes. The cloth rate was about 15,000 yards per hour. The cloth washed but still contained about 0.1% $H_2SO_4$ entering the bleach saturator. The nipped cloth was passed through a Textone solution containing 0.5% by weight of sodium chlorite and 0.5% by weight of sodium nitrate. (Textone is a commercial product containing approximately 80% of sodium chlorite.)

Caustic soda solution (20%) was added continuously to the saturator solution during the passage of the cloth to maintain the pH at between 3.2 and 3.7. The Textone concentration in the emerging cloth, nipped to 110% pickup, was about 0.4 to 0.56% by weight. The cloth passed to a stainless steel J-chute where it was held at about 210°F. for about one hour by the passage of steam. The cloth was then washed and dried. An excellent white was obtained and the goods were free of motes.

Example 2: About 3,000 yards of a Dacron-cotton blend, 45 inches wide, thread count 136 x 60 and running 3.58 yards per pound was continuously acid treated, stored and washed. Using 4 g. per liter of Textone in the saturator solution and maintaining a pH of 3 to 3.5 by the addition of caustic, the acidic goods were impregnated, squeezed to 100% pickup and steamed for about 1 hour in the J-box. After washing and drying, an excellent bleach was obtained.

Example 3: A 50-50 modified acrylic-cotton blend which turned dark when treated with alkaline peroxide was successfully bleached by the procedure of Example 2 using 10 g. per liter of Textone in the saturator solution at a pH of 3 to 3.5. The blend was 40 inches wide, thread count 136 x 60 and ran 3.17 yards per pound. Compared with a peroxide treatment under otherwise the same conditions, the tensile strengths using the chlorite bleach averaged 11% higher and the tear values 78% higher.

Example 4: About 7,500 yards of a regenerated rayon-cotton blend (45-55) 47 1/2 inches wide, thread count 90 x 66 and running 3.80 yards per pound was acid treated, stored and washed. It was bleached to an excellent white using 3 g. per liter Textone at a pH maintained between 3 and 3.5 by caustic addition. A peculiarity of this blend was that it suffered serious pinholing when bleached with peroxide. No such pinholing occurred using the chlorite bleach.

Example 5: About 25,000 yards of a nylon-Dacron-cotton blend constructed of nylon warp with twisted Dacron-cotton fill was continuously acid treated, stored, washed and bleached using 5 g. per liter of Textone, and 5 g. per liter of sodium nitrate at a pH of 3 to 3.5. The tensile strength of the nylon warp after the chlorite bleach was 160% higher than after a peroxide bleach.

# HYPOCHLORITES

## STABILIZED HYPOCHLORITES

### Amine Stabilized Hypochlorite

R.C. Davis and D.C. Wood; U.S. Patent 3,113,928; December 10, 1963; assigned to Whirlpool Corporation describe a process to prevent yellowing of cotton wash and wear fabrics by the addition of an amine to the laundry detergent and bleaching solution. The amine has a greater affinity for the chlorine than the resins used on the wash and wear fabrics such as the melamine-formaldehyde or the urea-formaldehyde resins.

The preferred amine is an alkanolamine with the alkyl radical being of relatively low molecular weight, preferably of 2 to 6 carbon atoms. Typical alkanolamines are monoethanolamine, diethanolamine, triethanolamine, N-methyl ethanolamine, benzyl dimethylamine, dimethylamine, phenyl ethanolamine, morpholine, N-methyl morpholine, N-(2-hydroxyethyl)morpholine, 2,6-dimethyl morpholine, N-ethyl morpholine, N-aminopropyl morpholine, N-aminoethyl morpholine and cyclic amines such as piperazine derivatives.

Typical amines are any of the aliphatic amines having an available reactive hydrogen atom. As a necessary characteristic of the amine or ether of this process is its ability to react with the chlorine from the chlorine containing bleach and is preferably a liquid at ordinary temperatures. The amine or ether may be incorporated with the chlorine containing bleach or bleaching solution or may be incorporated with the detergent. It may also be used in any other manner desired such as a separate additive to the laundry or bleaching water.

Example: A section of ordinary white wash and wear cotton fabric of a commercial type containing melamine-formaldehyde resin was washed with a solution containing 0.3% of a commercial laundry detergent, specifically sodium lauryl sulfate, 0.015% monoethanolamine and 1/2 ounce per gallon of 5.25% sodium hypochlorite solution, all percentages being by weight of the solution. Other detergents may be used if desired such as alkyl aryl sulfonate, nonyl phenol ethylene oxide condensate and the like. The fabric was washed and bleached for 10 minutes at a solution temperature of 125°F. The fabric was then rinsed at 100°F. for 5 minutes and finally tumble dried in a commercial dryer until dry to the touch. This same washing and bleaching followed by the same rinsing and drying was repeated. The fabric was hot ironed after each bleaching. Then, the fabric was soaked overnight in the same solution as described above except that the temperature was approximately 70°F.

In addition, a mustard stain was placed on the fabric to determine whether or not the chlorine of the bleach was readily available for efficient bleaching. The reflectance values of the fabric before and after each treatment were as follows:

| | Rd. | a | b |
|---|---|---|---|
| Original | 89.5 | -2.5 | +3.8 |
| After 1st Wash | 89.0 | +1.0 | +3.2 |
| After 2nd Wash | 91.0 | +0.6 | +2.6 |
| After Overnight Soak | 86.5 | +1.5 | +1.6 |

The reflectance data given above was obtained by the use of a Gardner color difference meter. This meter is a reflectance type color difference meter capable of determining the degree of whiteness, the degree of red or green, and the degree of blue or yellow of the fabric tested. An increasing degree of the "Rd." reading indicates an increasing degree of whiteness. The plus reading of the "a" readings indicates the fabric tested is on the "red" side while a minus reading indicates the fabric is on the "green" side. The "a" reading is not as important normally as the "b" reading. A positive "b" reading indicates the fabric is on the "yellow" side while a minus reading, which is normally desired for white fabrics, indicates the fabric is on the "blue" side.

The marked drop in yellowing as reflected by the decreasing "b" values is contrary to what would occur if the resin were not protected by the amine. The fact that the reduction in yellowing occurred showed that the amine protected against the bad effects of chlorine retention. Furthermore, the last "b" value shows that the mustard stain was almost completely eliminated, indicating that the bleach was still quite efficient in its operation. Also, the extensive overnight soaking shows that the amine is not consumed during the reaction but is still available for protection of the fabric.

## Hypochlorite with Surfactants

A.F. Steinhauer and J.C. Valenta; U.S. Patent 3,172,861; March 9, 1965; assigned to The Dow Chemical Company found that the bleaching and cleaning power of sodium hypochlorite bleaching solution is increased by the addition of surfactants, such as the sodium salt of an alkylated diphenyl oxide sulfonic acid.

The alkylated diphenyl oxide sulfonic acid alkali metal salts and their nuclearly mono and dichlorinated derivatives, suitable for use are those having from 8 to 22 carbon atoms in the alkyl chain and an average of from 1.8 to 2.3 sulfonate moieties per diphenyl oxide moiety.

## Hypochlorite, Optical Brighteners and Stabilizer

R.E. Zimmerer and W.I. Lyness; U.S. Patent 3,393,153; July 16, 1968; assigned to The Procter & Gamble Company describe a process to combine an alkaline, hypochlorite bleach solution with a compatible optical brightener and a stabilizer such as a styrene copolymer. This process provides a stable aqueous composition capable of rendering both chemical bleaching and optical brightening effects.

"Optical brighteners" (or "brighteners"), are defined as chemicals which are adsorbed by textile fibers and thereby impart to the textile an improved degree of whiteness or brightness (fluorescence) by means of their chemical ability to absorb ultraviolet radiation and reemit visible radiation. Optical brighteners have found widespread use as components of household detergent compositions. In fact, to achieve the degree of whiteness desired in the wash by most consumers, a combination of bleaching and optical brightening is generally required.

This requirement is usually met by using a brightener-containing detergent composition as the primary washing agent combined with (a) the subsequent addition of hypochlorite bleach to the wash water, or (b) prior use of a hypochlorite bleach in a separate step. It has long been deemed desirable to consolidate this bleaching/brightening effect into a single step process, e.g., by using a product which contains both a bleach and an optical brightener.

In general, this process comprises a stable aqueous composition having a pH ranging from 10.5 to 13.0 and consisting of: (a) from 1.0 to 10% alkali metal hypochlorite bleach; (b) from 0.002 to 2.0% hypochlorite bleach-compatible optical brightener; and, (c) from 0.1 to 2.0% stabilizing agent. The hypochlorite bleach component used in this process is present in an amount ranging from 1 to 10%, preferably from 3 to 7%, with 5.2% being especially desirable.

The optical brightener used in this composition is in an amount ranging from 0.002 to 2.0%, preferably from 0.01 to 0.1% with 0.05% being preferred. The optical brightener must be chemically stable and substantially insoluble, i.e., more than 99% of the amount employed remains undissolved in the aqueous hypochlorite-containing compositions, e.g., in solutions containing about 5% hypochlorite, but it must be sufficiently soluble, i.e., less than 25% remains undissolved at washing machine conditions, e.g., in aqueous solutions containing about 0.02% hypochlorite, so that it can be effectively deposited on desired textile substrata.

Optical brighteners that have the above outlined stability and solubility characteristics when incorporated into the compositions of this process are referred to as "hypochlorite bleach-compatible". The preferred optical brighteners are the stilbylmonotriazoles such as sodium 4-(2'H-naphtho[1',2'-d]triazol-2"-yl)-2-stilbenesulfonate and sodium 4-(2'H-5'-sulfonaphtho[1',2'-d]triazol-2"-yl)-2-cyanostilbene. Of these optical brighteners which are listed as sodium salts, other alkali metal salts thereof, e.g., potassium salts, can also be satisfactorily used.

The stabilizing agent is present in the compositions in an amount ranging from 0.1 to 2.0%, preferably from 0.5 to 1.5% with about 1.0% being particularly preferred. Stabilizing agents useful for the present process must be chemically nonreactive with the other ingredients of the composition, especially with the bleach and optical brightener, i.e., this component must be chemically stable in the presence of both the bleach and the optical brightener. Further, the stabilizing agent must be insoluble and dispersible in the bleach-containing solutions of the process. Additionally the stabilizing agent must be particulate, i.e., it must have a mean particle diameter ranging from 0.01μ to 40μ. Thus, the term "stabilizing agent" is defined as a particulate material that is hypochlorite bleach-stable, brightener-stable, insoluble, and dispersible in the compositions of this process. Preferably, in addition to exhibiting these necessary characteristics, the stabilizing agent is also capable of uniformly suspending the optical brightener component in the solutions of this process.

Moreover, certain stabilizing agents tend to make the solutions opaque; these stabilizing agents are preferred because of the desirability of opaque liquid laundry products, for appearance purposes, for visual ease in measuring and pouring, and for distinguishing liquid bleaches from other clear liquids such as water and vinegar. An example of a material useful as a stabilizing agent is colloidal (particulate) silica, preferably having a mean particle diameter ranging from 0.01μ to 0.05μ.

Particularly preferred materials useful as stabilizing agents in these compositions are copolymers of styrene with certain ester and acid monomers such as methyl or ethyl acrylate, methyl or ethyl maleate, vinyl acetate, acrylic, maleic or fumaric acid, and mixtures thereof, the mol ratio of ester and/or acid to styrene preferably being in the range of 1 ester and/or acid (monomer) unit per each 4 to 40 styrene units, said copolymers having a mean particle diameter ranging from 0.05μ to 1.0μ, and a molecular weight ranging from 500,000 to 2,000,000.

Examples of the above-defined polystyrene materials which are highly preferred as stabilizing agents in the compositions of this process are (a) a copolymer of styrene and methyl acrylate with a ratio of 1 methyl acrylate unit to 6 styrene units, and having a mean particle diameter of 0.2μ and an average molecular weight of 1,000,000, and (b) a copolymer of styrene, methyl acrylate, and acrylic acid with a ratio of 1.7 methyl acrylate units and 1 acrylic acid unit per 16 styrene units, and having a mean particle diameter of 0.1μ and an average molecular weight of 1,000,000.

The compositions can be prepared by physically intermixing the various components; it is convenient and desirable to start with readily available household bleach, which is generally an aqueous solution of 5.25% sodium hypochlorite, adjust the pH, and add the other components. However, particularly desirable results are achieved, both in the area of bleach-brightener compatibility and in general physical stability of the system, when the components of the composition are combined in accordance with a certain preferred method.

This method comprises the prior combination of the optical brightener and stabilizing agent before addition thereof to a hypochlorite-containing solution. Such a combination of optical brightener and stabilizing agent can be prepared, for example, by the addition of the optical brightener to an aqueous suspension of stabilizing agent with gentle heating, 40° to 95°C., and stirring.

Many of the polymeric materials useful herein as stabilizing agents are commercially available in latex form, i.e., an aqueous suspension containing 20% to 60% of the polymeric solids, and are generally referred to as "emulsion polymers" or, more generically they are referred to as "latex". This physical form of the stabilizing agent is very convenient and preferable for use in preparing the compositions of the process.

Example: In a 50 ml. flask, 15.0 g. of stabilizing agent suspension and 0.300 g. of optical brightener were combined by magnetically stirring and heating at 60° to 70°C. for about 30 minutes. The stabilizing agent suspension was a polystyrene-type latex (an aqueous emulsion containing about 40% of a copolymer of styrene and methyl acrylate with a ratio of 1 methyl acrylate unit to 6 styrene units, and having a mean particle diameter of about 0.2μ and a molecular weight of about 1,000,000); the optical brightener was sodium 4-(2'H-naphtho[1',2'-d]triazol-2"-yl)-2-stilbene-sulfonate.

This mixture was then added to 600 ml. of a commercially available household bleach (consisting of about 5.25% sodium hypochlorite, 4 to 5% sodium chloride, balance water) in which 0.6 g. of sodium hydroxide had been previously dissolved. The entire mixture was then agitated in a conventional laboratory blender for 5 minutes.

The resulting composition was stable, opaque and had a pH of 11.4. Visual examination under ultraviolet light revealed that the optical brightener was uniformly suspended throughout the composition. The fluorescence conferred on cotton by this composition was 135 GM units immediately after mixing and was 115 GM units after the composition had been stored for 7 days.

Stabilized Chlorinated Trisodium Phosphate and Sodium Hypochlorite

J.K. Stamm; U.S. Patent 3,364,147; January 16, 1968; assigned to W.R. Grace & Company describes a process to prepare a triple salt bleaching composition of the following formulae:

$$Na_3PO_4{:}0.25NaOCl{:}0.0005\text{-}0.32Na_5P_3O_{10}{:}10.5\text{-}11.5H_2O$$

$$Na_3PO_4{:}0.25NaOCl{:}0.0014\text{-}0.77Na_4P_2O_7{:}10.5\text{-}11.5H_2O$$

$$Na_3PO_4{:}0.25NaOCl{:}0.0003\text{-}0.018(NaPO_3)_6{:}10.5\text{-}11.5H_2O$$

It was found that alkali metal salts, in particular the sodium salts of condensed phosphates, can be combined chemically as triple salts with chlorinated trisodium phosphate to produce products which demonstrate chlorine stabilities distinctly better than those of conventional chlorinated trisodium phosphate. A remarkable improvement in chlorine stability is obtained if the secondary phosphate is added in an amount less than about 5% by weight of the final product.

Examples of such condensed sodium phosphates can be combined with trisodium phosphate and sodium hypochlorite to form a product having improved chlorine stability include sodium tripolyphosphate, $Na_5P_3O_{10}$ (commonly referred to as STP), tetrasodium pyrophosphate, $Na_4P_2O_7$ (commonly referred to as TSPP), or sodium hexametaphosphate, $(NaPO_3)_6$, either individually or in admixture with each other. The most favorable results are demonstrated when sodium tripolyphosphate ($Na_5P_3O_{10}$) is used.

The procedure comprises first forming a melt of trisodium phosphate and the condensed phosphate. A trisodium phosphate liquor can be formed by reacting a sodium hydroxide solution with phosphoric acid, and the condensed phosphate can be dissolved in the liquor in the molar proportions desired in the final product to form the melt. Suitable condensed phosphates include sodium tripolyphosphate, the preferred condensed phosphate, tetrasodium pyrophosphate, and sodium hexametaphosphate.

The condensed phosphate is preferably added in anhydrous form to prevent the water content of the solution from becoming excessive. The condensed phosphate may be added in hydrated form if the water content of the TSP is adjusted accordingly. The condensed phosphate concentration can range up to 5 weight percent based on the weight of trisodium phosphate dodecahydrate.

Above 5%, it becomes difficult or impossible to dissolve additional quantities of the condensed phosphate in the liquor, and this consideration therefore appears to establish the upper limit of the quantity of condensed phosphate which may be added.

The melt is then mixed with a sodium hypochlorite solution having a chlorine content of 10 to 23% in the proportion required to provide, in the final product, about 0.25 mol of sodium hypochlorite per mol of trisodium phosphate.

Example: This example shows the preferred procedure for forming the triple salt. 120 g. of NaOH was dissolved in 144 g. of water and cooled to room temperature. With good agitation 98 g. of $H_3PO_4$ was slowly added, the solution temperature rising to 90° to 100°C. during the acid addition. With continued agitation, 7.8 g. of $Na_5P_3O_{10}$ (2 wt. percent of the final composition) was added to the phosphate liquor and dissolved therein.

The temperature of the solution or melt was then permitted to drop 80° to 90°C., and 54.3 g. of NaOCl with an available $Cl_2$ of 16.0% was added to the melt with continued agitation. Agitation was continued as the melt was cooled to room temperature. The product was homogeneous crystals having the mol ratios of components as follows:

$$Na_3PO_4{:}1/4NaOCl{:}1/83Na_5P_3O_{10}{:}11H_2O$$

Using the same procedure, addition of 0.1, 1.0 and 5.0 wt. percent $Na_5P_3O_{10}$ to the liquor produces triple salts having condensed phosphate to trisodium phosphate mol ratios of 1/1708, 1/169, and 1/32 respectively.

Depending upon the chlorine content of the NaOCl solution, the chlorine content of the product may be from 1 up to 5.0%. Heretofore, commercial chlorinated TSP has been practically limited to a $Cl_2$ content of about 4%; surprisingly, products in accordance with this process have excellent chlorine stabilities even with chlorine contents of about 5%. Typically, products in accordance with this process both in the solid state and in solution, demonstrate chlorine stabilities which are considerably better than that of conventional chlorinated trisodium phosphate of lesser chlorine content.

Detergent Bleach Compositions with Stain Inhibitors

A.A. Rapisarda; U.S. Patent 3,551,338; December 29, 1970; assigned to Lever Brothers Company describes a stain inhibitor system which prevents or diminishes staining which occurs when textiles are laundered in a solution containing manganese ions and a bleaching agent which is capable of oxidizing the manganese. The inhibitor system comprises a water-soluble titanium compound and an inorganic condensed polyphosphate. This inhibitor system can be incorporated into a detergent formulation, a bleach formulation, or it can be added to the bath separately before the bleach is added. The process is of importance in communities where the water supplies contain significant quantities of manganese ions causing yellow or brown stains on textiles after bleaching. The term "bleaching" is intended to include both exposure to a bleaching agent in a bleaching bath per se or in a combined bleach-detergent bath. Bleaching agents, which release chlorine or oxygen, oxidize the manganese ions in the solution to form a manganese compound which is then deposited on the surface of the cloth.

As little as 0.1 part per million of manganese ion in the water will cause objectionable stains on cloth within 5 to 10 washes when the textile is exposed to the manganese ion and a bleaching agent which can oxidize the manganese. As little as 1 to 3 parts per million of manganese ion in the wash water along with about 15 parts per million of oxygen or about 200 parts per million of chlorine will severely discolor the cloth in a single wash.

The need for protection against discoloration is particularly required within the pH range generally encountered in the laundering of textiles, that is, at a pH between 5 and 10.5. It is within this range that manganese compound stains are found to deposit on the cloth and tenaciously adhere thereto.

The inhibitor compositions comprising a water-soluble titanium compound and an inorganic condensed polyphosphate have been found to be effective in preventing the discoloration of textile surfaces which occurs when the textiles are washed or bleached in solutions which contain manganese ions and a bleaching agent capable of oxidizing the manganese ions. Representative manganese-oxidizing bleaching agents include potassium persulfate, ammonium persulfate, sodium hypochlorite, calcium hypochlorite, sodium perborate, alkali metal dichloroisocyanurates, lauroyl peroxide, sodium peroxide, ammonium dipersulfate, hydrogen peroxide, and any other bleaching agents which are capable of oxidizing manganese ions to a higher valence state.

Suitable water-soluble titanium compounds include titanium dichloride, titanium fluoride, titanium nitrate, titanium oxalate, etc. Suitable condensed inorganic polyphosphates, which can be employed in combination with the above-described titanium compounds, include the alkali metal tripolyphosphates and pyrophosphates, e.g., sodium tripolyphosphate, potassium pyrophosphate, etc. Generally 0.1 to 50 ppm of titanium ion and 50 to 3,000 ppm of the polyphosphate in the wash or bleach solution is capable of providing adequate stain inhibition. It is preferred to employ from about 0.5 to 10 ppm of titanium ion and from about 100 to 1,000 ppm of the polyphosphate, based on the wash or bleach solution.

The following examples describe a test laundry procedure employed to facilitate the evaluation of the ability of the herein disclosed stain inhibitors to prevent manganese discoloration. In this procedure, swatches of white cloth were washed in a wash-water containing about 0.1 part per million of manganese ions, 180 parts per million, calculated as calcium carbonate, of calcium and magnesium ions in a molar ratio of 2 to 1, about 0.25 weight percent of a detergent formulation, as described, and about 200 parts per million of chlorine.

The cloths were washed, under conditions which simulated normal laundry procedure, in a small washing machine with the temperature of the wash solution at about 120°F. The wash procedure consisted of a wash cycle and a fresh water rinse cycle. After rinsing, the cloths were squeeze-dried. The cloths were washed a total of 10 times using each of the described detergent formulations. Reflectance measurements were made on each sample after the second, fifth, and tenth wash, using a General Electric Recording Spectrophotometer at a setting of 430 millimicrons. The reflectance for each measurement on the various formulations are given below.

Example: The following spray-dried nonionic detergent compositions were prepared:

| Components | Composition by weight | |
|---|---|---|
| | 1 | 2 |
| Water | 15.00 | 15.00 |
| Silicate solids (2.4 ratio R.U.) | 4.68 | 4.68 |
| Sodium carboxymethylcellulose | 0.40 | 0.40 |
| Sterox DJ [1] | 8.00 | 8.00 |
| Pluronic L-60 [2] | 1.00 | 1.00 |
| Pentasodium tripolyphosphate | 40.00 | 40.00 |
| Sodium sulfate | 30.57 | 30.41 |
| Miscellaneous (dyes, etc.) | 0.35 | 0.35 |
| Titanium trichloride (=0.05% Ti) | | 0.16 |
| Total | 100.00 | 100.00 |

[1] Dodecylphenol condensed with 10 moles (avg.) of ethylene oxide.
[2] Polyoxypropylene having a molecular weight of from 1,500 to 1,800.

The washing procedure outlined above was followed. The reflectance measurements on these samples are given below:

| Wash: | Composition number | |
|---|---|---|
| | 1 | 2 |
| 2 | 87.5 | 88.1 |
| 5 | 83.1 | 86.2 |
| 10 | 65.9 | 74.2 |

It can be seen from the above data that the detergent formulation containing titanium trichloride, i.e., Composition 2, is characterized by a significantly higher reflectance than the control. This increased reflectance indicates the substantial reduction of manganese staining on the cloths washed with the inhibitor-containing formulation.

### Zinc Sulfate Stabilizer for Hypochlorite Containing Amino Phosphonic Acid

T.M. King; U.S. Patent 3,629,124; December 21, 1971; assigned to Monsanto Company found that zinc sulfate and similar salts of copper, aluminum and cadmium prevent the interaction of hypochlorite and other chlorine releasing agents and amino phosphonic acids present in a bleaching composition. The amino phosphonic acids serve as sequestering or chelating agent in the composition for the calcium and magnesium ions in hard waters. The amino phosphonic acids have the general formula:

$$R_n-N \left( \begin{array}{c} X \\ | \\ -C-P=(OH)_2 \\ | \quad \| \\ Y \quad O \end{array} \right)_{3-n}$$

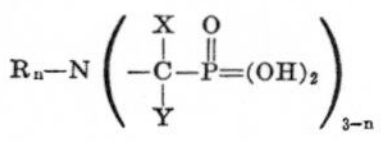

wherein: n represents an integer 0 or 1; X and Y represent hydrogen or alkyl; R represents hydrogen, aliphatic hydrocarbon, halo-substituted aliphatic hydrocarbon, hydroxy-substituted aliphatic hydrocarbon or:

$$-\left( \begin{array}{c} X \\ | \\ C \\ | \\ Y \end{array} \right)_m -N-Z' \atop | \atop Z$$

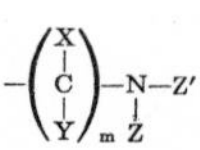

wherein: m represents an integer from 1 to 30; X and Y represent hydrogen or alkyl; Z represents hydrogen or

$$-\overset{X}{\underset{Y}{C}}-\overset{O}{\underset{}{P}}=(OH)_2$$

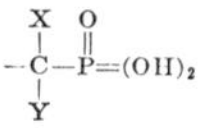

Z' represents

$$-\overset{X}{\underset{Y}{C}}-\overset{O}{\underset{}{P}}=(OH)_2 \text{ or}$$

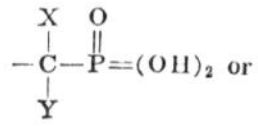

$$-\left(\overset{X}{\underset{Y}{C}}-\overset{}{\underset{Z}{N}}\right)_p \overset{X}{\underset{Y}{C}}-\overset{O}{\underset{}{P}}=(OH)_2$$

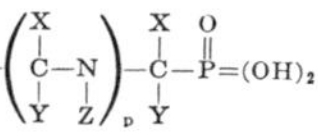

wherein: p represents an integer from 1 to 30. As used hereinafter the term "amino phosphonic acids" generically describes all of the foregoing. With respect to the foregoing general formula it should be noted that when R is either an aliphatic hydrocarbon, halo-substituted aliphatic hydrocarbon or hydroxy-substituted aliphatic hydrocarbon, it is preferably either the saturated or double-bonded unsaturated form containing from 1 to 30 carbon atoms with 6 to 30 carbon atoms being particularly preferred.

When either X or Y is an alkyl group it is preferred that the alkyl group contains from 1 to 30 carbon atoms with lower alkyl groups containing from 1 to 4 carbon atoms being particularly preferred. One of the preferred classes of the foregoing amino phosphonic acids are the amino tri(lower alkylidene phosphonic acids) or the water-soluble salts thereof and which acids have the general formula:

$$N-\left[\overset{X}{\underset{Z}{C}}-\overset{O}{\underset{OH}{P}}-OH\right]_3$$

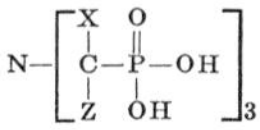

wherein X and Y are the same as defined above, i.e., either hydrogen or alkyl. The following compounds are illustrative of the amino phosphonic acids:

$$N(CH_2PO_3H_2)_3$$
$$N[C(CH_3)(CH_3)PO_3H_2]_3$$
$$CH_3N(CH_2PO_3H_2)_2$$
$$n-C_4H_9N(CH_2PO_3H_2)_2$$
$$(H_2O_3PCH_2)_2NCH_2CH_2N(CH_2PO_3H_2)_2$$

In place of the amino phosphonic acids, alkylene diphosphonic acids may also be used. These have the general formula:

$$(OH)_2=\overset{O}{\underset{}{P}}-\left(\overset{X}{\underset{Y}{C}}\right)_n-\overset{O}{\underset{}{P}}=(OH)_2$$

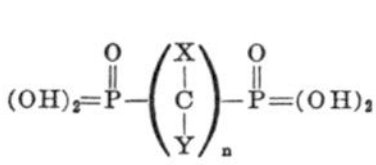

wherein n is an integer from 1 to 10, X represents hydrogen or lower alkyl (1 to 4 carbon atoms, such as methyl, ethyl, n-propyl, n-butyl and isomers thereof), and Y represents hydrogen, hydroxyl or lower alkyl (1 to 4 carbon atoms, such as methyl, ethyl, n-propyl, n-butyl and isomers thereof). Compounds illustrative of these alkylene diphosphonic acids include the following:

methylenediphosphonic acid, $(OH)_2(O)PCH_2P(O)(OH)_2$

ethylidenediphosphonic acid, $(OH)_2(O)PCH(CH_3)P(O)(OH)_2$

isopropylidenediphosphonic acid, $(OH)_2(O)PC(CH_2CH_3)P(O)(OH)_2$

1 hydroxy, ethylidenediphosphonic acid, $(OH)_2(O)PC(OH)(CH_3)P(O)(OH)_2$

Examples of stabilizing salts, in addition to zinc sulfate are: copper chloride, zinc chloride, aluminum nitrate, copper nitrate, cobalt nitrate, nickel nitrate, aluminum acetate and zinc acetate.

Example: For household dry bleaching the following additives within the ranges specified when incorporated with the chlorine-releasing agent give an effective formulation.

| | |
|---|---|
| Chlorine-releasing agent,[1] percent available chlorine per total weight of formulation | 5 to 10. |
| Additives, percent by weight: | |
| Stabilizing material (zinc chloride) | 0.01 to 15. |
| Threshold-sequestering amino phosphonic acid agent | 0.0001 to 50. |
| Inorganic phosphate [2] | 0 to 50. |
| Inert additive [3] | 30 to 75. |
| Organic anionic surfactant | 0 to 10. |

[1] Chlorinated trisodium phosphate, trichloroisocyanuric acid, dichloroisocyanuric acid, sodium dichloroisocyanurate, potassium dichloroisocyanurate or mixtures of these.
[2] Sodium or potassium -tripolyphosphate, -pyrophosphate, -orthophosphate or mixtures of these.
[3] Sodium or potassium -carbonates, -borates, -silicates, -metasilicates, -sulfates, -chlorides or mixtures of these.

The following dry composition (parts by weight) is especially adapted for use as a household dry bleach in an aqueous system at a concentration of 50 to 100 ppm available chlorine for bleaching and stain removal.

| | |
|---|---|
| Potassium dichloroisocyanurate | 13.0 |
| Sodium tripolyphosphate | 25.0 |
| Amino tri(methylene phosphonic acid) | 4.2 |
| Zinc chloride | 0.8 |
| Sodium sulfate | 55.0 |
| Sodium dodecyl benzene sulfonate | 2.0 |
| | 100.0 |

## SPECIALLY TREATED HYPOCHLORITES

### Calcium Hypochlorite Thickened with Sodium Silicate

D.J. Jaszka and R.W. Marek; U.S. Patent 3,036,013; May 22, 1962; assigned to Olin Mathieson Chemical Corporation describe a process to stabilize granules of calcium hypochlorite with a solution of sodium silicate. Other water soluble alkali metal salts such as borax, sodium metaborate, sodium carbonate, trisodium phosphate, disodium phosphate or potassium fluoride may also be used. Sodium silicate is the preferred soluble salt which provides an insoluble calcium silicate coating on the calcium hypochlorite granule. Sodium silicate is cheap, readily available and most effective since the resulting coating is tightly adherent, stabilizes the calcium hypochlorite with respect to loss of available chlorine on storage and prevents pinholing.

The sodium silicate solution used for the treatment of calcium hypochlorite granules is neutral or alkaline and has a concentration of from 5 to 50% $Na_2SiO_3$ but preferably from 15 to 25%. Using solutions in the preferred range of concentration, the time of contact with the calcium hypochlorite granules should be 0.1 to 1 minute at room temperature. The time should be increased for more dilute solutions and decreased for more concentrated solutions. It should be increased at lower temperatures and decreased at higher temperatures.

The treating solution may be applied by spraying or any other suitable means but immersion is ordinarily most convenient. Under these conditions the calcium hypochlorite granules are coated only on the surface and the product has the desired improved properties including enhanced stability and inability to cause pinholing in normal use. Products prepared under the preferred conditions are composed of 80 to 95% of a core of calcium hypochlorite with a coating of calcium silicate amounting to 5 to 20%.

In the treatment of calcium hypochlorite granules with sodium silicate solution, a reaction occurs in which the calcium is converted to calcium silicate forming sodium hypochlorite as a by-product. Most of the relatively unstable sodium hypochlorite is removed with the liquor since it is very soluble in water and any residual sodium hypochlorite is largely decomposed in drying. The resulting coated granules, consisting largely of calcium hypochlorite with a coating of calcium silicate, have enhanced stability and freedom from pinholing.

Example: A saturated solution of borax made up from approximately 20 grams of $Na_2B_4O_7 \cdot 10H_2O$ and 40 grams of water was poured over 40 grams of granular high test calcium hypochlorite contained in a Buchner funnel. The solution was immediately removed by suction and the residue was vacuum dried to a powder containing 58.3% available chlorine. From the available chlorine content, the approximate composition of the granule is 80% calcium hypochlorite covered with 20% of calcium borate. When tested the product caused no pinholing or damage to the cotton fabric. When stored at room temperature in a glass stoppered bottle for 70 days, the available chlorine dropped from an initial 66.2% to a final 47.3%, a loss of 0.27% per day. This is sufficiently low to be satisfactory for commercial use. In a bleaching test of tea stained cotton cloth, reflectances were raised from 57 to 84 using the product of this example compared with a bleach from 58 to 85 using untreated "HTH." The bleaching effectiveness of the calcium hypochlorite was thus not significantly affected by the treatment.

## Packaged Calcium Hypochlorite for Household Laundry

H.L. Robson and W.H. Sheltmire; U.S. Patent 3,154,495; October 27, 1964; assigned to Olin Mathieson Chemical Corporation describe a process to release controlled amounts of hypochlorite in household laundry by packaging calcium hypochlorite and sodium carbonate in envelopes of nonwoven fabrics made of thermoplastic fibers resistant to hypochlorite. On immersion in water, the sodium hypochlorite formed is leached from the envelope, while the precipitated calcium carbonate is retained in the envelope.

Example: An envelope was made up from a laid Dacron (polyethylene terephthalate) fabric weighing approximately 1 1/2 ounces per square yard having a porosity of 3.0. A sheet approximately 8 inches in length and 4 inches in width was folded in half to form an envelope about 4 inches square. Into the envelope was placed approximately one ounce of an intimate mixture of Ad-Dri bleach (product of U.S. Patent 2,959,554) containing about 8% available chlorine. The envelope was then heat-sealed along the two opened edges and at the seam where the ends over-lapped, thereby forming a completely sealed envelope.

Approximately 10 pounds of white goods was placed into a Kenmore automatic washing machine and the wash cycle was set to high water level, 12 minutes' duration and hot temperature. After the correct water level was reached and the water shut off, a synthetic detergent and at the same time the Dacron envelope with Ad-Dri bleach were introduced. The washing and rinsing then proceeded normally. These observations were noted: (1) The introduction of the Dacron envelope containing Ad-Dri bleach had no visible adverse effects on the cloth. (2) The envelope tended to remain on the surface at all times even after the envelope was purposely placed far down into the washing machine. (3) All of the bleach mixture appears to dissolve in 3 to 4 minutes. (4) At the end of the 12 minute wash cycle, there was no residue left in the bleach envelope nor was there any physical damage of any kind to the envelope. (5) An available chlorine determination by colorimetric methods showed a slight available chlorine content at the end of the wash cycle but none at the end of the first rinse cycle.

## Preparation of Dry, Stable Lithium Hypochlorite

G.J. Orazem, R.B. Ellestad and J.R. Nelli; U.S. Patent 3,171,814; March 2, 1965; assigned to Lithium Corporation of America, Inc. describe a process to produce dry and stable granules of lithium hypochlorite suitable for use in home laundry detergent and bleaching compositions. The starting material is an aqueous lithium sulfate solution containing essentially lithium sulfate, sodium sulfate, and potassium sulfate. Such solutions are derived from the production of lithium sulfate by a procedure which includes decrepitating spodumene, roasting the decrepitated spodumene with sulfuric acid or under such conditons as to convert the lithium values to lithium sulfate, leaching out the lithium sulfate with water, neutralizing excess acid in the resulting solution with limestone or the like generally to a pH of about 6, filtering, purifying by removal of calcium and magnesium, and concentration of the filtrate by evaporation of water therefrom. Such resulting aqueous solutions may contain from 120 to 270 g./l. of lithium sulfate; from 10 to 150 g./l. of sodium sulfate; from 0 to 60 g./l. of potassium sulfate. The starting solutions preferably contain from 210 to 230 g./l. of lithium sulfate, and from 80 to 100 g./l. of sodium sulfate.

In this process the starting solution is first admixed with aqueous sodium hydroxide in approximately stoichiometric quantity to react with the lithium sulfate in the starting solution to convert it to lithium hydroxide and thereby form additional sodium sulfate. While any dilution of aqueous sodium hydroxide can be used in this step, it is desired, for reasons of economy and for reductions in volume of liquids to be handled, to employ strong aqueous solutions of sodium hydroxide, advantageously aqueous solutions containing from about 40 to 50% of sodium hydroxide. The addition of the aqueous sodium hydroxide to the starting solution results in the formation of a slurry or solution in which the concentration of sodium sulfate is very substantially increased. In order to remove excessive amounts of sodium sulfate from the slurry or solution, the latter is cooled down to a low temperature. Before doing so, however, it is desirable in certain cases, dependent on the composition of the starting solution, to dilute the slurry or solution by the addition of water thereto, in sufficient quantity so that the mixture when cooled down to a low temperature, advantageously to about 32°F., or somewhat above or below said temperature, for instance, 30° to 35°F., in a crystallizer, results in the optimum quantity of Glauber's salt ($Na_2SO_4 \cdot 10H_2O$) being crystallized out.

The mass is filtered and the filter cake is washed with water and the water washings are added to the filtrate. The washed Glauber's salt may, if desired, be purified and represents a salable by-product. The recovered solution, comprising the filtrate and water washings, now contains essentially lithium hydroxide, sodium sulfate and potassium sulfate. To the recovered solution, there is then added, in at least a stoichiometric proportion, aqueous sodium hydroxide and a stoichiometric proportion of chlorine, to convert the lithium hydroxide to lithium hypochlorite in accordance with the following equation:

$$LiOH + NaOH + Cl_2 \longrightarrow LiOCl + NaCl + H_2O$$

As in connection with an earlier step in the process, it is desirable to use strong aqueous solutions of sodium hydroxide, advantageously aqueous solutions containing about 40 to 50% sodium hydroxide. It is desirable to utilize a slight excess of sodium hydroxide over stoichiometric proportions in the formation of the lithium hypochlorite from the lithium hydroxide in order to enhance the stability of the lithium hypochlorite. This slight excess of sodium hydroxide, which is desirably of the order of about 1/2%, serves to inhibit loss of available chlorine from the lithium hypochlorite in the lithium hypochlorite solution prior to the spray drying step, as well as subsequently during storage and shipping of the dry lithium hypochlorite compositions.

The lithium hypochlorite-containing solution is now ready for spray drying. In a typical illustration of the nature of such solution, it will contain about 12% lithium hypochlorite, 12 to 13% sodium chloride, 4 to 5% sodium sulfate, 0.7 to 0.8% potassium sulfate, very slight amounts of lithium hydroxide or sodium hydroxide, and the balance, of the order of about 70%, water.

Various types of spray driers can be used but it is particularly desirable to utilize the so-called parallel flow spray drier. The lithium hypochlorite-containing solution is advantageously fed into the spray drier at room temperature. The inlet air temperature should be between 380° to 420°F. and the outlet air temperature should be between 200° and 240°F.

<u>Example:</u> 320 parts of a 50 weight percent aqueous sodium hydroxide solution was added to 1,200 parts of an aqueous lithium sulfate solution taken from a process stream, said 1,200 parts of solution containing 220 parts of $Li_2SO_4$, 95 parts of $Na_2SO_4$, 20 parts of $K_2SO_4$, and 865 parts of water. The mixture was diluted with 220 parts of water, and then cooled to 32°F., resulting in the crystallization of Glauber's salt ($Na_2SO_4 \cdot 10H_2O$), which was removed by filtration. Then 310 parts of 50 weight percent aqueous sodium hydroxide solution was added to the filtrate, followed by chlorination with 275 parts of gaseous chlorine. The chlorinated solution

was spray dried in a spray dryer, using inlet and outlet temperatures of 400°F. and 220°F., respectively, and the spray dried composition was rapidly cooled to below 120°F. The resulting dried composition showed, on analysis, 5% moisture and 40% available chlorine. Illustrative of particularly satisfactory dry lithium hypochlorite compositions produced in accordance with this process are those containing approximately 35 to 40% available chlorine.

The dry lithium hypochlorite compositions are in the form of white particles whose average particle size varies but which, in typical instances, will lie in the range of 60 to 80 microns and whose bulk density will commonly lie in the range of 38 to 40 pounds per cubic foot. They may be admixed with surfactants, such as benzene-, toluene-, xylene-, dodecyl-, tridecyl-, tetradecyl and pentadecyl benzene sodium sulfonates, and said surfactants, which may be utilized in varying amounts, generally up to about 1.5 to 2% by weight of the dry lithium hypochlorite compositions, may be added before or after the spray drying step.

In general, those compositions which contain surfactants will tend to have somewhat greater average particle sizes and lesser bulk densities than the corresponding compositions without the surfactants. The lithium hypochlorite compositions, either with or without surfactants, can be compacted, if desired, to provide granules of particular character and bulk densities for convenience in handling and packaging.

The dry lithium hypochlorite compositions made in accordance with this process have shown excellent bleaching properties when used for bleaching cotton and other fabrics and with about the same results as to fiber strength and on various types of dyes as sodium hypochlorite. They are compatible with various substances such as sodium sulfate, sodium phosphates and polyphosphates, abrasives, sodium silicates, optical brighteners, and the like.

They may, thus, be diluted therewith or with various diluents to produce products, for home laundry use, scouring powders, and the like, containing, say, from 4 to 8% available chlorine or less. In general, the lithium hypochlorite dry compositions should be kept in closed containers to avoid undue deterioration. For bulk shipping, it is desirable to use fiber drums with polyethylene liners.

## Manufacture of Javel Extract by Fluidized Bed Process

J. Aigueperse and J. Barjhoux; U.S. Patent 3,287,233; November 22, 1966; assigned to Societe d'Electro-Chimie, D'Electro-Metallurgie et des Acieries Electriques d'Ugine, France describe a fluidized bed process using the sodium chloride crystals formed as part of the fluidized bed to manufacture concentrated sodium hypochlorite solutions (Javel extract) of between 70 to 140 chlorometric degrees. The concentrated hypochlorite solution is then diluted to 30 to 70 chlorometric degrees. This dilution may be carried out at the point of usage.

The process comprises continuously reacting chlorine and sodium hydroxide in a suspension of sodium chloride crystals maintained in a fluidized bed and under such conditions that precipitation of sodium chloride effected by the reaction is essentially made with increase in size of the crystals which constitute the fluidized bed, and preferentially with a precipitation as separated fine particles. According to the process, a continuous current of a hypochlorite solution of the final desired grade is introduced at the bottom of a vertical reactor used as crystallizing apparatus and containing a suspension of sodium chloride crystals of a suitably chosen granulometry. This solution rises up in the crystallizing reactor by maintaining in suspension in the fluidized bed the crystals on which takes place the greatest part of the sodium chloride precipitation. Gaseous chlorine, diluted or pure, and a quantity of a sodium hydroxide solution, that substantially stoichiometrically required, are injected into the bottom of the reactor. From the top of the reactor the desired clear solution is recovered, and on the other hand, the necessary solution to provide the continuous current of the hypochlorite solution which is injected into the bottom of the reactor after its passage through a cooling apparatus.

Precipitation of the sodium chloride resulting from the reaction continuously takes place in the vicinity of the solid crystals which constitute the bed, and preferably upon these crystals, which increase in size. The crystals which have reached a sufficient dimension travel progressively down towards the bottom of the bed, and it is easy to withdraw from this area the big grains or crystals of regular sizes. After draining, these grains include only small quantities of mother-waters, less than 3%.

To avoid formation of fine particles in excessive quantities, the whole of the volume of the reactor is filled by the fluid bed. Additionally, the surface of the crystals which form the fluidized bed must be large enough to allow the greatest part of the sodium chloride which precipitates to cooperate and increase the size of the grains or crystals while taking into account the speed of increase in size of the sodium chloride crystals.

The average granulometry of the fluidized bed is not critical, but sizes of grains between 0.5 and 4 mm. are particularly recommended. The cooling apparatus, which is traversed by the solution taken from the top of the reactor and introduced again at the bottom, is used to eliminate calories or heat produced by the reaction. The temperature is maintained under 35°C., preferably under 30°C. in the whole of the liquid circuit.

Example: The process is carried out in the apparatus schematically shown in Figure 3.1. This apparatus comprises a crystallizing reactor (1) at the bottom of which a recycled hypochlorite solution of 100 chlorometric degrees is introduced by conduit (2), gaseous chlorine is added by a line (3) and a sodium hydroxide solution is delivered by a line (4). The large sodium chloride crystals which have formed are withdrawn through pipe (5). Near the top of the reactor, the finished product is extracted through pipe (6) and the solution to be recycled is removed by a line (7). A pump (8) effects extraction of the solution to be recycled and sends it back through a cooling apparatus (9) and then to the reactor (1) through the conduit (2).

The reactor had a diameter of 25 cm. and a height of 1.20 m. The sodium chloride crystals in suspension in the reactor had an average granulometry of about 1 mm. The output of the solution introduced at the bottom of the reactor was 1 $m.^3/h.$ and the reactives were introduced continuously at the rate of 10 kg./h. for the chlorine and of 26.5 l./h. for the sodium hydroxide solution which has a concentration of 430 g./l. The solution to be recycled enters the cooling apparatus at 19°C. and exits therefrom at 14°C.

31 l./h. of Javel extract of 100 chlorometric degrees and 4.3 kg./h. of sodium chloride crystals of a granulometry of about 2 mm. were produced.

FIGURE 3.1: JAVEL EXTRACT REACTOR

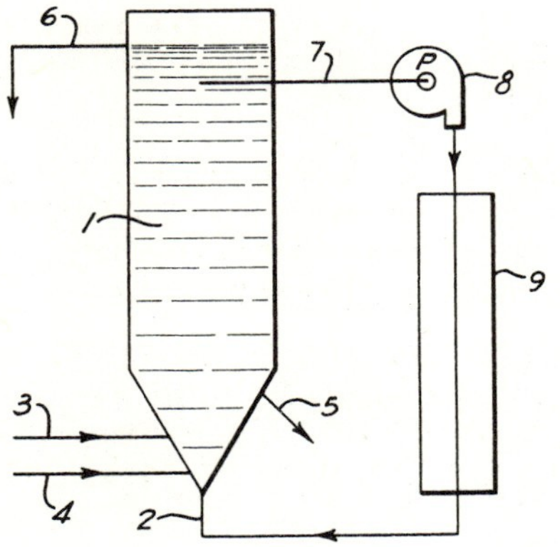


The dilution of the obtained extract was made in a tank equipped with an agitator. Mixing of one volume of Javel extract of 100 chlorometric degrees with one volume of water furnished a Javel extract of 50 chlorometric degrees. This Javel extract of 50 chlorometric degrees produced had a density of about 1.16. Its sodium hypochlorite content was 166 g./l. and the ratio NaOCl/NaCl was 2.25.

## Removal of Iron from Hypochlorite Solutions

M.J. Skrypa, F.R. Baran and W.W. Low; U.S. Patent 3,557,010; January 19, 1971; assigned to Allied Chemical Corporation describe a process for the production of hypochlorite solutions containing less than 0.1 part per million of iron which comprises adding calcium chloride and an alkali metal carbonate to the hypochlorite solutions containing iron impurities and permitting the precipitation of calcium carbonate which results in hypochlorite solutions containing less than 0.1 part per million of iron.

Example 1: The preparation of sodium hypochlorite is as follows. About 2,700 ml. of caustic soda solution prepared by diluting 50% mercury cell caustic soda to 14% with water was poured into a 4 l. beaker equipped with a glass cooling coil and a propeller stirrer. The stirred solution was cooled to a temperature of 15° to 20°C. before the addition of any chlorine.

The chlorine was introduced through a glass sparger with the apparatus set up to introduce liquid chlorine into the sparger. Essentially all of the chlorine was vaporized in the sparger before passing into the caustic soda solution. The temperature was maintained below 20°C. and chlorination was continued until an excess of up to 0.15% caustic remained. The sodium hypochlorite solution contained about 13% by weight NaOCl. Dilute solutions are prepared by mixing the concentrated bleach with water to the desired concentration.

Example 2: This example demonstrates that treatment of a sodium hypochlorite bleach with sodium carbonate and calcium chloride not only reduces the iron content significantly, but prevents any post-filtration precipitation. The sodium hypochlorite bleach was that prepared in Example 1, and contained 13% NaOCl. The calcium chloride was added in the form of a dilute aqueous solution containing approximately 3.5 to 20 weight percent $CaCl_2$ and which was prepared by dissolving the desired amount of calcium chloride in 20 ml. of water. This solution was then poured into approximately 1,500 ml. of the bleach.

The sodium carbonate previously had been added to the bleach as a powder and dissolved. The treated bleach solution was separated from the precipitate of calcium carbonate by filtration. The treated bleach was analyzed for its iron content after which it was allowed to stand for observation of the formation of any precipitate.

It was readily seen that not only was the iron content of the bleach reduced to below acceptable levels, but there was also no post-precipitation. Further, it was demonstrated that it is preferred to maintain the sodium hydroxide content in the bleach below about 0.07% by weight to achieve the most effective removal of iron.

## Liquid Hypochlorite Bleach Thickened with Clay

B.J. Zmoda; U.S. Patent 3,558,496; January 26, 1971 describes a process to thicken liquid hypochlorite or other bleach solutions from chlorinated organic compounds, by the addition of two clay dispersions, one charged positively and the other negatively. The thickened bleach may be applied to any, including vertical, surfaces. It will remain in contact with these surfaces to perform bleaching function.

The clay carrying a negative charge may be exemplified by clays such as a refined hectorite. Also applicable is the modified clay disclosed in U.S. Patents 3,109,847 and 2,974,108. The clay carrying a positive charge may be exemplified by materials such as a colloidal alumina which is composed of boehmite alumina fibrils. The tiny crystals are grown from solution.

Baymal is a porous material with a specific surface area of 275 square meters. Acetate groups absorbed on the surface of the micro crystals constitute about 13% of the product. Also applicable is the positively charged alumina coated silica disclosed in U.S. Patent 3,007,878 and the chemically modified alumina disclosed in U.S. Patent 3,031,418.

Example:

| | Weight percent | Actives percent |
|---|---|---|
| Baymal, 10% aqueous solution | 10.0 | 1.0 |
| Macaloid, 5% aqueous solution | 10.0 | 0.5 |
| 5.25% sodium hypochlorite solution | 40.0 | 2.0 |
| Water | Balance | |
| Total | 100.0 | |

The composition is prepared by reshearing the sols of the Baymal and the Macaloid in the balance of water. After this is performed, the sodium hypochlorite solution is added. Sodium hydroxide is added to adjust the pH to 12 to 12.5 in order to obtain better stability of the sodium hypochlorite solution. The Baymal and the Macaloid have a synergistic action in thickening the sodium hypochlorite. The composition thus formed is thixotropic and has a low shear, yet will remain in place for extended periods when poured upon any surface. Available chlorine is 2.0%. After 12 weeks' aging at room temperature in an uncolored bottle, 1.8% chlorine remained.

Stable, Dry Calcium Hypochlorite Bleach for Washing Machines

A. Long and D.L. Sawhill; U.S. Patent 3,639,284; February 1, 1972; assigned to Olin Corporation describe a process to prepare granular, free flowing bleaching composition for washing machines consisting of calcium hypochlorite, sodium tripolyphosphate and a diluent salt such as sodium metasilicate.

The process provides a solid bleaching composition in the form of discrete particles at least 90% of which pass 10 mesh and are retained on 40-mesh U.S. Standard screens, substantially each of said particles consisting essentially of calcium hypochlorite, sodium tripolyphosphate and at least one solid, anhydrous diluent alkali metal salt having a cation selected from the group consisting of sodium and potassium and an anion selected from the group consisting of metasilicate, carbonate, bicarbonate, chloride, pyrophosphate, acid pyrophosphate, phosphate, acid phosphate and mixtures thereof;

(a) the weight ratio of the sodium tripolyphosphate to the calcium hypochlorite being from 2.9:1 to 8.6:1;

(b) the weight ratio of said diluent salt to calcium hypochlorite being from 0.4:1 to 4:1; and

(c) the weight ratio of the total of sodium tripolyphosphate plus diluent salts to calcium hypochlorite being from 2.9:1 to 13.1.

Commercial calcium hypochlorite as generally sold contains at least 70% available chlorine and usually contains 71 to 73%. The commercial HTH brand of calcium hypochlorite (granular) typically has a minimum (loose packed) density of about 0.79 and a maximum (shaken) density of about 0.83. Commercial sodium tripolyphosphate, suitably in the form of spray-dried beads, granular or powdered is used in preparing the compositions. The diluent salts used include metasilicates, chlorides, carbonates and phosphates of both sodium and potassium.

The test procedures for washings are as follows. All washes were carried out in a Mark XII Whirlpool top-loading machine having a total capacity of 16 gallons. Water temperature was 130° to 135°F. in all tests. In each test, 20 pieces of unbleached broadcloth 36 x 48 inches in size were used. To the water was added the cloth, one ounce of a commercial alkylaryl sulfonate type detergent Tide and 120 g. of the dry bleach composition. The wash cycle consisted of 12 minutes and the total cycle for fill, wash, rinse and spin was 40 minutes. After each washing, the cloths were dried in a Mark XII Whirlpool electric clothes dryer at 140°F. for 30 minutes.

The test procedures for brightness and whiteness are as follows. Bleaching efficiency was determined by measurements commonly used in the textile industry. A piece of cloth after the washing, bleaching and drying operations was ironed lightly with a steam iron and subjected to light-reflectance reading with a Model 610 Photovolt Reflection Meter. Readings in percent reflectance were taken with blue, amber and green filters.

Each recorded reading was an average of three or four individual readings taken on each swatch backed by three similar swatches. The reflectance readings were taken automatically at 45° to the incident light beam. From the percent reflectance readings, brightness and whiteness were calculated as follows.

Brightness = (blue and amber and green filter readings)/3

Whiteness = 4 blue filter readings - 3 green filter readings

The brightness and whiteness figures are purely empirical and have no units. They closely represent what the human eye perceives. The larger the number, the brighter or whiter the fabric. Brightness values over 80 are considered good and whiteness values over 70 are considered to be good.

Example: The following mixture of 200 lbs. HTH granular, 1,100 lbs. sodium tripolyphosphate granular, and 100 lbs. sodium metasilicate granular was blended for 30 minutes in a ribbon blender and Chilsonated at a pressure of 10,000 lbs. per lineal inch on the rolls of a Chilsonator Model 7 LX 10D under the following conditions:

| | |
|---|---|
| Chilsonator roll speed | 18 rpm |
| Chilsonator motor load | 10 - 15 amp. |
| Chilsonator roll pressure | 1,200 - 2,200 lbs. total (pulsating) |
| Horizontal-feed screw speed | 160 rpm |
| Vertical-feed screw load | 4 amp. |
| Horizontal-feed screw load | 1.4 amp. |

The product was crushed and screened to separate a fraction in the -10 +40 mesh range. It contained 9.4% available chlorine. The pH of the wash water containing 120 g. of this product with 1 ounce of Tide and 20 test cloths was 10.6. Twenty washings, bleachings and dryings were performed as described above and the bleaching efficacy was determined as before with the following results:

| | Brightness | Whiteness |
|---|---|---|
| Original | 73.5 | 41.5 |
| After 20 washes | 86.5 | 75.3 |

The bleach was effective and no pinholing was observed.

## Calcium Hypochlorite Resistant to Ignition

J.P. Faust; U.S. Patent 3,669,894; June 13, 1972; assigned to Olin Corporation describes a process to prepare calcium hypochlorite (70%) with higher resistance to decomposition when heated and specifically resistant to ignition from a lighted cigarette. In this process a calcium hypochlorite containing 85 to 90% calcium hypochlorite is exposed to an inert gas, e.g., air or nitrogen, having a humidity of 80 to 100% at temperatures of 80° to 110°F. for a time sufficient to form a hypochlorite product containing 6 to 12% water.

The granular (0.05 to 3 mm.) calcium hypochlorite product contains from 75 to 82% calcium hypochlorite. It is not only safened with respect to ignition and exothermic decomposition but also is marketable as a guaranteed 70% calcium hypochlorite having a suitable margin for loss during shelf storage.

Example 1: A sample of Eimco filter cake containing 39.42% calcium hypochlorite and 48.8% water was taken from an operation producing commercial 70% calcium hypochlorite. It was further pressed in a porous bag to reduce the water content and then dried in a vacuum oven. The resulting dried material contained 86% calcium hypochlorite and 1% water. The dried material was exposed to air at 100°F. having 90% relative humidity for 1 hour. It then contained 76.24% calcium hypochlorite and 11.28% water. When a lighted cigarette was laid on the product, it produced no self-sustaining reaction although the 86% material reacted vigorously in a self-sustaining reaction until all of that product was decomposed.

Example 2: Calcium hypochlorite filter cake from a commercial operation for preparing 70% calcium hypochlorite and containing about 25% water was vacuum dried to 4.65% water and 85.3% calcium hypochlorite. This starting material was reactive toward lighted cigarettes and burning matches.

Part of the above-described starting material was placed in a desiccator, the bottom of which contained water at 104°F. The desiccator was flushed thoroughly with nitrogen to remove all the air. The product remained in the atmosphere of wet nitrogen for 20 minutes. A portion was removed and found to be unreactive toward lighted cigarettes and matches. The product was found to contain 82% calcium hypochlorite and 9.3% water.

Example 3: Another portion of the starting material described in Example 2 was exposed to air at 100°F. having 90% relative humidity for 10 minutes. A sample was found to be unreactive toward lighted cigarettes and matches. It contained 10.4% water and 80.85% calcium hypochlorite.

## Thickened Hypochlorite Bleaching and Cleaning Composition

B.M. Hynam, J.L. Wilby and J.R. Young; U.S. Patent 3,684,722; August 15, 1972; assigned to Lever Brothers Company found that increasing the viscosity of aqueous hypochlorite solutions by the addition of a $C_8$ to $C_{18}$ alkali metal soap and hydrotropes such as amine oxides and betaines improves the bleaching and disinfecting properties of the solutions, particularly if used as cleaners for lavatories.

Of these additives, the amine oxides have the structural formula:

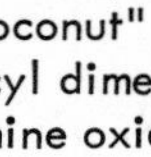

wherein R is a $C_8$ to $C_{18}$ alkyl group, preferably a $C_{10}$ to $C_{16}$ alkyl group, and $R_2$ and $R_3$ can be short-chain alkyl groups, such as methyl, ethyl, n-propyl and isopropyl. $R_2$ and $R_3$ will generally be the same, but can differ if this is desired. Typical amine oxides suitable are lauryl dimethylamine oxide, myristyl dimethylamine oxide, cetyl dimethylamine oxide, "coconut" dimethylamine oxide, "hardened tallow" dimethylamine oxide, hexadecyl dimethylamine oxide, lauryl diethylamine oxide, and "coconut" diethylamine oxide.

Long-chain substituted betaines suitable for use are compounds having the structural formula:

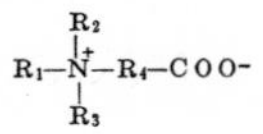

wherein $R_1$ is a $C_8$ to $C_{18}$ alkyl group, preferably a $C_{10}$ to $C_{16}$ alkyl group, and $R_2$, $R_3$ and $R_4$ are $C_1$ to $C_3$ alkyl groups. Specific examples of suitable betaines are octyl, decyl, dodecyl, tetradecyl, hexadecyl and octadecyl dimethyl betaines in which $R_4$ is an alkylene group with 2 or 3 carbon atoms.

The alkali-metal salt (soap) of a fatty acid will usually be the sodium salt, but potassium or lithium salts can also be used. The fatty acid can be any natural or synthetic C8 to C18 fully saturated fatty acid, such as caprylic acid, capric acid, lauric acid, myristic acid, palmitic acid and stearic acid, and the mixtures of fatty acids, suitably hardened, derived from such natural sources as tallow, coconut oil, groundnut oil and babassu oil can be used.

These soaps will not disperse in hypochlorite solution without the aid of a hydrotrope, and the presence of the hypochlorite-soluble surface active agent meets this requirement. A preferred soap is an alkali-metal laurate. Soaps with a shorter carbon chain have been found to be less effective thickeners, and soaps of a higher carbon chain length are less soluble in hypochlorite solution, the insolubility increasing with increasing chain length.

A suitable way of incorporating the thickening mixture is to form a premix of hydrotrope and soap in warm water, and then add hypochlorite solution to the premix to obtain the desired final composition. The alkali-metal hypochlorite will usually be sodium hypochlorite, but potassium hypochlorite and lithium hypochlorite may be used if desired. The alkali-metal hypochlorite will generally comprise from 1 to 10% by weight of the bleaching composition.

Although not an essential ingredient of a bleaching composition of the process, a caustic alkali is generally incorporated into liquid compositions of this type, and hence a bleaching composition can include a caustic alkali, preferably in an amount from 0.5 to 2% by weight, if desired. The caustic alkali will usually be sodium hydroxide, but potassium hydroxide and lithium hydroxide may also be used.

The increased viscosity of the composition due to the presence of the thickening agent improves the cleaning and disinfecting properties of the composition by increasing its tendency to adhere to treated surfaces, for example the internal surfaces of lavatory pans. The preferred viscosity for this purpose is between 10 and 100 centistokes (cs.) at 25°C., as measured using an Ostwald viscometer.

| Hydrotrope | Soap | Hydrotrope:soap ratio |
|---|---|---|
| Ammonyx LO (lauryl di-methylamine oxide). | Lithium caprylate | 80:20 |
| Ammonyx LO | do | 60:40 |
| do | Sodium caprate | 80:20 |
| do | do | 60:40 |
| do | Sodium laurate | 80:20 |
| do | Lithium laurate | 80:20 |
| Ammonyx MO (myristyl dimethylamine oxide). | Lithium caprate | 20:80 |
| Ammonyx MO | do | 60:40 |
| do | do | 80:20 |
| Ammonyx CO (cetyl dimethylamine oxide). | Sodium caprate | 80:20 |
| Ammonyx CO | do | 60:40 |
| Ambiteric D | Sodium myristate | 80:20 |
| do | do | 60:40 |
| do | Lithium myristate | 80:20 |
| do | Sodium palmitate | 80:20 |
| do | do | 60:40 |
| do | Lithium palmitate | 20:80 |
| do | do | 60:40 |
| do | do | 80:20 |
| do | Sodium laurate | 80:20 |
| do | Lithium laurate | 80:20 |

The preceding table lists some examples of the hydrotrope/soap compositions of the process.

In all of these examples the solution thickened was a sodium hypochlorite solution with 10% available chlorine. Samples were prepared at three different concentrations of free alkali in each case (sodium hydroxide concentrations 0.5, 1.0 and 2.0%). The concentration of thickening mixture employed in all cases was 0.5%.

Example: A bleaching composition according to this process is made by admixture of the following components:

| Component | Percent |
|---|---|
| Hydrotrope* | 0.72 |
| Sodium laurate | 0.28 |
| Perfume** | 0.08 |
| Caustic soda | 0.05 |
| 15% aqueous sodium hypochlorite solution | 66.6 |
| Water to 100% | |

*A betaine sold as Ambiteric D.
**A perfume blend of 10% terpene hydrocarbons, 15% cineole, 30% linalol and 10% camphor.

This composition was stable, and retained its perfume and viscosity on storage. Its viscosity was 15 cs.

## HYPOCHLORITES MODIFIED BY STYRENE POLYMERS

### Fabric Brightening Styrene Polymer

W.J. Park; U.S. Patent 3,606,989; September 21, 1971; assigned to Purex Corporation, Ltd. describes a process using styrene-methacrylic acid copolymer additives to hypochlorite bleach to improve the brighteners of natural or synthetic fabrics. The composition used in this process includes dilute aqueous solution of sodium hypochlorite. In common with conventional bleach solution this solution will contain sufficient free alkali to have a pH of at least about 10 and preferably sufficient alkali to have a pH of 11.5 or more e.g. 0.1 to 1.0% by weight free alkali.

Percent concentrations of sodium hypochlorite will be above 2.5% and generally be less than 10%, most solutions being in the range of 3 to 7% by weight sodium hypochlorite. Other bleaching agents may be used including other alkali metal salts generating bromine or chlorine in aqueous solution.

The second nonaqueous component is the water insoluble synthetic organic polymer which is of a particle size and constitution permitting suspension in the aqueous hypochlorite solution. Particle size will generally range between 0.05 and 5 microns, with smaller sizes being less visible and larger sizes less stable in suspension in typical compositions within the process. Concentrations of the polymer will range from an effective amount for imparting added brightness through deposition on fabric to that amount forming a deposit visible as such on fabric. Convenient concentrations are between 0.3 to 3% by weight. It is further characteristic of preferred polymers that they exhibit an index of refraction more than 0.05 unit greater than the index of the aqueous hypochlorite solution. As such the polymer particles are perceptible as a component of the composition although not individually visible, i.e., an opacity is imparted to the bleach composition.

The synthetic organic polymer is any chemical species substantially insoluble and nonreactive in the bleach solution. Numerous polymers will have the requisite chemical stability, i.e., resistance to oxidation. In addition the polymer should be easily suspendible in the bleach composition, i.e., be physically stable in the bleach. This property is realized in polymers having polar groups along the polymer chain, e.g., hydroxyl, carboxyl, carbonyl, peroxy, ester, amide, nitrile and similar hydrophilic groups. With such groups, the polymer chain per se can be hydrophilic or hydrophobic.

Particularly effective are copolymers of the styrenes with between 5 and 25% and preferably up to 15% by weight of an acid group containing comonomer such as, among the organic acid groups, a carboxyl, ester, and nitrile group and among the inorganic acid groups a phosphonic, sulfonic or sulfonate group some of which com monomers are listed above. These polymers are preferred to have specific gravities above about 1.0 at 25°C. and will generally be between 0.95 and 1.2 in specific gravity. Where

Where the acid group is an ester group it may have the structure A—R in which A is an acid radical particularly selected from oxycarbonyl, sulfo or phosphinyl groups and R is an alkyl group containing 1 to 4 carbon atoms. The term "acid group" has reference to the capacity of the group to act as a proton donor in aqueous media.

Example: All parts are by weight. Washing solutions containing 16 gallons of 150 ppm hardness water, a cup of 6% bleach solution (243 ppm available chlorine) and 0.2% Tide detergent were prepared. In the example the bleach solution contained 0.3% (solids basis) of styrene-methacrylic acid emulsion copolymer (10% acid) having a particle size of about 0.3 microns as an opacifier brightener. In the Control the opacifier was omitted from the bleach solution.

Swatches of cotton, nylon and polyester cloth were washed successively 5 times at 120°F. for 15 minutes. Percent reflectance for six swatches of each cloth was measured after odd-numbered washings using a Gardner Color Difference Meter and Hunter D-40 Reflectometer. In all instances the reflectance values for the opacified bleach exceeded those of the nonopacified conventional (Control) bleach. Measurements for blue-red and yellow-green values showed consistently superiority or equivalence for the opacified bleach washed cloths.

A panel of 100 persons was requested to compare nylon and cotton swatches produced in the above washings. 80% of those inspecting the cloths selected the opacified bleach (example) whitened cloth, nylon or cotton, as whiter than the corresponding nonopacified bleach (Control) cloths. Duplication of the above experiment with particulate depositable polymer contents of 0.1, 0.5, 0.75, 1.0, 1.5, 3 and 5% (solids basis) provides equivalent results. Softness or "hand" of the fabric was not deleteriously affected by the deposit of polymer. Conventional anionic and nonionic surfactants may be used in the composition of this process.

Hypochlorite Encapsulated with Styrene-Acrylic Copolymer

R.A. Robinson and B.R. Briggs; U.S. Patent 3,655,566; April 11, 1972; assigned to Purex Corporation, Ltd. describe a process to prepare stable household hypochlorite bleach solution containing optical brighteners finely dispersed and carried by particles of styrene-methacrylate polymer as well as encapsulated by the same polymer of somewhat different monomer ratio. Emulsion polymerization is used to prepare the copolymers, both for the carrier as well as the encapsulating part of the system.

The brightener is protectively carried in the bleach solution by a finely particulate synthetic organic polymer carrier which is dispersed through the solution, e.g., in amounts between 0.05 and 5% by weight based on the total weight of the composition. The carrier comprises polymer particles typically between 0.5 and 2 microns in average particle size and having an inner portion consisting essentially of the optical brightener compound and a styrene-acrylic polymer and an outer portion forming an encapsulating layer over the inner portion and consisting essentially of styrene polymer free of the brightening compound.

The particles comprise per 100 parts by weight of polymer from 65 to 98 parts of a styrene monomer having 8 to 12 carbon atoms and conversely from 2 to 35 parts of methacrylic or acrylic acid monomer copolymerized therewith. The inner portion polymer contains a major proportion but less than 90% by weight of the styrene and at least a major weight proportion of the acid monomer, with the outer portion of the particle polymer containing the balance of these monomers.

From 0.5 to 25 and preferably from 5 to 10 parts per 100 parts by weight of the polymer, of a hydrophilic comonomer may be employed in replacement of a like amount of styrene monomer, e.g., monomer selected from the hydroxyester, ether, amide and cyano derivatives of acrylic or methacrylic and/or a vinyl sulfonate monomer having the formula $R{-}CH{=}CH{-}SO_3Me$ in which R is hydrogen or an aromatic or alkyl radical having up to 10 carbon atoms and Me is an alkali metal.

The particles typically contain from 0.5 to 5% by weight of the optical brightening compound, based on the weight of the styrene in the polymer. The optical brightening compound may be selected from derivatives of 4,4'-diaminostilbene-2, 2'-disulfonic acid, dibenzothiophene-5,5-dioxide, azole, coumarin, pyrazine and 4-aminonaphthalimides.

The polymer particles thus distributed in bleach have been found to be highly substantive to fabric to carry the optical brightener through the wash cycle onto the fabric where it remains to give an appearance of brightness to the fabric. The aqueous solution of bleaching agent typically has a pH of at least 10 and may include as the bleaching agent typically a hypochlorite ion generating compound such as those generally used for bleaching, e.g., a heterocyclic N-chlorimide or sodium hypochlorite, and the like, in amounts between 1 and 10% by weight.

The liquid household bleach opacified in accordance with this process may be any hypochlorite ion containing solution containing sufficient free alkali to have a pH of 10 and preferably 11.5 and higher, typically from 0.1 to 1.0% by weight free alkali. Percent concentrations of hypochlorite ion will range between 1 and 10% by weight with a practical minimum being 2.5%. Most bleaches contain between 3 and 7% hypochlorite ion, and this concentration is most suited to use in the process.

Various other bleaching agents may be used including the heterocyclic N-chlorimides such as trichlorocyanuric acid, dichlorocyanuric acid and salts thereof such as the alkali metal salts, e.g., sodium and potassium dichlorocyanurates. Other imides are hypochlorite ion-generating also in aqueous solution and may be used, e.g., N-chlorosuccinimide, N-chloromalonimide, N-chlorophthalimide and N-chloronaphthalimide.

Example: (a) Polymer Preparation — Dissolve 5 parts of dioctyl ester of sodium sulfosuccinic acid and 2 parts of tetrasodium N-(1,2-dicarboxyethyl)-N-octadecylsulfosuccinamate in 700 parts of water. To this add 50 parts of methacrylic acid followed by 30 parts of a 25% aqueous solution of sodium vinyl sulfonate. Dissolve 5 parts of Calcofluor ALF (optical brightener compound) in 250 parts of styrene. Emulsify the styrene solution in the aqueous phase and carry out the polymerization by heating the emulsion to 130°F. and adding one part of sodium persulfate catalyst.

The temperature rises to about 190° to 200°F. from the exothermic heat of reaction. Cool to 130°F. and add 30 parts of a 25% aqueous solution of sodium vinyl sulfonate followed by 250 parts of styrene. Stir for 30 minutes holding at 130°F. and then add catalyst and polymerize the second monomer addition with the heat of the exotherm and cool.

(b) Bleach Composition Preparation — Add 0.5 percent of the latex obtained in Part (a) to a 5% aqueous solution of sodium hypochlorite.

(c) Stability – Fluorescence is evaluated at periodic intervals. The bleach shows fluorescence initially and for 3 months of 70°F., the normal shelf life for hypochlorite bleach. Fabric washed with the bleach shows greater whitening (brightening) both on visual and instrumental inspection.

Control — Duplicate Example 1 but mix the methacrylic acid first with styrene and then add to the aqueous surfactant solution, followed by heating to polymerize. In parts (b) and (c) the emulsion which had similar size particles and the same milky appearance prior to addition to the bleach as the Example emulsion, shows immediate physical deterioration and quickly demulsifies and settles as a flocculated mass in the holding vessel.

## Styrene Terpolymer Latex Opacifier for Hypochlorite

B.R. Briggs; U.S. Patent 3,663,442; May 16, 1972; U.S. Patent 3,666,680; May 30, 1972; U.S. Patent 3,689,421; September 5, 1972; all assigned to Purex Corp., Ltd. adds a styrene-vinyl acid-vinyl sulfonate latex to aqueous hypochlorite bleach. The opacifying polymer is a particulate form of styrene copolymer formed in a critical manner. Most polymer latices prepared from a styrene, a vinyl acid and a vinyl sulfonate with or without a fourth monomer such as a vinyl acid ester are not stable in highly alkaline, liquid bleach compositions.

Surprisingly the same proportions of monomers provide vastly different opacifying benefits depending on the polymerization technique employed. To get useful latices of particulate styrene copolymer for opacifying bleach it has been found essential to first form a solution of the vinyl acid monomer and vinyl sulfonate monomer in water and the noncationic surfactant prior to adding the styrene monomer to the reaction vessel. In this manner a latex having styrene terpolymer finely dispersed therethrough is obtained which is stable for long periods even at elevated temperatures in liquid household bleach. As the polymer components there may be employed styrene per se or vinyl benzene or a substituted styrene such as vinyl toluene or butyl styrene, i.e., alkyl substituted styrenes in which the alkyl groups contain from 1 to 4 carbon atoms such that the styrene monomer contains from 8 to 12 carbon atoms, inclusive. Or the styrene monomer may be monohalogen ring substituted such as chlorostyrene or bromostyrene. The water soluble vinyl sulfonate monomer has the formula

$$R{-}CH{=}CH{-}SO_3Me$$

in which R is a hydrocarbon radical free of aliphatic unsaturation having up to 10 carbon atoms, e.g., an aromatic radical such as tolyl, benzyl or phenyl radical; an alkyl radical such as methyl, ethyl, propyl, butyl, isobutyl, pentyl, neopentyl, hexyl, heptyl, octyl, 2-ethyl hexyl, nonyl and decyl, or hydrogen and Me is an alkali metal, e.g., sodium, potassium, lithium and cesium.

The vinyl acid component may be described generically as a water soluble $\alpha,\beta$-ethylenically unsaturated monocarboxylic acid, i.e., acrylic or methacrylic acids. The proportions of styrene monomer, vinyl sulfonate monomer and vinyl acid monomer are from 65 to 97.5 parts styrene from 0.5 to 25 parts vinyl sulfonate and from 2 to 34.5 parts of the vinyl acid per 100 parts by weight of the terpolymer.

Polymerization is carried out to provide polymer particles ranging in size between 0.5 and 2 microns. It is often desirable to incorporate a fourth monomer in the polymer. This fourth monomer will be a hydrophilic acrylic monomer such as a derivative of methacrylic or acrylic acid and free of carboxyl groups.

Thus, such derivatives as the ester, hydroxyester, ether, amide or cyano derivatives of acrylic or methacrylic acids may be used in amounts of from 0.5 to 25 parts by weight and preferably between 5 and 10 parts by weight, in substitution for an equal weight amount of the styrene monomer, per 100 parts of the final opacifying polymer. Specific termonomers of choice include the hydroxyalkyl esters of methacrylic acid in which the alkyl group contains from 1 to 4 carbon atoms and particularly hydroxyethyl and hydroxypropyl methacrylate, and acrylamide, methacrylamide, acrylonitrile and methyl vinyl ether.

The monomers described are emulsion polymerized using conventional catalysts, oxidizers or reducers, temperatures and pressures but with the critical step of first dissolving the water soluble vinyl acid and vinyl sulfonate in water, suitably with the emulsifying surfactant, prior to addition of the styrene. If used, the vinyl acid derivative fourth monomer should also be predissolved in water prior to introduction of styrene into the reaction vessel. Apart from the just-mentioned sequence of reactant introduction, the preparation of the polymers useful as opacifiers herein is carried out as for any other exothermic emulsion polymerization.

Thus an aqueous solution of a suitable surfactant is mixed with the water soluble vinyl acid. Thereafetr, the water insoluble styrene reactant is mixed in and agitated until emulsified as the oil phase. The emulsion is then maintained at an elevated temperature through exothermic and/or added heat in admixture with a suitable catalyst, e.g., and preferably water soluble persulfates such as ammonium and sodium and potassium persulfate and peroxides, e.g., hydrogen peroxide; and also catalysts such as t-butyl perbenzoate and t-butyl hydroperoxide, as well as other oil soluble materials such as bisazobutyronitrile and cumene hydroperoxide. Following reaction for the required period and at temperatures between 100°F. and boiling the reaction mixture is cooled and neutralized with alkali. The latex may be spray or otherwise dried without loss of dispersibility or stability in liquid household bleach. Suitable surfactants for effecting emulsion polymerization and for suspending the finely particulate polymer in bleach are the noncationic types, i.e., anionic, nonionic or amphoteric. Various of these surfactants will show greater or less tolerance for the harsh environment of liquid household bleach, depending on the concentration and pH thereof.

Example 1: (a) Polymer Preparation — Dissolve 1 part of dioctyl ester of sodium sulfosuccinic acid in 750 parts of water. Add and dissolve 50 parts of methacrylic acid and 10 parts of sodium vinyl sulfonate. Following achieving complete solution, add 500 parts of vinyl toluene. Apply heat to raise the mixture temperature to 105°F. Add one-half part sodium persulfate (initiator) and one-tenth part sodium formaldehyde sulfoxylate (reducing agent) were added.

Increase the temperature to 200°F. Thereafter reaction is completed and the reaction mixture cooled, neutralized with potassium hydroxide and filtered. Polymer particles were in the range of 0.5 to 2 microns in particle size. The product may be spray dried for subsequent incorporation into dry bleach products, or the aqueous suspension may be added to liquid bleach.

(b) Liquid Bleach Opacification — One part of the latex obtained in (a) was mixed with 100 parts of a 5% aqueous solution of sodium hypochlorite having a pH of 11.5 to opacify the same. After 3 months at 90°F. the bleach was still opacified, and the loss of available chlorine was comparable to a nonopacified bleach.

Control 1 — The methacrylic acid was mixed first with styrene and then added to the aqueous surfactant solution, followed by heating to polymerize. The emulsion which had similar size particles and the same milky appearance prior to addition to the bleach as the emulsion (b), showed immediate physical deterioration and quickly demulsified and settled as a flocculated mass in the holding vessel.

Control 2 — The procedure is duplicated without addition of sodium vinyl sulfonate. A fine latex is obtained which is useful in opacifying bleach. Increased quantities of the latex, relative to those used, are required to obtain equivalent opacity. Stability at 90°F. is less than two months. The procedure was again duplicated but the amount of methacrylic acid was reduced to 10 parts (0.5%), well below the minimum required. The latex obtained was stable for only two weeks.

Example 2: Example 1 is duplicated but reducing the styrene to 400 parts and substituting 100 parts of hydroxyethyl methacrylate. The latex obtained is highly stable in bleach.

Whitening Tests — Whitening tests were made with aqueous bleach containing 5.25% by weight sodium hypochlorite and 0.4% by weight of opacifying latex prepared in Example 1 (a). Swatches of nylon, cotton and Dacron were washed in a conventional home automatic washer. Other swatches of the same cloths were washed in the same bleach without added opacifier. Evaluation of whitening was made by a panel of persons after the first, third and fifth washing. In each instance the panel found superior whitening in the swatches washed with bleach containing opacifier.

## HYPOCHLORITES IN TEXTILE MILL PROCESSING

### Hypochlorite for Crease-Proofed, Wash and Wear Textiles

D.D. Gagliardi and M.W. Pollock; U.S. Patent 3,099,625; July 30, 1963; assigned to Argus Chemical Corporation describe a process to wash and bleach crease-proofed textiles without damage by incorporating in the detergent and bleaching composition a nitrogenous compound containing at least one NH group. The synthetic polymers commonly used for crease-proofing contain a significant proportion of NH groups in the resin molecule, either in a straight chain or in a cyclic structure.

Typical polymers of these types include the polyamides, such as nylon, polyacrylonitrile, melamine-formaldehyde resins, open-chain urea-formaldehyde resins and cyclic urea-formaldehyde resins. However, these polymers are so sensitive to hypochlorite that they are usually sold with instructions that they are not to be used with such bleaches. The bleaching of textiles treated with these resins thus has presented a serious problem.

After an almost predictable number of washings, the fabric will acquire a yellowish color, and will then develp holes or even disintegrate when subjected to high temperature, as in a dryer or during ironing. Apparently, the hypochlorite attacks the NH groups on the polymer molecule, and converts these to N—Cl groups, which are retained during subsequent washings. When N—Cl groups are heated to a sufficiently high temperature, HCl is liberated, and this scorches the fabric and attacks the textile. After a sufficient number of N—Cl groups have been formed in this way, it is quite easy for the fabric to becomes severely damaged when heated.

It has been proposed that this problem could be overcome by using polymers containing N-substituted groups, such as the N-substituted cyclic ureas, thus eliminating free NH groups. However, it is very difficult to effect complete replacement of hydrogen in NH groups, and cyclization in the cyclic ureas may not be complete. Thus, some NH groups almost invariably remain, which can react with the hypochlorite bleach to produce N—Cl groups.

Also, cyclic ureas are sensitive to alkaline hydrolysis, which can take place in the bleaching solution, opening the ring and thus forming NH groups. which then react to form N—Cl groups. After about 20 washings, fabrics crease-proofed with polymeric cyclic ureas are no more resistant to the action of hypochlorite bleach than are open chain urea-formaldehyde resins.

The problem can be avoided if nonchlorine-containing bleaches are used, or if the bleaching compound is sufficiently unreactive so it does not form N—Cl groups. In the latter case, however, the bleaching effect is not as great as might be desirable. In the former, the expense of the bleaching is considerably increased because such bleaches are more expensive than hypochlorites. These problems are avoided by bleaching in the presence of a nitrogenous organic compound (chlorine acceptor) having at least one NH group whose hydrogen atom is labile, and readily replaced by chlorine. This process is effective to protect natural and synthetic fibers sensitive to chlorination or oxidation, and to protect the NH-containing resinous material applied to them for the purpose of crease-proofing, etc. A variety of organic nitrogenous compounds can be employed. Any organic compound dispersible or soluble in the bleaching bath and having an NH group whose hydrogen is sufficiently labile to be reactive with chlorine under the bleaching conditions to form an N—Cl group can be employed in the process. It is preferred that the NH groups of the compound be more reactive with chlorine than the NH groups on the polymer, but a relatively slow rate of reaction, or low reactivity, compared to the polymer, can be compensated for by adding more of the compound.

The compounds can have an open chain or cyclic structure, which can be saturated, unsaturated or aromatic. In the preferred compounds, the NH group is adjacent a carbonyl C=O, thiocarbonyl C=S, imino C=NH, thiono S=O or phosphoro P=O group, and forms a linking nucleus of one of these types:

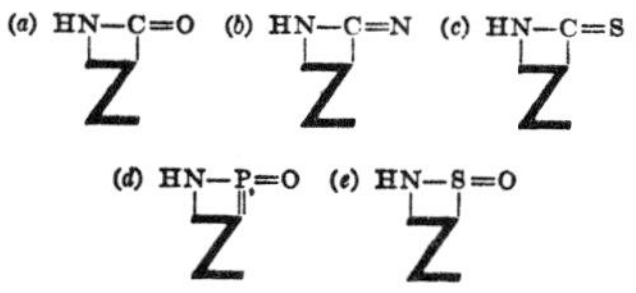


The free valences on the nucleus are satisfied by the radical Z, which can be one, two or three radicals, depending on the valence of the radical, taken in sufficient number to satisfy the valences of the linking nucleus, including groups forming a heterocyclic ring with the linking nucleus.

Preferably, Z is selected from the group of monovalent, bivalent and trivalent radicals consisting of hydrogen, alkyl, alkylene, aryl, amino and saturated and unsaturated mono- and polynuclear alicyclic and heterocyclic rings condensed therewith, having from 1 to 30 carbon atoms, all of which may bear heterocyclic atoms such as oxygen, nitrogen and sulfur and substituents such as hydroxyl, halogen, amino, alkyl and aryl groups.

Typical Z groups are methyl, ethyl, amyl, ethylene, propylene, butylene, 2-ethyl hexyl, dodecyl, isobutyl, $NH_2$—, t-butyl, phenyl, benzyl, xylyl, tolyl, cyclohexyl, naphthyl, pyridyl, —$CH_2CH_2$—O—$CH_2$—, —$CH_2$—NH—$CH_2$—, and —$CH_2$—S—$CH_2$—.

The chlorine acceptor can be incorporated directly in the bleaching bath or in the anhydrous bleaching composition. The acceptors are stable in the presence of the hypochlorite bleach in the absence of appreciable amounts of water. In the presence of sufficient water, N— Cl groups may be formed, diminishing or destroying the effectiveness of the acceptor. Dry powdered bleaching compositions can be prepared and the acceptors are readily incorporated in such mixtures by conventional methods, such as by simple mixing of the dry ingredients.

If, however, the bleach is prepared in aqueous solution it is preferable to package the chlorine acceptor in a separate package, for addition to the bleaching bath at the time of use. Compartmented packages are readily devised to meet the need to supply prescribed amounts of the acceptor. A supply of capsules or tablets of the acceptor can, for example, be included in the package with the bleaching composition, one or more tablets to be added to each bath, depending upon the amount of bleach that is added. Compartmented packages enclosed by materials soluble in the bleaching or washing bath can be used, containing the correct amount of each ingredient for one wash or bleach in an automatic washer, for example. Quite large amounts of chlorine acceptor are required, to supply enough NH groups per unit weight of detergent used in a wash-and-bleach operation. At least 7.5% by weight of the detergent composition is necessary to obtain a noticeable lessening in attack on the fabric, and preferably 10% or more is used.

The chlorine acceptors of this process can be mixed with soap or nonsoap base detergents, including cationic, anionic or nonionic detergents.

Example 1: A series of washings were carried out on an 80 x 80 white desized and bleached cotton fabric which had been crease-proofed with a methylated urea-formaldehyde resin. The cloth was padded with an aqueous bath of 8% methylated urea-formaldehyde resin and 0.9% magnesium chloride hexahydrate. The resin was applied in a laboratory padder, with one dip and one nip, at an 80% wet pickup.

The fabric was then frame dried and cured in one operation at 300°F. for 10 minutes. The resulting fabric was then tested by the AATCC accelerated chlorine damage test No. 92-1958 with and without various chlorine acceptors. The tensile strength of the fabrics tested was determined before and after the washings. The samples were air dried after washing for 4 hours at room temperature, and then pressed on a flat bed ironer, cotton setting, for 30 seconds. The following results were obtained:

| Chlorine Acceptor Added to NaOCl* | Percent** |
|---|---|
| None | 0 |
| Dimethyl hydantoin | 45 |
| Acetylene diurea | 49 |
| Urea | 91 |
| Melamine | 94 |

*One molar equivalent of NH groups present in the chlorine acceptor for each molar equivalent NaOCl.
**Percent strength retained after bleach and scorch.

It is apparent that the four acceptors tested gave considerable protection against the loss of strength during the bleaching operation. The best compounds tested were urea and melamine. The other compounds would have been more effective if used in a larger amount.

Example 2: A resin-impregnated cotton fabric was prepared from an 80 x 80 white desized and bleached cotton print cloth, using an aqueous padding bath containing 8% methylated melamine formaldehyde resin and 1.9% magnesium chloride hexahydrate. The resin was applied using a laboratory padder, one dip and one nip, at 80% wet pickup.

The impregnated samples were frame dried and cured in one operation at 300°F. for 10 minutes. The effectiveness of various nitrogenous chlorine acceptors in preventing the yellowing and scorching of the fabric was determined using the AATCC test measuring whiteness of the fabric against the control. The results obtained were as follows:

| Chlorine Acceptor Added* | Percent** |
|---|---|
| Control (one) | 67 |
| Melamine | 74 |
| Dimethoxy acetylene diurea | 74 |
| Triazone | 76 |
| Ethylene urea | 76 |
| Dicyandiamide | 76 |
| Levulinic acid hydantoin | 76 |
| Methyl acetylene diurea | 77 |
| Dimethyl hydantoin | 78 |
| Dihydroxy ethylene urea | 78 |
| Urea | 78 |

*One molar equivalent of NH groups per mol of NaOCl.
**Percent of whiteness (untreated fabric equals 8% whiteness), photovolt reflectance: MgO = 100%.

The control showed a reflectance of 67% after the bleaching, due to the canary yellow color formed. The degree of yellowing was materially decreased in the presence of the chlorine acceptor. Whiteness in this test has a value of 80, so that it is evident how close the treated fabrics are to the norm, in contrast to the control. All of the compounds tested clearly minimized chlorine attack of the fabric during the bleaching.

## Hypochlorite Bleaching of Regenerated Cellulose Sponge Web

A. Politzer, O.V. Drtina and J.S. Bedoch; U.S. Patent 3,473,884; October 21, 1969; assigned to Nylonge Corporation describe a continuous process to bleach regenerated cellulose sponge web with sodium hypochlorite. The freshly produced, sodium sulfate containing webbing is carried in zig-zag fashion, through rollers above six catch basins with guide bars and a pair of pressure rollers to squeeze out the solutions from the web at the end of each basin.

Suitable spray heads above the two first catch basins apply wash water to remove the sodium sulfate. The spray over the third basin applies an aqueous sodium hypochlorite solution at a concentration of between 0.5 and 5 g./l. Next antichlor is applied, sodium bisulfate or sodium thiosulfate solution at a concentration of 2.4%. A fungicide solution is sprayed on the web as it passes over the sixth basin.

## Hypochlorite Bleaching of Jute for Light-Fastness

A.B.S. Gupta and S.K. Majumdar; U.S. Patent 3,521,991; July 28, 1970; assigned to Indian Jute Industries' Research Association, India describe a process for improving the light-fastness of jute and reducing surface hairiness by contacting the jute with moist chlorine gas, aqueous chlorine solution or aqueous solution of hypochlorous acid and thereafter extracting the jute with aqueous solution of alkali metal hydroxides, phosphates, sulfites or bisfulfites.

Example: Five strips of a good quality jute fabric, each 10-in. wide and 20-ft. long (approx. 550 gm.), were separately treated with acid hypochlorite solution, using a liquor ratio of 10 to 1, in a laboratory open-type jigger, the time of traverse of fabric from one end to the other being 30 seconds. The acid hypochlorite solution was prepared by acidifying the alkaline sodium hypochlorite solution with dilute hydrochloric acid until the pH of the solution was brought down to 6, and its average chlorine concentration was adjusted to 0.7% by dilution. The treatment with hypochlorite solution was carried out at room temperature (25° to 30°C.) by allowing the fabric to run, back and forth, for 7 minutes, when the pH of the solution decreased to about 3, and 95% of the average chlorine was consumed by the fabric.

On allowing the liquor to drain off, the fabric was washed twice with cold water, in each washing operation the material being run twice in each direction. The individual samples were next extracted separately with 1% solution of different chemicals with a liquor ratio of 10 to 1 at 80° to 85°C. for 15 minutes. The bath was dropped, fabric washed twice with cold water as before and bleached with 1 volume hydrogen peroxide solution, containing 1% sodium silicate and 0.25% trisodium phosphate, using a liquor ratio of 10 to 1, at 70° to 75°C. for 1 hour. The bleached material was washed twice with tap water, once with water neutralized with acetic aicd and dried in the air.

The extent of improvement in color of the samples was ascertained by measuring the brightness index in a Photovolt Reflection Meter, using a tristimulus green filter. The light-fastness was determined on exposing the samples in the Fade-Ometer and comparing the extent of discoloration with that of standard dyed samples according to AATCC method. The results obtained are given below.

Effect of Extraction of Chlorinated Material with Different Chemicals (1%) on the Bleached Fabric

| Chemical Used in Extraction | Brightness Index | Light-fastness (Grade) |
|---|---|---|
| Sodium sulfite | 56 | 4-5 |
| Sodium bisulfite | 42 | 4-5 |
| Caustic soda | 53 | 5 |
| Trisodium phosphate | 56 | 4-5 |
| Tetrasodium pyrophosphate | 50 | 4-5 |

The results show that extraction of the chlorinated fabric with sodium bisulfite (1%) produces on bleaching a shade much inferior to those prepared by extraction with other chemicals. Light-fastness of the samples prepared by extraction with different chemicals are more or less the same, while that of the sample bleached only by the conventional hydrogen peroxide method was found to be much less (of the order of 1-2). The use of sodium sulfite is, however, preferred as an extracting chemical for higher retention of the tensile strength.

The sulfite extracted material showed a loss in dry strength of about 10% and in wet strength of about 44%, while on using other alkali metal salts, the loss in dry strength was of the order of 13 to 20% and that of wet strength of 40 to 58%. The loss in wet tensile strength of the untreated grey fabric was found to be about 15%.

## Hypochlorite Bleaching of Fiber Glass Fabrics

A.J. Jinnette; U.S. Patent 3,671,179; June 20, 1972; assigned to Burlington

Industries, Inc. found that a heat treatment of fiber glass fabrics, in the greige condition, at 600° to 1000°F. followed by aqueous sodium hypochlorite bleaching solution and drying, yields clean and white fabrics ready for finishing. The heating step is at a lower temperature than the conventional coronizing process carried out at 1000° to 1200°F.

In this process the heating step may be carried out in an oven of conventional design. Oven temperatures of about 600° to 1450°F. and fabric temperatures of about 600° to 1000°F. are suitable. Dwell time in the oven will vary with fabric weight and oven temperature. At high oven temperatures as little as 6 seconds dwell time may suffice.

Then the fabric, which conventionally would pass directly to the resin-dye bath or other industrial finishing solutions such as Volan or a silane, is treated at room temperature with a bleach solution, preferably by padding. An aqueous solution of sodium hypochlorite of between 0.3 and 15% available chlorine is preferred, although aqueous solutions of calcium hypochlorite, hydrogen peroxide, potassium permanganate, sodium perborate or sodium chlorite may be used.

Following the application of the concentrated bleach solution the wet fabric is passed to a drying oven which heats the fabric to 180° to 250°F. Retention time in the oven and the oven temperature required to achieve this cloth temperature and to dry the fabric will vary with fabric weight and style and are such that the action of the liberated chlorine decolorizes the brown carbonaceous residues.

# PEROXIDES

## HYDROGEN PEROXIDE, ACTIVATED AND/OR STABILIZED

### Peroxide Stabilized with Silicic Acid Esters

K. Dithmar; U.S. Patent 3,003,910; October 10, 1961; assigned to Deutsche Gold- und Silber-Scheideanstalt vormals Roessler, Germany found that silicic acid esters which are at least partially soluble in water can be used as stabilizers to retard the rapid decomposition of the peroxidic compounds, such as hydrogen peroxide, sodium peroxide and salts of perborates when used in bleaching cellulose products such as cotton or wood pulp or synthetic textiles.

The slow decomposition of the silicic acid ester avoids turbidity in the bath as well as a deposition of silicates on the fibers or textiles. Furthermore, it has also been found that alkoxysilanes can be used with excellent results as the silicic acid esters. These alkoxysilanes are soluble in water and exhibit excellent stabilizing characteristics. Typical alkoxysilanes, for example, are tetrakis(2-methoxy-ethoxy)-silane, methyl-tris(2-methoxy-ethoxy)-silane and dimethyl-bis(2-methoxy-ethoxy)-silane. The stabilizing effect of the silicic acid esters can be increased if, in addition, water-soluble alkali metal salts or particularly water-soluble alkaline earth metal salts, such as salts of magnesium or of calcium, are added to the bleaching baths.

The treatment of the textiles and fibers in the presence of silicic acid esters is preferably conducted in an alkaline medium, for example, at a pH between 7.5 and 11.5. For the attainment of the desired pH the usual buffering salts, such as phosphates and the like can be employed. Particularly good results can be obtained if the bleaching operation is conducted at temperatures over 60°C., preferably between 75° and 100°C. In addition, the usual chemical textile aids can be used in the bleaching bath according to the process, such as the sodium salt of dodecyl sulfate, oxyethylated fatty alcohols with 12 to 16 C-atoms or the condensation products of proteins with aliphatic acids.

Example: 1 kg. of fabric having a cotton warp and woof, which had previously been boiled in an alkaline medium, was treated for 2 hours at 95°C. in a peroxidic bleaching bath. The bath to fabric ratio was 12:1 and the chemical composition of the bath expressed in grams per liter of soft water was as follows:

| | |
|---|---|
| Hydrogen peroxide (35% by weight) | 10 |
| $MgSO_4 \cdot 7H_2O$ | 0.1 |
| Tetrakis-(2-methoxy-ethoxy)-silane | 1.8 |
| Sodium hydroxide | 0.3 |
| Oxyethylated dodecyl alcohol | 0.3 |

After this treatment the cotton fabric was rinsed and dried in the usual manner. The traces of cotton seed shells, which always make the fabric unattractive, had been so sufficiently dissolved and loosened by the treatment in the bath that they fell out when the fabric was stretched. When tested, the whiteness content of the treated cotton fabric was found to be 84% of the MgO reference plate of a Zeiss-Elrepho whiteness meter. The untreated cotton fabric had a whiteness content of 54% when similarly tested. The active oxygen content of the bleaching bath in this example which initially amounted to 1.8 g./liter, was lowered to 0.89 g./liter due to the bleaching operation conducted on the cotton fabric. Simultaneously, the pH dropped from an initial value of 9.8 to 9.2. A fabric having a polyester fiber warp and woof when bleached under the same conditions exhibited an equally good increase in whiteness effect.

## Peroxide Bleach with Dipersulfate for Cotton

G.T. Gallagher and N. Weinberg; U.S. Patent 3,026,166; March 20, 1962; assigned to FMC Corporation found that the absorbency of cotton fibers bleached with hydrogen peroxide is improved by the addition of alkali metal or ammonium dipersulfates to the bleach solution. The bleach solution comprises a typical hydrogen peroxide cold-bleach solution, containing as an added, essential ingredient, about 3 to 15 g./liter, and preferably about 5 to 7 g./liter, of a dipersulfate of ammonia or an alkali metal. The aqueous alkaline hydrogen peroxide bleach solution normally will contain, in addition to the dipersulfate, about 0.75 to 5% by weight of hydrogen peroxide, about 5 to 25 g./liter of sodium hydroxide, or an equivalent amount of another alkali such as potassium hydroxide or an alkali phosphate, and a stabilizing amount of a typical bleach bath stabilizer such as sodium silicate or a phosphate.

A stabilizing amount of the silicate is about 15 to 50 g./liter, and of the phosphate is about 2 to 15 g./liter. Preferably this solution will also contain a wetting agent of the type normally used in bleaching solutions, preferably a nonionic or anionic wetting agent such as an alkyl aryl sodium sulfonate or an alkyl aryl polyether alcohol, in the amount of about 1 to 5 g./liter. Hydrogen peroxide and alkali may be introduced into the bleaching solution in the form of sodium peroxide, which acts in the bleaching solution to provide both hydrogen peroxide and caustic. The dipersulfate containing bleach liquor will bleach cotton fibers more effectively than bleach baths containing somewhat less hydrogen peroxide. That is, the inexpensive dipersulfate, which contains active oxygen, serves to replace part of the hydrogen peroxide for bleaching, as well as to improve its effect on the absorbency of the fibers.

It has been found that the bleaching solution containing the dipersulfate should not undergo prolonged heating above about 100°F. before application to the fibers, as such preheating reduces the effectiveness of the dipersulfate in improving the absorbencies and brightnesses of the fibers in bleaching. Heating the dampened fibers containing the bleach to about 120°F. or even higher, however, does not have this deleterious effect. The brightnesses of the bleached fibers were determined on a Gardner Automatic Multi-Purpose Reflectometer. Brightnesses of 80% or better are considered satisfactory.

The absorbencies of the fibers were measured by placing a sample of the cloth in taut condition on a horizontal frame, dropping one drop of distilled water from a height of 3/8" onto the cloth, and measuring the time for disappearance of a specular reflection from the water. Time values of 20 to 30 seconds or less are considered satisfactory.

Fluidities, a measure of the extent of damage to the fibers, were measured by the ASTM Cuprammonium Method, D-539-53, with the results being stated as rhes units. Rhes values on the order of about 6 to 7 or lower are considered satisfactory. The fibers also were tested for ash, for their contents of water-extractable materials, their contents of enzyme-extractable materials, and for their contents of fats, oils and waxes by standard techniques. These additional determinations were made to compare the general properties of fibers cold-bleached with solutions containing, and those not containing, the dipersulfate. The ash determinations were made by charring samples of cloth in a platinum crucible using a Meker burner, and muffling to constant weight. Ash values on the order of 0.20 to 0.25% or lower are considered satisfactory. The percent enzyme-extractable values were determined with a solution of Rapidase, a proteolytic enzyme.

Example: Swatches of cotton drill cloth, having a weight of 1.85 yards per pound and a reflectance of 63% was immersed in the following solutions A and B, and passed through rollers to adjust the content of bleach solution on the cloth to 100%. The samples were then packed into a vessel and covered with a further cloth containing the bleach solution, and permitted to stand for 19 hours at room temperature, about 75° to 85°F. The treated fibers were then washed free of bleach solution by three water washings and dried in an air circulating over an 200°F. The bleached samples were then tested, with the results reported under the heading "Cloth Analysis".

| Ingredients | Grams per Liter<br>Solution A | <br>Solution B |
|---|---|---|
| Hydrogen peroxide* | 18 | 16.8 |
| Sodium silicate** | 48 | 48 |
| Sodium hydroxide | 18 | 18 |
| Detergent*** | 1.2 | 1.2 |
| Potassium dipersulfate | 0 | 7.2 |
| Water | Balance | Balance |

*Hydrogen peroxide was introduced as a 35% by weight solution in water. The amount reported represents the amount of 100% hydrogen peroxide.

**The amount of 42° Bé solution employed.

***Alkyl aryl sodium sulfonate.

Cloth Analysis

| Value | Sample A | Sample B |
|---|---|---|
| Reflectance (percent) | 81.6 | 83.0 |
| Absorbency | 1 minute | 2.5 seconds |
| Fluidity (Rhes) | 4.7 | 4.8 |

(continued)

| Value | Sample A | Sample B |
|---|---|---|
| Water extractables, (percent) | 0.48 | 0.48 |
| Enzyme extractables, (percent) | 0.69 | 0.61 |
| Fats, oils and waxes (percent) | 0.6 | 0.28 |
| Ash, (percent) | 0.19 | 0.21 |

## Oxidation of Polyacrylonitrile Fibers with Peroxide or Chlorite

F.J. Lowes; U.S. Patent 3,127,233; March 31, 1964; assigned to The Dow Chemical Company describes a process to make acrylonitrile polymeric fibers or films more flexible and more readily dyeable by mild oxidation with hydrogen peroxide or sodium chlorite solutions. Homopolymers of acrylonitrile or its copolymers with up to 20% of vinyl acetate, methyl acrylate, methacrylonitrile, acrylamide and similar monomers.

The aqueous alkaline mild oxidizing agent of the process should ordinarily be substantially free from discoloring metal ions, especially those of colored polyvalent metal ions which form chelates with and become bound to the polymer. The most economical agents to use have as the only metal ions present the colorless alkali metal or alkaline earth metal ions. The alkalinity of the aqueous oxidizing agent is preferably provided by sodium or potassium hydroxide. The oxidizer is preferably a peroxide or a hypochlorite, illustratively hydrogen peroxide or sodium hypochlorite or calcium hypochlorite or hypochlorite-chloride.

The concentration of alkali in the treating solution should not be significantly over 2.5 Normal, as much higher concentrations lead to different results. Concentrations of sodium hydroxide or of potassium hydroxide in the treating solution ranging from about 0.2 to about 10%, by weight, are quite satisfactory. The concentration of hydrogen peroxide or of hypochlorite used as the mild oxidizer is not critical, but is conveniently in the range from about 1.5 to 10% of the weight of solution.

Many of the known methods of processing polymers of acrylonitrile in an aqueous system seem to result in a tendency for the polymeric product to turn yellow when heated. This tendency, apparently related to the generation in the polymer molecule of randomly distributed amide groups, can be avoided by means of this process. In one illustration of this process, an alkyline solution of hydrogen peroxide was prepared by dissolving 0.2 g. KOH in each 100 ml. of aqueous 3% hydrogen peroxide. Several samples of polyacrylonitrile aquagel, 0.01 inch thick, were immersed in the alkaline peroxide at 26° to 27°C. Samples were removed periodically, rinsed, with water, and dried to destroy the aquagel condition.

The so-dried samples were then immersed in water and their wet flexibility was evaluated qualitatively. Each sample was then subjected to a standard dyeing procedure, using Irgalan Blue GL, an acid dye (Color Index Acid Blue 166), and the color yields were compared, as were the depths of penetration of the dye.

TABLE 1

| Immersion Time, hours | Wet Flexibility | Penetration of dye |
|---|---|---|
| 0.25 | No swelling; some brittleness | None. |
| 1.0 | Slight swelling; less brittle | Some; incomplete. |
| 4 | Swells, flexible | Thorough. |
| 8 | do | Do. |
| 24 | do | Do. |

Using treating solutions like those in the preceding example, other samples of the same polymer aquagel were immersed for the times and at the temperatures given in Table 2. The samples were removed, washed, dried irreversibly, and subjected to dyeing as described.

TABLE 2

| Temp., °C. | Time, Hours | Wet Flexibility | Penetration of dye |
|---|---|---|---|
| 35 | 2 | Swells; flexible | Thorough. |
| 35 | 1 | Sl. Swelling; flexible | Incomplete. |
| 35 | 0.5 | Sl. Swelling; Sl. Brittle | Do. |
| 50 | 0.5 | Swells; flexible | Thorough. |
| 50 | 0.25 | do | Do. |
| 70 | 0.25 | do | Do. |
| 70 | 0.10 | do | Do. |
| 90 | 0.10 | do | Do. |
| 90 | 0.05 | do | Do. |

An alkaline hypochlorite solution was prepared by dissolving 100 g. of NaOH in water, diluting to 1 liter, and bubbling chlorine into the solution until the weight had increased by 71 g. The resulting solution contained about 1 gram mol of NaOCl and about 0.5 gram mol of free NaOH per liter. The aqueous alkaline mild oxidizing agent was used in place of the alkaline hydrogen peroxide of the previously reported tests, to treat samples of polyacrylonitrile, both as aquagel and as previously dried filaments. In each case, the treated product, after exposures like those reported above, and when washed and dried, was found to have wet flexibility, and to be more receptive to dyes than is the normal hydrophobic polyacrylonitrile.

An aqueously processed polyacrylonitrile which had been polymerized in and wet spun from 60% zinc chloride solution, in the form of continuous filament tow, turned unacceptably yellow when heated for an hour in an oven at 140°C. Another specimen of the same tow was immersed for 1 hour in aqueous sodium hypochlorite (5% available chlorine) at room temperature, rinsed with water, and exposed for the same time to the same temperature in the same oven without discoloration.

Peroxide with Oxime or Hydroxamic Acid Stabilizer

K. Dithmar and E. Naujoks; U.S. Patent 3,153,565; October 20, 1964; assigned to Deutsche Gold- und Silber-Scheideanstalt vormals Roessler, Germany found that damage to polyamide textile fibers, such as nylon or Perlon, during peroxide washing and bleaching can be prevented by oximes or polyamic acids. Examples of these protective stabilizers are oxaldihydroxamic acid, benzaldoxime, propionic hydroxamic acid, benzhydroxamic acid, succinic dialdehyde dioxime, succinic diamide dioxime, and cyclohexanone oxime.

The compounds referred to develop their protective effect in the bleaching bath in the presence of peroxides by preventing the attack of the active oxygen on the polycarbonamide fibers. The bleaching or washing baths can contain active oxygen up to a maximum amount of 0.3 mol of $H_2O_2$ or 0.3 g. atom of active oxygen per liter of of bleaching liquid. The protective agents described are added in amounts of 0.01 to 0.05 mol per liter. However, certain individual agents are effective even in 1/10 of this concentration. When used according to instructions, peroxide detergents yield maximum concentrations of 0.25 g. or 0.016 gram atom of active oxygen/liter.

In this case, 0.001 to 0.005 mol of the protective agents per liter of peroxide detergent bath is sufficient, while once again individual products provide complete protection of the fibers when used with 1/10 of this concentration. They may also be directly incorporated with washing powders to provide peroxide detergent mixtures which in practice do not reduce the tensile strength of polyamide fibers, even when the treatment is repeated several times. The bleaching and washing baths may as customary also contain stabilizers (water glass, magnesium salts or phosphates) and also means for adjusting the pH value and surface active substances.

The protective agent can also be products applied in aqueous or alcoholic solution to the polycarbonamide fibers and then dried. The strength of textiles made of polycarbonamide fibers and treated in this manner is scarcely reduced, even after subsequent repeated peroxide treatment in a bleaching or washing bath which does not contain any protective agent. Finally, the products may also be used by incorporating them in the Perlon or nylon substance itself from which the fibers are spun.

Example 1: A stranded yarn material consisting of ε-caprolactam filaments with a titer of 60 deniers was treated in the bath ratio of 1:50 with occasional manipulation, for 2 hours at a temperature of 60°C. in a bath which contained:

4.8 g./l. of active oxygen in the form of 35% hydrogen peroxide (0.3 mol)
0.1 g. crystalline magnesium sulfate
1.0 cc/l. of commercially available water glass 38° Bé.

This strong bleaching treatment in the absence of a protective agent caused a 48% reduction in the tensile strength of the polycarbonamide yarn. However, when 1.2 grams of oxaldihydroxamic acid per liter were added to the bleaching bath the tensile strength of the bleached yarn was only reduced 7.9%.

Example 2: ε-Caprolactam Perlon fabric having a warp and weft both consisting of 60 denier ε-caprolactam silk was washed for 2 hours at 90°C. in a washing solution which contained peroxide and to which 0.3 g. of benzaldoxime was added as protective agent per liter of the solution. The tensile strength of the ε-caprolactam fabric dropped after treatment from 45.5 Rkm. to 43.83 Rkm., this corresponding to a strength loss of 3.7%. The bath had the following composition per liter:

1.0 g. of sodium perborate
5.0 g. of soap flakes
0.2 g. of dry water glass
1.5 g. of calcined soda
0.7 g. of sodium bicarbonate
1.6 g. of sodium pyrophosphate
0.3 g. of α-benzaldoxime

Rkm. = the number of kilometers of the yarn which will be sustained by the yarn before breaking.

## Stabilization with Phosphonic Acid Derivatives

R.R. Irani; U.S. Patent 3,234,140; February 8, 1966; assigned to Monsanto Co. describes a process to stabilize aqueous peroxy solutions with phosphoric acid derivatives, including their sodium salts. These compounds are stable in alkaline bleaching solutions, at relatively high temperatures and are compatible with optical whiteners. The acid, which in the form of its salt can be used as a stabilizer, has the following general formula:

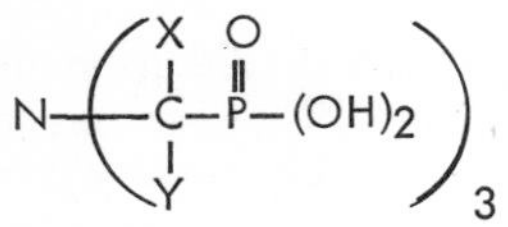

wherein X and Y represent hydrogen or a lower alkyl group (1 to 4 carbon atoms). Compounds used in the process are: amino tri(methylphosphonic acid), amino tri(ethylidenephosphonic acid), amino tri(isopropylidenephosphonic acid), amino mono(methylphosphonic acid)-di(ethylidene phosphonic acid), and amino tri(butylidene phosphonic acid). While the free acids can be used, the water-soluble salts are preferred, especially the sodium salts of amino tri(lower alkylidenephosphonic acids) and in particular the penta sodium salt has proven to be quite effective. Other alkali metal salts, such as potassium, lithium and the like, as well as mixtures of the alkali metal salts may be used. In addition, any water-soluble salt, such as the ammonium salt e.g., $N[CH_2PO_3(NH_4)_2]_2(CH_2PO_3HNH_4)$, and the amine salts:

$$N\{CH_2PO_3[N(CH_3)_2]_2\}_2[CH_2PO_3HN(CH_3)_2]$$

which exhibit the characteristics of the alkali metal salt may be also used. The stabilizing agents of the process exhibit, in addition to their stabilizing ability, the highly beneficial properties of being highly water-soluble and hydrolytically stable, that is, having a substantial resistance to hydrolysis or degradation under various pH and temperature conditions.

Peroxy solutions which are capable of being stabilized in addition to hydrogen peroxide and its addition compounds, such as the peroxide of sodium and the superoxide of potassium, include urea percompounds, perborates, persulfates, and the peracids such as persulfuric acid, peracetic acid, peroxy monophosphoric acid and their water-soluble salt compounds such as sodium, potassium, ammonium and organic amine salts.

Depending upon the particular peroxy compound used, the pH of the aqueous peroxy solution is usually adjusted with inorganic alkali metal basic materials, such as sodium hydroxide, sodium carbonate, sodium silicate, di- and trisodium phosphates and the like, including mixtures of these as well as the potassium forms of the foregoing materials, to a pH of between about 7.5 and about 12.5. Usually if the pH is higher than about 12.5 rapid bleaching occurs and the peroxy compounds rapidly decompose so that it is difficult to control a proper bleaching rate without undue damage to the fibers. At pH values lower than about 7.5, the rate of bleaching in most cases is slow to the extent of being uneconomical for bleaching.

The concentration of peroxy solutions can vary depending upon the type of peroxy compound, pH, temperature, type of bleaching desired and the like, however, normal concentrations, i.e., from about 0.01 to about 5% can be used with concentrations from about 0.2 to about 3% being preferred. The stabilizing agents may be dissolved in the peroxy solution which is ready for use or may be incorporated in a concentrated peroxy solution, such as a 35% solution of hydrogen peroxide, which is usually further diluted to form the peroxy solution for bleaching. In addition, the stabilizing agent can be incorporated in dry bleach compositions, such as perborate compositions, by admixing therewith, and the resulting composition dissolved in the aqueous system immediately preceding its end use application. In any event, the stabilizing agent is intended to be used with the peroxy solution at the time of its use for bleaching purposes. The concentration of the stabilizing agent of the process in the peroxy solution can vary from about 0.001 to 5% with from about 0.1 to 1% being especially preferred.

Example 1: Into a conventional 3-necked, 3-liter flask fitted with a reflux condenser, stirrer and thermometer was added 600 g. of diethyl phosphite and 127.5 g. of 29% aqueous ammonia solution. The flask was placed in an ice bath and after the mixture had become cooled to about 0°C., 325 g. of 37% aqueous formaldehyde solution was added. The flask was removed from the ice bath and heated with the reaction occurring at above 100°C. After the reaction was completed, the flask was allowed to cool to room temperature and the reaction products were extracted with benzene and separated by fractional distillation. Hexaethyl amino tri(methylphosphonate) distilled between 190° to 200°C. at a pressure of 0.1 mm. and was obtained in a quantity of 184 g.

The free acid amino tri(methylphosphonic acid), $N[CH_2P(O)(OH)_2]_3$, was prepared by hydrolysis of a portion of the foregoing prepared ester. In a flask similar to that described above 40 g. of the ester was refluxed with about 200 ml. of concentrated hydrochloric acid for a period of about 24 hours. The free acid, a sirupy liquid, crystallized on prolonged standing (about 1 week) in a desiccator. The yield was 20 g.

Example 2: Penta sodium amino tri(methylphosphonate):

$$N{-}[CH_2P(O)_3Na_2]_2{-}[CH_2P(O)_3HNa]$$

was prepared by dissolving the free acid obtained in Example 1 in 140 ml. of 10% NaOH solution and evaporating the aqueous solution to dryness at about 140°C. with the anhydrous form of the salt being formed.

In order to illustrate the stabilizing ability of the compounds, the following tests were made with the indicated results. The testing procedure used consisted of kinetic runs carried out in a suitable flask, stirred by a vibrating stirrer and thermostated at about 85°C. The flask initially contained 1 liter of solution of the following typical composition: 1.0% sodium silicate, 0.35% $H_2O_2$, stabilizing agent [amino tri(methylphosphonic acid)] and 2.5 x $10^{-5}$% $Cu^{++}$ (as $CuSO_4$), and pH adjusted to 10.0. The run was started by adding 10 ml. of concentrated peroxide solution to 990 ml. of solution. At intervals, 10 ml. aliquots of solution were withdrawn by pipette, quenched in 100 ml. $H_2O$, acidified with 1 ml. concentrated $H_2SO_4$ and the residual $H_2O_2$ titrated with 0.1 N $KMnO_4$.

It was found that a stabilizing agent was effective in stabilizing the peroxy solutions at very low concentrations (from 0.001 to 0.01%) over periods of time from about 20 minutes up to about 2 hours, while the peroxy solution without the stabilizing agent exhibited no bleaching ability after about 15 minutes. In order to illustrate the bleaching ability of a peroxy solution stabilized with the compounds of the process, the following tests were made with the indicated results. Four 5" x 6" swatches of unbleached desized sheeting were prewet with distilled water and placed in a suitable stirrer flask containing 1 liter of a bleaching solution of the following initial composition: 0.35% $H_2O_2$, 1% sodium silicate, 0.25 ppm Cu(II) (as $CuSO_4$)

and stabilizing agent as indicated. The temperature was thermostated at about 90°C. At intervals of about 15 minutes, 10 ml. aliquots of solution were withdrawn by pipette and residual $H_2O_2$ determined by permanganate titration. Cloth swatches were withdrawn after 15, 30, 60 and 120 minutes; rinsed well in distilled water, and air dried. The swatches were pressed and then reflectance measured versus the original unbleached cloth on a Gardner Reflectometer. Averages of four readings at different cloth orientations are reported. Reflectance values for blue light were measured relative to a white ceramic plate as 100%. The test swatches were then cut into 1" strips and measured for tensile strength according to ASTM Designation D-39-49, revised 1955 "Standard General Methods of Testing Woven Fabrics", A Breaking Strength, — Raveled Strip Method. The following tables illustrate the results of the test.

TABLE 1

| Stabilizing Agent* Concentration, (%) | Bleaching Times, (min.) | Residual $H_2O_2$, (%) | Reflectance (Blue Light), % |
|---|---|---|---|
| 0.01 | 0 | 100 | 69 |
| | 15 | 95 | 87 |
| | 30 | 94 | 90 |
| | 60 | --- | 92 |
| | 90 | --- | 93 |
| 0.1 | 0 | 100 | 69 |
| | 15 | 95 | 86 |
| | 30 | 92 | 88 |
| | 60 | 88 | 89 |
| | 90 | 86 | 91 |
| | 120 | 80 | 93 |

*Amino tri(methylphosphonic acid).

TABLE 2

| Stabilizing Agent* Concentration, (%) | Bleaching Times, (min.) | Breaking Strength, (lbs.) |
|---|---|---|
| Control (unbleached) | --- | 47.7 |
| 0.01 | 15 | 50 |
| | 30 | 50.8 |
| | 60 | 47.4 |
| | 120 | 46.5 |
| 0.1 | 15 | 45.8 |
| | 30 | 46.1 |
| | 60 | 46.5 |
| | 120 | 45.9 |

*Amino tri(methylphosphonic acid).

The above results indicate that the stabilized peroxy bleach solutions raised the brightness of the bleached cotton fabrics from about 69% to about 86 to 93%, without substantial degradation of the fabric.

## Hydrazine Antistaining Agent for Peroxide

H. Baier; U.S. Patent 3,348,903; October 24, 1967; assigned to Deutsche Gold- und Silber-Scheideanstalt vormals Roessler, Germany describes the use of hydrazine hydrate or its salts in peroxide bleaching solutions for textiles to prevent damage to the fibers due to the catalytic action of iron particles. One of the main sources of the presence of iron dust during bleaching is the grinding dust formed during the grinding of the steel needles of the carding machines or during grinding of the drums of such machines. However, iron from other sources can also reach the bleaching apparatus.

It was found that the occurrence of so-called catalytic injury can be prevented during the bleaching of fibers, yarns, woven fabrics, knitted fabrics and textile goods, especially those of cellulose or containing cellulose, with aqueous bleaching liquors containing peroxidic bleaching agents, when bleaching liquors are employed which contain additions of hydrazine, hydrazine hydrate or its salts, hexitols, such as, sorbitol, alkali metal or ammonium acetate or alkali metal or ammonium chromate or bichromate, either individually or in admixture. When peroxidic bleaching liquors containing such substances are employed in the presence of iron and steel particles catalytic injury only occurs to a very slight extent or is completely prevented.

Only very small quantities of such additional substances suffice to achieve the desired effect. In general, with the addition of hydrazine or its compounds 0.025 to 0.25 g. per liter of bleaching liquor suffice. As a rule, 1 to 3 g. of the hexitols or 1 to 5 g. of the acetates per liter of bleaching liquor suffice. In the case of the chromates and bichromates, additions of only 0.05 to 0.1 g. per liter suffice. In many instances it suffices if the additional substances are not added to the bleaching liquors themselves but rather to the pretreating liquors which, for example, are used for desizing, wetting, washing and the like.

The peroxidic compounds employed in the preparation of peroxidic bleaching liquors are those usually employed, such as, for example, hydrogen peroxide, sodium peroxide, other alkali metal or alkaline earth metal peroxides, persulfuric acid or its salts, perboric acid or its salts, percarbonic acid or its salts, perphosphoric acid or its salts or similar compounds which form hydrogen peroxide in an aqueous acid medium. Such peroxidic bleaching compounds, for example, can be employed in such quantities as to provide 0.5 to 7.0 g. of active oxygen per liter of bleaching bath. The following example illustrates the process.

Example: A woven cotton fabric, which in spots was strongly contaminated with iron dust originated in the grinding of cast iron drums of the carding machine was first desized by placing 100 g. of the fabric in 1 liter of an aqueous desizing solution containing 1 g. Degomma DK (pancreas-diastase), 1 g. NaCl and 1 g. hydrazine hydrate (24%) per liter and held in such solution at 55°C. overnight. As a control, a further 100 g. sample of the same fabric was desized in the same manner, except that the hydrazine hydrate was omitted from the desizing solution. On the next day both samples were taken out of the desizing solutions, squeezed out and thoroughly washed with water. The iron dust associated with the fabric which was desized in the solution containing the hydrazine hydrate showed no rusting phenomena whereas that associated with the fabric which was desized in the solution from which the hydrazine hydrate was omitted was strongly rusted.

Both samples were then given a peroxide bleaching by the impregnating-steaming process in which such samples were first impregnated with an aqueous bleaching liquor containing 33 cc water glass 38°/40° Bé., 8 g. of solid caustic soda, 35 cc $H_2O_2$ (35% by weight) and 2 g. of sodium dodecyl benzene sulfonate per liter, then squeezed out to 100% moisture content, and rolled flat on a glass tube and steamed in this condition at about 95° to 100°C. by direct supply of steam thereto for 2 hours. After such bleach the samples were rinsed with water. A good bleaching effect was achieved and they were completely free from seed shells. Neither rust formation nor catalytic injury was observed on the bleached sample which had been desized with the solution containing the hydrazine hydrate whereas the bleached sample which had been desized without the addition of hydrazine hydrate exhibited strong rust formation at the spots to which the iron dust had adhered and also exhibited catalytic injury to the fibers which in some instances led to the formation of holes.

## Hydrogen Peroxide Detergent Bleach with Aminoxides

K. Lindner and E. Eichler; U.S. Patent 3,388,069; June 11, 1968; assigned to Henkel & Cie GmbH, Germany found that surface active aminoxides stabilize aqueous concentrates of hydrogen peroxide or adducts of hydrogen peroxide to urea, melamine, alkali borates, alkali ortho- or polyphosphates. The hydrogen peroxide content of these concentrates varies between 10 to 40% $H_2O_2$. The aminoxides suitable as stabilizers contain 14 to 16 carbon atoms and are derived from tertiary amines containing 1 or more saturated higher alkyl radicals. Especially valuable aminoxides are those which, in addition to a total of two methyl and/or ethyl radicals and/or oxyethyl and/or monoxypropyl or dioxypropyl radicals, also contain an alkyl radical of high molecular weight, so that the sum of the carbon atoms in the alkyl radicals amounts to at least 10, and preferably 14 to 16.

Examples of aminoxides, include lauryl dimethyl aminoxide, lauryl diethyl aminoxide, myristyl dimethyl aminoxide, myristyl diethyl aminoxide, cetyl dimethyl aminoxide, cetyl diethyl aminoxide, or aminoxides which, instead of the above-named alkyl radicals of high molecular weight, contain 8 to 10 or 18 carbon atoms, or, alternately, contain one methyl or one ethyl radical, instead of the two low molecular weight methyl or ethyl radicals. Aminoxides, however, can also be used which contain two high molecular weight or only one low molecular weight alkyl radical such as dihexyl ethyl aminoxide, dioctyl methyl aminoxide or the like, providing the aminoxides are sufficiently water-soluble for effective use as stabilizers of aqueous hydrogen peroxide compounds, or compositions. Either the methyl and/or ethyl radicals of these compounds, however, can be replaced by oxyethyl, oxypropyl or dioxypropyl radicals.

The aminoxides of this process are manufactured by the condensation of high molecular weight alcohols, such as lauryl alcohol, myristyl alcohol, coconut fatty alcohols, cetyl alcohol or mixtures of such alcohols with dimethylamine or diethylamine, followed by oxidation. Water-solubility of these compounds is effected by the hydrophilic nature of the aminoxide group and by any free hydroxyl group or groups that may be present. The surface active aminoxides have especially pronounced stabilizing effects in an acid-to-neutral milieu, i.e., at a pH from 1 to 7. For example, the aminoxide may be used in a strongly acid (pH of 2 to 4), or weakly acid (from 4 to 6.8) environment or vary from 3 to 8. Stabilization, however, may also be achieved from several days to several weeks in a weakly alkaline milieu, i.e., at a pH from 7.2 to 8.

If the shelf life requirements are less stringent, as, for example, in the laundry industry, the pH of the environment may be made more strongly alkaline. For example, a pH up to about 10 is suitable for this purpose. The shelf life of the concentrates, however, decreases as the alkalinity increases, so that pH values over 10.5 do not give optimum results with respect to stability, however, concentrates having higher pH values may also be used.

These concentrates may be stored as a weakly acid, or neutral, or weakly alkaline mixture, which are adjusted to the desired higher pH values by the addition of alkaline or acidic compounds upon being used. Conventional alkaline reagents may be used for increasing the pH of the milieu. Alkaline salts are especially useful, such as those commonly used as additives in bleaching and washing, for example, the alkaline or alkali carbonates, bicarbonates, ortho-phosphates, pyro-phosphates, and poly-phosphates such as tripolyphosphates, borates, silicates, and the like, and various mixtures thereof. If the composition supplying the active oxygen comprises the perhydrates of the abovementioned inorganic salts, the alkaline materials will generally be present in sufficient quantities, so that addition of other alkaline compounds will not generally be necessary.

Acidic substances may also be added to the compositions of this process in order to lower the pH value to 9 or less. If condensed phosphates are to be used, the weakly acid glassy phosphates of the Graham salt type have proven to be particularly well suited for this purpose. Nonreducing inorganic or organic acids or acid salts are used for adjusting the pH values in the acid range, if necessary. Phosphoric acid as well as the polyphosphoric acids, as mentioned above, are suitable in addition to sulfuric acid, citric acid, malonic acid, tartaric acid, ascorbic acid, and the like, or various mixtures thereof.

Example: A concentrate comprising approximately 10% by weight of hydrogen peroxide is prepared by mixing 33.3 parts by weight of 30% hydrogen peroxide, 6.7 parts by weight of lauryl dimethyl aminoxide and 60 parts of distilled water. This concentrate is divided into three portions, and the pH of each is adjusted to 3.4, 6.0 and 7.7, respectively, by the addition of either a dilute caustic soda solution or dilute sulfuric acid, according to the case. The temperature of the three solutions is maintained at 20° to 22°C. The oxygen losses during storage in polyethylene containers is determined by titration with potassium permanganate.

The stability of the concentrates thus prepared improves as the pH is lowered. Whereas the alkaline pH results in a concentrate that remains somewhat stable for several days, the acid pH systems are kept stable for weeks. Concentrates prepared having a high pH are less suitable for applications where relatively long storage periods are encountered, without danger that the concentrate will lose active oxygen. Commercial laundry preparations, can thus be prepared of acid pH and may be kept stable for weeks and used as needed after adjusting to the desired pH value.

## Strontium Carbonate for Control of Hydrogen Peroxide Decomposition Rate

W.E. Helmick and B.O. Pray; U.S. Patent 3,437,599; April 8, 1969; assigned to PPG Industries, Inc. describe a process to control the rate of decomposition of hydrogen peroxide in bleaching baths by the addition of alkaline earth metal carbonates such as strontium carbonate. Concentrations of 0.3 to 0.8% by weight were found effective.

## Borax-Sodium Silicate Stabilizer

R.N. Suiter; U.S. Patent 3,449,254; June 10, 1969; assigned to Allied Chemical Corporation found that a combination of borax and sodium silicate stabilized hydrogen peroxide solution used for cotton knit goods without the disadvantages of the use of sodium silicate as the stabilizer, such as scaling of the equipment and damage of the fabric. The bleaching efficiency was determined by the brightness of the bleached cotton, the degradation of the cotton cellulose and the removal of motes.

The brightness obtained when employing borax-sodium silicate mixtures is not a linear function of the amount of sodium silicate present in the solution. As illustrated in the drawing, Figure 4.1, wherein 42° Bé. sodium silicate is employed, the brightness obtained, excluding the optical dye effect, drops one unit as the 42° Bé. sodium silicate content is reduced from 100 to 75%. Then from 75 to 25% 42° Bé. sodium silicate there is a plateau in the curve. Below 25% 42° Bé. sodium silicate the brightness falls off rapidly. The most desirable properties and minimum scaling and deposition result when proportions are employed corresponding to the plateau of the curve. While the reasons for the properties received when operating within this plateau are not known, the improved properties are not the result of the additive effect of the properties of the two stabilizers. The explanation appears to best lie in some coaction or cogeneric effect between borax and sodium silicate within the recited range which does not occur when employing larger amounts of sodium silicate.

FIGURE 4.1: BRIGHTNESS RESULTS OBTAINED WITH VARIOUS COMBINATIONS OF BORAX-SILICATE STABILIZER

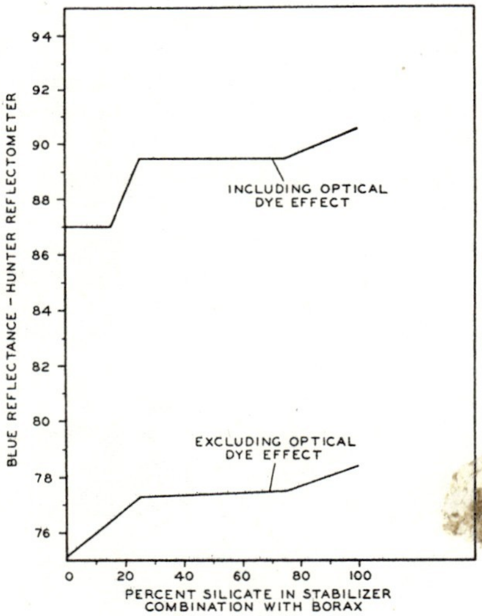


Source: R.N. Suiter; U.S. Patent 3,449,254; June 10, 1969

The brightness of the bleached cotton knit goods was determined using a Hunter D40 Reflectometer using the blue filter. The capillary absorbency is determined by the following procedure: 1/2-inch strips of conditioned cotton goods, cut in wale direction, are fastened without stretching between prongs projecting from the edge of two disks, spaced about 6 inches apart by means of a supporting rod passing through the center of each disk. The assembly is then placed in a vertical glass cylinder containing 1 to 1 1/2 inches of distilled water. At the end of 5 minutes as determined by a stopwatch started simultaneously with placing the assembly in the cylinder, the height of the water absorption in the cotton strips is measured by means of a transparent scale attached to the outside of the cylinder by flexible bands. The zero point is adjusted to the water level and each 1/16 inch rise of water in the fabric is equivalent to 1 point of absorbency. The absorbency test is carried out at 73°F. and a relative humidity of 50%. A capillary absorbency of 25 to 30 is acceptable and absorbency above 30 is rated excellent.

## Solid, Stabilized Hydrogen Peroxide-Polyvinyl Pyrrolidone Compositions

D.A. Shiraeff; U.S. Patent 3,480,557; November 25, 1969; assigned to GAF Corp. describes a process to prepare a stable, solid bleaching composition by evaporating to dryness an anhydrous aqueous solution containing hydrogen peroxide and a water-soluble polymeric N-vinyl heterocyclic compound, such as polyvinyl pyrrolidone.

The following water-soluble polymeric N-vinyl heterocyclic compounds are suitable for use as stabilizing agents: (a) polymeric N-vinyl lactams, i.e., polymeric organic ring compounds containing in their ring an acyl group

$$\overset{\displaystyle O}{\overset{\|}{(-C-)}}$$

attached to a nitrogen atom, such as the polymers of N-vinyl pyrrolidone, preferably poly-N-vinyl-2-pyrrolidone, N-vinyl-2-piperidone and N-vinyl-2-caprolactam, polymeric N-vinyl-2-oxazolidones, polymeric N-vinyl-3-morpholinones; (b) polymeric N-vinyl-imidazoles and water-soluble copolymers including graft copolymers prepared from the above N-vinyl-heterocyclic and dissimilar vinyl monomers such as vinyl acetate, isopropenyl acetate, vinyl laurate, vinyl stearate, vinyl oleate, vinyl benzoate, vinyl chloride, 2-chloro propene, methyl vinyl ether, ethyl vinyl ether, isopropyl vinyl ether, isobutyl vinyl ether, 2-oxethyl vinyl ether, phenyl vinyl ether, methyl acrylate, ethyl acrylate, methyl methacrylate, 2-ethylhexyl acrylate, hexyl acrylate, styrene, methoxystyrene, ethylstyrene, chlorostyrene or the like.

Homopolymers, random type copolymers and graft copolymers of the above described types having molecular weights from about 1,000 to 500,000 are preferred in this process. Generally polyvinylpyrrolidone having molecular weights of from about 300,000 to 400,000 is preferred. These preferred polyvinylpyrrolidone compounds may also be defined in terms of their Fikentscher K values and include polyvinylpyrrolidone having Fikentscher K value ranging from 15 to 100. For some applications it may be advantageous to use mixtures of the above described polymeric materials having different molecular weight ranges. When copolymers are used, the weight ratio of the component N-vinyl heterocyclic monomer to that of the dissimilar vinyl monomer in the copolymer will determine its properties. Choice of the weight ratio will depend on the particular properties desired in the copolymer. Preferred weight ratios of N-vinyl heterocyclic monomer to dissimilar vinyl monomers include ratios from 40:60 upward. The stabilized, solid compositions have a wide range of utility in both the commercial and industrial fields. For example, they may form the basis of washing concentrates, bleaching agents, disinfecting agents, sterilization agents, etching agents and cosmetic agents. Moreover, they may be utilized in various bleaching operations, such as the bleaching of wool and human hair or they may be used as clarification agents for beverages such as beer, whiskey, wine and other alcoholic and fermented beverages.

In addition, such compositions may be used to provide a source of oxygen which can be released at a controlled rate for sterilization and may be applied directly to wounds or form the basis of a permanently antiseptic material by first spraying its aqueous counterpart on bandages and gauze and then allowing the materials to dry. Moreover, the stabilized compositions may be used as catalysts in, in situ, polymerizations requiring a free radical source. The stabilized compositions may be formulated with the usual additives, for example, pH modifiers, detergents, sunscreen agents, emollients, brighteners and the like, depending on the particular use desired.

## Biguanidines to Protect Polyamide Fibers in Peroxide Bleaching

K. Dithmar and P. Koblischek; U.S. Patent 3,628,906; December 21, 1971; assigned to Deutsche Gold- und Silber-Scheideanstalt vormals Roessler, Germany describe a process to protect polyamide fibers during peroxide bleaching by the addition of biguanide or alkylbiguanide compounds. Better whiteness and freedom from discoloration of the fibers, due to the presence of iron or manganese salts, is obtained with these compounds.

$$\mathrm{RNH{-}\underset{\underset{NH}{\|}}{C}{-}NH{-}\underset{\underset{NH}{\|}}{C}{-}NHR_1}$$

wherein R and $R_1$ are hydrogen, alkyl groups having 1 to 6 carbon atoms, cyclohexyl and substituted cyclohexyl group, a substituted alkyl group having 1 to 6 carbon atoms, and salts thereof. If the unsubstituted biguanide is used, a good filament protective action and a good white effect are found:

$$\mathrm{NH_2{-}\underset{\underset{NH}{\|}}{C}{-}NH{-}\underset{\underset{NH}{\|}}{C}{-}NH_2} \quad \text{(biguanide)}$$

If substitution is made in the first position or in the fifth position with a methyl, ethyl, propyl, isopropyl, butyl, heptyl, hexyl, and/or cyclohexyl group, a good filament protective action and a good degree of whiteness are observed:

$$\mathrm{Alkyl{-}NH{-}\underset{\underset{NH}{\|}}{C}{-}NH{-}\underset{\underset{NH}{\|}}{C}{-}NH_2} \quad \text{(1-alkylbiguanide)}$$

A good fiber protection and a good white value also result if each alkyl or cycloalkyl group is in the first and fifth positions of the biguanide molecule:

$$\mathrm{Alkyl{-}NH{-}\underset{\underset{NH}{\|}}{C}{-}NH{-}\underset{\underset{NH}{\|}}{C}{-}NH{-}Alkyl} \quad \text{(1,5-dialkylbiguanide)}$$

Several alkyl substituents on one of the end position amino groups influence the fiber protective action according to the process detrimentally. If this kind of substitution is present with further alkyl groups, the fiber protecting action ceases, as for example with 1,5-tetraalkylbiguanide. In general, it can be stated that mono- or disubstituted derivatives and their salts (analogously each biguanide unit in macro-biguanides) are effective. On their part, the alkyl groups can also be substituted, thus, for example, consisting of $—CH_2Cl$ groups. Also, the fiber protective action and the white effect are not detrimentally influenced to any extent thereby. The alkyl groups can on their part also, for example be substituted by a further biguanide group.

If the biguanide molecule is substituted in second position, however, the fiber protective action will then be nullified whereby the substances will become useless as additives to peroxide-containing bleaching baths for polyamide fibers. The imine and and secondary amino groups must therefore be unsubstituted if an optimum effect is to be obtained. If instead of alkyl or cycloalkyl groups, the end position N-atoms are substituted with phenyl groups, then the fiber protective action will not be detrimentally affected, but the degree of whiteness of the polyamide fibers bleached with this additive will be decreased.

The aliphatic or cycloaliphatic groups being in first position and/or in fifth position can in turn be substituted, for example, by chlorine or also by a further biguanide grouping so that for example, a dibiguanide results. These agents can suitably be used together with peroxide compound or active oxygen producing compounds and the additives commonly useful in bleaching or washing, such as pH regulators, wetting, emulsifying, stabilizing, finishing and optical brightening agents. But they can also be applied as pretreatment to the fibers or threads prior to the action of the bleaching or washing baths.

In order to take into consideration high-hydrogen peroxide concentrations common in recent times for the impregnation bleach and in order to obtain plainly recognizable differences in fiber protection effect, very strong hydrogen peroxide baths with 30 cc commercial 35% $H_2O_2$ per liter bleaching bath were brought into action on bleach sensitive polyamide fibers in the following examples and comparisons. Thus, also the dose of the fiber protective agent, of which as little as 1/10 of a gram per liter suffice with only 10 cc 35% hydrogen peroxide for the polyamide fiber protection, had to be increased to 1 gram fiber protective agent per liter. The abovementioned compounds furnishing fiber protection are bases. Instead of the bases, their salts can also be employed with success, for example, preferably their salts with sulfuric acid or hydrochloric acid, but also the salts with tartaric acid or short chain organic acids with up to 4 carbon atoms. By the use of these salts of the biguanide charge weight in the following analyses was adjusted to 1 g./l. of the free base.

The stabilization of the baths was carried out in the usual manner with water glass and magnesium sulfate. The fiber protective effect obtained was measured through determination of the resistance to tearing of the bleached polyamide fiber textiles measured before and after bleaching, whereby, also, the strength of the polyamide fiber textiles that had been bleached without fiber protective agent under otherwise equal conditions was measured and compared. The degree of whiteness of the textiles (fabric) was always distinguished with the Zeiss-Elrepho step photometer with filter 6 and white standard 565 as percent on the whiteness degree scale.

To a part of the bleaching baths, 1 mg. manganese or iron in salt form was added per liter each time. The bleaching damage to the polyamide fibers without the addition of fiber protective agent is very considerable in samples from all origins, but nevertheless varied in degree here and there. Therefore, one sample was chosen from each of the two fiber types 6 and 6,6-polyamide of high-bleach sensitivity and also high-strength loss in $H_2O_2$ bleach. Optical brighteners were not used in the tests. For the examples 6,6-polyamide was preferably used because this according to experience is somewhat more difficult to bleach than 6-polyamide (Perlon).

Example: The fiber protection effect of the biguanides mentioned below was ascertained on 6,6-polyamide fabrics in a bleaching bath of the following composition: 0.1 magnesium sulfate, 1.0 cc water glass 38° Bé., 30.0 cc hydrogen peroxide 35% by weight, 1.0 g. fiber protective agent, all per liter distilled water. The pH value was adjusted with soda lye to 10.2.

Bleaching Conditions

| | |
|---|---|
| Active oxygen content | 5.6 g./l. |
| Temperature | 90°C. |
| Bleaching time | 2 hours |
| Bath ratio | 1:50 |
| Test material | 6,6-polyamide (nylon, 40/13 mat) |
| Tearing strength (untreated) | 181.5 g. |

The tearing strength was ascertained in standard climate at 20°C. and 65% rel. air moisture and is the average of 10 individual values. The percentages indicated in the table below state the loss relative to the initial value.

| Fiber protective agent | Tear strength, g. | Loss, percent |
|---|---|---|
| According to this process: | | |
| (a) Biguanide | 165.8 | 8.7 |
| (b) 1-methyl-biguanide | 166.8 | 8.1 |
| (c) 1-n-propyl-biguanide | 167.9 | 7.5 |
| (d) 1-isopropyl-biguanide | 160.1 | 11.8 |
| (e) 1-n-hexyl-biguanide | 167.7 | 7.6 |
| (f) 1-cyclohexyl biguanide | 163.7 | 9.8 |
| (g) 1,5-dimethyl biguanide | 153.3 | 15.5 |
| For comparison: | | |
| None | 59.3 | 67.3 |
| 1-phenyl biguanide | 165.9 | 8.6 |
| 1,1-dimethyl biguanide | 128.2 | 29.4 |
| 1,2-dimethyl biguanide | 110.1 | 39.9 |

## Hydrogen Peroxide Bleach Activated with Peroxydiphosphate

R.E. Yelin, L.A. Sitver and R.F. Villiers; U.S. Patent 3,649,164; March 14, 1972; assigned to FMC Corporation found that the addition of up to 0.5% tetrapotassium peroxydiphosphate to aqueous alkaline hydrogen peroxide bleaching solution improves the whiteness of cellulose textiles.

In bleaching with hot alkaline hydrogen peroxide, the desized and scoured fabric is caused to absorb an aqueous alkaline hydrogen peroxide solution, which generally contains about 0.5 to 1.5% of hydrogen peroxide, based on weight of material (percent OWM), and sufficient alkali to get the pH up to the range of about 9 to 14, generally by a combination of caustic soda and sodium silicate. Wetting agents are conventionally used to insure fast wet-out of the textile; they may be anionic, cationic or nonionic. Most mills use chelating agents to prevent interference by trace metals during processing. The textiles are preferably padded with a hot solution, at temperatures of 140° to 160°F. The saturated fabric is then heated to the boiling point. In a J-box, this is of course 212°F.; bleaching time is about 1 minute to 2 hours. Higher temperatures can be obtained in high pressure apparatus, such as the Vapor-lock in a typical 45 psi treatment at 275°F., time required is about 10 seconds to 5 minutes. After bleaching, the chemicals are rinsed out of the textile, and the product further treated as required.

In accordance with this process, about 0.05 to 0.5% OWM of a water-soluble peroxydiphosphate is added to the bleach, to get an increase in brightness which cannot be obtained by adding the same amount of active oxygen as hydrogen peroxide, together with a decrease in the noncellulosics (fats, oils, waxes and sizing) present in the textile material. In general, an increase of about 2 points in reflectance is obtainable with about 0.2 to 0.3% of peroxydiphosphate; at the same time, the noncellulosics are reduced to about 60 to 80% of the values obtained without the peroxydiphosphate.

Example 1: A sample of 100% cotton Oxford was bleached with 2.1% (50%) hydrogen peroxide, 0.15% sodium hydroxide, 1.0% silicate (sodium) and 0.1% Triton X-100 (wetting agent) at 212°F. for 60 minutes. The reflectance obtained was 86.3%. Percentages are on weight of fabric. Fabric was saturated to 100% wet pickup at room temperature and steamed. A similar sample of cotton Oxford was bleached with the same formulation, and under the same conditions as above, but 0.2% tetrapotassium peroxydiphosphate was added to the bleach solution. An increase in reflectance of 2.3 points, to 88.6% was obtained.

Example 2: A sample of 100% cotton Oxford was bleached as in Example 1, but with 1.05% (50%) hydrogen peroxide, and the reflectance obtained was 85.9%. The addition of 0.2% tetrapotassium peroxydiphosphate to this bleach formulation gave an increase of 1.7 reflectance points to 87.6%.

## Hydrogen Peroxide and Tripolyphosphate Bleach and Water Softener

A. Smeets; U.S. Patent 3,686,126; August 22, 1972; assigned to Citrex, SA, Belgium found that hydrogen peroxide has a solubilizing effect on sodium tripolyphosphate (TPP) and that aqueous solutions of these components together with a sequestering agent give an improved bleaching and water-softening composition. The sodium tripolyphosphate, $Na_5P_3O_{10}$, used may be in any of its crystallographic forms and in any state of hydration including the anhydrous compound $Na_5P_3O_{10}$, and the hexahydrate compound $Na_5P_3O_{10}\cdot 6H_2O$. The hydrogen peroxide used may be in the form of conventional commercial aqueous solutions having an $H_2O_2$ concentration of from 5 to 50% by weight, preferably about 30% by weight.

The organic sequestering agent used as stabilizer must have the property of inhibiting the decomposition of the hydrogen peroxide. Any organic sequestering agent may be used having this property and which is sufficiently resistant to the action of the hydrogen peroxide, a preferred sequestering agent being an aminopolycarboxylic acid, such as ethylenediamine tetraacetic acid and in particular diethylenetriamine pentaacetic acid (hereinafter referred to as DPTA). The bleaching composition may be prepared as follows. The required quantity of organic sequestering agent is added to an aqueous solution containing a predetermined proportion of $H_2O_2$, which is cooled to about 15°C. The TPP is added to the solution at the same temperature during agitation. To achieve this effect it is necessary further to cool the solution as during the dissolution of the TPP an increase in temperature is obtained. After agitating for about 1 hour at the same temperature, the solution is filtered, the pH is adjusted with sulfuric acid preferably to a value of from 5.5 to 6.5, and the solution allowed to reach ambient temperature. The solution is then ready for use. When packaging the solution care should be taken that storage vessels are absolutely clean, as impurities catalyze the decomposition of the hydrogen peroxide components.

## CALCIUM, UREA AND BUTYL PEROXIDES

### Calcium Peroxide and Sodium Bicarbonate for Bleaching Wood Pulp

J.R. Moyer; U.S. Patent 3,251,780; May 17, 1966; assigned to The Dow Chemical Company describe a process utilizing a thermally stable mixture of calcium peroxide and sodium bicarbonate which when mixed with water form a solution with a pH of 12.13 to 9.60 and with a complete release of the peroxygen value of the mixture. The composition is particularly suitable for bleaching wood pulp.

pH's ranging from 12.13 to 11.5 may be obtained using a weight ratio of sodium bicarbonate of calcium peroxide ranging from about 1.5 to about 1.8 parts sodium bicarbonate to each part calcium peroxide with various amounts of water, for example, from 1 to 250 parts of water per part of the dry mixture. The desired pH value in this range is more easily attained by the addition of small amounts of either a standard base, e.g., sodium hydroxide, or an acid, e.g., hydrochloric acid. At a pH of 12.13 and below, a substantially complete release of the peroxygen value of the calcium peroxide is obtained. Above a pH of 12.13, however, there is a significant reduction in the peroxide release. The weight ratios required to obtain the desired pH within the above range are determined by reference to the graph in the annexed drawing. The graph in Figure 4.2 is a plot of the pH values obtained in the aqueous mixtures of sodium bicarbonate and calcium peroxide in the weight ratios shown.

FIGURE 4.2: pH VALUES vs. $NaHCO_3$ ADDITIONS

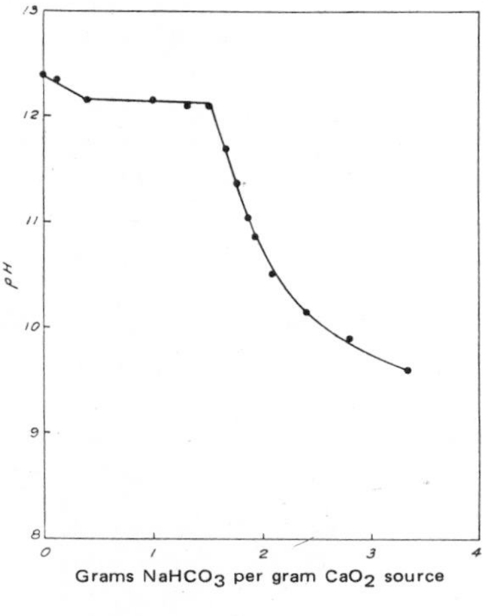


Source: J.R. Moyer; U.S. Patent 3,251,780; May 17, 1966

In obtaining the pH values plotted on the graph, aqueous slurries were prepared by adding 1 g. of 75% pure calcium peroxide and an amount of sodium bicarbonate to calcium peroxide represented by the heavy points on the graph to 100 ml. of water. The pH of each of the so-prepared aqueous mixtures was determined by a Leeds and Northrup pH indicator in combination with a Beckman glass electrode. The peroxygen value of each mixture was determined iodometrically using starch as an indicator. In the mixtures which registered a pH below 12.13, a complete release of the peroxygen value of the calcium peroxide was noted. However, in the mixtures having a pH above 12.13, the release of the peroxygen value was reduced as the pH increased.

A particular advantage of the bleaching composition is that the aqueous bleaching mixture prepared may have a pH of not lower than 9.60, if desired. Thus, materials such as wood and wood pulp can be effectively bleached with such an alkaline solution. In carrying out the method, an aqueous mixture of sodium bicarbonate and calcium peroxide in a weight ratio within the range stated above is prepared. The mixture is dissolved in that quantity of water needed to provide the desired pH applicable to the material being bleached. The material to be bleached is then contacted with the aqueous mixture for a time sufficient to obtain the desired degree of bleaching, followed by removal of the so-bleached material from the aqueous mixture and rinsing.

The sodium bicarbonate ingredient of the bleacing composition provides the acid needed to form hydrogen peroxide, the actual bleaching agent, from the calcium peroxide according to the following reaction:

$$CaO_2 + NaHCO_3 + H_2O \longrightarrow CaCO_3 + H_2O_2 + NaOH$$

If other grades of pure calcium peroxide are employed in the bleaching composition the weight ratios of sodium bicarbonate to calcium peroxide as used in obtaining the points on the aforementioned graph will be substantially the same as those using commercially pure materials and will require only slight modification, as determined by testing, to produce the pH in solution as indicated in the graph. To further illustrate the utility of the process as applied to bleaching wood pulp, aqueous bleaching mixtures were prepared each having the composition and pH as indicated in the table.

A batch of redwood kraft pulp was prebleached with 8% $Cl_2$, 4% NaOH and 2% NaOCl and was found to have a brightness after these three stages of 74.8 air dry and 70.5 oven dry (1 hour at 105°F.). The so-treated pulp was immersed in each of the so-prepared aqueous mixtures, maintained at a temperature of about 160°F. for 180 minutes and air dried or oven dried at 105°F. for 1 hour. Magnesium sulfate and sodium silicate were added to the solution as a stabilizer to prevent peroxide decomposition. This is a conventional step in the wood pulp bleaching art. The brightness of each of the so-bleached batches was next determined by using a Photovolt Reflectometer with a Hunter green tristimulus filter, with MgO as a standard with a percent brightness of 100. The results of these tests are recorded in the table.

In addition, as blanks for comparison purposes, two aqueous sodium peroxide solutions, a well-known bleach for wood pulp, were prepared equivalent or greater in released peroxygen value to the above aqueous mixtures prepared under the same process. Pieces of pulp of the same grade and type as above were also bleached using the same procedure and under the same conditions. The so-bleached pulp was tested for brightness with the results also recorded in the table shown on the following page.

Comparison of the Effects of $CaO_2$ and $NaHCO_3$ with $Na_2O_2$ in Bleaching Redwood Kraft Pulp

| Example | Source of Active Oxygen | Percent Active Oxygen [1] | Percent 41 Bé. Sodium Silicate [1] | Percent $MgSO_4$ [1] | Final pH | Temperature (° F.) | Brightness [2] (Percent) | |
|---|---|---|---|---|---|---|---|---|
| | | | | | | | Air Dry | Oven Dry |
| 1 | $NaHCO_3+CaO_2$ [3] | 0.205 | 2.0 | 0.025 | 11.4 | 160 | 82.0 | 79.2 |
| 2 | $NaHCO_3+CaO_2$ [3] | 0.205 | 1.0 | 0.025 | 11.6 | 160 | 82.0 | 80.2 |
| 3 | $Na_2O_2$ | 0.205 | 2.0 | 0.025 | 10.2 | 190 | 84.2 | 82.5 |
| 4 | $Na_2O_2$ | 0.205 | 1.0 | 0.025 | 10.2 | 190 | 83.2 | 81.5 |

[1] As weight percent of dry pulp.
[2] Of dry pulp after bleaching.
[3] One gram $CaO_2$ to 1.75 grams $NaHCO_3$/100 ml. $H_2O$.

## Calcium Peroxide and Citric Acid for Bleaching Wool and Rayon

J.R. Moyer; U.S. Patent 3,259,584; July 5, 1966; assigned to The Dow Chemical Company found that a stable bleaching composition is obtained by mixing calcium peroxide with citric acid or citric acid monohydrate. When the composition is mixed with water, a complete release of the peroxygen value is obtained. pH's ranging from about 9.5 to 5 may be obtained using a weight ratio of citric acid monohydrate to calcium peroxide ranging from 1.6 to 1.7 parts citric acid monohydrate to each part calcium peroxide with various amounts of water, for example, from 1 to 250 parts of water per part of the dry mixture.

Where citric acid is substituted for the monohydrate, weight ratios of the citric acid to the peroxide ranging from about 1.46 to about 1.55 may be used to obtain pH's of from 9.5 to 5. The desired pH value in this range is more easily attained by the addition of small amounts of either a standard base, e.g., sodium hydroxide or an acid, e.g., hydrochloric acid. At a pH of 9.5 and below, a substantially complete release of the peroxygen value of the calcium peroxide is obtained. Above a pH of 9.5, however, there is a significant reduction in the peroxygen release.

## Bleaching Cottons with Urea Peroxide

H. Papini; U.S. Patent 3,350,161; October 31, 1967 found that urea peroxide is an efficient bleach in aqueous solutions for cotton and other textiles. Nonionic detergents and sodium chlorite additions are used in some formulations.

Example 1: Bleaching of Cotton Yarn on Beams — In the bleaching of cotton yarns on beams, a beam of 253 pounds with 13/1 carded yarn is placed in a machine which is filled with cold water, with the machine first run 5 minutes inside-out. Then the 25% of aqueous detergent solution containing 10 ounces of nonionic detergent is run 5 minutes outside-in and 15 minutes inside-out. Next there is added a 3% solution containing 7 pounds and 10 ounces of urea peroxide and a 2% solution containing 5 pounds of caustic soda flakes run outside-in over a 5-minute period. The temperature is then raised slightly over a 20-minute period, with a series of 5-minute runs, the first 5-minute run being inside-out at 140°F., the next run being for 5 minutes inside-out at 160°F. and the last run being for 5 minutes outside-in at 180°F. The next succeeding step then involves two successive runs at 180°F. and a 10-minute run inside-out at 180°F. Then the machine is drained with a hot running wash being carried on for 10 minutes inside-out.

The machine is refilled with hot water at 180°F. and a 0.67% solution containing 2 pounds of acetic acid of 84% concentration is added, with a 5-minute run outside-in and a second 5-minute run inside-out. Then the machine is drained, followed by a cold running wash for 10 minutes. Finally the beam is extracted if it is not to be dyed and any strings are removed. This entire operation, which should not take over 90 minutes, will give a very bright, light cotton yarn having a smooth handle and unaffected as far as strength is concerned, with a saving of time from 2 to 3 or more hours and with the production of a superior product. It will be noted that the urea peroxide was only used for about 20 minutes to achieve this result, at varying temperatures from 120° to 180°F., in the presence of caustic soda and with a pH of 8 to 11.

Example 2: Bleaching Cotton Knit Goods with Urea Peroxide — The knit goods are placed in a bleach tub with a liquor ratio of 20:1 and with all percentages based on the weight of the goods. The bath is then boiled for 20 minutes with 5% caustic soda and 3% sodium hydrosulfite, followed by rinsing with warm water. Then the bath is treated for 30 minutes at 180° to 190°F. with 2% of urea peroxide, 1% sodium chlorite and 0.4% of oxalic acid. Then the pH is adjusted to 10 or above with caustic soda and the bath is run for 30 minutes longer, followed by rinsing, souring, rinsing, and then with another run of 20 minutes at 160°F. with 0.025% of an ultraviolet brightener. The resultant knit goods had a superior hand and texture and were not at all unduly affected as far as strength was concerned.

Example 3: Bleaching Cotton Duck — 650 yards of cotton duck were placed in a bath containing by weight of the bath 1% of sodium chlorite, 1/2% of urea peroxide, 1/4% oxalic acid, 1/4% nonionic detergent. Six ends were run at 190°F. for 45 minutes and then 2 gallons of 33% caustic soda were added to 100 gallons of the bath to give a pH of about 14 and 10 ends were run for 40 minutes. A superior duck finish was achieved.

Example 4: White Wash Laundering — One load of 300 pounds of white goods was processed in a prosperity type stainless steel washer wheel. The goods were given 3 successive 8-minute suds treatments with flush, soap and draining between the suds treatments. Then 4 suds treatments consisting of 600 cc of a solution of 4 pounds of urea peroxide and 5 gallons of water containing 2 pounds of soda ash. All of these 4 cycles were run at 150°F., and finally the goods were given 4 rinses, 1 at 8 minutes at 180°F. followed by dumping; a second of 8 minutes at 160°F. followed by dumping; a third at 80°F. followed by dumping; and a fourth of 8 minutes including a blue and sour cold treatment. A most satisfactory whitening of the goods was achieved without loss in strength.

## Bleaching Cotton with Butyl Hydroperoxide

R.M. Lincoln and J.A. Meyers III; U.S. Patent 3,574,519; April 13, 1971; assigned to Atlantic Richfield Company describe a process for the bleaching of cotton fabrics in the vapor phase and at 70°C. by tertiary butyl hydroperoxide. The process is also suitable for mixed cotton-polyester fabrics. Since by the method of this process, organic hydroperoxides, in particular tertiary butyl hydroperoxide, can be used in the same amounts as hydrogen peroxide to give equally good bleaching of cotton materials and since the hydroperoxides can be produced at a considerably lower cost than the hydrogen peroxide there is a distinct advantage for utilizing the hydroperoxides.

In addition to the economic advantage for the hydroperoxides they have the additional advantage of having greater stability as compared with hydrogen peroxide. Hydrogen peroxide unless special precautions are followed, decomposes even at ordinary temperatures to give water and oxygen. This type of oxygen evolved does not result in bleaching. Special precautions are not necessary with organic hydroperoxides since they decompose only at elevated temperatures. For example, tertiary butyl hydroperoxide starts to decompose only at about 90°C.

Other organic hydroperoxides in addition to tertiary butyl hydroperoxide are also available commercially, for example, cumene hydroperoxide and the amyl hydroperoxides. These hydroperoxides can be utilized to the same advantage as the tertiary butyl hydroperoxide. Bleaching by the method of this process provides important advantages over the hydrogen peroxide process. The hydroperoxides require no special handling to avoid spontaneous oxygen release. Solutions of hydrogen peroxide must be prepared daily, since oxygen continuously evolves from bleach solutions containing hydrogen peroxide. Hydroperoxide bleach solutions maintain a constant active oxygen content for many weeks or months. In order to make a direct comparison with hydrogen peroxide bleaching, identical runs were carried out. These runs showed that the vapor phase bleaching with tertiary butyl hydroperoxide gives equally good results as with hydrogen peroxide.

## Pollution-Free Bleaching of Wood Chips by Tertiary Butyl Hydroperoxide

R.M. Lincoln and J.A. Meyers III; U.S. Patent 3,707,437; December 26, 1972; assigned to Atlantic Richfield Company describe a process for the pulping and bleaching of wood chips in a single stage using tertiary butyl hydroperoxide in an aqueous alkaline medium. This method prevents atmospheric pollution as well as wastewater pollution.

The raw material for the example which follows was prepared from the two peeled quaking aspen bolts which were chipped into about 5/8-inch chips in a 4-knife, 38-inch Carthage chipper. The chips were screened on a Sweco screen and the large knots and slivers were removed from the on-2-mesh fraction, the accepted chips were broken down further in a 12-inch Sprout-Waldron refiner fitted with B-2975-A spiked plates set at a near-zero gap. The chips, after this reduction, were put through a Sweco screen and the through-4-on-10-mesh fraction were used in the pulping runs. These chips were about 1/10 to 1/16 inch cross-section and approximately 5/6-inch in length. Solutions for treating the chips were prepared by mixing the required weights of aqueous solutions containing 10 weight percent tertiary butyl hydroperoxide and 20 weight percent sodium hydroxide, then further diluting with water to the specified liquor-to-wood ratio.

The chips to be cooked were placed on a polyethylene sheet and turned over while being sprayed with a calculated amount of the appropriate solutions to achieve the different liquor compositions used. After they had been sprayed, the chips were transferred to vessels of about 400 ml. capacity and because of the relatively low liquor to wood ratios there was usually no free liquor. The vessels were capped and kept in a controlled-temperature oil bath which had been preheated sufficiently above the selected temperature so that the heat flow to the vessels caused the temperature to fall to that for the reaction, which required about 3 minutes. To cook the times included the 3 minutes for the vessel to become heated to reaction temperature. At the end of the reaction time the vessels were withdrawn and quenched in cold water. Known amounts of water were added to facilitate removal of the product from the vessels. The chips and diluted spent liquor were left overnight to equilibrate at ambient temperature and at about 12% consistency before withdrawing the liquor for pH determination. For yield and lignin determinations the cooked chips were collected and washed with several 1 to 2-hour soakings in warm distilled water until the filtrate had a pH of 7 to 7.5 and was clear. Klason lignin analysis were determined according to the method of The Institute of Paper Chemistry. Method number 428 which is essentially a modification of TAPPI Standard Method T 222 except that the precipitate was collected on glass fiber filters instead of crucibles. In the example which follows, the cooking and related data on the products obtained are set forth.

Example: Several runs were made in accordance with the foregoing description in order to determine pulp yield and lignin remaining in the pulp, utilizing a reaction temperature of 100°C. but varying the other reaction conditions. The Klason lignin content of the original chips was 21 weight percent. The conditions and results are set forth in the table below.

| | Run number | | | | | | | |
|---|---|---|---|---|---|---|---|---|
| | 1 | 2 | 3 | 4 | 5 | 6 | 7 | 8 |
| t-Butyl hydroperoxide, wt. percent [a] | 18 | 18 | 15.3 | 15.3 | 0 | 0 | 15.3 | 15.3 |
| Sodium hydroxide, wt. percent [a] | 0 | 0 | 8 | 16 | 8 | 16 | 16 | 16 |
| Acetic acid, wt. percent [a] | 4 | 0 | 0 | 0 | 0 | 0 | 0 | 0 |
| Liquor:wood, wt. ratio | 1.2:1 | 1.2:1 | 2:1 | 2.4:1 | 2:1 | 2:1 | 2.4:1 | 2.4:1 |
| Time in bath, min | 60 | 60 | 60 | 60 | 60 | 60 | 30 | 15 |
| Final pH | 4.3 | 5.3 | 9.6 | 12.7 | 11.6 | >13 | 12.7 | >13 |
| Pulp yield, wt. percent [a] | 96 | 96 | [b] 89 | [b] 76 | [c] 84 | [c] 76 | [b] 79 | [b] 84 |
| Klason lignin, wt. percent O.D. pulp | 21 | 21 | 18 | 14 | 21 | 21 | 15 | 18 |
| Klason lignin removed, wt. percent [d] | 4 | 4 | 25 | 49 | 19 | 20 | 43 | 30 |

[a] Percentages on oven-dried wood basis unless noted otherwise.
[b] Product at least as light in color as chips.
[c] Product darkened to a tan color.
[d] Based on Klason lignin content of chips, 21 wt. percent.

In order to understand the relationship between pulp yield and weight percent Klason lignin on dry pulp basis and the weight percent Klason lignin removed an explanation of these determinations is necessary. Thus, in Run 4, for example, for each 100 g. of chips, since the pulp yield was 76%, there would be 76 g. of pulp and 24 g. of the chips removed as solids in the liquor. Since the lignin content of the pulp was 14%, there would be 10.7 g. of lignin in the pulp and 10.4 g. of lignin in the solids of the liquor. Thus, in the pulp, 10.7 g. divided by 21 g. times 100 gives 51% of the original lignin and accordingly 49% of the original lignin had been removed. Runs 1 and 2, wherein no sodium hydroxide was employed, show that substantially no delignification is obtained. Runs 4 and 7 show high delignification, and in addition excellent color properties of the delignified pulp. Runs 5 and 6 show that in the absence of tertiary butyl hydroperoxide the pulp had very poor color and delignification was inferior.

## Brightening or After-Bleaching Kraft Pulp with Tertiary Butyl Hydroperoxide

R.M. Lincoln and J.A. Meyers III; U.S. Patent 3,707,438; December 26, 1972; assigned to Atlantic Richfield Company describe a method for replacing one or more of the chlorine or chlorine-compound bleaching steps of kraft pulp by the use of tertiary butyl hydroperoxide as a brightening agent in an alkaline medium, in particular aqueous sodium hydroxide or sodium carbonate. The amount of tertiary butyl hydroperoxide and alkaline medium required depends upon the extent of the prior bleaching of the kraft pulp.

Kraft brown stock, as obtained from the pulping process, requires extensive treatment with chlorine and chlorine-compounds in order to obtain a bleached and brightened pulp. A large number of paper mills are designed to carry out this treatment in four stages. The first stage normally consists of treating with chlorine and the second stage, which follows, is an extraction with aqueous sodium hydroxide, these stages being designated C and E, respectively. Following the extraction stage the pulp is treated with hypochlorite, the H stage; the final brightness is obtained by treating with chlorine dioxide, the D stage. This four-stage sequence, CEHD, gives kraft pulps having a GE brightness of 86 to 87. In order to obtain higher brightness levels numerous five-stage process have also been designed and used, for example, CEHHD, CEHDH, CEHED, CHDED and CEDED, wherein the letters refer to the same stages as described in the four-stage sequence.

Six and eight stage bleaching processes are also used, for example, CEHDED, CHEDED, and CEHCHDED. The primary reason for employing various sequences such as those described is to obtain the maximum amount of bleaching or brightness with minimum amount of chemical utilization and chemical cost. Another important reason for employing such sequences is to minimize the pulp yield loss since in obtaining bleaching by the use of these chlorine compounds some lowering of yields results by formation of water-soluble chlorinated compounds of cellulose and hemicellulose.

These methods, however, are subject to many disadvantages. The most important are: All of the sequences employ chlorine or chlorine-compounds which give both serious wastewater and air pollution problems. The compounds employed are expensive and thus must be used in exactly the optimum amounts in order to minimize the chemical cost and to avoid over-bleaching with attendant color and yield loss. This method completely avoids the pollution problems associated with the prior art method which fact is of exceedingly great importance since the public and many governmental agencies are now greatly concerned with these types of pollution and are requiring either stringent curbing or their entire elimination.

In the method kraft pulps either as obtained in the form of the brown stock from the pulping stage of manufacture, or as obtained from any of the stages of conventional bleaching are treated with a mixture of tertiary butyl hydroperoxide and an alkaline compound such as sodium hydroxide or sodium carbonate in aqueous solution to provide a brightened pulp having a GE brightness equal to or greater than that obtained by the conventional methods which have been described. The method is applicable to both hard wood and soft wood types of pulp, and in a particular important embodiment, the alkaline tertiary butyl hydroperoxide treatment is used to replace conventional chlorine dioxide treatment.

Example: In the runs which are shown in the table commercial kraft pulp at different stages in the kraft bleaching process for both hardwood type (aspen) pulp (H), and softwood (coniferous) pulp (S) were treated at 85°C. for 2 hours with brightening solutions consisting of an aqueous solution of tertiary butyl hydroperoxide and sodium hydroxide. In each run, an 18 g. sample of pulp (consisting of 4 g. of pulp, on an oven dried basis, and 14 g. of water) was treated with a brightening solution consisting of water containing 10 weight percent tertiary butyl hydroperoxide and 4 weight percent sodium hydroxide. In Run 1 the amount of the brightening solution was 16 ml. so that the amount of tertiary butyl hydroperoxide employed was 1.6 g. or 40 weight percent of the weight of oven dried pulp and the amount of sodium hydroxide was 0.64 g. or 16 weight percent.

In Runs 2, 3 and 4 the amounts of brightening solution were 8, 4 and 2 ml. respectively. The amounts of tertiary butyl hydroperoxide based on the oven dried pulp were 20, 10 and 5 weight percent respectively. Handsheets of the brown stock, chlorinated pulp (C), chlorinated and extracted pulp (CE), chlorinated, extracted and hypochlorite pulp (CEH), and chlorinated, extracted hypochlorite, and chlorine dioxide pulp (CEHD) as received were made in the conventional manner.

The Photovolt reflectances (compared to a standard magnesium oxide block as 100) were measured and compared with handsheets made from the pulps in Runs 1, 2, 3, and 4. The results obtained are set forth in the table. In this example the handsheets were made in conventional manner by transferring the pulp to a Buchner funnel provided with a filter paper. The brightening solution was removed by suction and the pulp washed with several volumes of water to assure complete removal of the solution. The wet handsheet was oven dried at 100°C. under vacuum. Unreacted tertiary butyl hydroperoxide and sodium hydroxide although known to be present at significant levels in the wash waters, were not recovered but could be recycled.

Photovolt Reflectances

| Pulp | Pulp as received | Run 1, 40% TBHP,[1] 16% NaOH | Run 2, 20% TBHP, 8% NaOH | Run 3, 10% TBHP, 4% NaOH | Run 4, 5% TBHP, 2% NaOH |
|---|---|---|---|---|---|
| Brown stock (H) | 34 | 80 | 79 | 65 | 54 |
| Brown stock (S) | 30 | 82 | 66 | 55 | 46 |
| C (H) | 45 | 88 | 87 | 73 | 68 |
| C (S) | 33 | 86 | 78 | 65 | 53 |
| CE (H) | 51 | 90 | 88 | 83 | 76 |
| CE (S) | 40 | 85 | 87 | 80 | 74 |
| CEH (H) | 79 | 94 | 95 | 92 | 91 |
| CEH (S) | 74 | 92 | 96 | 92 | 91 |
| CEHD (H) | 91 | - | - | - | - |
| CEHD (S) | 91 | - | - | - | - |

[1] TBHP is tertiary butyl hydroperoxide.

It was found that the pH of the brightening solution and the pulp at the start of the treatment was 10 or slightly higher and after treating the pH had dropped to about 9. It will be seen from these data that a 5% tertiary butyl hydroperoxide treatment gives superior improvement in brightness compared with the corresponding prior art treatment. For example, when hard wood brown stock is chlorinated the Photovolt reflectance is raised from 34 to 45, whereas in Run 4 it will be seen that the reflectance is raised from 34 to 54. Similarly, it will be seen from Run 4 that a CEH pulp can be brightened with a 5% tertiary butyl hydroperoxide treatment to the same brightness obtained with conventional chlorine dioxide treating of the CEH pulp. Run 2 shows that a 20% tertiary butyl hydroperoxide treatment gives superior results on a CEH pulp to anything that can be obtained commercially with chlorine dioxide.

Because of the particular interest in replacing chlorine dioxide in kraft bleaching 3 additional runs (5, 6 and 7) were carried out at levels of 2.5%, 1.0% and 0.5% tertiary butyl hydroperoxide with 1.0%, 0.4% and 0.2% respectively of sodium hydroxide in the same manner as has been described in obtaining the data in the table. For the hard wood CEH pulps the reflectances obtained were respectively 87, 84 and 85 and the soft wood reflectances were 85, 81 and 80 respectively. These experiments show that with as little as 0.5% tertiary butyl hydroperoxide and 0.2% sodium hydroxide brightening can be obtained.

## Bleaching Wood Pulp with Tertiary Butyl Hydroperoxide and Sulfite Treatment

R.A. Lincoln and J.A. Meyers III; U.S. Patent 3,709,778; January 9, 1973; assigned to Atlantic Richfield Company found that tertiary butyl peroxide buffered with an alkali is an effective bleaching agent for high lignin wood pulp. The bleaching is followed by a sulfite after-treatment. The amount of tertiary butyl hydroperoxide utilized to treat the pulp is in the same range as that utilized by the industry for hydrogen peroxide treating. In general, amounts ranging between 0.1 to 2 weight percent based on the weight of dry pulp can be used, although larger amounts can be employed. The amount of sulfite, for example, sodium bisulfite for the after-treatment may range between 1 and 6% although this is not critical and lower or higher amounts may be employed. If $SO_2$ is employed as equivalent amount is used.

Sodium silicate is the recommended buffer for hydrogen peroxide bleaching and it is known that hydrogen peroxide is less effective in the absence of a buffer. It was found, however, that tertiary butyl hydroperoxide requires no strong alkali. This is desirable since the possibility of cellulosic breakdown which can occur under alkaline conditions is completely avoided. In order to determine the bleaching effectiveness of tertiary butyl hydroperoxide as compared with hydrogen peroxide, tertiary butyl hydroperoxide as a bleaching agent was compared directly with hydrogen peroxide. Laboratory tests were standardized and based on TAPPI Routine Control test methods. The pulp employed was a commercial balsam plus pine groundwood. The pulp as received has a Photovolt reflectance (brightness) of 65.

It is well recognized in this art that an acceptable color improvement by hydrogen peroxide or similar competitive bleaching agents is 10 units, i.e., a Photovolt value of 75. Since such an improvement is difficult to achieve under the most carefully controlled conditions, improvements in excess of 10 units are very unusual and are deemed extremely difficult. The test method consists of first conditioning the pulp. The pulp as received contains approximately 50% water. A slurry of this pulp in 9 weights of water is allowed to stand overnight, then mixed for 10 minutes with a wire beater and filtered in a Buchner funnel for 15 minutes using a rubber dam. The resulting cake is used as stock for the bleaching test. The cake, in general, contains about 20 to 25% dry pulp.

A sufficient amount of the stock prepared as described above is mixed with water in a polyethylene bag to give a 10% consistency, i.e., 10 g. of dry pulp and 90 g. of total water. Under these conditions most of the water is absorbed by the pulp and there is little free liquid. The bleaching agent and buffer, if any, is added in an aqueous solution (generally about a 1.0 weight percent solution of the bleaching agent) in an amount such that the desired quantity based on dry pulp of the bleaching agent and buffer is added to the pulp solution in the bag.

The bag is kneaded to mix the pulp and bleach solution and is then immersed in a water bath at 170°F. for 1 hour. On removal from the bath the sulfite after-treatment agent is mixed by kneading and after 15 minutes the pulp is poured onto a Buchner funnel where the superficial liquid is removed through a filter paper. After removal of the superficial liquid the pulp on the funnel is covered by a second piece of filter paper and placed between layers of felt. The pulp pad covered by the filter paper and felt is squeezed between rubber rollers and then hung in an air current to air dry for 12 to 16 hours with the filter paper covers in place.

Following air drying the handsheet is dried in an oven at 110°C. for 1 hour. Following the oven drying step the amount of bleaching obtained is determined by measuring the reflectance by a Model 610 Photovolt reflectometer and comparing it with a handsheet made by the same procedure without bleaching or after-treatment solution.

The Photovolt reflectometer for this purpose uses a blue filter and the brightness values thus obtained can be converted to the TAPPI official General Electric Brightness values. The Photovolt reflectometer is adjusted so that a standard magnesium oxide block reads 100% reflectance.

Example 1: A series of runs were made utilizing the procedure set forth above. The results of these tests are set forth in the table below.

| Run No. | Bleach (wt. percent of dry pulp) | Buffer (wt. percent of dry pulp) | After-treat (wt. percent dry pulp) | Photovolt reflectometer, percent |
|---|---|---|---|---|
| 1 | None | None | None | 65 |
| 2 | 0.5 $H_2O_2$ | 2.5 sodium silicate | 1.5 $NaHSO_3$ | 66 |
| 3 | 1.0 $H_2O_2$ | 5.0 sodium silicate | 3.0 $NaHSO_3$ | 71 |
| 4 | 2.0 $H_2O_2$ | 10.0 sodium silicate | 6.0 $NaHSO_3$ | 74 |
| 5 | 0.5 TBHP [2] | None | 1.5 $NaHSO_3$ | 73 |
| 6 | 1.0 TBHP | do | 3.0 $NaHSO_3$ | 75 |
| 7 | 2.0 TBHP | do | 6.0 $NaHSO_3$ | 75 |
| 8 | 1.0 $H_2O_2$ | 1.0 STPP [3] plus 1.0 Det.[4]. | 2.0 $NaHSO_3$ | 75 |
| 9 | 2.0 $H_2O_2$ | 2.0 STPP [3] plus 1.0 Det.[4]. | 2.0 $NaHSO_3$ | 77 |
| 10 | 1.0 TBHP | 1.0 STPP [3] plus 1.0 Det.[4]. | 2.0 $NaHSO_3$ | 76 |
| 11 | 2.0 TBHP | 2.0 STPP [3] plus 1.0 Det.[4]. | 2.0 $NaHSO_3$ | 77 |
| 12 | None | None | 2.0 $NaHSO_3$ | 71 |
| 13 | do | do | 4.0 $NaHSO_3$ | 71 |

[1] Photovolt reflectance, blue filter, MgO=100%.
[2] THBP=tertiary butyl hydroperoxide.
[3] STPP=sodium tripolyphosphate.
[4] Det.=detergent (linear alkylbenzene sulfonate sodium salt, alkyl group 11 to 14 carbon atoms average, approximately 12 carbon atoms).

These data demonstrate that the tertiary butyl hydroperoxide without silicate buffering gives superior bleaching results compared to hydrogen peroxide utilizing sodium silicate buffering, compare Runs 2, 3 and 4 with Runs 5, 6 and 7. The data also show that tertiary butyl hydroperoxide gives equally good bleaching when a buffering agent such as sodium tripolyphosphate and an alkylbenzene sulfonate is used, as compared with silicate buffered hydrogen peroxide, compare Runs 8 and 9 with Runs 10 and 11. Runs 12 and 13 show that the sodium bisulfite alone does not give the required improvement in brightness.

The data also show that a wide range of amounts of bleach, buffer and after-treating agent may be employed and that, in general, slightly better results are obtained with the larger quantities, although it will be recognized, of course, in accordance with well-known principles in the art that excessive amounts of bleaching agent should not be used. Tertiary butyl hydroperoxide, however, if used in excess of that required can be recovered for reuse while hydrogen peroxide or the peracids cannot be recovered. Amounts of bleaching agent ranging between 0.1 and 5 weight percent based on the dry pulp are satisfactory and amounts of buffering agent, if used, can range between 0.5 and 10.0%.

## HYDROGEN PEROXIDE FOR DRYCLEANING

### Bleaching with Hydrogen Peroxide in Drycleaning

R.E. Keay, H.M. Castrantas and D.G. MacKellar; U.S. Patent 3,635,667; January 18, 1972; assigned to FMC Corporation found that yellowing of white garments in drycleaning can be avoided by the use of hydrogen peroxide and a small amount of ammonium hydroxide.

It is possible by this process to get acceptable whitening of white fabrics, approaching and equaling that of typical laundering with bleach, without damage to the fabric, in drycleaning operations employing a solvent and detergent, by using hydrogen peroxide, water and a volatile alkali in controlled proportion relative to the weight of fabric being cleaned (expressed as percent of weight of fabric). To get these results, it is essential to use in the drycleaning bath between about 0.25 and 2.5% wof of hydrogen peroxide, most preferably 0.5 to 1.5% and preferably 2.0 to 10.0% wof of water, using at least twice as much water as hydrogen peroxide; at least 2.0% wof of water plus peroxide; and sufficient volatile water-soluble alkali to bring the aqueous phase to a pH of preferably 8.5 and at least 7.0, generally equivalent to about 0.004 to 0.025% wof of 100% $NH_4OH$. Within these limits, reasonably good bleaching is obtained, approaching or equaling that of commercial laundering within the preferred limits, and with less damage to the cellulosic fabrics being cleaned than obtained in typical launderings.

Damage observed on fabrics drycleaned previously in the presence of hydrogen peroxide is due to the fact that cellulose fibers will selectively adsorb water and hydrogen peroxide from drycleaning baths, so that the fabric is exposed not to very dilute peroxide, as is the case in aqueous bleaching, but to hydrogen peroxide in concentrations dependent only on the ratio of hydrogen peroxide to the water in the systems; if the common 50% peroxide of commerce is added to the bath, it is 50% which is present on the fabric, and which acts on it.

To minimize this action, at least twice as much water as hydrogen peroxide is used in the bath, and sufficient volatile alkali to make the aqueous phase at least neutral, and preferably alkaline, so that the peroxide acts far more rapidly than when on the acid side. Since the pickup is not immediate, the fabric is in contact with at most 33% peroxide for a very brief period, the loss of active oxygen in the bleaching action increases the amount of water, so that the peroxide becomes more and more dilute on the fabric during the operation; hence, fabric damage is minimized.

This is borne out by a series of tests run on swatches of cotton, and 65/35 polyester/cotton blends, using perchloroethylene containing 4% of an anionic detergent in a ratio of 15 parts by weight of solvent to 1 of fabric. Tea-stained swatches are used, of the sort used in bleaching tests in aqueous wash systems to test bleaches, where a single wash with a good active oxygen bleach will give about 40 to 50% stain removal; fabric degradation was measured by AATCC method 82-1961, with results measured in fluidity expressed as rhes, higher numbers expressing greater degradation.

The results show that even at low concentrations of peroxide, insufficient to do any substantial bleaching, fabric damage in nonetheless marked. At fairly high concentrations, there is some bleach effect, much below that of normal bleach laundering, but fabric damage is still objectionable. Mere dilution of the peroxide gets somewhat improved bleach, still with more fabric damage than is desirable. But by inducing rapid bleach with volatile alkali, without increasing peroxide levels, both reduction in fabric damage to below that of a typical laundering, and bleach effectiveness approaching that of a typical laundering operation is obtained. The amounts of hydrogen peroxide which are used depend on the degree of bleach desired; at least about 0.25% wof is necessary if any really substantial bleach effect is to be obtained. From about 0.5 to 1.5% wof of hydrogen peroxide will give fair to good bleaching without unacceptable cellulose degradation. Somewhat better bleaching, with some but not excessive cellulose degradation, is obtained between 1.5 and 2.5% wof of peroxide. Above about 2.5%, cellulose degradation is excessive. Peroxide should be used in the range of 0.5 to 1.5%. At least 2 parts of water per part of peroxide should be used.

Example 1: A 100-pound load of naturally soiled garments was composed of 50/50 by weight of colored and white shirts and colored slacks in both 65/35 Dacron/cotton blends and 100% cotton. The drycleaning bath contained 500 pounds of solvent, in this case perchloroethylene, 4 pounds of detergent, in this case the isopropylamine salt of dodecylbenzene sulfonate, 2 pounds of 50 weight percent of aqueous hydrogen peroxide and 1 pound of 2.0 weight percent of aqueous $NH_4OH$. The soiled garments were cleaned for a period of 15 minutes in a typical drycleaning machine and then forced hot air dried. These garments were noticeably cleaner and colors brighter than comparable garments cleaned under similar drycleaning conditions but without the combination of hydrogen peroxide, water, and $NH_4OH$.

Example 2: In this example, the conditions of testing were the same as those of Example 1 except that 4 pounds of the sodium salt of a monophosphate ester of 1-dodecanol was used as the detergent in place of the isopropyl amine salt of dodecylbenzene sulfonate. The garments cleaned in this example were also cleaner and brighter than those cleaned via conventional techniques.

Example 3: In this example, the conditions of testing were the same as those of Example 1 except that 500 pounds of Stoddard solvent (saturated hydrocarbons) was used in place of perchloroethylene as the solvent. The garments cleaned in this example again exhibited a cleaner, brighter appearance than those cleaned in a similar system but without hydrogen peroxide, water and $NH_4OH$.

Hydrogen Peroxide Dispersed in Drycleaning Fluids

C.L. Cormany and J.A. Spotts, Jr.; U.S. Patent 3,679,590; July 25, 1972; assigned to PPG Industries, Inc. describe a process to clean fabrics with an aqueous hydrogen peroxide solution dispersed by means of an alkyl phosphate in an organic solvent. The process includes a final rinsing step to remove the residual peroxide from the fabrics by water containing up to 10% of a water-soluble alcohol such as ethyl alcohol or isopropanol. The process is suitable for use on cotton or cotton-polyester permanent press type fabrics.

According to this process, a laundering or cleaning bath is provided which is capable of treating fabrics soiled with both oleophilic and hydrophilic stains or soils by single contact therewith. This is accomplished by the provision of a cleaning bath which has as primary cleansing components both an organic, substantially water-immiscible cleaning solvent and an aqueous phase containing a water-soluble bleach. Laundering baths herein contemplated thus contain as a principal component methylchloroform, perchloroethylene, or like halogenated organic cleaning solvent along with an aqueous solution of a water-soluble bleaching agent such as aqueous hydrogen peroxide or a water-soluble hypochlorite such as sodium hypochlorite. These bleaching agents are substantially insoluble in the organic cleaning solvent of the bath.

Example 1: A solution was prepared by passing 3,600 ml. of methylchloroform containing 29.9 g. (0.5 weight percent of methylchloroform) of Alkapent 6TD (a long chain alkyl dihydrogen phosphate) and 47.8 ml. (1.0% by weight of methylchloroform) of distilled water through 50 cc of sodium carbonate disposed in a glass column 1 in. in diameter. The treatment with soda ash converted the Alkapent 6TD (a long chain alkyl dihydrogen phosphate) to long chain alkyl sodium phosphates.

From this solution a series of 150 ml. bleach baths were provided. 10 g. scoured cotton panels were placed in the solution for 2 minutes, then removed. Thereafter, the specified amount of hydrogen peroxide (as 70% hydrogen peroxide) was rapidly added and the solution mixed. A cleaning solution thus resulted in which an immiscible aqueous hydrogen peroxide phase was dispersed in the methylchloroform. The panels were returned to the bath at room temperature (about 25°C.) for 10 minutes, and then removed, rinsed thoroughly with distilled water then acetone following which they were air dried. The panels were ironed and their blue reflectance values measured. Also, the fluidity values of the treated cloth was determined by the American Association of Textile Chemists and Colorists standard test method AATCC 82-1961. The results are listed in Table 1 below.

TABLE 1

| | Weight percent of 70% $H_2O_2$ added[1] | Blue reflectance, percent | Cuam fluidity (rhes) |
|---|---|---|---|
| Liquor-to-goods ratio: | | | |
| 20:1 | 5 | 71.9 | 3.0 |
| 20:1 | 10 | 74.6 | 3.0 |
| 20:1 | 15 | 74.8 | 3.0 |
| 20:1 | 20 | 75.6 | 3.0 |
| 20:1 | 25 | 80.6 | 3.4 |
| 20:1 | 30 | 81.2 | 3.4 |
| No treatment | | 55.7 | 3.4 |

[1] By weight of cotton panels.

Example 2: A 10 g. neutral cotton panel was treated for 10 minutes at room temperature with 150 ml. portions of a stock solution prepared by adding 1.7 g. of sodium carbonate, 10 g. of Alkapent 6TD and 10 ml. of distilled water to 3,000 ml. of perchloroethylene and to which portion 0.5 ml. of 70% hydrogen peroxide was added. After removal from the treating bath, the panel was rinsed 6 times using each time 250 ml. of a perchloroethylene rinse composition containing 10% by volume 2-propanol and 1 ml. of distilled water. Following this rinsing, the panel was dried in a forced air oven at about 70°C. until all solvent vapors had been expelled, then ironed. Its residual hydrogen peroxide content was 0.09% and it had a Cuam fluidity of 3.8 rhes. Removal of stains by the cleansing composition of this process is illustrated in the following example.

Example 3: Samples of fabric each with mustard, ball point ink, grape juice, coffee or chocolate milk stains were prepared and then placed for 20 minutes at room temperature in a bleach bath of methylchloroform to which had been added (by weight of the methylchloroform) 6.5% of 70% hydrogen peroxide, 0.85% sodium carbonate, 5.0% water and by weight percent of the goods 0.25% Alkapent 6TD, the amount of such bleach bath providing a 20:1 weight ratio of solvent to goods. The fabrics were removed, their residual hydrogen peroxide destroyed, and their reflectances determined with the results shown in Table 2 on the following page.

Organic cleaning solvents other than methylchloroform, as exemplified by the use of perchloroethylene in Example 2 are effective. Usually these cleaning solvents are essentially water-immiscible. The more notable are halogenated hydrocarbon solvents, having 1 to 3 carbons. Other solvents include petroleum drycleaning solvents such as the Stoddard solvent.

TABLE 2

| | Initial reflectance, percent | | | Final reflectance, percent | | |
|---|---|---|---|---|---|---|
| | Blue | | Green | Blue | | Green |
| Stain | Inc.[1] | Exc.[2] | Inc.[1] | Inc.[1] | Exc.[2] | Inc.[1] |
| | 100% COTTON FABRICS | | | | | |
| Mustard | 37.4 | 36.9 | 79.2 | 60.0 | 58.8 | 85.4 |
| Coffee | 43.4 | 43.5 | 62.1 | 61.7 | 61.9 | 74.3 |
| Ink | 77.2 | 77.3 | 57.7 | 81.9 | 81.6 | 84.7 |
| Grape juice | 27.9 | 26.6 | 45.5 | 56.9 | 56.9 | 75.2 |
| Chocolate milk | 56.7 | 56.6 | 69.4 | 66.0 | 64.3 | 75.4 |
| | 50/50 DACRON-COTTON FABRICS | | | | | |
| Mustard | 50.1 | 45.9 | 77.1 | 68.4 | 60.0 | 80.6 |
| Coffee | 43.6 | 37.9 | 58.5 | 73.2 | 65.4 | 77.6 |
| Ink | 84.6 | 73.7 | 65.3 | 87.5 | 76.4 | 79.9 |
| Grape juice | 39.6 | 36.4 | 50.6 | 82.4 | 72.3 | 80.4 |
| Chocolate milk | 62.9 | 55.1 | 69.5 | 81.5 | 71.7 | 79.9 |

[1] Inc. = Including fluorescence.
[2] Exc. = Excluding fluorescence.

## PEROXIDES IN TEXTILE MILL PROCESSING

### Continuous Bleaching of Cross-Linked Cellulose Fabrics

L.V. McMackin; U.S. Patent 3,104,152; September 17, 1963; assigned to The Springs Cotton Mills describes a process to bleach with hydrogen peroxide, cross-linked, wash and wear type, cellulose fabrics in two stages. In the first stage, the fabric is saturated with a bleaching composition containing hydrogen peroxide, and is then heated to substantial dryness, i.e., to a moisture content of about 2 to 3% of the weight of the fabric. The thus treated fabric is then wet with an aqueous solution following which the fabric is heated again to substantial dryness. This procedure markedly improves the efficacy of the hydrogen peroxide and the bleaching of the fabric to desired whiteness, suitable for use in men's white shirts, etc.

Figure 4.3 is a diagrammatic view of a portion of a bleaching range illustrating generally the equipment used in the two-stage operation of the process. According to the process, open width, prebleached, woven, white, dry, cotton shirting is passed continuously at a speed of 90 yards per minute through guide rolls and into a pad box containing the following composition at 100°F., for imparting wash and wear effects, all parts by weight, and which may be compounded by mixing together the following components, in the order listed.

| | Lbs. |
|---|---|
| 39% 2,2'-sulfonyldiethanol in water | 720 |
| Soda ash (99% $Na_2CO_3$) | 55 |
| Borax ($Na_2B_4O_7 \cdot 10H_2O$) | 7 |
| 25% polyethylene in aqueous dispersion | 70 |
| Wetting agent (1,1,3,3-tetramethyl-phenoxy-polyethoxyethanol) | 5 |

The shirting is then passed through squeeze rolls adjusted to allow the fabric to retain about 65 to 70% of its own weight of the sulfonyldiethanol composition and then passed by means of a clip tenter frame for 90 feet through an area of blown hot air at 225° to 265°F., whereby the fabric is dried to about 5% moisture content, and then passed over a three-roll calender. The fabric is then passed through a gas-fired curing oven continuously containing 150 yards of fabric and heated to 355°F. The curing temperature is critical in that below 345°F. the sulfonyldiethanol will not react with the cotton shirting to cross-link the cellulose molecules, and at temperatures of more than 365°F., a cross-shading of the fabric occurs that cannot be removed by subsequent bleaching. The fabric is then passed over three cooling cans, through which are circulated water at tap temperature, and then collected on a take up roll, at which time it may be stored indefinitely or taken immediately to the bleaching operation. It will be noted that the previously white cotton shirting is discolored to a cream color upon emergence from the curing oven, and that it is the thus discolored fabric that has proven so difficult to bleach. All of the above operations have been carried out in a continous process with the fabric traveling at a rate of 90 yards per minute.

FIGURE 4.3: TWO-STAGE BLEACHING PROCESS

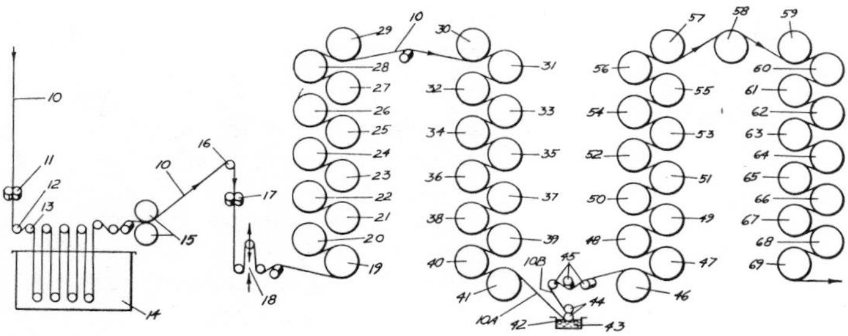

Source: L.V. McMackin; U.S. Patent 3,104,152; September 17, 1963

The cotton shirting is next put through a scouring operation prior to the bleaching. The scouring is important for satisfactory subsequent bleaching. The scouring is accomplished by passing the fabric in open width continuously at a speed of 150 yards per minute over guide means and into a pad box containing, at 165°F., 1 gallon of concentrated acetic acid (84%), 10 pounds of detergent (33% aqueous dispersion of sodium-N-methyl-N-oleoyl laurate), and water to bring the final volume to 250 gallons, and then through squeeze rolls and into 5 successive wash boxes, each equipped with squeeze rolls at its entry and exit and containing 250 gallons of water at 165°F. The exit squeeze roll of the last box is adjusted to allow the fabric to retain about 50% of its weight in water. The fabric passes out onto a conventional scray to interrupt the tension.

The fabric (10) is then passed over guide means (11) to (13) and, while still retaining about 50% water, into a pad box (14) containing the aqueous bleaching composition. The bleaching composition is maintained at 100°F. and is continuously fed chemicals so that it has the following constant composition.

| | Lbs. |
|---|---|
| 50% hydrogen peroxide in water | 62 |
| Wetting agent (1,1,3,3-tetramethylbutyl-phenoxy-polyethoxyethanol) | 1 |
| Soda ash (99% $Na_2CO_3$) | 3 |
| Sodium tripolyphosphate | 50 |
| $H_2O$ q.s. to give 250 gallons final volume | |

The fabric is passed through squeeze rolls (15) adjusted to allow it to emerge from the aqueous bleaching solution with about 36% of its own weight of water and aqueous bleaching solution. The fabric (10) is next passed over guides (16) and (17) and over a conventional adjustable tension device (18), and is then serpentined through a network of 23 conventional steam heated, cylindrical, rotating drying cans (19)-(41) in contact with the surfaces thereof. Each drying can is 72.3 inches in circumference and is maintained at 45 pounds per square inch internal steam pressure. The fabric remains on the above drying pans (19)-(41) for about 15 seconds, where the fabric is partially bleached, but not to the degree of whiteness desired. The fabric (10a) emerges from drying can (41) containing about 2 to 3% of its weight in residual bleach and water, about 90 to 94% of the moisture content prior to can drying having been driven off of the fabric and at which moisture content rapid bleaching has ceased for all practical purposes.

The bleaching treatment is then renewed by passing the fabric (10a) into a trough (42) containing water (43) at tap temperature. This rewetting renews the action of the hydrogen peroxide and the wetted fabric then passes through squeeze rolls (44) adjusted to allow about 35% water and residual bleach retention based on the weight of the goods. While some of the bleaching composition may be removed from the fabric during the rewetting step, enough residual bleach remains on the fabric to provide the essential subsequent bleaching. The fabric (10b) is passed over a series of expanders (45), and is then serpentined through a network of 21 steam-heated, cylindrical, rotating drying cans (46)-(66), in contact with the surfaces thereof.

Each drying can is 72.3 inches in circumference and is maintained at 45 pounds per square inch internal steam pressure. The fabric remains on the drying cans (46)-(66) for about 13 to 14 seconds and emerges from drying can (66) substantially dry and in fully and evenly bleached bright white condition. The fabric is passed over 3 cooling cans (67)-(69), each of which is 72.3 inches in circumference and contains therein circulating water at tap temperature. The fabric is collected on a take up roll, and may be taken immediately into a scouring and top finishing operation or may be stored for a period of time.

The fabric is next scoured by passing it continuously in open width, at the rate of 150 yards per minute, over guide means and into a pad box containing, at 165°F., 1 gallon of concentrated acetic acid (84%), 10 pounds of detergent (33% aqueous dispersion of sodium-N-methyl-N-oleyl laurate), and water to bring the final volume to 250 gallons, and then through squeeze rolls and into 5 successive wash boxes, each equipped with squeeze rolls at its entry and exit, and containing water at 165°F. The fabric is then dried to substantial dryness by passing it over 20 conventional steam heated, cylindrical, rotating, drying cans, in the above described manner, each being 72.3 inches in circumference and having an internal steam pressure of 4 to 10 pounds per square inch.

The dry fabric is passed over guide means and into a padder containing a top softener and fluorescent brightener composition at 110°F. The composition may be formed by mixing together the following components in the order listed.

| | |
|---|---|
| 25% aqueous dispersion of polyethylene | 120 lbs. |
| Blue tint (Indanthrene Blue GP Powder Fine, New Color Index No. Vat Blue 4) | 5 av. oz. |
| Fluorescent brightener (25% aqueous dispersion of bis(2-anilino-para-sodium sulfonate-4-monoethanolamino-6-triazinyl)-4,4'diamino-stilbene-2,2'-disodium sulfonate) | 30 fl. oz. |
| Nonionic wetting agent (sodium dicapryl-sulfosuccinate) | 10 lbs. |
| $H_2O$ q.s. to give final volume of 250 gallons | |

The fabric is passed through squeeze rolls adjusted to allow the fabric to retain 65 to 70% of its own weight of the top softener and fluorescent brightener composition. The fabric is next dried to 5% moisture content by passing it on a clip tenter frame under a zone of hot air at 250° to 300°F., and finally the fabric with 5% moisture content is passed through a compressive shrinkage machine equipped with a rubber belt. It will be noted that the fabric, after being collected from the curing oven, has been processed while traveling at a rate of 150 yards per minute. The cotton shirting produced according to the above process has outstanding bright whiteness, excellent crease resistance in both the wet and dry states, is free from odor, and will shrink less than 0.75%. The wash and wear finish is permanent since the sulfonyldiethanol has reacted chemically to cross-link the cellulose molecules.

## Bleaching Cotton Knits in Alkaline Hydrogen Peroxide

S.M. Rogers; U.S. Patent 3,142,531; July 28, 1964; assigned to Allied Chemical Corporation describes a process to bleach gray cotton knit goods in alkaline hydrogen peroxide solution under strictly controlled temperatures for definite time periods, to obtain optimum brightness, absorbency, low fluidity and freedom from motes. In this process the cotton knit goods are immersed in a bath of an aqueous wetting-out solution at a temperature within the range of 160° to 180°F. and containing, preferably 0.5 to 1 g./l. of an optical brightening agent and 1.5 to 3 g./l. of a wetting agent of the groups consisting of nonionic and cationic surface active wetting agents.

The goods are kept immersed in the bath for a time of from about 2 to 4 minutes at temperatures within the aforesaid range and the goods are preferably stirred while in the bath. Then, without rinsing the goods, the goods are squeezed to 175 to 200% saturation, followed by immersing the goods in an aqueous alkaline hydrogen peroxide bleaching solution free of alkali metal silicate, containing from about 8 to 9 grams per liter of hydrogen peroxide, from about 3 to 4 g./l. of sodium hydroxide and preferably also borax ($Na_2B_4O_7$) in amount from about 1 to 1.5 g./l. The goods are kept immersed in the peroxide bleaching solution at a temperature within the range of about 170° to 190°F., for a period of about 280 to 320 minutes, and the goods are then washed first with hot water, then with cool water, thereafter with aqueous sodium bisulfite solution and again with water.

The brightness of the bleached cotton knit goods was determined using a standard General Electric Reflectometer. A GE brightness of 86% or above is good for cotton.

The absorbency of the cotton knit goods is determined by the following procedure. 1/2 inch strips of the cotton goods are fastened without stretching between prongs projecting from the edge of two disks, spaced about 6 inches apart by means of a supporting rod passing through the center of each disk. The assembly is then placed in a vertical glass cylinder containing 1 to 1 1/2 inches of distilled water. At the end of 5 minutes as determined by a stopwatch started simultaneously with placing the assembly in the cylinder, the height of the water absorption in the cotton strips is measured by means of a transparent scale attached to the outside of the cylinder by flexible bands. The zero point is adjusted to the water level and each 1/16 inch rise of water in the fabric is equivalent to 1 point of absorbency. The absorbency test is carried out at 73°F. and a relative humidity of 50%. An absorbency of 25 to 30 is acceptable and absorbency above 30 is rated excellent.

Degradation or modification of the cotton fibers due to the chemicals used in the processing is very effectively and accurately measured by determining the fluidity of the cotton cellulose dissolved in cuprammonium hydroxide according to standard procedure. The fluidity values attached are largely independent of the fabric construction, and the test is a very useful one inasmuch as the degradation of the fibers can be used to predict the effect of chemical treatments on the tensile strength of the fabrics and their wearing qualities. The fluidity values are measured in rhes, i.e., reciprocal poises, with values about 6 indicative of the degradation of the cotton fibers and values below 4 indicating superior strength of the fibers. The mote content was determined by careful visual inspection of the cotton knit goods.

Example 1: A 50 g. sample of gray tubular cotton knit goods was immersed and stirred for 3 minutes at 180°F. in 500 ml. of an aqueous solution containing 3.0 g./l. of an isooctyl phenyl polyethoxy ethanol (Triton X-100) and 0.69 g./l. Calcofluor ST White solution [4,4'-bis(4-anilino-6-diethanolamine-s-triazin-2-ylamine)-2,2'-stilbene-disulfonic acid, disodium salt] brightener. Without washing, the goods were squeezed to a 200% solution saturation and piled in a glass kier. 60 ml. of an aqueous solution containing 8.75 g./l. $H_2O_2$, 1.0 g./l. $Na_2B_4O_7$ and 3.0 g./l. NaOH and 10 cc of water were added to the kier, circulated and heated to 180°F. and held for 5 hours. The kier was drained and the sample given 5 minute circulating rinses with separate portions of hot water, cool water, 2.0 g./l. $NaHSO_3$ and 0.7 g./l. $NaHSO_3$ after which the goods were removed from the kier and washed. The goods were then cut into pieces of about equal size which were placed between blotters to remove water, then pressed at 600 psi between fresh blotters and dried. The results were:

| | |
|---|---|
| Brightness | 90.1 |
| Fluidity | 1.9 |
| Absorbency | 61.0 |
| Mote removal | Excellent |

Example 2: The procedure of Example 1 was repeated but the isooctyl phenyl polyethoxy ethanol of the pretreatment wetting-out solution was reduced to 0.75 g./l. in the wetting-out solution of Example 1. The results obtained were:

| | |
|---|---|
| Brightness | 88.5 |
| Fluidity | 1.8 |
| Absorbency | 52.0 |
| Mote removal | Excellent |

Rapid Bleaching of Wool with Hydrogen Peroxide

B.K. Easton and N. Weinberg; U.S. Patent 3,148,018; September 8, 1964; assigned to FMC Corporation describe a process to bleach wool in alkaline hydrogen peroxide solution. The wool is first saturated with 50 to 120% of its weight with an aqueous alkaline hydrogen peroxide solution having a pH of about 7.5 to 9.5, and containing about 3 to 9 g. of hydrogen peroxide on a 100% basis, per liter of solution, and containing about 0.1 to 0.3% of an alkali metal salt of a molecularly dehydrated phosphoric acid, steaming the goods saturated with this aqueous alkaline hydrogen peroxide solution at a temperature of about 190° to 212°F. for a controlled time of about 1 to 3 minutes, and thereafter drying the goods by application of drying heat at a temperature of 150° to 200°F. until the goods contain no more than 10% free water.

The process is readily conducted in well-known wool-treating equipment, known to the trade as a "scouring train". This equipment comprises a series of 4 or 5 or more vats or bowls laid end-to-end, through which the wool passes. The first two or three bowls may contain a scouring solution consisting of synthetic detergents or soaps. If the scour is employed, the next one or two bowls will contain water for rinsing these scouring materials from the goods. The final bowl in the series contains the hydrogen peroxide bleaching solution. The time the goods spend in each bowl is about 2 to 3 minutes.

As the wool stock leaves the scouring train saturated with aqueous, alkaline hydrogen peroxide solution, it passes onto a conveyor within a housing, or is dropped into a chute, or passes through any other equipment which will provide a holding time of about 2 minutes and at the same time permit the wool to be steamed. It is then passed through a hot air dryer. The process advantageously carried out in the above scouring train, provides a means for continuously scouring, bleaching and drying wool stock in a matter of only about 15 minutes. This time consists of about 6 to 12 minutes of scouring, about 1 to 3 minutes of steaming, and about 2 to 5 minutes of drying. The wool produced by this process is as white as that obtained by conventional means in 2 to 3 hours or longer, and the alkali solubility value or degradation of the continuously bleached wool of the process is normally lower than that obtained by such conventional methods.

The wool stock bleached in the examples was scoured in a standard scouring train, by being treated for 2 1/2 minutes in each of three vats containing: in the first vat, 1,500 gallons of water at 90°F.; in the second vat, 18 pounds of an alkyl phenol type nonionic detergent in 2,200 gallons of water at 140°F.; and in the third vat, 8 pounds of the same detergent in 2,000 gallons of water at 140°F., and then water rinsed until substantially all of the detergent was removed. The wool employed was a wool stock of the type to be used in carpeting, and had a percent reflectance after scouring and rinsing of 40.5% and an alkali solubility after such treatment of 12.4%.

Example 1: 100 g. of the wool scoured by the above method was bleached in 1 liter of the following solution:

| | |
|---|---|
| Hydrogen peroxide, 35% | 26.1 g. |
| Tetrasodium pyrophosphate (anhydrous) | 2.4 g. |
| 28% $NH_4OH$ | 1.6 g. |
| Water to make 1 liter | |

The wool was held in the solution for 2 1/2 hours at a temperature of 135°F., following which it was rinsed with water until it exhibited a neutral pH, and dried at about 150°F. for 15 minutes until it contained 8% of free moisture. The bleached wool had a reflectance of 65%, and had an alkali solubility of 24.9.

Example 2: 100 g. of the wool scoured by the above method was soaked for 2 minutes in 1 liter of the following solution at room temperature:

| | |
|---|---|
| Hydrogen peroxide, 35% | 40 g. |
| Tetrasodium pyrophosphate (anhydrous) | 2 g. |
| 28% $NH_4OH$ | 4.5 g. |
| Water to make 1 liter | |

The wool was squeezed until it contained its own weight of the above solution, and was steamed for 2 minutes at 210°F. It was then dried at 185°F. in an oven for 15 minutes until it contained about 8% of free moisture. The wool had a reflectance of 68%, and its alkali solubility value was only about 15.3.

Example 3: 100 g. of the wool scoured by the above means was treated as in Example 2, except that the following solution was used:

| | |
|---|---|
| Hydrogen peroxide, 35% | 19.2 g. |
| Tetrasodium pyrophosphate (anhydrous) | 2.0 g. |
| 28% $NH_4OH$ | 4.5 g. |
| Water to make 1 liter | |

The whiteness of this wool was 62%, or slightly less than that obtained with the conventional immersion bleach of Example 1; its alkali solubility, however, was only 14.7.

Example 4: The procedures of Examples 2 and 3 were followed with the exception that the ammonium hydroxide was omitted from the respective bleach solutions of these examples. The wool samples were bleached to reflectances of 68% and 61%, respectively, and their alkali solubilities were 19.6 and 12.1, respectively. The alkali solubility test employed above was conducted according to ASTM-D1283-53T. Reflectances were determined with a Gardner Multipurpose Reflectometer using a blue filter.

## Continuous Alkaline Scouring and Peroxide Bleaching for Cotton

G.T. Gallagher and E.J. Elliott; U.S. Patent 3,148,019; September 8, 1964; assigned to FMC Corporation describe a process to continuously scour and bleach desized cotton textiles in a single treating zone, such as a J-box, to obtain goods with high reflectance and good absorbency. It has been found that desized gray cotton cloth can be bleached within a single treating zone to give a cloth having high reflectance and good absorbency by passing the cloth into an alkaline solution having a concentration of from about 1.5 to 5.0% by weight of an alkali metal hydroxide, wetting the cloth with from 50 to 150% of its weight of the alkaline solution, passing the wet cloth into a single treating zone, steaming the cloth at from about 200° to 212°F. for from about 15 to 45 minutes at one end of the treating zone, passing the cloth containing the alkaline solution directly into an aqueous hydrogen peroxide solution maintained at the other end of the single treating zone, heating the cloth within the aqueous hydrogen peroxide solution for about 15 to 45 minutes at temperatures of about 150° to 210°F. and maintaining the concentration of the alkali metal hydroxide in the aqueous hydrogen peroxide solution from about 1.0 to about 1.5% by weight. Figure 4.4 is representative of the equipment and means used for carrying out the chemical and physical treating operations.

FIGURE 4.4: CONTINUOUS SCOURING AND BLEACHING EQUIPMENT

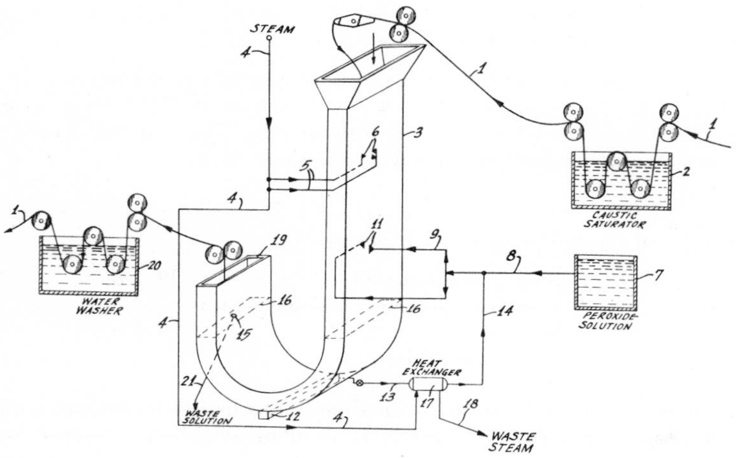


Source: G.T. Gallagher and E.J. Elliott; U.S. Patent 3,148,019; September 8, 1964

In Figure 4.4, (1) represents a cloth which has been treated for removal of any sizing or finishing agents. The cloth is passed into a caustic saturator (2) where it is impregnated with 50 to 150% of its weight of a 2 to 5% caustic soda (NaOH) solution. The caustic soda-impregnated cloth is then passed into the top of a J-box (3). Steam from conduit (4) passes through conduits (5) and into the J-box through openings (6). The cloth is steamed while moving slowly through the upper section of the J-box for from 15 to 45 minutes. After being steamed in the upper portion of the J-box, the cloth proceeds downward towards the U-shaped portion of the J-box and is sprayed through spray heads (11) with a mixture of fresh hydrogen peroxide solution and recirculating hydrogen peroxide solution.

The fresh make-up hydrogen peroxide solution from tank (7) proceeds through conduit (8) and mixes with recirculating hydrogen peroxide carried through conduit (14). This mixture flows through conduits (9) and (10) and is injected into the J-box through spray heads (11). The cloth then proceeds into the aqueous hydrogen peroxide bleach bath maintained at the base of the J-box up to levels (16). At the base of the J-box, a collecting trough (12) removes hydrogen peroxide from the base of the J-box and passes it via conduit (13) into heater (17). Heater (17) is heated by means of steam from conduit (4) and waste steam is removed via conduit (18). The hydrogen peroxide bleach solution which enters heater (17) through conduit (13) is heated to about 180°F. and passed via conduit (14) into conduit (8) to mix with the fresh hydrogen peroxide make-up solution from tank (7). Excess solution depleted of its aqueous hydrogen peroxide is removed as waste through orifice (15) located at a point opposite the points of entry of the hydrogen peroxide solution in order to remove spent solution containing a minimum of hydrogen peroxide.

Orifice (15) also acts to regulate the level (16) of aqueous hydrogen peroxide bleach solution which is present in the J-box, the excess being discharged through line (21). The cloth leaves the J-box through exit opening (19) in a clean and bleached state and is then passed into a water washer (20) for removal of residual caustic, hydrogen peroxide and other chemicals.

Example: Cotton fabric 80 x 78 threads per inch, weighing about 5 yards per pound and free of sizing materials was saturated with an aqueous solution containing by weight 3.0% caustic soda, 0.3% tetrasodium pyrophosphate and 0.1% Tergitol NPX, maintained at a temperature of from 170° to 180°F. The cloth was permitted to retain about 100% of its weight of the solution. The cloth was then passed into the straight section of a J-box and steamed for about 25 minutes at a temperature of about 212°F. At the termination of the steaming step, the unwashed cloth passed into the curved section of the J-box which contained a bleach solution having the following ingredients: 1.05% hydrogen peroxide, 2.0% sodium silicate, 0.3% tetrasodium pyrophosphate, 1.0% caustic soda, 0.1% Versenex 80, 0.1% epsom salt and 0.1% Tergitol NPX.

The bleach solution was maintained at a temperature of from 170° to 180°F. The cloth was passed through the J-box at a rate sufficient to maintain it in the bleach solution for about 35 minutes, and was then washed and dried. The bleaching solution at the base of the J-box was maintained at a 1.0% caustic soda level by the introduction of hydrogen peroxide make-up liquor containing 1.75% hydrogen peroxide at a rate of 0.3 gallon per minute. The bleached cloth was then tested for reflectance (whiteness) by a Hunter Lab Reflectometer. A high value indicates high reflectance and a good bleach. Values above 82% are generally acceptable.

The fluidity of the bleached cloth was tested according to the American Association of Textile Chemists and Colorists Tentative Test Method 82-1954, using cuprammonium hydroxide. This test is designed to test damage to the fiber after bleaching. A high value indicates high damage. Values below 10 signify no excessive chemical degradation of the cellulose, and are generally acceptable. The absorbency of the bleached cloth was tested in accordance with Tentative Test Method 79-1954 of the American Association of Textile Chemists and Colorists. This test is designed to test the time required for the cloth to absorb a given amount of water. High numerical values indicate poor absorbency. Values below 2 are considered acceptable.

The cloth was found to have a reflectance of 87.1%, a fluidity of 2.8 and an absorbency of 0.9 second. Following the washing and drying step, the bleached cloth was dyed in a continuous dyeing range with a blue vat dye. This method of dyeing and dye color are very sensitive to poorly bleached cloth and gives poor results unless the cloth possesses good absorbency and other desirable dyeing properties. The result was an excellent even dyeing through the entire length of cloth without evidence of resist marks, stains or other dye defects.

## Cotton Piece Goods or Yarns Bleached by Alkaline Peroxide Solutions

R.R. Currier; U.S. Patent 3,150,918; September 29, 1964; assigned to Pittsburgh Plate Glass Company describes a process of bleaching cotton and cotton-synthetic textiles or yarns using an alkaline peroxide solution in a batch process. The goods are first thoroughly saturated in an alkaline peroxide solution, followed by steaming at 210° to 212°F. The alkaline hydrogen peroxide solution used has a pH of 10 to 12, a peroxide concentration between 0.3 to 0.7% and contains about 0.5% of $[Na_2O(SiO_2)_{2.5}]$.

Example 1: A Turbo Rotary Dyer of 100 pound (dry goods) capacity was employed to prepare the bleaching solution. 50 gallons of water were introduced into the machine through a liquid intake and the water was heated by steam coils located in the bottom of the machine to 90°F. To the water contained in the machine was added 0.8 pound of flake caustic soda (NaOH), 2.5 pounds of 42° Bé. sodium silicate $[Na_2O(SiO_2)_{2.5}]$ and 8.3 pounds of 35% hydrogen peroxide. The inner basket of the machine was rotated during the addition of chemicals to assist in the dissolving of the chemicals. The rotation of the basket was stopped and to the inner basket was added a batch of gray cotton socks weighing 100 pounds. The inner basket was closed and the outer shell also closed and the basket again rotated for 5 minutes to thoroughly saturate the socks.

This provided a solution pick-up of 300% basis the weight of the dry socks. The excess solution was drained from the machine by opening the drain valve in the bottom of the shell encompassing the basket. Leaving the drain open, live steam was admitted to the machine through a steam inlet at a temperature of 212°F. for a period of 20 minutes. Hot water (160° to 180°F.) was then introduced through the liquid inlet and the basket again rotated to thoroughly wash the goods. After washing for 5 to 10 minutes, the rotation of the basket was discontinued, the machine drained, and the socks were bleached to a full white.

Example 2: Cotton yarn mounted on spools and placed in a package dyer such as a Gaston-County Package Dyer (AATCC Technical Manual reference) may be bleached in a similar manner within the purview of the process. Thus, in treating cotton yarn, the bleaching solution is admitted to the dyer in quantities sufficient to give the desired solution pick-up on the yarn. The solution, after the yarn has been saturated to the desired degree, is then removed from the dyer and steam is admitted at temperature of usually 210° to 212°F. for 20 minutes. After steaming, the yarn is washed with water, preferably hot (160° to 180°F.) and the yarn is bleached to a full white.

## Simplified Bleaching Process for Surgical Cotton

H. Grunert; U.S. Patent 3,198,597; August 3, 1965; assigned to Bohme Fettchemie GmbH, Germany describes a process for the preparation of surgical cotton of high absorbency and silk-like scroop. Hydrogen peroxide or persalts are used as the bleaching agents. The process consists of treating the raw cotton material with an aqueous alkaline bath containing a wetting agent at a temperature of 40° to 80°C., preferably about 60°C., adding to the bath bleaching and scrooping agents, heating the resulting bath stepwise at intervals up to a temperature of about 120°C., cooling the bath to below 100°C., rinsing the cotton fibers with warm water and then cold water, and drying the cotton to obtain the surgical cotton. The treatment with the wetting agent can be performed before or after the addition of the bleaching and scrooping agents. This method of treatment represents a very significant simplification of the previously known process in that a kier-boiling or boiling-off treatment can be entirely omitted.

The preliminary wetting bath is an aqueous alkaline solution containing 1 to 3 g./l. of a wetting and dispersing agent, preferably a nonionic, surface active compound. Examples of suitable nonionic, surface active agents are alkyl polyglycol ethers formed by the condensation of an alkyl alcohol having 8 to 18 carbons with an alkylene oxide, such as ethylene oxide, and alkyl phenol-polyglycol ether formed by the condensation of an alkyl phenol wherein the alkyl radical has 4 to 12 carbon atoms with an alkylene oxide such as ethylene oxide. The alkaline solution is preferably an aqueous solution containing 3 to 10 g./l. of an alkali metal hydroxide such as sodium hydroxide.

The bleaching agents are alkali metal perborates, persulfates, etc., and hydrogen peroxide which is preferred. Concentrations of 1 to 10 g./l. of the bleaching agents give a satisfactory whiteness. Examples of scrooping agents are polyglycol ethers of high molecular weight compounds having exchangeable hydrogen atoms such as hydroxyl, carboxylic acid, and amino compounds such as oleyl amine, stearic acid, oleyl alcohol. Especially useful are mixtures of unsaturated fatty alcohols having 12 to 18 carbon atoms and alkyl polyglycol ether condensation products of 1 mol of fatty alcohols having 12 to 14 carbon atoms with 6 to 12 mols of ethylene oxide. Concentrations of 1 to 3 g./l. of the scrooping agent give satisfactory results.

The stepwise heating of the bath is preferably effected in a closed pressure vessel in 2 to 5 steps, preferably 3 steps, for intervals of 10 to 30 minutes each. For example, the steps may be effected at temperatures of about 60°, 80° and 120°C. for time intervals of 10 to 30 minutes each. After the stepwise heating, the bath is cooled to below 100°C. and the cotton fibers are rinsed with warm water and then cold water, and then dried in the usual manner. Excess scrooping agent is removed from the cotton fibers by the rinsing.

Example 1: 35 kg. of cotton linters were subjected to a preliminary wetting treatment in a solution of 1,000 l. of water, 1 kg. of a condensation product of 1 mol of fatty alcohols having 12 to 14 carbon atoms and 8 to 10 mols of ethylene oxide and 3 kg. of 38° Bé. sodium hydroxide for about 20 minutes at 60°C. Then 4 kg. of 35% hydrogen peroxide and 1 kg. of a mixture of 5 parts of unsaturated sperm oil fatty alcohols having 14 to 18 carbon atoms and 1 part of a condensation product of 1 mol of fatty alcohols having 12 to 18 carbon atoms and 8 to 12 mols of ethylene oxide were added to the solution. The cotton was treated in this solution for 10 minutes at 60°C. Thereafter, the temperature of the solution was increased to 80°C. over a period of 10 to 15 minutes, maintained for 20 minutes at 80°C., raised over 20 minutes to 120°C. in the closed vessel, and maintained at that level for an additional 20 minutes. Subsequently, the solution was cooled to 90°C., the cotton was removed and rinsed first in warm and then in cold water. After drying, an absorptive cotton fiber having a silk-like scroop was obtained.

Example 2: 50 kg. of raw cotton were subjected to a preliminary wetting treatment in a solution of 1,000 l. of water, 5 kg. of 38° Bé. sodium hydroxide and 2 kg. of the condensation product of 9 mols of ethylene oxide with 1 mol of nonyl phenol for about 20 minutes at 60°C. Then, 5 kg. of 35% hydrogen peroxide and 3 kg. of the condensation product of 6 mols of ethylene oxide with 1 mol of stearic acid were added to the solution. The cotton was treated in this solution for 10 minutes at 60°C. Thereafter, the temperature of the solution was raised to 80°C. over a 10 to 15 minute period, maintained there for 20 minutes, raised to 120°C. within 20 minutes in the closed vessel and held at that level for an additional 20 minutes. The solution was then cooled to 90°C., and the cotton was rinsed in first hot and then cold water, and then dried to obtain a highly absorptive cotton fiber with a silk-like scroop.

## Continuous Bleaching of Keratinous Animal Fibers

A.E. Davis and A.M. Sookne; U.S. Patent 3,350,160; October 31, 1967; assigned to Gillette Research Institute, Inc. describe a continuous process to bleach wool, cashmere and other animal fibers in aqueous hydrogen peroxide. A ferrous sulfite mordant bath is the first and an iron removal bath is the last step in the process. The process will be more particularly described with reference to the drawing, Figure 4.5 diagrammatically illustrating the process and apparatus for the process.

FIGURE 4.5: CONTINUOUS BLEACHING OF ANIMAL FIBERS

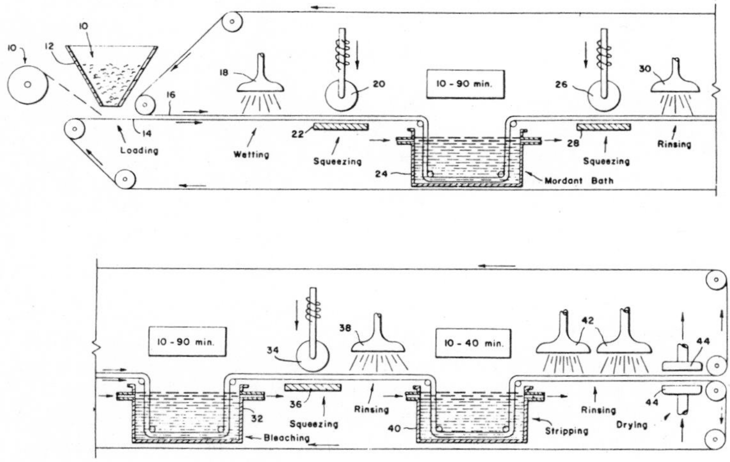


Source: A.E. Davis and A.M. Sookne; U.S. Patent 3,350,160; October 31, 1967

In accordance with the process, keratinous fibrous material either in the form of a woven or felted textile fabric illustrated by roll (10) or loose fibers maintained in the feed hopper (12) are fed between a pair of porous conveyor belts (14) and (16). The confined fibrous material is thoroughly wetted in passing below wetting means (18) connected to a suitable source of wetting liquid which may contain a wetting agent in addition to water. Following the spray wetting, the fibrous material is squeezed by, for example, roll (20) and cooperating platen (22) and the fibrous material is passed through a tank (24) containing an aqueous mordant solution of a ferrous salt containing sulfuric acid maintained at a temperature of from about 100° to 200°F. The concentration of the aqueous mordant solution is maintained substantially constant by continuously circulating the solution from a storage source as indicated by the directional arrows. Further, conventional heating means may be maintained in the tank (24) or at the outlet from the source of mordant solution. In the mordant step, the pH range of the solution should be between about 1 and 3. By maintaining the mordant bath at 150° to 170°F., the mordant treatment can be accomplished in a relatively short time, that is, from about 10 to about 90 minutes although a time period up to 150 minutes may be used.

The liquor-to-stock ratio of the mordant bath may be varied substantially. For example, a liquor-to-stock ratio of 5 to 1 to 100 to 1 or more will provide useful results. The concentration of the ferrous sulfate may also be varied substantially and useful results may be obtained in the range of from about 1 to about 20% as ferrous sulfate heptahydrate ($FeSO_4 \cdot 7H_2O$) based on the weight of the bath. Further, other ferrous salts may be used in equivalent amounts; for example, ferrous acetate and ferrous ammonium sulfate have provided satisfactory results. As the animal fibers leave the bath tank (24), the excess mordant solution is squeezed therefrom by squeeze roll (26) cooperating with platen (28). Then, the fibers carried by the endless belts (14) and (16) are passed under spray head (30) and the fibers are thoroughly rinsed and cooled. The temperature of the fibers following the rinsing step should not be greater than, for example 40° to 80°F. to minimize oxidation of the ferrous ion.

Following the rinsing step, the belts (14) and (16) carry the fibers into tank (32) containing the bleaching solution. The bleaching step comprises the treatment of the fibers with a solution of hydrogen peroxide containing suitable buffering materials. It is important that the pH of the bleach bath be maintained within the range of about 7.5 to 8.5. Preferred buffering materials for the bleach bath are tetrasodium pyrophosphate and sodium tetraborate used in combination. The amount of hydrogen peroxide in the bleach solution may be varied from about 0.5 to about 5% calculated as 100% $H_2O_2$ based on the weight of the bath.

The amount of tetrasodium pyrophosphate may be varied from about 0.5 to 5% also based on the weight of the bath while the sodium tetraborate may vary from 0 to 5% based on the weight of the bath and a preferred mixture of buffers is 1 part tetrasodium pyrophosphate to 2.5 parts of sodium tetraborate. The temperature of the bleaching bath is critical. For best retention of desirable fiber properties, the temperature of the bath should not be permitted to exceed 125°F. while a good working range is from about 100° to about 125°F. The time required to achieve the desired color reduction in the animal fibers in the bleach tank (32) may range from about 10 to 90 minutes.

After the fibers have been bleached the required amount in the bleach tank (32), the endless belts (14) and (16) carrying the fibrous material are passed below the squeeze roll (34) which cooperates with the fixed platen (36) to remove excess bleach from the belt. Thereafter the belt is passed below the spray head (38) and the fibers are thoroughly rinsed in a water solution maintained at a temperature of from about 100° to 150°F. to remove solubilized pigment, soluble iron and residual alkali and hydrogen peroxide. After the rinsing step, the fibers are carried through tank (40) containing the strip bath, comprising an aqueous solution of an iron removal agent. The strip bath may be maintained at a temperature in the range of from about 125° to 150°F., at a pH in the range of about 3.5 to 4.5. Where the iron removal bath contains sodium formaldehyde sulfoxylate, a concentration of from about 0.2 to 4.0 on the weight of the bath has been found to provide a very satisfactory results. The time employed for thorough removal of any residual iron in the fibers may be in the range of from about 10 to 40 minutes. Other iron removal agents may be employed in the continuous process of the method including, for example, oxalic acid. Following the stripping step, the fibers are thoroughly rinsed while passing beneath spray heads (42). The final rinsing must be thorough to remove residual chemicals which may affect future processing of the fibers if the chemicals are allowed to remain on the fibers. Following the thorough rinsing, the bleached animal fibers are dried while passing through dryer means (44).

Example: The following solutions were prepared:

Bath A — 0.05% Triton X 100 in water at 150°F.;
Bath B — 2.27% sulfuric acid and 3.1% ferrous sulfate heptahydrate in water at 150°F., pH 1.7;
Bath C — 1% tetrasodium pyrophosphate, 2.5% sodium tetraborate and 1.25% hydrogen peroxide in water at 115°F., pH 8.3;
Bath D — 0.3% sodium formaldehyde sulfoxylate in water adjusted to pH 3.5 with sulfuric acid at 140°F.

Specimens of Black/Brown Awassi carpet wool were arranged on a polypropylene netting (1/4 inch square openings) in thin layers, covered with an identical netting to hold the fibers in a fixed position and conveyed through the baths in the following manner.

At any instant the amount of fiber in any bath was such to produce a 24:1 bath:fiber ratio. The fiber-netting assembly was passed through Bath A to wet the fiber, squeezed through rollers at 60 lbs. pressure, then passed slowly through Bath B with an up-and-down motion for a period of 60 minutes during which time the pH was maintained at 1.7, the ferrous sulfate heptahydrate concentration was maintained at 3% and the temperature was maintained at 150°F. After 60 minutes of immersion in Bath B, the assembly passed through squeeze rollers at 60 lbs. pressure. The expressed liquor was returned to Bath B while the assembly proceeded through a forced spray of water at 80°F. for 2 minutes, followed by squeezing through rollers at 60 lbs. pressure, and then into Bath C which it moved through for 30 minutes. During the 30 minutes the hydrogen peroxide concentration was maintained at 1.25% and the pH maintained between 8.0 and 8.5 with additions of tetrasodium pyrophosphate and borax in the ratio of 1:2.5. The temperature was controlled at 115°F.

From Bath C the assembly passed through squeeze rolls at 60 lbs. pressure, the expressed liquor being returned to Bath C, through a force spray of water at 125° to 150°F., through squeeze rolls at 60 lbs. pressure into Bath D through which it passed in 20 minutes with the temperature maintained at 140°F., pH at 3.5 and sodium formaldehyde sulfoxylate concentration at 0.3%. After 20 minutes the assembly was squeezed to 60 lbs. pressure, force sprayed with water at 150°F., squeezed at 60 lbs. pressure and air dried. The treated fiber was cream color and uniformly bleached at the end of the processing train. The finished fibers had an ultimate elongation of 30% as compared with 29% for the untreated and a breaking strength of 1.25 g. per denier as compared with 1.27 g. per denier for the untreated fibers.

## Bleaching of Dyed Cotton Fabrics with Hydrogen Peroxide

H.G. Smolens; U.S. Patent 3,280,039; October 18, 1966; assigned to Pennsalt Chemicals Corporation and U.S. Patent 3,343,906; September 26, 1967; assigned to Pennsalt Chemicals Corporation describes a process for the simultaneous bleaching of dyed cotton fabrics, dyed by different dyestuffs. The goods may contain some dyed with naphthol dyes which have the tendency to bleed during the bleaching process, also some dyed with vat dyes and some undyed fabrics. The bleeding of the naphthol dyed goods is prevented by balanced and closely controlled alkalinity obtained with sodium carbonate and sodium bicarbonate without the use of a silicate. The process is suitable for continuous or batch operations.

In bleaching fabrics containing naphthol dyed yarns, three factors are most important to the prevention of staining and mark-off. These are: peroxide concentration (which should be as low as possible), pH (which should be maintained in all cases at or preferably below 10.0) and total alkalinity as measured by the use of phenol red indicator (which should be as high as possible). By total alkalinity is meant the amount of alkali present as measured by titrating a measured quantity of the solution with a standardized dilute acid solution until a phenol red end point is reached. The result may be expressed in terms of the grams per liter of NaOH which that amount of acid would neutralize, or alternatively in terms of normality, that is, the number of mols of alkali per liter multiplied by the number of $OH^-$ groups in the molecule.

Example 1: Approximately 300 yards of cotton goods having 1 inch blue and green stripes of vat dyed colored yarn alternating with white background is sewn to a length of approximately 300 yards of naphthol dyed red and white striped cotton cloth. The dry cloth is first desized and then is fed to a bleaching saturator containing a solution of 100 gallons of water, 0.5 pound of stabilizer, 5 pounds of sodium carbonate, 5 pounds of sodium bicarbonate, 8 pounds of 35% hydrogen peroxide. This saturator is fed with a solution from a head tank containing 100 gallons water, 35 pounds sodium carbonate, 56 pounds 35% hydrogen peroxide, and 35 pounds sodium bicarbonate.

This solution has a pH of about 9.9 and a total alkalinity of 0.25 normal. The squeeze rolls on the exit of the saturator are set to yield a cloth to bleaching solution ratio of 1:1. The cloth is run through a steam heated J-box so that each portion of the cloth receives an exposure of 1 hour to the 150° to 160°F. temperature in the interior of the J-box. The cloth is fed to washers and all bleaching solution is removed. After drying, both the naphthol dyed and the vat dyed colored yarn fabrics which have been treated identically throughout the bleaching process, show excellent whiteness and excellent brightness of color without evidence of running, staining, or mark-off.

Example 2: Approximately 300 yards of cotton goods having 1 inch blue and green stripes of vat dyed colored yarn alternating with white background is sewn to a length of approximately 300 yards of naphthol dyed red and white striped cotton cloth. The cloth is first desized and then fed to a bleaching saturator containing a solution (all parts by weight) composed of: 1,000 parts water, 12 parts sodium silicate, 1.2 parts sodium carbonate and 9.6 parts 35% $H_2O_2$. The saturator is fed from a head tank containing a solution (all parts by weight) composed of: 1,000 parts of water, 84 parts sodium silicate, 8.4 parts sodium carbonate and 67.2 parts 35% $H_2O_2$.

The solution in the saturator has a total alkalinity of 0.06 normal, a pH of 10.5 and is at room temperature. The cloth is run into a steam heated J-box slowly and is slowly withdrawn so that each portion of the cloth receives an exposure of about 1 hour to the 150°F. temperature in the interior of the J-box. The bleached cloth is fed to washers and all bleaching solution residue is removed. After drying, the cloth is examined. Bleaching quality of vat dyed goods is satisfactory but the naphthol dyed goods show, particularly in the white areas, a large amount of staining and mark-off. The naphthol dyed goods are not suitable for sale.

Example 3: Again using similar quantities and types of fabrics, identical equipment and similar desizing, steaming, washing and drying operations, the process of Example 2 is repeated using the following bleaching compositions:

| Head Tank | Parts |
|---|---|
| Water | 1,000 |
| Sodium silicate | 0 |
| Sodium carbonate | 84 |
| Sodium bicarbonate | 84 |
| 35% $H_2O_2$ | 67.2 |
| Saturator | |
| Water | 1,000 |
| Bleaching solution stabilizer | 0.5 |
| Sodium silicate | 0 |
| Sodium carbonate | 12 |
| Sodium bicarbonate | 12 |
| 35% $H_2O_2$ | 9.6 |

Here the total alkalinity in the saturator rises to 0.16 normal, a very desirable characteristic for the bleaching solution. The pH is 9.8 and the saturator remains at room temperature. After steaming, washing and drying as before, the goods are examined. The vat dyed fabrics again show good to excellent bleaching, and in addition the naphthol dyed goods show excellent bleach with no staining and no mark-off.

The head tank concentration represents approximately 7 times that used in the saturator. This desirable concentration ratio is obtained with no clogging of pipes or precipitation in J-boxes due to gelling of the solutions. For maximum economy and bleaching efficiency it is generally desirable to add 0.05 to 1.0 g./l. of stabilizer to the bleaching solution or in a 0.35 to 7.0 g./l. concentration to the head tank solution. Stabilizers may be added and other special purpose ingredients may be found advantageous, but none of these is essential to the process.

The Examples 1 and 3 with no silicate and with pH below 10 show no bleeding of the fabrics with naphthol dyes, nor any clogging of the pipes. Example 2 gives unsatisfactory results.

## Desizing and Bleaching Cotton with Hydrogen Peroxide

H.L. Potter and N.J. Stalter; U.S. Patent 3,353,903; November 21, 1967; assigned to E.I. du Pont de Nemours and Company describe a multistage process to desize and bleach greige cotton fabrics in preparation for dyeing. The process consists of three stages.

Stage 1: The fabric is dampened or impregnated with an aqueous solution containing 6 to 13%, preferably 8 to 11% caustic soda or caustic potash. Although not essential, the solution will advantageously also contain 0.2 to 2%, preferably 0.4 to 0.7% of a water-soluble molecularly dehydrated phosphate and/or also a wetting agent. The dampened fabric is subjected to the direct action of saturated steam at about atmospheric pressure for a time of from about 1 to 2 minutes, then rinsed with water.

Stage 2: The fabric from Stage 1 is impregnated or dampened with an aqueous solution containing: (a) 1 to 5% caustic soda or caustic potash, preferably 1 to 3% when two stages are to be used and 2 to 4% when three stages are to be used; (b) 0.2 to 2.5%, preferably 0.3 to 1.5% hydrogen peroxide ($H_2O_2$); and (c) a peroxide stabilizer at a concentration effective to prevent excessive decomposition of the hydrogen peroxide. Advantageously, although not essential, the solution will also contain a wetting agent and/or an organic heavy metal sequestering agent. The dampened fabric is subjected to the direct action of saturated steam at about atmospheric pressure for a time of from about 1 to 2 minutes, then rinsed with water.

Stage 3: If a further treatment stage is needed or desired, the fabric from Stage 2 is impregnated or dampened with an aqueous solution containing (a) 0.8 to 3%, preferably 1 to 2% caustic soda or caustic potash; (b) 0.2 to 2.5% preferably 0.7 to 1.6% $H_2O_2$; and (c) a peroxide stabilizer at a concentration effective to prevent excessive decomposition of the hydrogen peroxide. A wetting agent and/or an organic heavy metal sequestering agent may also be present. The fabric dampened with the treating solution is then subjected to the direct action of saturated steam at about atmospheric pressure for from about 1 to 2 minutes, then finally washed.

It has been found that many woven greige cotton fabrics can be rapidly and satisfactorily conditioned for dyeing by being subjected sequentially to the treatments of only Stages 1 and 2. However, in some instances, a subsequent treatment in accordance with the above Stage 3 may be indicated, particularly when a whiteness somewhat higher than that obtained by the two-stage treatment is found to be desirable. Whether practicing either the two-stage or the three-stage embodiment of the process, the fabric is handled throughout in its open width and extended condition, so that no storage of the fabric in piled or bulk form is involved and the problem of creasing and rub marks is entirely eliminated.

In each stage, a continuous length of the fabric in open width and extended state is dampened with the treating solution for that stage, the dampened fabric is passed through a steamer in which an atmosphere of saturated steam at substantially atmospheric pressure is continuously maintained, and the steamed fabric is then rinsed. The fabric is continuously advanced in open width through all stages to be employed, and the rate of advancement will be such as to provide a residence time in the steamer of each stage of about 1 to 2 minutes. The method produces excellent results, even when applied to heavy (at least 2 yds./lb.), tightly woven fabrics, such as twills, poplins, sateens, drills and jeans, in the greige (undesized and unscoured) state. The method, in both its two and three-stage embodiments, has been successfully applied to a wide variety of all-cotton fabrics and blended cotton fabrics, e.g., cotton-nylon and cotton-polyester blends, all in the greige state.

Example: In Stage 1, a 1.8 yd./lb. greige cotton twill fabric was immersed in an aqueous solution containing 10% caustic soda, 0.5% sodium pyrophosphate

($Na_4P_2O_7 \cdot 10H_2O$) and 0.2% sodium lauryl sulfate. The temperature of the solution was 180°F. Excess solution was expressed from the fabric so as to leave it dampened with an amount of the solution equal to the fabric weight. The dampened fabric was heated in an atmosphere of saturated steam at atmospheric pressure (the temperature being approximately 212°F.) for 2 minutes, then rinsed with water. In Stage 2, the fabric was dampened, steamed, then rinsed as previously, except that the treating solution used was one containing 2% caustic soda, 0.5% sodium pyrophosphate, 0.2% sodium lauryl sulfate, 1% sodium silicate (commercial 42° Bé. solution) and 0.35% $H_2O_2$.

In Stage 3, the dampening, steaming and rinsing operations were again repeated, but this time employing as the treating solution one containing 1% caustic soda, 2% sodium silicate, 0.1% sodium salt of diethylenetriaminepentaacetic acid, 0.1% sodium lauryl sulfate and 1.4% $H_2O_2$. The above three-stage treatment gave a prepared fabric having a whiteness of 83.0%, as measured by the Hunter Reflectometer, and an absorbency of 0.5 second, as determined by the standard AATCC drop test. It was substantially free of motes and in excellent condition for dyeing.

## Continuous Mercerizing and Bleaching Process for Yarns

T.E. Westall; U.S. Patent 3,370,911; February 27, 1968; assigned to The American Thread Company describes a continuous and combined process for mercerizing and bleaching cotton yarn, using an alkaline hydrogen peroxide solution. The process is carried out in a short period of time. The yarn moves forward continuously through the following major process steps: boiling out, mercerizing (caustic), caustic washing, steaming, bleaching, steaming, washing and finishing.

Example 1: A warp of yarn is made up by bringing 378 ends of cotton yarn together after the yarns have been individually passed through a flame to remove lint and fuzz. The yarn ends are brought together in side-by-side parallel relationship with all of the yarn ends lying in a single phase. Eighteen warps made in this manner are continuously advanced in parallel side-by-side relationship through the sequence of steps illustrated in Figure 4.6. The velocity of the yarn is 20 yards per minute throughout the process from the beginning until the yarn reaches steam heating chamber (30). The velocity increases to about 20.1 yards per minute in steam heating chamber (30), and remains at this level throughout the remainder of the process.

The sheet of yarn (10) is first contacted with boiling water in boiling-out tank (11) to remove grease and impurities. The yarn is then cooled to about room temperature in cold water tank (12), and is then passed into mercerizing tank (13) containing an aqueous 23% caustic soda solution at room temperature. The yarn is then washed in caustic wash tank (14) containing an aqueous solution of 7 to 9% by weight of caustic soda. The pressure of rolls (46) through which the warp of yarn is passed after it leaves caustic wash tank (13) is set so that the yarn contains 3% by weight of caustic solution as it enters first steaming chamber (15).

The yarn passes through steaming chamber (15) in a plurality of passes. Wet steam is continuously supplied to steaming chamber (15), and the temperature in that chamber is maintained at 190° to 200°F. The residence time of yarn in steaming chamber (14) is approximately 1 minute. The yarn is then passed through a second caustic wash solution in tank (20). This caustic wash solution also contains 1% by weight of caustic soda and is at a temperature of 140° to 180°F.

FIGURE 4.6: CONTINUOUS MERCERIZING AND BLEACHING EQUIPMENT

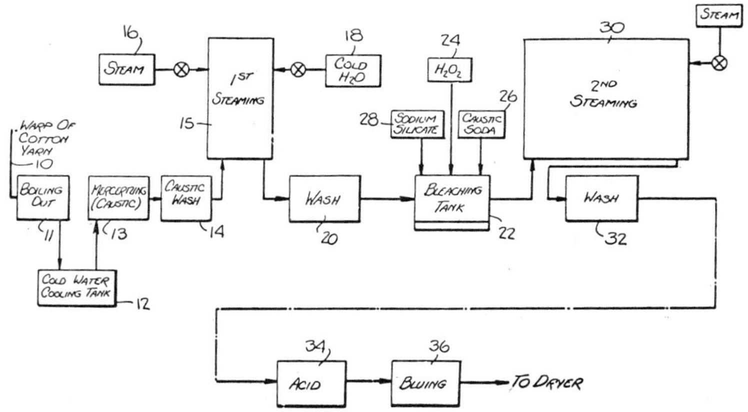


Source: T.E. Westall; U.S. Patent 3,370,911; February 27, 1968

After caustic washing in tank (20), the yarn enters bleaching tank (22). The yarn is contacted in bleaching tank (22) with a bleaching solution containing 4% by weight of hydrogen peroxide, 1% by weight of caustic soda, 2% by weight of sodium silicate (water glass, $Na_2O \cdot 3.3SiO_2$), 1/2 ounce per gallon of epsom salts (magnesium sulfate), and 1/2 ounce per gallon of diethylenetriaminepentaacetic acid chelating agent (Versenex 80). The temperature in bleaching tank (22) is 120°F. Make-up solutions are continuously supplied to bleaching tank (22) from hydrogen peroxide storage tank (24) containing 50% aqueous hydrogen peroxide, caustic soda, sodium silicate solution tank (28) containing 81% by weight of sodium silicate (water glass) having the approximate formula $Na_2O \cdot 3.3SiO_2$, epsom salts make-up tank containing 1 pound of epsom salts per gallon of water, and chelating agent make-up tank containing 1 pint of Versenex 80 per gallon of water. The flow rates of these make-up solutions in gallons per hour are as follows:

| | |
|---|---|
| Hydrogen peroxide | 15 |
| Sodium hydroxide | 3 |
| Sodium silicate | 4 |
| Epsom salts | 0.25 |
| Versenex 80 | 0.25 |

The total amount of make-up solution is approximately equal to the amount of solution carried out by the yarn. From bleaching tank (22) the yarn goes to second steaming chamber (30). Dry saturated steam is introduced into this chamber at 212°F. and 1 atmosphere, and the steam is superheated to about 218°F. by means of steam heaters within chamber (30). The speed of the yarn as it enters is 20 yards per minute. The yarn passes alternately up and down through a number of passes in chamber (30), and is kept under tension at all times. Due to the tension and the high temperature in chamber (30), the velocity increases by about 0.5% to 20.1 yards per minute at the exit. The yarn passes alternately over a driven roll and an idler roll. The roll speed of driven rolls progressively increases as the warp of yarn advances through chamber (30), so that there is no slip between the yarn and any of the rolls. The residence time of yarn in chamber (30) is approximately 5 minutes. The yarn upon leaving second steaming chamber (30) is washed with hot water at 190°F. in water wash tanks (32) and (32a), and then is contacted with dilute aqueous acetic acid at a temperature of about 140° to 160°F. in acid wash tank (34). After acid washing, the yarn is contacted with a bluing agent in bluing tank (36), which is maintained at 160°F. From bluing tank (36) the yarn passes through a drier. After drying, the yarn is taken up on a pallet.

## Continuous Desizing, Scouring and Bleaching for Cotton and Polyester

G. Cerana; U.S. Patent 3,377,131; April 9, 1968; assigned to Roberto Cerana, SpA, Italy describes a continuous process for desizing, scouring and hydrogen peroxide bleaching of cotton or mixed fabrics. The following are the steps in the process:

(1) The impregnation of the fabric, in full width (the fabric could be dry or wet), in the cold, is made with an aqueous solution containing hydrogen peroxide at a concentration of from 30 to 80 g./l., e.g., 40 g./l. For example, a solution available in the market known as ROCESA can be used.

(2) Placement of the fabric, in full width for about 15 to 20 minutes, in an impregnated condition (according to the nature, weight, quality of the fabric to be bleached) and at a temperature comprised between 90° and 100°C. This step can be carried out with saturated steam obtained through automatic saturation adjustment devices, and has the purpose of removing the residues of substances used in the spinning and weaving (starch, fecula, lubricants, etc.).

(3) Placement of the fabric, in full width, in hot water (about 90°C.) wherein it is immersed so as to activate the bleaching liquor, that is so as to have it acting on the fabric.

(4) Steaming, in full width, for about 2 to 3 minutes in steam at 100°C. for making the action of the preceding step completed.

(5) Washing of the fabric for a very short time (about 2 to 3 seconds) in water at about 80°C.

(6) Washing of the fabric, for a very short time (about 2 to 3 seconds) in water at about 70°C.

(7) Washing of the fabric in cold water at ambient temperature for about 2 to 3 seconds.

(8) Squeezing at a high squeezing efficiency, for example, up to about 60% of the weight of the fabric and subsequent folding (on a truck) or rolling (on a cloth beam) of the treated fabric.

The advantage of this process is compared to conventional (1) hypochlorite bleaching, (2) chlorite bleaching, and (3) this process as related to the treatment of a conventional cotton fabric is given in the following comparison. The article is a Madapolam fabric, 70 pieces of 8,400 meters total length, width 90 cm., grams per square meter 130 approximate, 100% cotton. (1) Man-hours, about 24; workers, 9 men. (2) Man-hours, about 17, workers, 3 men. (3) Man-hours, about 6; workers, 2 men.

## Peroxide Bleaching of Cellulose Textiles at Higher Temperatures

P. Ney and W. Kuhnmünch; U.S. Patent 3,397,033; August 13, 1968; assigned to Deutsche Gold- und Silber-Scheideanstalt vormals Roessler, Germany describe a peroxide bleaching process for cotton textiles using higher storage and steaming temperatures. Sodium perborate, sodium chlorite or peracetic acid may also be used in this process. The goods to be bleached are impregnated with the aqueous bleaching solution, and the impregnated goods are stored at room or moderately raised temperatures, for instance up to about 40°C. for several hours, for example, 3 to 10 hours, in a tightly packed state, and after such storage are treated in a pressure steamer at temperatures between 120° and 200°C. for 15 to 120 seconds. Preferably, such pressure steaming treatment is effected at temperatures between 130 and 150°C. at corresponding pressures for 30 to 60 seconds.

The tests were carried out on desized Reutlinger nettle cloth which still contained seed shells. The steaming apparatus employed was an Obermaier HT apparatus. The nettle cloth was impregnated with an aqueous bleaching solution which per liter contained:

| | |
|---|---|
| Water glass 38/40° Bé. | 30 ml. |
| NaOH | 10 g. |
| $H_2O_2$ (35% by weight) | 40 ml. |
| Lamepon A (condensation product of oleic acid chloride and sodium lysalbinate) | 2 g. |
| Wetting agent (lauryl sulfate) | 2 g. |

In all instances the cloth after impregnation was squeezed out to a 100% bleaching solution content and then subjected to the action of the retained bleaching solution at room temperature for the periods indicated in the following table. The steam was at a temperature of 130°C. which corresponds to a gauge pressure between 2.8 to 3.2. After completion of the bleach the cloth samples were each rinsed successively with hot, warm and cold tap water. The degree of whiteness, the time required for submerging when placed on water and the degree of seed shell removal was determined for each bleached sample. The results are given in the following table.

| Test No. | Room Temp. Bleach (hrs.) | Pressure Steaming (sec.) | Degree of Whiteness (%) | Submerging Time | Seed Shell Removal |
|---|---|---|---|---|---|
| 1 | 18 | -- | 77 | Over 2 min. | Shell-free |
| 2 | 3 | 60 | 80 | 25 sec. | Shell-free |
| 2a | 3 | -- | 72 | Over 2 min. | Still contained shells |
| 3 | 6 | 60 | 81.9 | 2 sec. | Shell-free |
| 3a | 6 | -- | 73 | 90 sec. | Still contained shells |
| Starting Material Desized | -- | -- | -- | Over 10 min. | --- |

As can be seen from the table, tests 2 and 3, which were according to the process, resulted in goods which were seed shell free, and had good wettability and a higher degree of whiteness than were obtained in tests 1, 2a and 3a. Analogous improved results were obtained when the above bleaching solution was replaced by aqueous bleaching solutions which per liter contained the ingredients given on the following page.

Bleaching solution a:

| | |
|---|---|
| Sodium perborate | 75 g. |
| Lamepon A | 2 g. |
| Lauryl sulfate | 2 g. |

Bleaching solution b:

| | |
|---|---|
| Water glass | 30 ml. |
| $Na_2O_2$ | 10 g. |
| $H_2O_2$ (35% by weight) | 30 ml. |
| Lamepon A | 2 g. |
| Lauryl sulfate | 2 g. |

## Simultaneous Scouring and Bleaching with Alkaline Hydrogen Peroxide

B.C. Lawes and H.L. Potter; U.S. Patent 3,514,247; May 26, 1970; assigned to E.I. du Pont de Nemours and Company found that textile fabrics containing portions dyed with a sensitive dyestuff, such as a vat or a naphthol dye, are simultaneously scoured and bleached by saturating the fabric with a bleaching solution containing 0.7 to 3% hydrogen peroxide and sufficient alkali to bring the pH of the solution to 10.5 to 13, then heating the saturated fabric to a temperature of 180° to 220°F. for 0.5 to 4 minutes. This process is suitable for cotton, rayon and mixed fabrics. It is sufficiently short so as not to require the stacking of fabrics into large piles.

In addition to hydrogen peroxide and sodium or potassium hydroxide, the bleach solution may also contain other materials such as stabilizers for the hydrogen peroxide, wetting agents, and heavy metal impurity sequestering agents such as the well-known polycarboxyamino compounds and their water-soluble salts, for example, ethylenediamine tetraacetic acid, diethylenetriamine pentaacetic acid, nitrilotriacetic acid, and their water-soluble salts. The bleaching solutions will also preferably contain a peroxide stabilizer which is effective under alkaline conditions. Effective stabilizers are the water-soluble salts of molecularly dehydrated phosphoric acid, examples of which are the sodium and potassium pyrophosphates, sodium hexametaphosphate and sodium tripolyphosphate. These will generally be used as concentrations of 0.2 to 2%. Sodium silicate, e.g., the 42° Bé. solution (approximate composition: 10% $Na_2O$, 25% $SiO_2$ and 65% $H_2O$) commonly employed in alkaline peroxide bleach solutions, is also effective, e.g., at concentrations of 0.5 to 3%.

The following abbreviations are used in some of the examples to designate certain constituents of the bleach solutions employed: DTPA designates a commercial diethylenetriaminepentaacetic acid sequesterant; DODS designates a commercial surfactant which is the sodium salt of dodecylated oxydibenzene sulfonate; and TSPP designates tetrasodium pyrophosphate decahydrate.

Example: Separate portions of a tightly braided cotton-polyester shirt fabric containing red stripes formed of yarns dyed with a naphthol red dye and black stripes formed of yarns dyed with a black vat dye woven into the fabric, were bleached using the bleaching solutions for Runs A, B and C whose compositions are indicated in the tabulation in the table. The fabric samples prior to the bleaching had been first given a mold scouring treatment involving saturating the fabric with an aqueous solution containing 0.83% TSPP and 0.5% soda ash, and then heating the saturated fabric at 150°F. for 60 minutes. The results are shown in the tabulation following.

| Run | Bleach | Bleach Conditions | Bleeding or Mark-off |
|---|---|---|---|
| A | 0.7% $H_2O_2$, 1.25% sodium silicate, 0.25% $NaCHO_3$, 0.25% sulfonated castor oil, 0.2% DTPA, 0.02% epsom salt, pH 10.2% | 1 hour at 210°F. | Considerable |
| B | 0.8% peracetic acid, 0.4% TSPP, pH 5.8 (NaOH) | 30 min. at 210°F. | Visible but less than in A |
| C | 1.4% $H_2O_2$, 0.9% NaOH, 2.0% sodium silicate, 0.05% epsom salt, 0.1% DTPA, 0.1% DODS, 0.5% TSPP, pH 11.4 | 2 min. at 210°F. | None visible |

The above data show that of the three bleaching runs, only Run C (in accordance with the process) effected bleaching without bleeding of the naphthol red dye from the red or black pin stripes woven into the fabric.

# PERACIDS AND PERSALTS

## NITROGEN BASE ACTIVATORS AND STABILIZERS

### Bleaching Nylon with Acylimide and Perborate

M.M. Baevsky; U.S. Patent 3,061,550; October 30, 1962; assigned to E.I. du Pont de Nemours and Company describes a process to bleach nylon and other textile fabrics with N-acylimide compounds such as N-benzoyl succinimide and sodium perborate or other peroxygen compound. The N-acylimide compositions have the general formula:

$$R{-}\overset{\overset{\displaystyle O}{\|}}{C}{-}R_1$$

wherein R is aromatic or substituted aromatic and $R_1$ is an imide radical, preferably derived from a cyclic imide. The above compositions are used to formulate superior home bleaches for nylon that are highly effective in removing discoloration from nylon fabrics without causing excessive strength losses, and that may be conveniently and effectively used alone or in combination with household soaps or detergents for bleaching purposes.

Although it has been customary to judge fabric color by eye, it is much more satisfactory to use precise instrumental methods whereby a numerical value can be obtained. In the following examples, color measurements are made using a differential colorimeter as described by L.B. Glasser and D.J. Troy in the Journal of the Optical Society, 42, 652-660 (1952). These authors describe a method whereby reflectance measurements can be converted into the color coordinates, L for lightness, a for greenness-redness, and b for yellowness-blueness. Using this system, negative (-) values of b represent blue shades, while positive (+) values represent yellow shades. The depth or intensity of color will be proportional to the numerical magnitude of the b value. Also described is a method for determining the total color difference ($\Delta$E) between a sample and a white standard when the sample contains an mixture of colors which cannot be described in terms of any single color coordinate.

In the following examples, bleaching tests are carried out on fabrics which have been discolored by either dye transfer or heat yellowing, since these are the most common sources of garment discoloration encountered in actual use. The dye transferred fabrics are discolored by laundering white fabrics in the presence of blue, red, and yellow fabrics which had been dyed to saturation with certain dyes of poor washfastness. The fabrics used for this purpose in the following examples are dyed with a blue dye, Anthraquinone Blue SWF (Prototype No. 12, "Technical Manual and Yearbook," AATCC, 1955), a red dye, Celanthrene Red 3BN (Prototype No. 234), and a yellow dye, Celanthrene Fast Yellow GL (Prototype No. 534).

The color of the dye transferred fabrics is numerically described in the examples in terms of the total color difference ($\Delta E$) when compared to the same fabrics before discoloration. Thus, the lower the total color difference ($\Delta E$), after treatment with an oxidizing agent, the greater the color removal and the more effective the bleach. The heat-yellowed fabrics described in the following examples are discolored by exposure to excessive heat. The color of these fabrics is described in terms of the +b value (yellowness). The effectiveness of the bleaching action of the various compounds given in the examples can be clearly determined by comparison of the magnitude of the +b values of a fabric, before and after bleaching.

Example: Nylon taffeta fabrics, discolored by exposure to heat and by laundering in the presence of dyed fabrics, are immersed in an aqueous bath buffered to pH 6 with citric acid-disodium phosphate, the weight of the bath being 50 times the weight of the fabric. A mixture of N-benzoyl succinimide and anhydrous sodium perborate is added to the bath and the fabrics allowed to soak for 1 hour at a temperature of 45°C.

The fabric samples are then removed from the bath, rinsed and dried, and the total color difference ($\Delta E$) before and after bleaching is determined with a differential colorimeter as described above. The table below shows the results obtained with various molar ratios of sodium perborate to N-benzoyl succinimide and with various concentrations of the bleaching composition. The concentration of bleaching composition is expressed as percent based on the weight of the fabric sample. Total color measurements before and after bleaching are shown for each of the samples. Also shown, for comparison, is the result obtained when sodium perborate is used alone under these conditions. Measurement of the Mullen bursting strength of the fabric before and after bleaching showed a strength loss of less than 2.0% in all cases.

| Molar Ratio | Concentration, % | ΔE Before | ΔE After |
|---|---|---|---|
| Sodium perborate alone | 0.2 | 9.4 | 9.1 |
| 2.5 | 0.4 | 9.4 | 1.3 |
| 2.5 | 0.6 | 9.4 | 1.5 |
| 6.0 | 0.34 | 8.8 | 1.0 |
| 6.2 | 0.7 | 9.4 | 1.2 |
| 12.4 | 0.6 | 9.4 | 1.2 |

When the buffering agents, citric acid and disodium phosphate, are mixed with the dry bleaching composition and the weight of bleaching composition increased sufficiently to provide the same concentration of N-benzoyl succinimide and sodium perborate in the bleaching bath, similar results are obtained.

Perborate with Hydrazine Activators

W.J.C. Viveen and C.U. Kloosterman; U.S. Patent 3,163,606; December 29, 1964; assigned to Koninklijke Industrieele Maatschappij vorheen Noury & van der Lande NV, Netherlands describe a process to improve the low temperature bleaching effect of perchlorates by the use of nitrogen containing activators, having at least two acyl groups attached to the same nitrogen atom. Effective bleaching is obtained on cotton or synthetic textiles, including nylons at temperatures between 15° and 85°C. with lower loss in tensile properties of the fibers.

As activators for oxygen releasing compounds according to this process there are included: N-diacylated amines, e.g., diacetylmethylamine, diacetylethylamine; N-diacylated ammonia, e.g., diacetamide, dipropionamide; N-diacylated amides, e.g., N-formyldiacetamide, N-acetyldiacetamide, (triacetamide), N-propionyldiacetamide, N-butyryldiacetamide, dicarboxylic acid acyl imides, e.g., N-acetylphthalimide, N-acetylsuccinimide; N-diacylated urethanes, e.g., N,N-diacetylethylurethane, N-acetyl, N-propionylethylurethane; N-diacetylated hydrazines, e.g., triacetylhydrazine; N-diacylated alkylenediamines, e.g., triacetylmethylenediamine, tetraacetylmethylenediamine, the N-diacyl compounds of semicarbazide, thiosemicarbazide and dicyanodiamide.

The efficiency of the activator is further improved if it is covered with a solid material soluble in the washing and/or bleaching liquid. As such, stearic acid, polyethyleneglycol such as, e.g., Carbowax 4000 and 6000 condensation products of ethylene oxide and propylene oxide, polyvinyl alcohol, carboxymethylcellulose, cetyl alcohol and fatty acid alkanolamides, are especially suitable.

It is recommended to granulate the activator before covering. Covering may be effected in coating pans suitable to the purpose. The covering material dissolved in water or in an organic solvent, is sprayed on the activator in finely divided form, after which the covered material is dried. Solutions of stearic acid, Carbowax or $C_{12}$ to $C_{14}$ isopropanolamide in isopropanol as well as solutions of polyvinyl alcohol and carboxymethylcellulose in diluted ethanol are very suitable. As to the quantity of the covering to be used, it is in general sufficient to apply 1 to 30% by weight on the activator. As activators for the active oxygen releasing compound in a washing and/or bleaching agent there have been found especially effective compounds with the general formula:

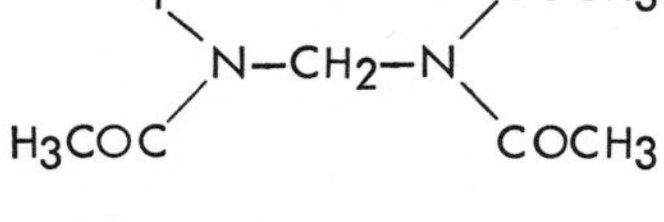


in which $R_1$ represents a hydrogen atom or a $—COCH_3$ group. These compounds may be obtained by either of the following two methods. Thus it is possible to prepare N,N,N',N'-tetraacetylmethylenediamine by allowing to react ketene on a mixture of N,N'-diacetylmethylenediamine and an inert solvent in the presence of an acid-reacting catalyst. This reaction may be executed in benzene as an inert solvent at a temperature of 30° to 80°C., preferably however at a temperature of 50° to 65°C., in the presence of 0.5 to 3%, preferably of 2% by weight of p-toluenesulfonic acid calculated on N,N'-diacetylmethylenediamine. The N,N,N',N'-tetraacetylmethylenediamine formed may be recovered from the reaction mixture in any conventional way. N,N,N',N'-tetraacetylmethylenediamine may, however, also be obtained by acetylating N,N'-diacetylmethylenediamine with the aid of acetic acid anhydride while eliminating continuously the acetic acid formed during the reaction, e.g., by distillation.

If during the reaction between diacetylmethylenediamine and acetic acid anhydride, the acetic acid liberated is not eliminated from the reaction mixture, mainly N,N,N'-triacetylmethylenediamine is formed. When only part of the quantity of acetic acid that theoretically could be formed during the acetylation is eliminated from the reaction mixture, mixtures are obtained that appear to contain triacetylmethylenediamine and tetraacetylmethylenediamine with the quantity of the former increasing as less acetic acid is eliminated.

Example: Pieces of cotton fabric were soiled with the juice of black currants, by evenly applying to them 2.7 cc, of black currant juice per 100 $cm.^2$ of surface of cotton fabric by means of a brush. After being dried in the air, the lightness of the fabric was measured with a photoelectric remission meter. It was 29% of the lightness of magnesium oxide. Parts of the pieces of material were washed for 15 minutes at 60°C. in suds of the following composition:

| | Grams per Liter |
|---|---|
| Sodium salt of dodecyl benzene sulfonic acid | 0.85 |
| Lauryl isopropanolamide | 0.10 |
| Tetrasodium pyrophosphate | 0.75 |
| Pentasodium tripolyphosphate | 0.60 |
| Alkaline water-glass ($Na_2O/SiO_2$ = 1/2) | 0.20 |
| Na CMC (100%) (sodium carboxymethylcellulose) | 0.04 |
| Sodium sulfate | 0.80 |
| Sodium perborate-tetrahydrate (10.2% active oxygen) | 0.50 |

The percentage of lightness of the washed pieces of material amounted to 64 (lightness of magnesium oxide = 100 units), which is 13 units higher than the percentage of lightness of the pieces of material after thorough rinsing with water, which was 51.

Other parts of the material were washed in three samples of suds of the same composition, to which respectively 0.1, 0.4 and 0.5 g./l. of tetraacetylhydrazine was added. These suds contained per atom of active oxygen from the perborate 0.63, 2.50 and 3.14 acetyl groups as tetraacetylhydrazine, respectively.

The percentages of lightness of the thus washed material were respectively 20, 31 and 34 units higher than the percentage of lightness of the material which was only thoroughly rinsed with water. The relative improvements by adding tetraacetylhydrazine in the concentrations mentioned were therefore 54, 138 and 162%.

## Perborate with Amide Activators

S.C. Bright and F.R.M. McDonnell; U.S. Patent 3,177,148; April 6, 1965; assigned to Lever Brothers Company found that some acyl organoamides activate bleaching solutions based on perborates and other persalts. A test procedure was developed to determine the suitability of the amides as activators in this process. It has been found that those amides which possess the necessary properties are simply and conveniently characterized by a test in which a persalt and the amide are heated together in water and the vigorous bleaching agent thus produced is estimated by its ability to liberate iodine from potassium iodide solutions at 0°C. This test is designated as the "Peracid Formation Test." A test solution is prepared by dissolving the following materials in 1,000 ml. distilled water:

| | Grams |
|---|---|
| Sodium pyrophosphate $Na_4P_2O_7 \cdot 10H_2O$ | 2.5 |
| Sodium perborate $NaBO_2 \cdot H_2O_2 \cdot 3H_2O$ (having 10.4% available oxygen) | 0.615 |
| Sodium dodecylbenzene sulfonate (the dodecyl group being that derived from tetrapropylene) | 0.5 |

To this solution at 60°C. is added an amount of amide such that for each atom of available oxygen present one molecule of amide is introduced. A water-soluble amide or one which is liquid at 60°C. or will disperse into the solution easily is added directly to the test solution, but other amides are first dissolved in 10 ml. ethyl alcohol and then added to a test solution prepared using 990 ml. distilled water.

The mixture obtained by addition of the amide is vigorously stirred and maintained at 60°C. After 5 minutes from the addition, a 100 ml. portion of the solution is withdrawn and immediately pipetted on to a mixture of 250 grams cracked ice and 15 ml. glacial acetic acid. Potassium iodide (0.4 gram) is then added and the liberated iodine is immediately titrated with 0.1 N sodium thiosulfate solution with starch as indicator until the first disappearance of the blue color. An amide which is suitable as an activator for hydrogen peroxide (or persalt) in low temperature bleaching is one which in this test gives a titre of at least 1.5 ml. Accordingly a "reactive" acyl organoamide means an acyl organoamide which has a titre of at least 1.5 ml. 0.1 N sodium thiosulfate in this test.

By "acyl organoamide" is meant an amide derived on the one hand from an organic carboxylic acid (forming an acyl radical) and on the other from an organic-substituted ammonia compound containing one hydrogen atom and at least one acyl radical attached to the nitrogen atom. The acyl organoamide is accordingly a compound of the formula $RCONR_1R_2$ in which RCO is a carboxylic acyl radical, $R_1$ is a second acyl radical and $R_2$ is any suitable radical as determined by the behavior of the corresponding compound in the Peracid Formation Test described above.

It has been found that suitable radicals for $R_2$ may generally be selected from the group consisting of alkyl, aryl, carboxylic acyl, carbamyl and sulfonic acyl radicals. Any two of the radicals RCO, $R_1$ and $R_2$ may be joined together in the form of a divalent radical which, with the nitrogen atom, will form a heterocyclic ring. Preferably, the radical RCO is aliphatic, especially one containing from 2 to 4 carbon atoms, for instance the acetyl radicals. Preferably also the radical RCO is not linked with either of the radicals $R_1$ and $R_2$ to form a divalent radical. A list of some reactive acyl organoamides is given below, together with the titre which each gives in the test described above.

| | Titre |
|---|---|
| N,N-diacetylaniline | 6.05 |
| N,N-diacetyl-p-toluidine | 6.65 |
| N,N-diacetyl-p-chloroaniline | 5.85 |
| N,N-dibutyrylaniline | 5.2 |
| Dibenzanilide | 3.1 |
| N-acetyl caprolactam | 4.5 |
| N,N'-diacetylbarbitone | 2.8 |
| N-acetyl phthalimide | 4.35 |
| N-acetyl saccharin | 5.6 |

Othe suitable amides are N-acetyl anthranil, and N,N-diacetyl-5,5-dimethylhydantoin. The amides concerned are normally solids and are in practice mixed with the persalt in a finely divided form in order to ensure that when the bleaching composition thus obtained is used, it is easily dispersed in the bleaching bath. As the amides in general show a tendency to be decomposed by the persalts, the compositions should in practice be prepared just before use.

Example 1: A detergent bleaching composition was prepared using a spray dried detergent powder base containing the following materials in parts by weight.

| | |
|---|---|
| Sodium dodecylbenzene sulfonate (the dodecyl group was that derived from tetrapropylene) | 21.9 |
| Anhydrous alkaline sodium silicate | 6.6 |
| Tetrasodium pyrophosphate | 13.1 |
| Pentasodium tripolyphosphate | 21.9 |
| Anhydrous sodium sulfate | 24.4 |
| Water | 12.1 |

To 10 parts by weight of this detergent powder base was added 0.865 parts of sodium perborate tetrahydrate (containing 10.4% available oxygen) and 1.03 parts of N,N-diacetyl-p-toluidine. A 0.44% by weight aqueous solution of the total composition (containing 0.37% of the detergent powder base, 0.00335% of available oxygen and 0.0382% of N,N-diacetyl-p-toluidine: that is 1.07 atoms of available oxygen for each molecule of the amide) was maintained in a thermostat at 60°C. A tea stained white cloth was immersed in the solution for 10 minutes and continuously stirred (the ratio of washing liquid to cloth being 20 to 1 parts by weight) after which it was taken out and rinsed in warm distilled water, and the cloth was finally dried and ironed.

The increase in percent reflectance of the cloth achieved by washing was measured with a Hunter reflectometer, using a blue filter; the increase in reflectance was 16.1%. When a similarly stained cloth was washed in the same way with a solution of the same composition except that no amide was used an increase in reflectance of 10.9% was observed, indicating that the amide conferred better bleaching properties at 60°C.

Example 2: A detergent bleaching composition was prepared as in Example 1 except that N,N-diacetyl-p-chloroaniline (1.14 parts) was used as amide. A 0.44% by weight aqueous solution of the total composition (containing 0.37% of the detergent powder base, 0.00335% of available oxygen and 0.0423% of N,N-diacetyl-p-chloroaniline: that is, 1.06 atoms of available oxygen for each molecule of the amide) was used to wash tea stained white cloth in exactly the same way as that described in Example 1. The introduction of the amide in this instance led to an increase in reflectance of 17%.

## Perborate plus Sulfonamides

E.A. Matzner; U.S. Patent 3,245,913; April 12, 1966; assigned to Monsanto Co. found that perborates with sulfonamide additions can be used to bleach and whiten textiles at temperatures as low as 50°C. The perborate-sulfonamide compositions are stable for prolonged periods of time under ordinary storage conditions and may be included in household bleaching compositions. The sulfonamides used have the following general formula:

$$R(SO_2-\overset{\overset{R''}{|}}{N}-\overset{\overset{O}{\|}}{C}-R')_n$$

wherein R and R' are organic radicals and R" is an organic radical and n is an integer of from 1 to 3. In the formula, n is preferably 1 and the organic radicals may be any of a wide variety of substituted or unsubstituted aliphatic or aromatic radicals. Thus, R, R', and R" may be like or dissimilar aliphatic or aromatic radicals and may be any combination of a wide variety of substituted and unsubstituted aliphatic or aromatic radicals.

The aliphatic radicals preferably contain from 1 to 10 carbon atoms in the aliphatic group. Although such radicals may contain more than 10 carbon atoms, compounds containing them often have limited solubility. Thus, the unsubstituted aliphatic hydrocarbon radicals in the above structure may include, for example, alkyl radicals having a straight or branched chain, e.g., methyl, ethyl, n-propyl, isopropyl, n-butyl, isobutyl, sec-butyl, t-butyl, n-amyl, isoamyl, n-hexyl, isohexyl, n-heptyl, isoheptyl, n-octyl, isooctyl, 2-ethyl-hexyl groups or radicals.

The substituents of substituted aliphatic radicals may be halogen atoms, sulfo, nitro, carboxy, methoxy, carbethoxy, amino, ethyl, carboxyl groups or radicals to provide the aforementioned substituted aliphatic radicals. The unsubstituted aromatic radicals may be pheny, pyridyl, benzyl, alpha- and beta-naphthyl, quinolyl, anthryl, phenanthryl, benzquinolyl and the like. The substituents of substituted aromatic radicals include halo, nitro, sulfo, and alkyl substituted groups or radicals and the alkyl substituted aromatic radicals may contain from 1 to 20 carbon atoms in the alkyl group.

A particularly advantageous method for preparing the acyl sulfonamides comprises a two step reaction, the first step of which comprising reacting one mol of a sulfonyl chloride having the formula $R-SO_2Cl$ wherein R is an organic radical as hereinbefore described with at least one mol of ammonia or an amine having the formula $R''NH_2$, where R" is an organic radical, as hereinbefore described, to form the acyl sulfonamides. The equations shown on the following page exemplify these reactions:

$$R{-}SO_2Cl + NH_3 \longrightarrow RSO_2{-}N(H){-}H + HCl$$

primary sulfonamide

$$R{-}SO_2Cl + R''NH_2 \longrightarrow RSO_2{-}N(R''){-}H + HCl$$

secondary sulfonamide

The reactions are usually exothermic and are carried out in a liquid alkaline medium for example in an organic medium such as pyridine or in an inorganic liquid medium such as a 2 to 20% by weight aqueous solution of a water-soluble alkaline metal salt. The exotherm may be controlled by standard procedures, such as by cooling, using well-known cooling techniques or by the controlled addition of the reactants. The acyl sulfonamides may then be readily obtained by the second step which comprises acylating the primary and secondary sulfonamides and refluxing and reacting the primary or secondary sulfonamide in an excess of an acyl halide or an acid anhydride, the excess acyl compound being appropriately removed by distillation or extraction. By so proceeding primary sulfonamides can be suitably mono- or diacylated and secondary sulfonamides can be suitably monoacylated.

The primary and secondary sulfonamides may be suitably acylated to obtain the acyl sulfonamides useful in the compositions of this process by heating and reacting 1 mol of either the primary or secondary sulfonamide with at least 1.1 mol of an acyl halide in at least 1.2 molar equivalents of an alkaline solvent such as pyridine at 100° to 120°C. for at least 4 hours. At the end of the reaction the reaction product is preferably taken up in an organic solvent such as chloroform and the hydrohalide, formed by the reaction of the acid produced in the alkaline solvent, is extracted with an aqueous acid solution. If any unacylated product remains such unacylated material may be extracted with an aqueous alkaline solution of a metal hydroxide.

For commercial laundry bleaches, the compositions consist of sodium perborate and from 0.1 to 2.0 mols, per mol of sodium perborate, of N-methyl, N-benzoyl-dodecylbenzene sulfonamide; N-methyl, N-benzoyl-para-toluenesulfonamide, N-methyl, N-benzoyl-para-nitrobenzenesulfonamide or N-ethyl, N-benzoyl-para-toluenesulfonamide or a mixture of such sulfonamides.

The compositions may also contain sodium phosphates and silicates, as for example, from 3 to 15% by weight of sodium perborate and from 0.1 to 2.0 mols per mol of sodium perborate of an acyl sulfonamide such as N-methyl, N-benzoyl-para-toluenesulfonamide, from 10 to 60% by weight of sodium tripolyphosphate or a mixture of such phosphate and sodium silicate and the remainder consisting essentially of sodium sulfate.

Example: Dry mixed compositions containing the ingredients in the percentages given in Table 1 on the following page, were prepared. The bleaching capacity of each of Compositions 1 through 10 was determined by dissolving 0.25% by weight of each composition in 1 liter of water in separate cylindrical receptacles. The receptacles were provided with mechanical agitation and the solutions therein were maintained at a temperature of 60°C. Each solution had an available oxygen concentration of 11.5 parts per million. The solutions containing Compositions 1 through

TABLE 1

| Composition Number | 1 | 2 | 3 | 4 | 5 | 6 | 7 | 8 | 9 | 10 |
|---|---|---|---|---|---|---|---|---|---|---|
| Ingredient: | | | | | | | | | | |
| Sodium perborate | 6 | 6 | 6 | 6 | 6 | 6 | 6 | 6 | 6 | 6 |
| N-methyl, N-benzoyl-dodecylbenzenesulfonamide | 15 | 0 | 0 | 0 | 0 | 15 | 0 | 0 | 0 | 0 |
| N-methyl, N-benzoyl-para-toluenesulfonamide | 0 | 11 | 0 | 0 | 11 | 0 | 11 | 0 | 0 | 0 |
| N-methyl, N-benzoyl-para-nitrobenzenesulfonamide | 0 | 0 | 15 | 0 | 0 | 0 | 0 | 0 | 0 | 0 |
| N-ethyl, N-benzoyl-para-toluenesulfonamide | 0 | 0 | 0 | 15 | 0 | 0 | 0 | 0 | 0 | 0 |
| Sodium tripolyphosphate | 40 | 30 | 30 | 40 | 40 | 40 | 30 | 40 | 30 | 25 |
| Sodium sulfate | 39 | 41 | 30 | 0 | 43 | 30 | 20 | 54 | 52 | 30 |
| Sodium carbonate | 0 | 0 | 19 | 27 | 0 | 0 | 10 | 0 | 0 | 25 |
| Sodium silicate | 0 | 0 | 0 | 0 | 0 | 0 | 10 | 0 | 0 | 14 |
| Sodium dodecylbenzene sulfonate | 0 | 12 | 0 | 12 | 0 | 9 | 13 | 0 | 12 | 0 |

7 respectively contained a mol ratio of sodium perborate to acyl sulfonamide of 1:1. Twenty 5" x 5" swatches of unbleached naturally yellowed muslin were analyzed for reflectance (Rd) and (a) + (b) color values on a Gardner Automatic Color Difference Meter. Two swatches were placed in each of the ten receptacles containing the dissolved compositions and washed for 10 minutes.

After this period the swatches were dried, pressed and again analyzed on the Gardner Automatic Colorimeter. The reflectance ΔRd (brightening) and bleaching efficiency Δ(a) and Δ(b) were calculated by subtracting the differences in the readings before and after the washing operation. The loss of available oxygen was also determined for each solution. The results are summarized in Table 2. The Gardner Automatic Color Difference Meter is a tristimulus colorimeter, that is, it contains three photocells which measure (1) reflectance (Rd); (2) green to red color, (a); and (3) blue to yellow color, (b).

TABLE 2

| Composition Number | Loss of available [1] oxygen (percent) | ΔRd[2] | Δ(a)[3] | Δ(b)[3] |
|---|---|---|---|---|
| 1 | 83 | +7.6 | −0.1 | −1.9 |
| 2 | 86 | +8.0 | −0.4 | −2.5 |
| 3 | 74 | +6.2 | −0.1 | −1.5 |
| 4 | 77 | +6.5 | −0.1 | −1.5 |
| 5 | 86 | +8.0 | −0.4 | −2.4 |
| 6 | 76 | +6.5 | −0.1 | −1.4 |
| 7 | 85 | +8.1 | −0.4 | −2.4 |
| 8 | 3 | +4.0 | +0.1 | −1.2 |
| 9 | 4 | +4.6 | +0.1 | −1.3 |
| 10 | 3 | +4.2 | +0.1 | −1.2 |

[1] Determined by iodometric titration of spent wash solutions.
[2] Positive values indicate degree of increase in reflectance or brightening.
[3] Negative values indicate degree of color disappearance or bleaching.

The above values for loss of available oxygen indicate that the acyl sulfonamides in Compositions 1 through 7 promote the release of the oxygen in the oxygen releasing compounds sodium perborate in aqueous solution and that such oxygen was consumed in bleaching and brightening the fabric. Also these results show that the acyl sulfonamides promote the bleaching and brightening of the yellow unbleached muslin. In contrast, solutions of Compositions 8 through 10 which contain sodium perborate, but no acyl sulfonamides, did not lose or release oxygen and did not bleach or brighten the muslin to any appreciable extent. Compositions similar to Compositions 1 through 7 were prepared except that urea peroxide was used in place of sodium perborate. These compositions when dissolved in water in concentrations of 0.25% effectively bleached and brightened unbleached muslin swatches when treated as in Example 1.

Perborate Activated with Trihydroxytriazine

J.H. Blumbergs, D.G. MacKellar and S. Berkowitz; U.S. Patent 3,332,882; July 25, 1967; assigned to FMC Corporation found that peroxygen compounds such as perborate and hydrogen peroxide can be activated to bleach, disinfect and sanitize more effectively when used in aqueous solutions at temperatures below boiling by incorporation of an aromatic or lower aliphatic acyl derivative of 2,4,6-trihydroxy-1,3,5-triazine. These compounds have the general formula:

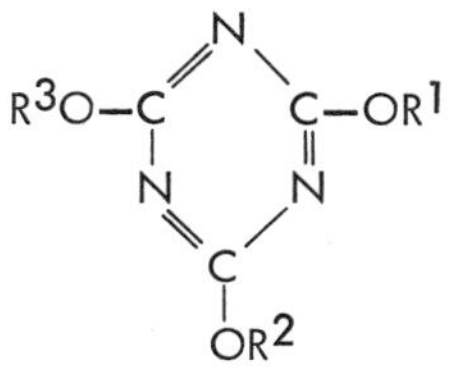

where at least one of $R^1$, $R^2$, or $R^3$ are acyl groups having the formula:

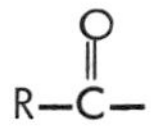

wherein R is either hydrogen, an aromatic group of an aliphatic group having from 1 to 6 carbon atoms, and any remaining $R^1$, $R^2$ or $R^3$ groups are hydrogen. The acyl derivative must be soluble in the bleach solution in amounts of at least 100 ppm to be effective. The monoacyl derivative has the formula:

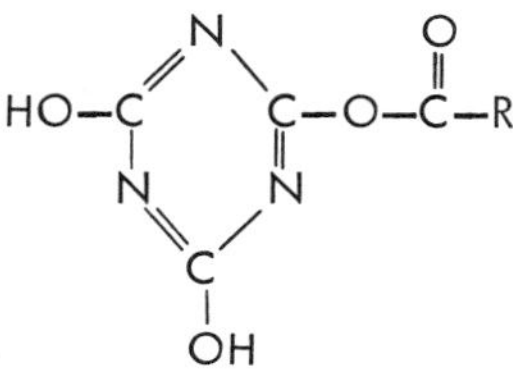

where R may be hydrogen, an aromatic grouping, or an aliphatic group having from 1 to 6 carbon atoms. The diacyl and triacyl derivatives are identical to the monoacyl structure shown above except that additional acyl groupings are added to the cyclic structure by acylating one or both of the remaining hydroxyl groups. The solid peroxygen compound is mixed together with a phosphate builder such as sodium tripolyphosphate in combination with one or more working ingredients. These include antiredeposition agents such as sodium carboxymethylcellulose, anionic or nonionic surfactants and anticorrosion agents such as sodium silicate.

The activator is added to mixture, preferably in an amount sufficient to have at least one acyl group present for every atom of available oxygen. Obviously, larger amounts of the activator can be used to assure complete activation of the peroxygen compound. Also, smaller amounts of the activator can be employed where a controlled bleaching effect is desired. The solid peroxygen compound is employed in amounts sufficient to achieve the desired degree of bleaching. For home bleaching applications amounts sufficient to supply from 10 to 100 ppm of active oxygen normally are employed in the wash solution. A preferred amount is 40 ppm. In the makeup of the final bleaching mixture it is desirable to keep the solid peroxygen compound and the activator out of contact with one another until placed in the wash water to avoid any possible reaction between these ingredients, even in the dry state. This can be done most readily when making up a single mixture by coating the activator with a water-soluble coating prior to mixing it with the remaining components of the bleaching mix. In this way the activator and the solid peroxygen compound can be mixed together and still remain out of contact with one another until the mixture is placed in the wash water. Another alternative is to separately package the solid peroxygen compound and the activator so that they are physically out of contact with one another in the consumer package.

The solid peroxygen compounds which are useful in the process as bleaching, disinfecting or sanitizing agents are those which liberate perhydroxyl anions readily when dissolved in an aqueous media. These include the alkali perborates such as sodium perborate and other alkali metal percompounds such as percarbonates, persilicates, perphosphates, and perpyrophosphates. In addition, such compounds as sodium peroxide, zinc peroxide, calcium peroxide, magnesium peroxide, urea peroxide, and others are included within the term solid peroxygen compound.

Example: Bleaching tests were carried out by bleaching tea, coffee and wine stained cotton swatches (5" x 5") with sodium perborate tetrahydrate alone and in the presence of various activators. The procedure used was as follows: 32 cotton swatches (5" x 5" desized cotton Indianhead fabric, 48 fill by 48 warp threads per inch, uniform in weave and thread count) were stained with tea, coffee and wine in the following manner.

Five tea bags were placed in a liter of water and boiled for 5 minutes. Thereafter, the swatches were immersed in the tea and the boiling continued for 5 minutes. Thirty-two additional swatches of the same cloth were coffee stained by boiling 50 grams of coffee in a liter of water, immersing the swatches in the coffee solution, and boiling for an additional 5 minutes. The wine stains were created by soaking swatches of the same cloth in a red wine at room temperature.

The stained swatches were then squeezed to remove excess fluid, dried, rinsed in cold water, and dried. Three of the stained cotton swatches were then added to each of a series of stainless steel Terg-o-Tometer vessels containing 1,000 ml. of a 0.2% standard detergent solution at a temperature of 120°F. Measured amounts of sodium perborate tetrahydrate bleach were then added to each vessel sufficient to correspond to an active oxygen content of 60 ppm. The pH of the solutions was adjusted to 9.0 using soda ash.

Cut up pieces of white terry cloth toweling were then added to provide a typical household wash water: cloth ratio of 20:1. The activators specified in the table below were then added to the solution in the ratio set forth. The Terg-o-Tometer was then operated at 72 cycles per minute for 15 minutes at a temperature of 120°F. At the end of the wash cycle, the swatches were removed, rinsed under cold tap water and dried in a Proctor-Schwartz skein dryer. The tests were run in triplicate and included detergent blanks. Reflectance readings of the swatches were then taken before and after the wash cycle with a Hunter Model D-40 Reflectometer and the readings were averaged. The percent stain removal was obtained according to the formula shown on the following page.

$$\text{Percent stain removal} = \frac{\text{reflectance after bleaching} - \text{reflectance before bleaching}}{\text{reflectance before staining} - \text{reflectance stained}} \times 100$$

The results are reported in the table below.

| Activator Added | Ratio Activator/Perborate by weight | Percent Stain Removal Tea | Coffee | Wine |
|---|---|---|---|---|
| None | -- | -- | -- | -- |
| Detergent blank | -- | 26.9 | 45.0 | 61.0 |
| None | -- | -- | -- | -- |
| Perborate blank | -- | 33.2 | 51.2 | 69.4 |
| Diacetyl derivative of 2,4,6-trihydroxy-1,3,5-triazine | 1/1 | 62.4 | 88.6 | 82.7 |
| Triacetyl derivative of 2,4,6-trihydroxy-1,3,5-triazine | 0.8/1 | 63.0 | 89.2 | 96.0 |
| Dibenzoyl derivative of 2,4,6-trihydroxy-1,3,5-triazine | 1.2/1 | 61.4 | 90.0 | 88.0 |

## Substituted Hydantoin Activator for Perborate

K. Dithmar and P. Koblischek; U.S. Patent 3,349,035; October 24, 1967; assigned to Deutsche Gold- und Silber-Scheideanstalt vormals Roessler, Germany found that 1-substituted-3-acylhydantoin is an activator for persalt, such as sodium perborate containing stable bleaching and washing compositions. These hydantoins have increased activating efficiency combined with optimum storage stability. They permit efficient bleaching at 60° to 70°C. The activators are 1-substituted hydantoins represented by the structure

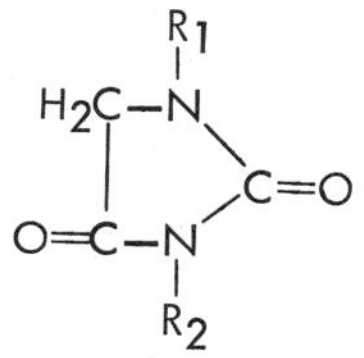

wherein $R_1$, in the 1 position, is an alkyl, e.g., lower alkyl, cycloalkyl, aryl, alkaryl or aralkyl group and $R_2$, in the 3 position, is preferably an acetyl group. These monoacetyl hydantoins in contrast to known materials, such as, N,N-diacetyl-5,5-dimethyl hydantoin, are not substituted in the 5 position, e.g., with a methyl group. Only these specific hydantoin compounds have both high activating effect and high stability during storage. The substituents given above in the 1 position increase stability during storage while substituents in the 5 position reduce stability during storage. Another advantage of the N-monoacetyl hydantoins which are not substituted in the 5 position is that they have smaller molecular weight and therewith higher effectiveness in addition to being cheaper to produce. These compounds do not decompose at room temperature in a powder mixture containing sodium perborate, sodium tripolyphosphate, sodium pyrophosphate and dry water glass and do not react until the powder mixture is dissolved and, if necessary, heated to medium wash temperature, e.g., 30° to 60°C. When dissolved and employed in peroxide washing agent solutions, the compounds effectively increase the bleaching effect of such solutions. The compounds containing an acetyl group are cheaper than similar compounds containing a benzoyl group and accordingly have an advantage over known benzoylamides.

The activating agents satisfy the prerequisite of effectiveness by releasing the acyl group which reacts with $H_2O_2$ to form monoperoxy acid when the agents are dissolved in the washing process, as well as the prerequisite of stability during storage at normal temperatures, and accordingly are satisfactory as a commercial, dry powder in admixture with customary auxiliary washing agents, although such washing agents are packed in paper packages or boxes which permit atmospheric moisture to come in contact with the powder.

These prerequisites are satisfied by acyl derivatives of hydantoins as the activating agents wherein an alkyl or aryl group is substituted in the 1 position, not in the 5 position as in known activators, and an acyl group is present in the 3 position. Initially this controlled substitution, i.e., substitution in the 1 and 3 position, enables use of the cheap acetyl group. If the acyl group is present in the 1 position, the compounds do not have satisfactory reactivity for reacting with $H_2O_2$ in aqueous solution to increase the bleaching potential.

Example 1: This example illustrates the stability during storage of the compositions of the process. The following wash powder composition was compounded containing sodium perborate, phosphates, silicate and 1-phenyl-3-acetyl hydantoin in the following ratio by weight:

| | Grams |
|---|---|
| Tetrapropylene benzylsulfonate | 37.50 |
| Sodium pyrophosphate | 18.75 |
| Sodium tripolyphosphate | 18.75 |
| Dry water glass | 10.50 |
| Sodium perborate | 18.00 |
| 1-phenyl-3-acetyl hydantoin (MP 146° to 147°C.) | 30.00 |

The mixture was stored for a year and tested for remaining active oxygen content and decomposition after about one month. Three times during this period tests were run to determine how well a washing bath prepared from this mixture could remove red wine stain from staple rayon fabric in comparison to a corresponding wash bath without added acylated hydantoins. Despite storage for one year there was practically no oxygen loss. Normal loss or decomposition of an active oxygen material during storage without addition of an activator amounts to more than 5% of the starting material per year.

Example 2: This examples illustrates the effectiveness of the compositions of the process in removing strong red wine stain. A piece of staple rayon fabric was boiled for an hour in red wine and then washed three consecutive times for 15 minutes with a wash solution containing 4.45 grams per liter of the wash powder described in Example 1 at 60°C. Subsequently the increase in brightness of the fabric was measured on a Zeiss-Elrepho whiteness measurer using a white standard 565 and filter 6. A piece of staple rayon fabric was soiled or stained in a like manner and washed in the same way in a wash solution containing the same wash powder without the addition of 1-phenyl-3-acetyl hydantoin. The following results were obtained on the day the washing compositions were made up:

| | Percent |
|---|---|
| (a) Whiteness of textile after staining with red wine before washing | 32.0 |
| (b) Whiteness of textile after staining with red wine and subsequent washing at 60°C. with washing agent containing activator | 62.1 |
| (c) Whiteness of textile after staining with red wine with subsequent washing at 60°C. with the same washing agent as employed in (b) but without added activator | 43.9 |
| (d) Difference in degree of whiteness of textiles, stained with red wine washed with and without an activator | 18.2 |

The following results were obtained employing the same washing composition 9 months later, i.e., after storage for 9 months:

| | Percent |
|---|---|
| (a) Whiteness of textile stained with red wine without washing | 32.0 |
| (b) Whiteness of textile staining with red wine and subsequent washing with washing agent containing an activator | 60.1 |
| (c) Whiteness of textile after staining with red wine and subsequent washing with the washing agent without added activator | 43.1 |
| (d) Difference in degree of whiteness of textiles washed with and without an activator | 17.0 |

The difference in degree of whiteness after 9 months was reduced from 18 to 17%, a reduction of only 1%, which is practically an insignificant loss. This demonstrates that 1-phenyl-3-acyl hydantoins in admixture with components of a washing agent are not changed by storage. Other activators proved effective in removing red wine stains are 1-alkaryl-3-acetyl hydantoins employing a washing composition and bath as above. These activators are likewise effective in increasing the bleaching effect.

## N-Halo-N-Alkyl-o-Sulfobenzamide Activator for Persalt

F.E. Hardy; U.S. Patent 3,652,660; March 28, 1972; assigned to The Procter & Gamble Company describes the use of sodium N-chloro-N-methyl-o-sulfobenzamide as an activator for perborate in washing composition and sodium N-bromo-N-methyl-o-sulfobenzamide which is a bleaching agent by itself. The general formula of these compounds is as follows:

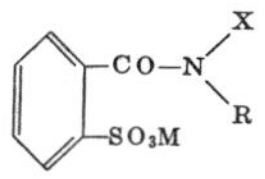

In the formula shown on the preceding page X is chlorine and bromine, R is an alkyl group and M is a cation. M is preferably an alkali metal or alkaline earth metal ion. The alkyl group R generally has from 1 to 20 carbon atoms, and preferably is either a short chain group of 1 to 4 carbon atoms, or, especially if surface-active compounds are desired, a relatively long chain group of 10 to 18 carbon atoms.

N-chloro compounds are valualbe in particular as activators for perborate and like compounds, so that these become effective bleaching agents at temperatures below 85°C., for instance in the range from 38° to 82°C. and especially in the range from 49° to 60°C. The N-chloro compounds generally have little or no bleaching activity on their own and cause very little or no chlorine type odor in use. These compounds can be used in combination with alkaline builder soaps, organic detergents, soaps and all types of synthetic surfactants, also in automatic dishwashing compositions.

Example 1: The preparation of sodium N-chloro-N-methyl-o-sulfobenzamide is as follows. N-methyl-o-sulfobenzamide was prepared by reacting o-sulfobenzoic anhydride with methylamine and treating the resulting amine salt with Amberlite IR 120 ($H^+$) resin. 32.6 grams of the N-methyl-o-sulfobenzamide were dissolved in 200 ml. of water and the solution was neutralized with sodium hydroxide. 480 ml. of 0.42 molar hypochlorous acid (from distillation of a mixture of commercial sodium hypochlorite and boric acid) were added and the mixture was stored for 3 hours at room temperature and then overnight at 4°C. On evaporation of the solution to a small volume the N-chloro body crystallized. This material (39.4 grams) was filtered and dried. On analysis there was found (percent): C, 33.25; H, 3.2; N, 4.75. $C_8H_7ClNNaO_4S \cdot H_2O$ requires (percent): C, 33.15; H, 3.1; N, 485.

Example 2: The preparation of sodium N-bromo-N-methyl-o-sulfobenzamide is as follows. Five grams of the acid N-methyl-o-sulfobenzamide as prepared in Example 1 were dissolved in 50 ml. of water and the solution was neutralized with sodium hydroxide. 150 ml. of 0.16 molar hydrobromous acid (prepared by treating bromine water with mercuric oxide and distilling the filtrate) were added and the mixture was stored at room temperature for 2 hours. Evaporation to dryness gave the N-bromo body (7.3 grams) in microcrystalline form. On analysis there was found (percent): C, 28.1; H, 2.65; N, 3.9. $C_8H_7BrNNaO_4S \cdot H_2O$ requires (percent): C, 28.75; N, 4.2.

Example 3: The preparation of sodium N-chloro-N-hexadecyl-o-sulfobenzamide is as follows. Three grams of o-sulfobenzoic anhydride were added to a solution of 3.9 grams of hexadecylamine and 2.5 ml. of triethylamine in 40 ml. of anhydrous benzene. The mixture was heated under reflux for 3 hours and stored at room temperature overnight. Evaporation to dryness gave triethylammonium N-hexadecyl-o-sulfobenzamide, which was recrystallized from ethanol. The yield was 7.9 grams.

This salt was dissolved in ethanol-water (1:9) and converted to the free sulfonic acid by treatment with Amberlite IR 120 ($H^+$) resin. Neutralization with sodium hydroxide followed by evaporation to dryness gave the corresponding sodium salt which was recrystallized from acetone. This material (5 grams) was suspended in 125 ml. of water, and 34 ml. of 0.45 molar hypochlorous acid (prepared as in Example 1) were added. The mixture was stirred at 35°C. until it became clear (3 hours). The N-chloro body was isolated by lyophilization as an amorphous acid. Yield 4.5 grams. On analysis there was found (percent): C, 57.45; H, 7.9; N, 2.95.

$C_{23}H_{37}ClNNaO_4S$ requires (percent): C, 57.35; H, 7.7; N, 2.9. Similar results are obtained when hypobromous acid is employed on an equimolar basis in place of hypochlorous acid in that sodium-N-bromo-N-hexadecyl-o-sulfobenzamide is obtained.

Example 4: A control washing solution A was prepared containing 0.5% by weight of the following detergent-bleaching composition (the proportions being by weight).

| | |
|---|---|
| Linear alkyl benzene sulfonate | 14.0 |
| Sodium toluene sulfonate | 1.4 |
| Sodium tripolyphosphate | 36.7 |
| Sodium silicate | 5.8 |
| Coconut monoethanolamide | 1.5 |
| Sodium perborate tetrahydrate | 20.5 |
| Sodium carboxymethylcellulose | 0.8 |
| Sodium sulfate | 9.5 |
| Perfume, brightener, minor components, etc. | 1.8 |
| Moisture | 8.0 |

A solution B contained in addition 5% and a solution C contained 20%, by weight of the detergent composition, of the compound of the formula:

$$\begin{array}{l} CON(CH_3)Cl \\ | \\ C_6H_4 \\ | \\ SO_3Na \end{array}$$

These proportions correspond to 14 and 56 molar percent respectively, referred to the perborate. Standard tea stained cloth swatches were treated in the solutions and the percentage stain removal measured. Three test conditions were employed: (1) 10 minutes agitation at 49°C., (2) 16 hours soaking beginning at 49°C. and allowing to cool naturally to room temperature, and (3) heating over a period of 1 hour from 49° to 82°C. and then agitating for 10 minutes. Percentage stain removal figures are given below:

| | Solution | | |
|---|---|---|---|
| Washing Condition | A | B | C |
| 1 | 36 | 47 | 54 |
| 2 | 66 | 83 | 87 |
| 3 | 82 | 91 | 93 |

Comparison of the results for solutions of compositions of the process B and C with the control solution A shows the activating effect of the compounds of the process, which is as would be expected more marked in conditions 1 and 2 than in condition 3 where the temperature approached that at which perborate alone is known to become fully effective.

## Puffed Borax Stabilizer for Peroxygen-Activator Systems

G.G. Corey and B. Weinstein; U.S. Patent 3,661,789; May 9, 1972; and U.S. Patent 3,671,439; June 20, 1972; both assigned to American Home Products Corp.

found that nonionic surfactants and puffed borax improve the storage stability of perborate bleach composition activated with benzoylimidazole. The following groups of nonionic surfactants are suitable stabilizers:

(1) Straight chain alkylphenoxypoly(ethyleneoxy) ethanols having the general formula:

wherein R is an alkyl radical and n is the number of mols of ethanol oxide in the molecule (Igepals).

(2) Ethoxylates of isomeric linear secondary alcohols having the general formula:

$$\begin{array}{l} CH_3-(CH_2)_n-CH_3 \\ \;\;| \\ \;\;O-(CH_2-CH_2O)_x-H \end{array}$$

wherein n is the number of mols of methylene and x is the number of mols of ethylene oxide in the molecule (Tergitols).

(3) Condensation products of ethylene oxide with a hydrophobic base formed by the condensation of propylene oxide with propylene glycol having the general formula:

$$\begin{array}{l} HO-(CH_2CH_2O)_a-(CHCH_2O)_b-(CH_2CH_2O)_c-H \\ \qquad\qquad\qquad\qquad\;\;| \\ \qquad\qquad\qquad\qquad CH_3 \end{array}$$

wherein a and c represent mols of ethylene oxide and b represents mols of propylene glycol (Pluronics).

(4) Addition products of propylene oxide to ethylene diamine followed by the addition of ethylene oxide having the general formula:

$$\begin{array}{ccc} H-(C_2H_4O)_y-(C_3H_6O)_x & & (C_3H_6O)_x-(C_2H_4O)_y-H \\ & \searrow N-CH_2-CH_2-N \swarrow & \\ H-(C_2H_4O)_y-(C_3H_6O)_x & \nearrow \quad\quad \nwarrow & (C_3H_6O)_x-(C_2H_4O)_y-H \end{array}$$

wherein x and y represent respectively the number of mols of propylene oxide and ethylene oxide in the molecule (Tetronics).

(5) Ethylene oxide adducts of straight chain alcohols having the general formula:

$$CH_3—(CH_2)_x—CH_2—(O—CH_2—CH_2)_y—OH$$

wherein x is the number of methylene groups (chiefly $C_{12}$ to $C_{18}$) and y is the number of mols of ethylene oxide present (Alfonics).

(6) Glycols are exemplified by such as propylene glycol, triethylene glycol and trimethylene glycol.

The nonionic surfactants or glycols are incorporated into the oxygen bleach-activator system in from 1/10 to about 5 times the weight of the oxygen bleach-activator used. Preferably the nonionic surfactant is used in equal weight amounts with the oxygen bleach-activator. Puffed borax may also be included in the composition. Found to be particularly useful are puffed boraxes having particle size distribution so that the major portion of the puffed borax is of a size within the U.S. sieve range of from -20 to +200. In more preferred forms over 90% of the particles of puffed borax are in the U.S. sieve range from 40 to 60. At the optimum compositions of the instant disclosed process, the bulk density of the puffed borax is 15 lbs./ft.$^3$.

The oxygen bleach-activator compositions may also include conventional additives for such compositions. These may include binders, other fillers, builders, optical brighteners, perfumes, colorings, enzymes, bacteriostats, etc., all of which may be added to provide properties required in any particular instance. Additionally, the stabilized oxygen bleach-activator compositions can be incorporated into cleaning compositions containing soap and/or synthetic organic detergents and formulated for use as heavy duty household detergents, fine fabric washing detergent systems or clothes washing formulations in general.

Example 1: Samples were made by mixing the activating system, BID (benzoyl-imidazole) with a number of fillers referred to hereinafter, and then combining the above mixture with sodium perborate monohydrate. The samples were then stored at 90°F./90% RH in open containers for 72 hours. The samples were removed and titrated for the amount of active oxygen present with the standard permanganate titration. On the table below, following column I specifies the filler tested. Column II to IV, respectively, set out the amounts of filler, BID and sodium perborate in the compositions tested. Column V lists the measured $O_2$ loss under the adverse storage conditions.

| I | II | III | IV | V |
|---|---|---|---|---|
| Filler | Grams of filler | Grams BID | Grams Na perborate $H_2O$ (15.4% active $O_2$) | Percent active oxygen loss [a] |
| Puffed borax [b] | 2.00 | 0.25 | 0.23 | 22 |
| Borax decahydrate [c] | 2.00 | 0.25 | 0.23 | 49 |
| Low density sodium carbonate (Flozan) [d] | 2.00 | 0.25 | 0.23 | 94 |
| Light density sodium tripolyphosphate [e] | 2.00 | 0.25 | 0.23 | 29 |
| Soap flakes [f] | 2.00 | 0.25 | 0.23 | 55 |

[a] After 72 hours at 90° F./90% R.H.
[b] 15 lbs./cu. ft.
[c] 52 lbs./cu. ft.
[d] 32 lbs./cu. ft.
[e] 33 lbs./cu. ft.
[f] 85% tallow/15% coconut oil ratio soap having an Iodine Value (Hanus method) of 38–42.

The table clearly delineates the unexpected stability of oxygen bleach-activator systems containing puffed borax and those containing standard fillers including light density fillers or soaps.

Example 2: A perborate activating system was prepared in the following manner. One gram of benzoylimidazole (BID) was solubilized in a nonionic or cationic surfactant or a glycol which was liquid at ambient room temperature. Then 0.05 gram of the solubilized BID mixture was mixed with 2.00 grams of puffed borax (15 lbs./ft.$^3$ density). This puffed borax activator system was admixed with sodium perborate monohydrate (15.2% active oxygen) giving a BID:Na perborate monohydrate ratio of 0.25:0.23. The sample was then stored in open containers at 90°F./75% RH for 72 hours, at which time active oxygen content was determined by the standard permanganate titration.

The results showed that (1) the use of puffed borax as a filler increases the active oxygen life of Na perborate; (2) the use of nonionic surfactants tends to increase the active oxygen life of Na perborate while cationics have an opposite effect; and (3) the various nonionics give varying results. In a further series of experiments delineating the activity of nonionic surfactants and glycols as stabilizers for oxygen bleach-activator systems, equal amounts of the activator (BID) and glycol or nonionic surfactant (0.25 gram) were obtained with a sodium perborate bleach composition. After aging under adverse storage conditions the amount of oxygen lost was ascertained using a standard permanganate titration. The results demonstrated that the combination containing activator and nonionic surfactant or glycol was more stable than the combination containing the activator alone.

## Perborate plus Poly(N-Tricarballylic Acid)Alkyleneimine Stabilizers

A. Werdehausen and P. Krings; U.S. Patent 3,686,128; August 22, 1972; assigned to Henkel & Cie GmbH, Germany found that aminopolycarboxylic acids and their salts increase the stability and cleaning properties of washing agents and stabilize the optical brighteners present in the composition. To the conventional perborate containing detergent bleaching compositions are added up to 50% by weight the salts of poly(N-tricarballylic acid)alkyleneimines having an average molecular weight of from 430 to 500,000 in which at least one third of the recurring substituted alkyleneimine groups have the following formula

(1)
$$-\text{N}-\text{CH}_2-\text{CH}-\\ \phantom{-\text{N}}|\phantom{-\text{CH}_2}| \\ \phantom{-\text{N}-}X\phantom{-\text{CH}_2-}R$$

and the remainder of the recurring substituted alkyleneimine groups have the following formula

(2)
$$-\text{N}-\text{CH}_2-\text{CH}-\\ \phantom{-\text{N}}|\phantom{-\text{CH}_2}| \\ \phantom{-\text{N}-}Y\phantom{-\text{CH}_2-}R$$

wherein, in both formulas, R represents a member selected from the group consisting of hydrogen and methyl, X represents

$$-\text{CH}-\!\!-\!\!-\text{CH}-\!\!-\!\!-\text{CH}_2\\ \phantom{-}|\phantom{-\!\!-\!\!-\!\!-}|\phantom{-\!\!-\!\!-\!\!-}| \\ \phantom{-}\text{COOH}\phantom{-}\text{COOH}\phantom{-}\text{COOH}$$

Y represents a member selected from the group consisting of hydrogen,

$$-\text{N}-\text{CH}_2-\text{CH}-\phantom{x}\text{and}\phantom{x}-\text{N}-\text{CH}_2-\text{CH}-\\ \phantom{-\text{N}}|\phantom{-\text{CH}_2}|\phantom{xxxxxxxx}|\phantom{-\text{CH}_2}| \\ \phantom{-\text{N}-}X\phantom{-\text{CH}_2-}R\phantom{xxxxxxx-\text{N}-}Y'\phantom{-\text{CH}_2-}R$$

and Y' represents a member selected from the group consisting of hydrogen and

$$-\text{N}-\text{CH}_2-\text{CH}-\\ \phantom{-\text{N}}|\phantom{-\text{CH}_2}| \\ \phantom{-\text{N}-}X\phantom{-\text{CH}_2-}R$$

The amino groups present in the polyalkyleneimines can be completely or partially substituted by tricarballylic groups. Completely substituted polyalkyleneimines contain only the recurring groups according to Formula (1) and have a linear structure of the following formula

(3)
$$X-\text{NH}-\text{CH}_2-\text{CH}-\left[\text{N}-\text{CH}_2-\text{CH}\right]_n-\text{NH}-X\\ \phantom{X-\text{NH}-\text{CH}_2-}R\phantom{-}|\phantom{\text{CH}_2-\text{CH}}|\\ \phantom{X-\text{NH}-\text{CH}_2-\text{CH}-[}X\phantom{-\text{CH}_2-}R$$

wherein X and R are defined above and n is an integer from 2 to 130, preferably from 3 to 85, corresponding to a molecular weight of from 430 to 15,000, preferably from 500 to 10,000. Such polymers in which only a part, for example, at least 33 1/3% and less than 100% of the amino groups carry a tricarballylic group are mixed polymers which are built from the groups of the Formulas (1) and (2). Usually they contain branched chains and are partially illustrated by the following formula

(4)
$$-\text{N}-\text{CH}_2-\text{CH}-\text{N}-\text{CH}_2-\text{CH}-\text{N}-\text{CH}_2-\text{CH}-\text{N}-\text{CH}_2\text{CH}-\text{N}=\\ |\phantom{x}|\phantom{xx}|\phantom{xx}|\phantom{xx}|\phantom{xx}|\phantom{xx}|\phantom{x}|\\ X\phantom{x}R\phantom{x}\text{CH}_2\phantom{x}R\phantom{x}X\phantom{x}R\phantom{x}\text{CH}_2\phantom{x}R\\ \phantom{xxxx}R-\text{CH}\phantom{xxxxxxxxx}R-\text{CH}\\ \phantom{xxxx}|\phantom{xxxxxxxxxxxxxxxx}|\\ \text{HN}-\text{CH}_2-\text{CH}-\text{NX}_2\phantom{xx}X-\text{N}-\text{CH}_2-\text{CH}_2\text{N}-\\ \phantom{\text{HN}-\text{CH}_2-\text{CH}}R\phantom{xxxxxxxxxxxxxxxx}R\phantom{xxxxxxx}X$$

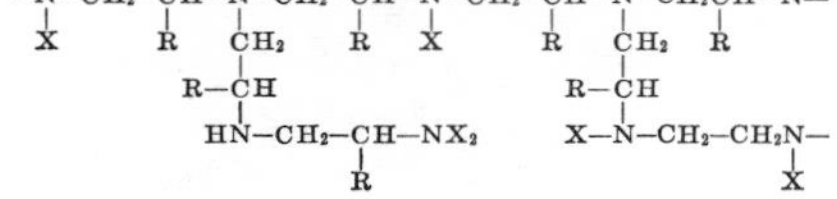

In formula (4) X and R are defined above. The branched and/or mixed polymers have average molecular weights of from 430 to 500,000. Preferably such alkylcarboxylic acid derivatives of polyethyleneimine or polypropyleneimine are used in which from 50 to 100% of the primary and secondary amino groups in the polyalkyleneimine molecule carry tricarballylic acid groups. Of particular interest are the N-tricarballylic acid derivatives of polyethyleneimines.

The poly(N-tricarballylic acid)ethyleneimines are amphoteric substances. They can, therefore, depending upon the alkalinity or acidity of the washing, bleaching, and cleansing agents, be present as salts of alkali metals and ammonium salts, especially salts of sodium and potassium, and as salts of organic ammonium bases, as inner salts, or as salts of strong acids, for example, mineral acids such as sulfuric acid and organic acids such as p-toluene sulfonic acid. The following recipes have proven particularly good in practice.

Powdery, Low-Foaming Washing Agent:

3 to 15% of a sulfonate basic washing component from the class of alkylbenzene sulfonates, olefin sulfonates and n-alkane sulfonates (sodium salts)
0.5 to 5% of an alkylpolyglycolether (alkyl $C_{12}$ to $C_{18}$) or alkylphenolpolyglycolether (alkyl $C_8$ to $C_{14}$) with 5 to 10 oxyethylene groups
0 to 5% of a $C_{12}$ to $C_{18}$ soap (sodium salt)
0.2 to 5% of foam inhibitors from the class of trialkylmelamines and saturated fatty acids with 20 to 24 carbon atoms, or their alkali metal soaps
10 to 50% of a condensed alkali metal phosphate from the class of the pyrophosphates or the tripolyphosphates
0.1 to 25% of poly(N-tricarballylic acid)alkyleneimine or its alkali metal salt
1 to 5% of sodium silicate
10 to 35% of sodium perborate tetrahydrate
0 to 5% of enzymes
0.05 to 1% of at least one optical brightener from the class of diaminostilbenedisulfonic acid or diarylpyrazoline derivatives
0.1 to 30% of an inorganic alkali metal salt from the class of the carbonates, bicarbonates, borates, sulfates and chlorides
0 to 4% of magnesium silicate
0.5 to 3% of sodium cellulose glycolate

Powdery Foaming Fine Washing Agent:

1 to 30% of a sulfonate basic washing component (sodium salt)
0.5 to 10% of alkylpolyglycolether sulfate (alkyl $C_8$ to $C_{16}$, 1 to 5 oxyethylene groups)
0 to 20% of an alkylpolyglycolether (alkyl $C_{10}$ to $C_{18}$) or alkylphenolpolyglycolether (alkyl $C_8$ to $C_{12}$), with 5 to 12 oxyethylene groups
0.2 to 25% of poly(N-tricarballylic acid)alkyleneimine or its alkali metal salt
0 to 5% of a higher fatty acid ethanolamide or diethanolamide
0 to 20% of sodium tripolyphosphate
0 to 1% of a brightener (diarylpyrazoline derivatives)
3 to 70% of sodium sulfate

Examples 1 and 2: A washing agent of the following composition was used.

| | Percent by Weight |
|---|---|
| Na n-dodecylbenzene sulfonate | 8 |
| Sodium soap of $C_{12}$ to $C_{22}$ fatty acids | 5 |
| Sodium salt of $C_{12}$ to $C_{22}$ fatty acids | 5 |
| Pentasodium triphosphate | 40 |
| Sodium silicate ($Na_2O \cdot 3.3SiO_2$) | 5 |
| Magnesium silicate | 2 |
| Sodium cellulose glycolate | 1 |
| Sodium perborate-tetrahydrate | 25 |
| Water | 8 |
| Brightener of the pyrazoline type | 0.8 |
| Brightener of the diaminostilbene type | 0.2 |

The brighteners had the following structures:

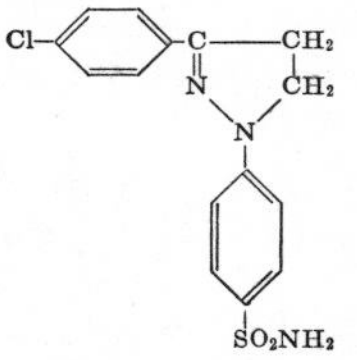

Pyrazoline type

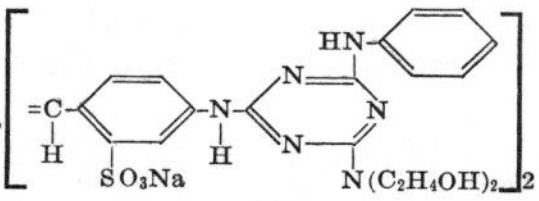

Diaminostilbene type

To this agent were added each time 2% by weight of the sodium salt of the poly(N-tricarballylic acid)ethyleneimine listed in the following table. For comparative purposes, a washing agent was used which, instead of the polymers according to the process, 2 % of sodium nitrilotriacetate (NTA) or 2% of Na ethylenediamine tetraacetate (EDTA) was added.

With these agents, textiles of polyamide fiber (Perlon, registered trademark) were washed in a laboratory washing machine where the washing liquor was heated from 20° to 60°C. within 15 minutes and was kept at this temperature for additional 15 minutes. The washing agent concentration was 5 g./l. and the weight ratio of textiles to liquor was 1:30. The water used had a hardness of 16° dH as well as a copper ion content of $10^{-5}$ mols per liter. The degree of whiteness of the four times rinsed and then dried wash was determined by photometer. The results demonstrated the superiority of the use of the polymers of this process.

| Example | Polymeric (N-alkylcarboxylic acid)-ethylenimine utilized (2%) | Molecular weight | Degree of whiteness after— 1 washing | Degree of whiteness after— 5 washings |
|---|---|---|---|---|
| 1 | Poly-(N-Tricarballylic acid)-ethyleneimine. | 2,040 | 109 | 117 |
| 2 | do | 3,500 | 110 | 116 |
| | NTA | | 100 | 104 |
| | EDTA | | 103 | 110 |

## ANHYDRIDE ACTIVATORS

### Persalt with Substituted Phthalic Anhydride

J. Malafosse and P.H. Gonse; U.S. Patent 3,298,775; January 17, 1967; assigned to l'Air Liquide, SA pour l'Etude et l'Exploitation des Procedes Georges Claude, France describe a process to activate sodium perborate bleaches by chloro- or nitrophthalic anhydrides. The bleaching power of mixtures based on solid anhydrides of dicarboxylic acids and per compounds is considerably increased if one or more hydrogen atoms in the anhydride molecule, are replaced by an electrophilic radical such as $NO_2$, Cl, Br, $HSO_3$, and $NaSO_3$. These substitutions increase the reactivity of the anhydrides in the bleaching baths, particularly at temperatures between 30° and 80°C., and at pH values from 7 to 10, and the dissolution of the corresponding anhydrides is accelerated.

The substituted phthalic anhydrides have the advantage of being solid and stable products which can be easily obtained from an industrial material. The solubility in water of the corresponding acids, particularly the solubility of nitrophthalic acids, is very much higher than that of phthalic acid. The anhydrides of organic diacids offering particular interest are the nitrophthalic anhydrides, preferably 3-nitrophthalic anhydride, and the halophthalic anhydrides, such as 4-chlorophthalic anhydride. These products are used individually or in admixture. They are obtained from the phthalic anhydride by nitration or chlorination by the conventional methods.

The substituted anhydrides react with the hydrogen peroxide, even in weakly alkaline medium, in order to form corresponding peracids much more completely than the unsubstituted phthalic anhydride. With the object of obtaining an oxidizing mixture which forms in aqueous solution, it is possible to mix beforehand, in the solid state, the substituted phthalic anhydride and a solid source of active oxygen, such as: monohydrated or tetrahydrated perborate, percarbonate, perpyrophosphate, urea peroxyhydrate and in general the mineral or organic peroxyhydrates, as well as the mineral or organic peroxides.

It is also possible to supply the substituted anhydride and the liquid hydrogen peroxide in the liquid form or in the form of peroxyhydrate separately to the bleaching solution at the moment of use. The active oxygen concentration of the bleaching bath is preferably between 20 and 500 mg. of active oxygen per liter. The material to be bleached is treated with a solution adjusted to a pH value between 7 and 11. There are used, per molecule of peroxide, 0.4 to 2 molecules of carboxylic acid anhydride substituted by at least one electrophilic group, preferably 0.8 to 1.5 molecules. The bleaching can be carried out at slightly elevated temperatures, for example, in the range from 30° to 80°C. and preferably at 60°C.

The comparative tests described in the different examples were carried out by measuring the decoloring power of the bleaching compounds on an aqueous solution of sulfur black, sold commercially under the trademark "Sulfanol M.B.S., " used in solubilized form. It has been found that a good correlation exists between the results of this test and those of the practical bleaching experiments with a commercial alkali washing mixture.

In the series of examples given below, a sulfur black solution with an optical density equal to 0.6 (green filter, thickness 10 mm.) is prepared; this solution is buffered to an alkaline pH value by a mixture of sodium carbonate and bicarbonate, or by any other buffering system such as those formed from sodium borate. This solution is brought to 60°C. and the per compound and the anhydride are added, separately or mixed before use, and the bleaching bath is kept at 60°C. for 20 minutes. The sample is then quickly cooled to normal temperature and its optical density is measured by colorimetry. The result is expressed as a percentage of decoloration, related to a comparison sample not provided with active oxygen and placed under the same operational conditions.

$$\text{Percent decoloration} = \frac{\text{Optical density of comparison sample} - \text{optical density of test specimen}}{\text{Optical density of comparison sample}}$$

The quantities of active oxygen which are used and the molecular ratios between the anhydride and active oxygen are different according to the tests, but in order to obtain comparable results, such as those described in the following examples, there are systematically employed 100 mg. of active oxygen per liter of solution and a pH value of 9.5 is used, the molar ratio between the buffering agent and the active oxygen being 5.

Example 1: Under the conditions described above, the operation takes place in the absence of organic anhydrides, using sodium borate in its monohydrated form $NaBO_3 \cdot H_2O$ or tetrahydrated form $NaBO_3 \cdot 4H_2O$. An average degree of decoloration of 5% is obtained. When the temperature is brought to 92°C., always under the same general conditions, the average degree of decoloration is 40%.

Example 2: A mixture of 3-nitrophthalic and 4-nitrophthalic anhydrides is used, obtained by the mixture of acids resulting from a phthalic anhydride nitration by means of a sulfonitric mixture being transformed into anhydride. For a molar ratio of 1 between the anhydride and the active oxygen, a degree of decoloration of 45% is obtained. For a molar ratio of 1.3, a degree of decoloration of 68% is obtained.

Example 3: Under the same conditions, the degree of decoloration obtained is compared, the source of active oxygen being on the one hand monohydrated sodium perborate with 15.7% of active oxygen and on the other hand sodium percarbonate with 11.7% of active oxygen, the anhydride in both cases being a sample of 3-nitrophthalic anhydride (melting point 160° to 163°C.), the molar ratio between the anhydride and the active oxygen being equal to 1.3. A degree of decoloration of 49% is obtained with the perborate and of 53% with the percarbonate. The preservation tests carried out on the mixtures of persalts with a low degree of hydration and of substituted anhydrides have been shown to be perfectly satisfactory.

## Activating Peroxygens with Mixed Acid Anhydrides

D.G. MacKellar, J.H. Blumbergs and H.M. Castrantas; U.S. Patent 3,338,839; August 29, 1967; assigned to FMC Corporation describe a process to activate liquid or solid peroxygen compounds by mixtures of carboxylic acid anhydrides. The peroxygen compounds, including both solid peroxygen compounds and liquids such as hydrogen peroxide can be activated to bleach, disinfect and sanitize more effectively when used in aqueous solutions at temperatures below boiling by incorporating as an activator, either (a) a mixed carboxylic acid anhydride having the formula:

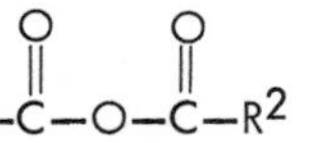

where $R^1$ is an aliphatic group and $R^1CO$ is derived from an aliphatic monocarboxylic acid having from 2 to 19 carbon atoms, and $R^2$ is an aromatic group and $R^2CO$ is derived from an aromatic carboxylic acid which may be a mono- or polycarboxylic acid in which any additional carboxylic acid substituents are not attached to adjoining carbon atoms of the aromatic ring, or (b) a mixture of about equal molar amounts of an aliphatic monocarboxylic acid anhydride derived from an aliphatic monocarboxylic acid having from 2 to 19 carbon atoms, and an aromatic carboxylic acid anhydride derived from either an aromatic mono- or polycarboxylic acid, in which any additional carboxylic acid groups are not attached to adjoining carbon atoms of the aromatic ring. The activator must be soluble in the aqueous treating solution in amounts of at least 100 ppm.

In practice for making up detergent mixtures, the solid peroxygen compound is mixed together with a phosphate builder such as sodium tripolyphosphate in combination with one or more working ingredients. These include antiredeposition agents such as sodium carboxymethylcellulose, anionic or nonionic surfactants and anticorrosion agents such as sodium silicate.

The activator is added to the mixture preferably in an amount sufficient to have one mol of anhydride present for every atom of available oxygen. Obviously, larger amounts of the activator can be used to assure complete activation of the peroxygen compound. Also, smaller amounts of the activator can be employed where a controlled bleaching effect is desired. The solid peroxygen compound is employed in amounts sufficient to achieve the desired degree of bleaching. For home bleaching applications amounts sufficient to supply from 10 to 100 ppm of active oxygen in the wash solution normally are employed. The preferred amount is that which will yield 40 ppm of active oxygen.

In the makeup of the final bleaching mixture it is desirable to keep the solid peroxygen compound and the activator out of contact with one another until placed in the wash water to avoid any possible reaction between these ingredients, even in the dry state. This can be done most readily when making up a single mixture by coating the activator with a water-soluble, or dispersable coating prior to mixing it with the remaining components of the bleaching mix. In this way the activator and the solid peroxygen compound can be mixed together and still remain out of contact with one another until the mixture is placed in the wash water. Another alternative is to separately package the solid peroxygen compound and the activator so that they are physically out of contact with one another in the consumer package.

The solid peroxygen compounds which are useful in the process as bleaching, disinfecting and sanitizing agents are those which liberate perhydroxyl anions readily when dissolved in an aqueous media. These include the alkali perborates such as sodium perborate and other alkali metal percompounds such as percarbonates, persilicates, perphosphates, and perpyrophosphates. In addition, such compounds as sodium peroxide, zinc peroxide, calcium peroxide, magnesium peroxide, urea peroxide, and others are included within the term solid peroxygen compound.

The mixed carboxylic acid anhydrides can be produced by a variety of processes. These include the reaction of an aliphatic anhydride such as acetic anhydride with an aromatic carboxylic acid, e.g., benzoic acid, at temperatures of from ambient to 70°C. The result of the above reaction is a mixed carboxylic acid anhydride having both aromatic and aliphatic moieties. A method for acetylating an aromatic carboxylic acid, e.g., benzoic acid, is by reacting ketene with the acid in an inert solvent. A mixed anhydride is formed, e.g., acetic-benzoic acid anhydride, as a product of the reaction. The preferred mixed anhydrides are those which are most easily prepared such as the benzoic-acetic anhydride, further these mixed anhydrides are more effective per unit weight than the substituted, more complex anhydrides.

Example: Bleaching tests were carried out by bleaching tea, coffee and wine stained cotton swatches (5" x 5") with sodium perborate tetrahydrate alone and in the presence of various activators. The procedure used was as follows: 32 cotton swatches (5" x 5" desized cotton Indianhead fabric, 48 fill by 48 warp threads per inch, uniform in weave and thread count) were stained with tea, coffee and wine in the following manner.

Five tea bags were placed in a liter of water and boiled for 5 minutes. Thereafter, the swatches were immersed in the tea and the boiling continued for 5 minutes. Thirty-two additional swatches of the same cloth were coffee stained by boiling 50 grams of coffee in a liter of water, immersing the swatches in the coffee solution, and boiling for an additional 5 minutes. The wine stains were created by soaking swatches of the same cloth in a red wine at room temperature. The stained swatches were then squeezed to remove excess fluid, dried, rinsed in cold water, and dried. Three of the stained cotton swatches were then added to each of a series of stainless steel Terg-o-Tometer vessels containing a 1,000 ml. of a 0.2% standard detergent solution at a temperature of 120°F.

Measured amounts of sodium perborate tetrahydrate bleach were then added to each vessel sufficient to correspond to an active oxygen content of 60 ppm. The pH of the solutions was adjusted to 9.0 using soda ash. Cut up pieces of white terry cloth toweling were then added to provide a typical household wash water:cloth ratio of 20:1. The activators specified in the table below were then added to the solution in the ratio set forth in the table. The Terg-o-Tometer was then operated at 72 cycles per minutes for 15 minutes at a temperature of 120°F. At the end of the wash cycle, the swatches were removed, rinsed under cold tap water and dried in a Proctor-Schwartz skein dryer. The tests were run in triplicate and included detergent blanks. Reflectance readings of the swatches were then taken before and after the wash cycle with a Hunter Model D-40 Reflectometer and the readings were average. The percent stain removal was obtained according to the following formula:

$$\text{Percent stain removal} = \frac{\text{Reflectance after bleaching} - \text{reflectance before bleaching} \times 100}{\text{Reflectance before staining} - \text{reflectance stained}}$$

The results are reported in the table below.

| Activator | % Stain Removal | | |
|---|---|---|---|
| | Tea | Coffee | Wine |
| Detergent blank | 26.9 | 45.0 | 61.0 |
| Perborate blank | 33.2 | 51.2 | 69.4 |
| Mixed acetic-benzoic anhydride | 72.3 | 94.0 | 92.6 |
| Mixture of benzoic anhydride plus acetic anhydride (1:1 mol ratio) | 74.0 | 94.3 | 92.8 |
| Mixed acetic-m-chlorobenzoic anhydride | 73.0 | 94.2 | 93.0 |
| Diacetic anhydride of isophthalic acid | 76.8 | 95.3 | 94.2 |

Persalt with Encapsulated Activator

W.G. Woods; U.S. Patent 3,532,634; October 6, 1970; assigned to United States Borax & Chemical Corporation describes a bleaching composition containing a persalt such as sodium perborate activated by an organic anhydride and a salt of a transition metal, with an atomic number of 24 to 29. The salts are the chlorides, sulfates, nitrates, perchlorates, water-soluble oxides of chromium, manganese, iron, cobalt, nickel and copper. The composition also contains a chelating agent. The activators improve the bleaching effect of the perborate to make it comparable to hypochlorites. The organic activator and the metal salt are kept from the rest of the composition by separate packaging, by encapsulation or by multilayer tableting.

Example 1: Washing tests using various activator compositions were conducted with Tide, a commercially available heavy duty detergent, and sodium perborate. The washing solutions contained 0.1% Tide and 0.125% sodium perborate. The tests were conducted using a Model 7243 Terg-o-Tometer apparatus. This apparatus consists of four 2,000 ml. stainless steel beakers immersed in a constant temperature water bath maintained at 120°F. Each beaker is fitted with a stainless steel agitator which is set to oscillate at 150 rpm in these tests.

The procedure involves adding 750 ml. of distilled water to the beaker and allowing the system to equilibrate at the desired temperature. Tide detergent (0.75 grams: 0.1%) then is added with agitation. After 1 minute of agitation, three swatches of cloth, 4" x 5", are added. With continuous agitation, the desired reagents are added at 1 minute intervals and the agitation continued for 15 minutes after the last reagent is added. The swatches then are removed from the wash solution, shaken with 400 ml. of distilled water, squeezed by hand, further rinsed in a running stream of distilled water on both sides, wrung by hand, and then ironed on the nonreflectance side.

Reflectance measurements were made before and after washing using Gardner Laboratories Multipurpose Reflectometer No. 1478 with a green filter. Standard tile Ser. No. R-2712-59 (64.1 reading) was used with unbleached muslin samples and standard tile No. R-2717-59 (14.5 reading) was used with the EMPA cotton Bleach Test Cloth No. 1. The percentage increase in reflectance was calculated as follows:

$$\text{Percent increase} = \frac{(\text{initial reflectance}) - (\text{final reflectance})}{(\text{initial reflectance})} \times 100$$

The increases are averaged for the three swatches in each run and reported as the mean. The increase in reflectance attributed to the activator system was calculated by taking the percent increase in reflectance obtained with the persalt activator, chelated metal ion, or combinations thereof, and subtracting the percent increase obtained by sodium perborate alone (15.5). The difference is the percent increase in reflectance attributed to the activator system and was expressed as Reflectance Increase. The results are recorded in the table in which the amounts represent the concentration of activator ingredients expressed as millimols per liter.

| No. | Activator system | Amount | Reflectance increase |
|---|---|---|---|
| 1 | Acetic anhydride | 8.14 | 23.7 |
| 2 | do | 4.07 | 7.6 |
| 3 | Succinic anhydride | 8.12 | −2.3 |
| 4 | Sodium p-acetoxybenzenesulfonate | 12.21 | 19.0 |

(continued)

| No. | Activator system | Amount | Reflectance increase |
|---|---|---|---|
| 5 | do | 8.14 | 13.4 |
| 6 | do | 1.63 | −1.5 |
| 7 | Tetraacetylhydrazine | 16.2 | 26.3 |
| 8 | do | 12.0 | 12.6 |
| 9 | do | 8.1 | 4.4 |
| 10 | Trisacetylcyanurate | 8.13 | 10.0 |
| 11 | Phthalic anhydride | 8.14 | 1.0 |
| 12 | Cobaltous chloride | 0.03 | 7.2 |
| | Picolinic acid | 0.25 | |
| 13 | Acetic anhydride | 8.14 | 115.7 |
| | Cobaltous chloride | 0.03 | |
| | Picolinic acid | 0.25 | |
| 14 | Succinic anhydride | 8.12 | 23.1 |
| | Cobaltous chloride | 0.03 | |
| | Picolinic acid | 0.25 | |
| 15 | Sodium p-acetoxybenzenesulfonate | 8.14 | 92.0 |
| | Cobaltous chloride | 0.03 | |
| | Picolinic acid | 0.25 | |
| 16 | Phthalic anhydride | 8.14 | 16.0 |
| | Cobaltous chloride | 0.03 | |
| | Picolinic acid | 0.25 | |
| 17 | Tetraacetylhydrazine | 16.2 | 149.8 |
| | Cobaltous chloride | 0.03 | |
| | Picolinic acid | 0.25 | |
| 18 | Tetraacetylhydrazine | 8.1 | 106.3 |
| | Cobaltous chloride | 0.015 | |
| | Picolinic acid | 0.25 | |
| 19 | Triacetylcyanurate | 8.13 | 96.4 |
| | Cobaltous chloride | 0.03 | |
| | Picolinic acid | 0.25 | |
| 20 | Ferrous sulfate | 0.03 | −1.5 |
| | 2,6-dicarboxypyridine | 0.25 | |
| 21 | Acetic anhydride | 4.07 | 12.8 |
| | Ferrous sulfate | 0.03 | |
| | 2,6-dicarboxypyridine | 0.25 | |
| 22 | Sodium p-acetoxybenzenesulfonate | 8.14 | 29.9 |
| | Ferrous sulfate | 0.03 | |
| | 2,6-dicarboxypyridine | 0.25 | |
| 23 | Cobaltous chloride | 0.03 | 1.0 |
| | 1,10-phenanthroline | 0.275 | |
| 24 | Acetic anhydride | 8.14 | 124.9 |
| | Cobaltous chloride | 0.03 | |
| | 1,10-phenanthroline | 0.275 | |
| 25 | Sodium p-acetoxybenzenesulfonate | 8.14 | 157.4 |
| | Cobaltous chloride | 0.03 | |
| | 1,10-phenanthroline | 0.275 | |
| 26 | Tetraacetylhydrazine | 12.0 | 134.3 |
| | Cobaltous chloride | 0.03 | |
| | 1,10-phenanthroline | 0.275 | |
| 27 | Sodium p-acetoxybenzenesulfonate | 1.63 | 101.5 |
| | Cobaltous chloride | 0.03 | |
| | 1,10-phenanthroline | 0.268 | |
| 28 | Sodium p-acetoxybenzenesulfonate | 12.21 | 157.4 |
| | Cobaltous chloride | 0.03 | |
| | 1,10-phenanthroline | 0.268 | |
| 29 | Cupric sulfate | 0.03 | 2.9 |
| | 2,2′-bypyridine | 0.25 | |
| 30 | Acetic anhydride | 8.14 | 35.5 |
| | Cupric sulfate | 0.03 | |
| | 2,2′-bipyridine | 0.25 | |
| 31 | Sodium p-acetoxybenzenesulfonate | 8.14 | 25.1 |
| | Cupric sulfate | 0.015 | |
| | 2,2′-bipyridine | 0.25 | |
| 32 | Cupric sulfate | 0.03 | −1.5 |
| | Piconlinic acid | 0.25 | |
| 33 | Sodium p-acetoxybenzenesulfonate | 8.14 | 25.0 |
| | Cupric sulfate | 0.03 | |
| | Picolinic acid | 0.25 | |
| 34 | Cupric sulfate | 0.03 | 6.4 |
| | 1,10-phenanthroline | 0.275 | |
| 35 | Sodium p-acetoxybenzenesulfonate | 8.14 | 39.3 |
| | Cupric sulfate | 0.03 | |
| | 1,10-phenathroline | 0.275 | |
| 36 | Manganous sulfate | 0.03 | 0 |
| | Picolinic acid | 0.25 | |
| 37 | Acetic anhydride | 8.14 | 71.9 |
| | Manganous sulfate | 0.03 | |
| | Picolinic acid | 0.25 | |
| 38 | Nickel acetate | 0.03 | 0.6 |
| | Picolinic acid | 0.25 | |
| 39 | Acetic anhydride | 8.14 | 43.5 |
| | Nickel acetate | 0.03 | |
| | Picolinic acid | 0.25 | |
| 40 | Cobaltous chloride | 0.03 | 1.4 |
| | 2,2′-bipyridine | 0.25 | |
| 41 | Sodium p-acetoxybenzene sulfonate | 8.14 | 151.3 |
| | Cobaltous chloride | 0.03 | |
| | 2,2′-bipyridine | 0.25 | |
| 42 | Trisacetylcyanurate | 8.13 | 137.1 |
| | Cobaltous chloride | 0.03 | |
| | 2,2′-bipyridine | 0.25 | |
| 43 | Acetic anhydride | 6.51 | 159.0 |
| | Cobaltous chloride | 0.024 | |
| | 2,2′-bipyridine | 0.2 | |
| 44 [1] | Acetic anhydride | 6.51 | 64.4 |
| | Cobaltous chloride | 0.024 | |
| | 2,2′-bipyridine | 0.2 | |

[1] Test No. 44 was in the absence of Tide.

Example 2: The procedure of Example 1 was followed except 2.02 grams of potassium monopersulfate were employed instead of sodium perborate. The results shown on the following page were obtained.

| | Activator system | Amount | Reflectance increase |
|---|---|---|---|
| Number: | | | |
| 45 | Acetic anhydride | 8.14 | 2 |
| 46 | Cobaltous chloride | 0.03 | 7 |
| | Picolinic acid | 0.25 | |
| 47 | Acetic anhydride | 8.14 | 13 |
| | Cobaltous chloride | 0.03 | |
| | Picolinic acid | 0.25 | |

Thus, as illustrated in the above examples, the synergized persalt bleaching compositions are far superior to persalt bleaches activated by the usual activators or combinations of metal salts with chelating agents. For example, No. 13 gave a reflectance increase of 115.7; yet, the sum of the increase from the ingredients (Nos. 1 and 12) was only 30.9. Furthermore, such superior performance is obtained at a relatively low water temperature of 120°F.

## Persalt Bleach for Dyed Cottons

J.R. Moyer; U.S. Patent 3,563,687; February 16, 1971; assigned to The Dow Chemical Company describes a process to provide a color-safe bleaching solution from a dry composition containing a persalt and an organic acid anhydride with added alkali builder to give a pH of 9.0 to 9.5. Examples of inorganic peroxygen compounds include, the inorganic perborates, peroxides, percarbonates, perphosphates and the like. Sodium perborate monohydrate has been widely used because of its low cost and availability. Potassium perborate and ammonium perborate are especially desirable because of their stability for sustained periods of time. Among other specific peroxygen compounds which may be included are urea peroxide, zinc peroxide, magnesium peroxide, and calcium peroxide. These peroxygen compounds may be either anhydrous or in the hydrated form as long as they are sufficiently free of uncombined water so as to be unreactive toward the organic acid anhydride in a dry mix prior to use.

Also included in the commonly used dry bleach compositions is an organic acid anhydride. Examples of suitable anhydrides include succinic anhydride, maleic anhydride, phthalic anhydride, glutaric anhydride and benzoic anhydride. Mixed anhydrides or derivatives of the aforementioned organic anhydrides are also included. Generally nontoxic organic acid anhydrides are employed; however, anhydrides of a toxic nature may be employed if desired.

The Figure 5.1 represents a comparison of the pH of a peroxybenzoic acid bleaching solution to color safeness and bleaching effectiveness. The dry bleach compositions usually contain equimolar quantities of the peroxygen compound and organic acid anhydride. However, the alkalinity of the prepared solutions may be varied to some degree by employing other than equimolar quantities of the ingredients. Thus, a more alkaline solution can be prepared if greater than an equimolar amount of an inorganic peroxygen compound is employed.

For example, a dry bleaching composition containing 1 gram-mol percent of sodium perborate monohydrate and 1 gram-mol percent of benzoic anhydride will produce an aqueous solution containing peroxybenzoic acid having a pH of 8.5. The pH is defined as the final pH after complete reaction of the organic acid anhydride and the peroxygen compound in an aqueous solution. Thus, the requisite pH may be maintained to some extent by employing other than equimolar amounts of these constituents.

FIGURE 5.1: COLOR DIFFERENCE AND WHITENESS AS FUNCTION OF pH

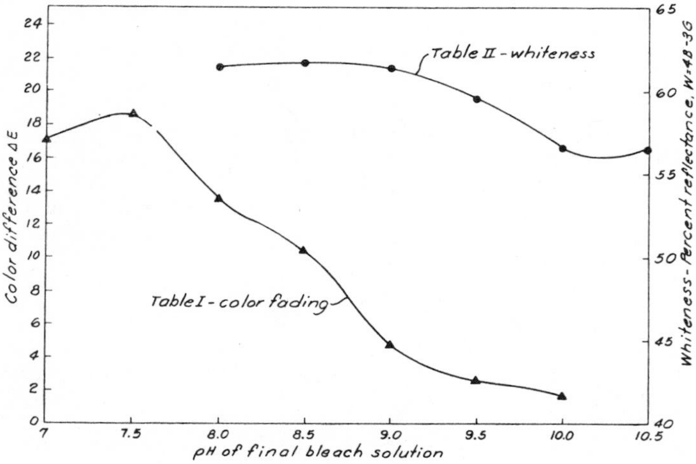


Source: J.R. Moyer; U.S. Patent 3,563,687; February 16, 1971

Example 1: A series of bleaching baths with final pH values ranging from 7 to 10 were prepared in 0.5 pH value increments in the following manner. A stainless steel agitating vessel was employed as a test vessel. To 1,000 ml. of water at 105°C. was added 2.0 grams of a commercial laundry detergent, 0.277 gram of benzoic anhydride and 0.128 gram of sodium perborate monohydrate. Hydrolysis of the anhydride was allowed to occur during which time the pH of the solutions fell. Either sulfuric acid or trisodium phosphate was then added to the bath to provide a predetermined pH between 7 and 10. The amount of sulfuric acid or trisodium phosphate was noted.

Following the preparation of the initial bleaching solutions a second series of solutions were prepared as test solutions. The solutions were prepared employing the same constituents and amounts as determined from the first series described directly hereinbefore and including the requisite amount of sulfuric acid or trisodium phosphate to provide a predetermined pH in the final bleaching solution. Four pieces of blue cotton cloth about 5 inches square were agitated in each of the so-prepared bleach baths for about 10 minutes at 100 cycles agitation per minute. The pieces of cloth were then rinsed in water and ironed dry.

The color of each piece of cloth after treatment was measured with a Hunter Model D-40 Color and Color Difference Meter. This color difference was compared to a control standard which consisted of a piece of the same material but not subjected to bleaching. The color difference between each sample was determined and the average color difference of the four pieces of cloth in each bath was calculated and is tabulated in following Table 1. Color difference of the treated fabric for each solution was expressed as ΔE, and Rd, a and b, are readings obtained from the color difference meter.

TABLE 1

| Run Number | pH Bleaching Solution | Rd | a | b | ΔE |
|---|---|---|---|---|---|
| Control | -- | 7.65 | 6.35 | -38.25 | -- |
| 1 | 7 | 9.25 | -2.15 | -23.85 | 17.00 |
| 2 | 7.5 | 9.70 | -3.10 | -22.75 | 18.56 |
| 3 | 8 | 8.80 | -0.35 | -26.80 | 13.46 |
| 4 | 8.5 | 8.85 | 0.75 | -29.75 | 10.45 |
| 5 | 9.0 | 8.10 | 3.75 | -34.40 | 4.74 |
| 6 | 9.5 | 8.00 | 4.55 | -36.35 | 2.71 |
| 7 | 10.0 | 8.00 | 5.00 | -37.10 | 1.91 |

Example 2: In this test a series of bleaching baths having predetermined pH values were prepared in a manner similar to that in Example 1. However, in this series of tests, undyed muslin cloth (greige goods) was subjected to the bleaching solutions at a temperature of 130°C. The whiteness of the bleached cloth was determined after bleaching by means of a Hunter lab Model D-40 reflectometer having filters to give both green and blue light. The percentage of light reflected is indicative of the whitening power of the solutions. A whiteness factor was calculated employing the equation:

$$W = 4B - 3G$$

wherein: W = whiteness, B = percent of blue light reflected and G = percent of green light reflected. This equation is well-known in the art as one means of comparing the whitening power of bleach solutions. The pH of the solution and reflectance of the blue and green light as well as the whitening factor is set forth in Table 2 including a control consisting of unbleached muslin cloth.

TABLE 2

| Run Number | pH | B, Reflectance | G, Reflectance | W, W = 4B - 3G |
|---|---|---|---|---|
| Control | -- | 66.4 | 72.2 | 49.0 |
| 1 | 8.0 | 75.9 | 80.7 | 61.5 |
| 2 | 8.5 | 75.5 | 80.1 | 61.7 |
| 3 | 9.0 | 74.5 | 78.9 | 61.3 |
| 4 | 9.5 | 73.1 | 77.6 | 59.6 |
| 5 | 10.0 | 71.4 | 76.3 | 56.7 |
| 6 | 10.5 | 71.0 | 75.7 | 56.9 |

The Figure 5.1 represents a graphic comparison of the effect of the final pH of the bleaching solutions on color stability and whitening power employing the results of the foregoing examples. It is evident that color safeness, noted as ΔE in the figure, i.e., resistance to fading, is increased markedly when employing a pH within the range described herein. The whitening power, i.e., W = 4B - 3G of the solutions, on the other hand, is not detrimentally affected thus indicating the value of employing higher pH values in organic monoperacid bleaching solutions.

## Peroxygen, Alkali Metal Silicate and Organic Acid Anydride

F.J. Donaghu; U.S. Patent 3,640,876; February 8, 1972; assigned to Kerr McGee Chemical Corporation describes a method to produce a high oxidation potential and an increased bleaching effect, by mixing to a persalt, such as sodium perborate, a synergistic mixture of a silicate and an organic acid anhydride. The synergistic reaction between the silicate, the acid anhydride and the peroxy compound produces oxidative species of higher oxidation potential than the peroxy compound alone or in combination with either the silicate or the acid anhydride. This high potential oxidative species is rapidly formed even at relatively mild temperatures, that is, below 60°C. and reaches its peak of greatest oxidation potential within a relatively short period of time.

Peroxy compounds which may be used, either singly or in combination, include hydrogen peroxide; alkali metal and alkaline earth metal peroxides; alkali metal and alkaline earth metal perborates; salts of peroxyacids such as peroxycarbonates, peroxyphosphates and the like; and peroxyhydrates such as phosphate peroxyhydrates, sodium carbonate peroxyhydrates and the like. The peroxy compound used must be relatively stable at room temperatures so that no significant loss of active oxygen content occurs during storage. Usually it is preferred to use peroxy compounds which are available as solids so that a stable, dry composition may be provided. The peroxy constituent preferably used is an alkali metal perborate and more particularly sodium perborate. Both the tetrahydrate and monohydrate form of sodium perborate may be used since both are available commercially in crystalline or powder form and have good storage stability.

The alkali metal silicate constituent of the composition preferably is sodium metasilicate, which may be used in either its anhydrous form or in a hydrated form, such as the pentahydrate or octahydrate. Other water-soluble alkali metal silicates may, of course, also be used. The acid anhydrides which may be used are organic acid anhydrides which are capable of reacting with the peroxy constituent of the composition in an aqueous medium to form a peracid of the acid anhydride. Suitable organic acid anhydrides which may be used either singly or in combination, include, for example, acetic anhydride, propionic anhydride, phthalic anhydride and succinic anhydride.

Example: In this example, a comparison was made of several compositions, one of which is typical of the improved oxidizing composition of this process. The compositions compared were: (A) the oxidizing composition containing 20 parts by weight of sodium perborate, 20 parts by weight of phthalic anhydride and 1 part by weight of sodium metasilicate; (B) 20 parts by weight of sodium perborate and 20 parts by weight of phthalic anhydride; (C) 20 parts by weight of sodium perborate and 1 part by weight of sodium metasilicate; and (D) sodium perborate alone. The compositions were formed by dry mixing the stated ingredients each of which was in dry, particulate form.

Compositions (A) through (D) were subjected to an electrometric test for measuring the potential (electromotive force) of oxidative species in solution. It has been found that, in general, the actual bleaching ability of a compound or composition is proportional to the oxidative shift in potential when the composition is put in solution. Thus, any compound or composition which produces a strong oxidative shift in potential, that is, electromotive force, without significant lowering of the pH, is considered to provide a good oxidative bleach. A standard procedure was followed in evaluating each of these compositions. Thus, a measured amount of the composition being tested was introduced into 200 ml. of a buffer solution of 0.01 M $Na_2B_2O_4$ in an amount sufficient to provide a concentration of 1 gram perborate per liter. The temperature of the solution was maintained at 50°C. throughout the test period. The electrochemical potential shift of the solution formed by the composition being tested was measured using platinum saturated calomel electrodes. In addition, a glass calomel electrode was provided in the solution and connected to the recording potentiometer so that changes in pH could be followed. The change in EMF for each composition was measured in this manner for 15 minutes.

The results of this test are illustrated in Figure 5.2 wherein a series of curves are shown. EMF measurements are plotted as the ordinate and time, in minutes, is plotted as the abscissa in the figure. The curves identified by A, B, C and D were obtained by testing compositions (A) through (D), respectively. The results of this test clearly show the synergistic effect of EMF of the oxidizing composition. Thus, the combination of sodium metasilicate with sodium perborate (curve C) provides no significant increase in oxidation potential over the use of sodium perborate alone (curve D). The combination of phthalic anhydride with the perborate (curve B) results in only a slight oxidative shift in potential.

FIGURE 5.2: OXIDATION POTENTIAL AS FUNCTION OF TIME

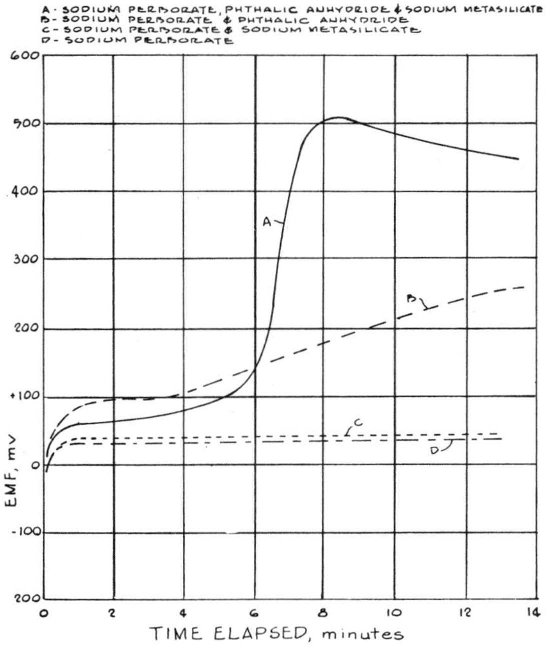


Source: F.J. Donaghu; U.S. Patent 3,640,876; February 8, 1972

The combination of the metasilicate and the acid anhydride with the perborate (curve A), the composition of this process provides relatively high oxidative shift in potential. This shift in potential is substantially greater than would be expected from the combination of the three constituents. The composition reached its greatest peak of oxidative potential in 8 minutes. This particularly advantageous since it is well within the 15 minute time period generally used in home laundry practices.

In addition, the pH of the solution to which composition (A) was added remained at 9.4 throughout the test period. These results clearly show that the oxidative potential of a peroxy compound may be increased to high value when an alkali metal silicate and an organic acid anhydride are added with the peroxy compound to an aqueous medium. Compositions (A) through (D) were then tested in a standard colorimetric test in order to compare the ability of the compositions in bleaching a standard colored solution. This series of tests clearly showed that composition (A) effectively bleached the colored solutions, even when the bleaching was carried out at mild temperatures over a relatively short period of time.

Peroxygen with Expanded Hydrated Salt Carrier Particles

R.C. Rhees; U.S. Patent 3,640,885; February 8, 1972; assigned to Kerr McGee Chemical Corporation describes a process to manufacture dry, free-forming sodium perborate bleach containing an organic anhydride, an alkali silicate, bonded to particles of expanded sodium tetraborate pentahydrate, as carrier particles. A variety of hydrated inorganic salts, suitable for use as the carrier in the bleaching composition may be expanded in this manner. Such suitable inorganic hydrate salts include, for example, potassium tetraborate pentahydrate, $K_2B_4O_7 \cdot 5H_2O$; sodium metaborate tetrahydrate, $Na_2B_2O_4 \cdot 4H_2O$; tetrasodium pyrophosphate decahydrate, $Na_4P_2O_7 \cdot 10H_2O$; sodium tetraborate decahydrate, $Na_2B_4O_7 \cdot 10H_2O$; and sodium tetraborate pentahydrate, $Na_2B_4O_7 \cdot 5H_2O$. The bulk density of the expanded, porous particles produced from these materials may be controlled within the range of from 10 to 40 lbs./ft.$^3$, and preferably between 15 to 30 lbs./ft.$^3$.

The solid peroxygen compound is bonded to the expanded porous salt particles by intimately mixing the two materials at moderately low temperatures under relatively dry conditions. The term "relatively dry" as used herein refers to bonding conditions under which at no time during mixing of the materials does the total water content in the mixture (that is, free water plus that combined as water of hydration) exceed the amount which can be completely bound as the highest hydrates of the compounds present in the mixture.

For example, in preparing compositions containing sodium tetraborate as the expanded, porous carrier and sodium perborate as the peroxygen compound, the amount of water present during formulation of the composition should at no time exceed the total of that found in the hydrates $Na_2B_4O_7 \cdot 10H_2O$ and $NaBO_3 \cdot 4H_2O$. While fully hydrated components may be used, those with lesser amounts of water of hydration are preferred. However, some free water to water which can exchange between compounds must be present during formulation to bond the components together. Thus, the bonding is carried out under relatively dry conditions.

The amount of water present in the mixture during formulation may be controlled in several ways. Thus, the solid peroxygen compound and the expanded, porous carrier particles may be mixed and moisture added to the mixture, with agitation, at a controlled rate. The free water thus added to the admixture is rapidly taken up by the porous particles and thereby bonds the peroxygen compound to the expanded porous carrier particles. The moisture may be added in any suitable manner such as spraying, atomizing, and the like.

Preferably, the amount of water present in the mixture is controlled by employing a higher hydrate of either the peroxygen compound or the expanded, porous carrier and then, in absence of added water, mixing and heating the two components up to that temperature (transition temperature) at which one of the components releases at least a portion of its water of hydration. It has been found that the released water of hydration redistributes among the constituents of the mixture and is rapidly taken up by the porous carrier particles, thereby effectively bonding the constituents in the mixture together.

It is normally expected that when a solid water-soluble hydrated peroxygen compound is heated above its transition temperature and starts to melt into the released water of hydration, decomposition of the peroxygen compound will be initiated. However, it has been discovered that when this heating is performed in the presence of the expanded, porous carrier particles substantially no decomposition of the peroxygen compound occurs. It is believed that this is due to the small amount of free water present in the mixture resulting from the rapid transfer of released water of hydration to the expanded, porous carrier. To facilitate this transfer of released water, it is generally preferred to use a higher hydrate of the peroxygen compound and a lower hydrate of the expanded inorganic salt. In addition, it is necessary to agitate the mixture during heating to provide a substantially uniform composition.

The temperature to which the mixture is heated will, of course, depend on the components of the mixture and the degree of hydration of these components. As noted above, the mixture must be heated to at least the temperature at which one of the hydrates in the mixture releases at least a portion of its water of hydration. Accordingly, sodium tetraborate pentahydrate is rapidly heated to a temperature between 400° and 500°C. to form expanded, porous particles containing between 2 and 4 mols of water per mol of tetraborate and having a bulk density of between 15 and 30 pounds per cubic foot.

If desired, the expanded porous sodium tetraborate may then be dried to reduce the water of hydration content even further, that is to 1 mol of water per mol of tetraborate. The use of such dried carrier particles is generally preferred for the dried carrier particles readily absorb water present in the mixture during preparation of the composition, thereby facilitating production of the composition. In addition, the dried particles absorb any free water which may be encountered during handling and storage of the composition, thereby preventing decomposition of the peroxygen compound prior to use.

The dried tetraborate particles are placed in a container and mixed with finely divided sodium perborate tetrahydrate ($NaBO_3 \cdot 4H_2O$). The mixture is then heated to 55° to 70°C. with continued agitation. In this temperature range, the sodium perborate releases some of its water of hydration, with the released water being transferred to the sodium tetraborate. This transfer of water from the hydrated perborate to the expanded tetraborate particles not only bonds the components together, thereby preventing segregation during handling and storage, but also provides a composition having increased stability, for the lower hydrates of sodium perborate are more stable than the tetrahydrate. The resulting product is dry, uniform and free-flowing. Since the perborate is bonded to the expanded tetraborate particles, segregation of these materials does not occur upon handling and storage.

Example: A dry, stable peroxygen bleaching composition having a relatively low bulk density is prepared in the following manner. A quantity of sodium tetraborate pentahydrate is expanded according to the procedure disclosed in U.S. Patent 3,454,357 to provide a quantity of expanded, porous particles which are then dried to reduce the water of hydration content thereof to 1.2 mols of water per mol of tetraborate. The dried particles thus produced have a bulk density of 19.1 lbs./ft.$^3$.

A measured quantity of the dried expanded, porous sodium tetraborate particles and an equivalent amount by weight of particulate sodium perborate tetrahydrate, having a bulk density of 43 lbs./ft.$^3$, are introduced into a container in a controlled temperature oven. The expanded sodium tetraborate and the sodium perborate are heated, while continuously agitated, to a temperature of between 65° to 70°C. for 10 minutes. At this temperature water of hydration is released from the perborate and the released water is rapidly taken up by the dried tetraborate particles to bond the perborate to the particles and provide a uniform, free-flowing, nonsegregating product. Agitation of the product is continued while the product cools. Since the perborate is bonded to the expanded porous tetraborate particles, separation of these two constituents is substantially prevented during handling and storage.

A decomposition inhibitor, sodium metasilicate, is included in the composition by mixing 22 grams of finely divided metasilicate with 886 grams of the bonded sodium perborate-sodium tetraborate product. The metasilicate in the resulting composition is not bonded to the tetraborate particles in the resulting composition but is physically mixed therewith. The bulk density of the resulting composition is 24 lbs./ft.$^3$. The available oxygen content of the bleaching composition thus prepared is determined after formulation and after 14 weeks storage. There is substantially no loss in available oxygen content after this prolonged period of storage, thereby demonstrating the storage stability of the composition.

The bleaching composition prepared in the manner described above is subjected to a series of test to compare the bleach activity of the composition of this process with that of an equivalent amount of fresh sodium perborate. In this series of tests a standard procedure is followed in which test cloths were washed in an agitator type washer under standard conditions of temperature, agitation rate and wash water. Tests are made using cloths stained with tea, ink, vacuum cleaner dust and chlorophyll.

The following test procedure is used: the percent reflectance of each test cloth is first determined. The test cloth is then washed for 15 minutes in wash water containing 0.70 gram sodium perborate per liter at a temperature of 194°F. Following the wash cycle the cloth is subjected to two 3 minute rinse cycles, after which it is dried, ironed and the percent reflectance again measured. The change in percent reflectance is determined for each test cloth and the results set out in the table shown on the following page. The results of this series of tests clearly show that the composition of this process is as effective a bleach as perborate alone. However, the bleaching composition of the process is superior to conventional peroxygen bleaches for this composition has excellent storage stability and has a controllable, moderately low bulk density. In addition, the sodium perborate, which is bonded to the expanded, porous carrier, does not separate or segregate from the composition so its concentration remains substantially uniform throughout the composition.

| Stained cloth washed | Change in percent reflectance | |
|---|---|---|
| | Sodium perborate | Sodium perborate bonded to sodium tetraborate |
| Ink | 40.8 | 40.8 |
| Tea | 14.3 | 14.7 |
| Vacuum cleaner dust | 7.8 | 6.4 |
| Chlorophyll | 11.3 | 11.3 |

## Maleic Anhydride Copolymer Stabilizers for Peroxygen Bleach

D. Du Bois, J. Heino, and A.F. Walden; U.S. Patent 3,663,443; May 16, 1972; assigned to The Procter & Gamble Company describe a process to add to a peroxygen bleaching composition a small amount of a water-soluble copolymer of maleic anhydride and a vinyl compound of the formula RCH=HCR, where one R represents a hydrogen atom and the other R represents a $C_1$ to $C_4$ alkyl ether radical or a hydrogen atom; or the alkali metal or ammonium salt of these copolymers. The two preferred copolymers are: maleic anhydride-vinyl methyl ether or maleic anhydride-ethylene copolymer.

Detergent and bleaching compositions, which are preferred, contain, by weight, from 5 to 20% of a noncationic detergent, from 15 to 60% of an inorganic builder salt (preferably, sodium tripolyphosphate), from 0.5 to 5% of active oxygen in the form of a water-soluble inorganic hydrogen peroxide addition compound, and from 0.5 to 5% of a water-soluble copolymer. The molecular weights of the copolymers used are related to their viscosities.

For the maleic anhydride-vinyl ether $C_1$ to $C_4$ alkyl copolymers, the specific viscosity varies preferably between 0.1 and 6.0, most preferably between 0.2 and 5.0; the specific viscosity is defined by measuring the viscosity of 1 gram of the copolymer in 100 cc methylethylketone in a Cannon-Fenske viscosity meter at 25°C. The viscosity of the maleic anhydride-ethylene copolymer varies preferably between 1.2 and 100 cp. when measured as an aqueous solution containing 2% of the copolymer, the solution being adjusted to have a pH of 10, in a Brookfield RTV viscosity meter at 10 rpm and at 25°C.

The hydrogen peroxide addition compounds, which are used are the perborates, e.g., the sodium perborate mono- and tetrahydrates. Other perborates which can be used are the potassium and ammonium or "true" perborates having the formulas $2KBO_3H_3O$ and $2NH_4BO_3 \cdot H_2O$, respectively. Other valuable hydrogen peroxide addition compounds are the carbonate peroxyhydrates, e.g., $2Na_2CO_3 \cdot 3H_2O_2$, and the phosphate peroxyhydrates. Although a great variety of sodium, potassium, ammonium and alkali earth metal phosphates can be used, sodium pyrophosphate peroxyhydrate ($Na_4P_2O_7 \cdot 2H_2O_2$) is preferred. The most suitable organic hydrogen peroxide addition compound which can be used in the process is the urea peroxide $[CO(NH_2)_2 \cdot H_2O_2]$, because it is one of the few free-flowing dry organic hydrogen peroxide addition compounds.

Test A: To five aqueous solutions (water hardness 3.4 mmol./l. $Ca^{++}$ and $Mg^{++}$, ratio $Ca^{++}$: $Mg^{++}$ 3:1) was added sodium perborate tetrahydrate, corresponding to 200 ppm active oxygen, and 0.5% by weight of a detergent composition consisting of 10% of sodium alkylbenzene sulfonate (average C atoms of alkyl radical: 11.8), 2% of a polyoxyethylene-polyoxypropylene condensate (molecular weight of polyoxypropylene: 1,750, polyoxyethylene: 80% by weight of total weight of condensate), 32% of sodium tripolyphosphate, 6% of sodium silicate, 2% of hydrogenated fish oil fatty acid (average molcular weight 285), 6% of sodium sulfate, and 42% of moisture (all percentages by weight). The first solution was used as a reference. To the second and third aqueous solution 1.0 and 5% respectively by weight, calculated on the weight of the detergent composition, of the water-soluble maleic anhydride-vinyl methyl ether copolymer were added. And to the fourth and fifth aqueous solution 1.0 and 5% respectively by weight, based on the weight of the detergent composition, of the water-soluble maleic anhydride-ethylene copolymer were added.

Each aqueous solution was heated to and maintained at 92°C. The active oxygen was determined after 5, 10, 15 and 20 minutes by the permanganate method. (See, for details, "Quantitative Inorganic Analysis," A. Vogel, 3rd ed., 1962, Longmans, London, p. 295; but whereby 50 cc of a sample solution was acidified with 50 cc of sulfuric acid (0.1 N), and titrated with a 0.1 N solution of $KMnO_4$, until permanent pink; 1 cc $KMnO_4$ = 16 ppm of active oxygen.) The results are presented in Table 1.

TABLE 1

| Solution | $H_2O_2$ decomposition controlling agent | Percent by weight[1] | Percent of active oxygen available after— 5 min. | 10 min. | 15 min. | 20 min. |
|---|---|---|---|---|---|---|
| 1 | | 0 | 78 | 65 | 56 | 52.5 |
| 2 | Maleic anhydride-vinyl methyl ether copolymer.[2] | 1.0 | 82 | 71 | 63 | 58 |
| 3 | do[2] | 5.0 | 90 | 85 | 80.5 | 77 |
| 4 | Maleic anhydride-ethylene copolymer.[3] | 1.0 | 88 | 69 | 59 | 56 |
| 5 | do.[3] | 5.0 | 92 | 89 | 85 | 82.5 |

[1] Percentage by weight of copolymer calculated on weight of detergent composition.
[2] Specific viscosity about 0.4 (1 g. in 100 cc. methylethylketone at 25° C.); ratio of monomers 1:1.
[3] Viscosity, 2 cps. in 2% aqueous solution at 25° C.; ratio of monomers 1:1.

Test B: The effectiveness of those copolymers when compared with a strong sequestering agent, such as, for example, ethylenediaminetetraacetic acid (EDTA) is shown in Table 2. To emphasize the effect, seven solutions were prepared with deionized water, containing 1 ppm of $Fe^{+++}$ ions, and 0.5% by weight of the detergent compositions described in Test A. Each solution was heated to and maintained at 92°C. To each solution was then added sodium perborate tetrahydrate, corresponding to 200 ppm of active oxygen, and at the same time 1.0 and 5% respectively by weight, based on the weight of the detergent composition, of ethylenediaminetetraacetic acid to solutions 2 and 3, of maleic anhydride-vinyl methyl ether copolymer to solutions 4 and 5, and of maleic anhydride-ethylene copolymer to solutions 6 and 7.

The active oxygen available at 5, 10, 15, and 20 minutes was determined by the permanganate method for solutions, 1, 4, 5, 6 and 7, and by the thiosulfate method for the solutions containing EDTA. (Thiosulfate method: a 10 cc sample solution is acidified with sulfuric acid 1.0 N; 30 cc of a 15% KI solution plus 100 cc distilled water is added, and the whole is stirred; after 10 minutes in the dark it is titrated with 0.01 thiosulfate; when yellow color fades, 0.5 gram starch is added and titration continued until colorless solution is obtained.)

TABLE 2

| Solution | $H_2O_2$ decomposition controlling agent | Percent by weight [1] | Percent of active oxygen available after— 5 min. | 10 min. | 15 min. | 20 min. |
|---|---|---|---|---|---|---|
| 1 | | 0 | 75 | 56 | 46.5 | 50 |
| 2 | EDTA | 1 | 66 | 54.5 | 45.5 | 33 |
| 3 | EDTA | 5 | 73.5 | 61.5 | 51 | 35.5 |
| 4 | Maleic anhydride-vinyl methyl ether copolymer.[2] | 1 | 80 | 56 | 46 | 39 |
| 5 | do.[2] | 5 | 83 | 76.5 | 72 | 66.5 |
| 6 | Maleic anhydride-ethylene copolymer.[3] | 1 | 79 | 65.5 | 56 | 48 |
| 7 | do.[3] | 5 | 86 | 86 | 83.5 | 80.5 |

[1] Percentage by weight of decomposition controlling agent, calculated on the weight of detergent composition.
[2] Specific viscosity, about 0.4; ratio of monomers 1:1.
[3] Viscosity, 2 cps. in 2% aqueous solution at 25° C.; ratio of monomers 1:1.

Comparing the percentages of available active oxygen in Table 2, EDTA at 1% concentration seems to enhance the decomposition of the hydrogen peroxide addition compound, while the difference in available active oxygen between solutions with and without 5% EDTA after 20 minutes is insignificant. The difference in available active oxygen in the solution containing 5% of maleic anhydride-ethylene copolymer and the control solution after 20 minutes is more than 40%.

Example 1: A granular bleaching and detergent composition described below, is prepared by slurrying the surface-active agents and builders, spray-drying the slurry, and mixing the perborate, copolymer and perfume with the spray-dried product. The end product consists of:

| | Percent |
|---|---|
| Sodium alkylbenzene sulfonate (average chain length 11.8) | 12 |
| Hydrogenated fish oil fatty acids (average MW 285) | 5 |
| Tallow alcohol-ethylene oxide (average ethylene units 11) | 3 |
| Sodium tripolyphosphate | 35 |
| Sodium silicate | 5 |
| Maleic anhydride-vinyl methyl ether copolymer (specific viscosity, 0.4) | 3 |
| Sodium perborate tetrahydrate | 30 |
| Sodium sulfate, CMC, perfume, moisture | Balance |

The maleic anhydride-vinyl methyl ether copolymer (ratio of monomers 1:1) can be replaced by the maleic anhydride-ethylene copolymer at an equal percentage basis, or by the disodium salts of one or both of them. The sodium perborate tetrahydrate can be replaced on an equal percentage basis by the carbonate peroxyhydrate, $2Na_2CO_3 \cdot 3H_2O_2$, or by the sodium pyrophosphate peroxyhydrate $Na_4P_2O_7 \cdot 2H_2O_2$.

Example 2: A detergent composition was prepared by slurrying the detergent and builders, spray-drying the slurry, and adding the copolymer, perfume and perborate to the spray-dried product. The finished product consisted of:

|  | Percent |
|---|---|
| Sodium 3-(N,N-dimethyl-N-tetradecylammonio)-2-hydroxypropane-1-sulfonate | 6 |
| Sodium tripolyphosphate | 40 |
| Sodium perborate tetrahydrate | 25 |
| Sodium sulfate, moisture, perfume | 15 |
| Maleic anhydride-vinyl methyl ether copolymer (specific viscosity, 0.4) | 5 |
| Sodium silicate | 9 |

The percentage of active oxygen available in a washing solution, containing 0.5% by weight of the detergent composition, heated at 92°C., was, according to the usual permanganate method, after 10 minutes, 77%, and after 20 minutes, 65%. Heavily soiled clothes, washed for 30 minutes with a solution of the detergent composition, were thoroughly cleaned and bleached.

## PERBORATES WITH MISCELLANEOUS ACTIVATORS

### Alkyl Phosphate Activators for Inorganic Persalt Bleaches

K. Dithmar and E. Naujoks; U.S. Patent 3,073,666; January 15, 1963; assigned to Deutsche Gold- und Silber-Scheideanstalt vormals Roessler, Germany found that the bleaching of natural or synthetic fibers with peroxides, perborates or perphosphates is activated and the desired degree of bleaching is obtained at a lower temperature by the use of alkyl phosphates. It was found that the reaction products of acylating agents, such as acetyl chloride with esters of phosphoric acid, are very excellent activators for per compounds. Such acylated esters, for example, have the following formula:

$$\begin{array}{c} R-O \\ \phantom{R-O}\diagdown \\ \phantom{R-O}P-O-\underset{\underset{O}{\|}}{C}-R_1 \\ \phantom{R-O}\diagup\;\| \\ R-O\phantom{\diagup}\;O \end{array} \quad \text{and} \quad \begin{array}{c} \phantom{R-O-P}O-\underset{\underset{O}{\|}}{C}-R_1 \\ \phantom{R-O-P}\diagup \\ R-O-P \\ \phantom{R-O-P}\|\diagdown \\ \phantom{R-O-P}O\phantom{\;}O-\underset{\underset{O}{\|}}{C}-R_1 \end{array}$$

wherein R, for example, can be the alkyl groups, such as ethyl, derived from the phosphoric acid ester and

$$-\underset{\underset{}{\|}}{\overset{O}{C}}-R_1$$

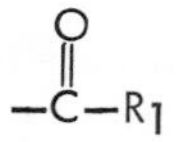

is the acyl group, such as an acetyl group, derived from the acylating agent. The acylated esters are produced by known methods and exhibit a very good activating action, sufficient stability and a pleasant odor. It was found that relatively small quantities of the activators provide a good activation in aqueous peroxidic baths. For example, quantities between 0.1 to 1.0 mol per mol of active oxygen calculated as $H_2O_2$ are sufficient. The bleaching and washing baths which contain peroxidic textile bleaching agents and the activators can in addition contain known stabilizers, buffering agents, wetting agents, detergents, protective agents for fibers and optical brighteners. The best action is obtained in neutral, weakly acidic and weakly alkaline baths or, in other words, at a pH between 3 and 11. The temperature of the baths preferably is between 60° and 80°C. The active oxygen content of such baths preferably is between 0.05 and 20 grams per liter. It is important for the action of the activators that they are dissolved directly together with the per compounds or shortly before or after such per compounds in order that coaction is ensured.

Example: A rayon staple fiber was boiled in red wine and its white content measured and found to be 31%. The measurement of such white content was with the Zeiss-Elrepho-whiteness degree measurer against white standard 565 and with filter 6. This stained fabric was washed for 45 minutes at 60°C. in a quiescent bath of the following composition per liter:

| | Grams |
|---|---|
| Lauryl sulfate | 1.250 |
| Sodium pyrophosphate | 0.625 |
| Sodium tripolyphosphate | 0.625 |
| Dry water glass | 0.350 |
| Sodium perborate | 0.600 |
| Diethyl monoacetyl phosphate | 0.980 |

The pH of such bath was 7.5 to 8. The treated fabric was rinsed and dried and then again its white content measured as above. After such treatment its white content was 69.4%. The same stained fabric when treated under the same conditions, except for the omission of the diethyl monoacetyl phosphate, only had a white content of 48.0%.

## Perborate Bleach with Heavy Metal Catalyst and Chelating Agent

J.O. Konecny and R.E. Meeker; U.S. Patent 3,156,654; November 10, 1964; assigned to Shell Oil Company describe a process to bleach cotton with perborate and the addition of cobalt sulfate and an aminocarboxylic acid chelating agent. With these additives perborate bleach can be used efficiently in home laundry equipment at 50° to 60°C.

The bleaching activity of peroxy compounds can be materially improved by using therewith heavy metal ions which catalyze peroxide decomposition together with a special type of chelating agent for the heavy metal. The chelating agent must be one which is not only itself stable and without undesirable catalytic effect on peroxide decomposition in the bleaching composition and bath in which it is to be used, but also should be one which forms a complex with the heavy metal ions which complex is soluble and similarly stable in the bleaching bath. The chelating agent used must furthermore be one which is not a stronger complexing agent for the heavy metal ions present than is the material to be bleached.

Representing the complexing agent by Q and the material to be bleached as R, one can write equations representing the equilibria which control the proportion of the heavy metal ions which will be available for promoting the desired bleaching and which will be held in inert complexed form in the solution so that loss of peroxide through undesirable decomposition away from contact with the material being bleached can be minimized. These are, using cobalt ions as an example of suitable heavy metal ions:

$$Co^{++} + (Q) \rightleftharpoons (CoQ)^{++}$$

$$Co^{++} + R \rightleftharpoons Co^{++} \text{ adsorbed on R}$$

$$(CoQ)^{++} \rightleftharpoons Co^{++} \text{ adsorbed on R} + (Q)$$

Thus besides its desirable synergistic effect in promoting the bleaching, the chelating agent should also preferably form sufficient of the chelate $(CoQ)^{++}$ to maintain essentially all the heavy metal ions not adsorbed on R in inert soluble complexed form so that undesirable peroxide decomposition is minimized without interfering with the required amount of adsorption of cobalt ions indicated in the second of these equations. Thus an important feature of the process is that it provides a simple, one step method of operation whereby bleaching is improved by heavy metal ions adsorbed on the material to be bleached so as to catalyze the decomposition of the peroxide at the precise location where bleaching is desired. The metal ions are believed to decompose hydrogen peroxide by a catalytic cycle illustrated, in the case of copper ions, by the equations:

$$Cu^{++} + O_2H^- \longrightarrow Cu^+ + O_2H$$

$$Cu^+ + H_2O_2 \longrightarrow Cu^{++} + OH + OH^-$$

This localized production of OH radicals is from the peroxide initially present as the peroxide bleaching agent, either by direct catalytic decomposition as when using hydrogen peroxide or by catalytic decomposition of such peroxide formed from the starting peroxide bleaching agent, for instance by hydrolysis of a peracid or persalt such as a perborate or the like. These OH radicals are among the strongest oxidizing agents known and probably contribute to the observed improvement in the bleaching. The improvement in bleaching obtained by using the indicated combination of chelating agent and heavy metal ions depends on the amount of heavy metal ion adsorbed on the fabric or other material being bleached, e.g., $Co^{++}$ adsorbed on R.

It has been found that with a constant amount of chelating agent such as pyridine-2-carboxylic acid, e.g., the bleaching increases with increasing amounts of cobalt sulfate or the like. With a constant amount of the latter and increasing amounts of the chelating agent, sufficient in all cases to provide an appreciable excess and thus prevent precipitation of heavy metal hydroxide which would promote undesirable peroxide decomposition, the bleaching decreases, the decrease being coincident with decreased adsorption of heavy metal ions by the fabric. The total concentration of heavy metal in the bleach bath can be varied widely without detrimentally affecting the bleach provided the heavy metal to chelating agent ratio is maintained at a proper constant ratio.

It is not desirable to use peroxide decomposition catalysts such as heavy metal ions alone. A small improvement in bleaching can indeed be obtained by adding copper or cobalt sulfate to a sodium perborate bleaching bath, for example. But the increase in bleaching effect is small and accompanied by extensive peroxide loss through decomposition both by the metal ions and precipitated hydroxides. For example, when bleaching of a standard cotton test cloth for 15 minutes with sodium perborate (10 grams per liter of bleach solution) at 60°C., the addition of 0.015 millimol of copper sulfate per liter, results in an increase of reflectance (ΔR) of the cotton cloth of 9.4 units compared with 6.1 units under the same conditions without added copper sulfate, but the loss of peroxide through decomposition is increased from 10 to 32%. However, when the catalysis of peroxide decomposition by copper is localized at the surface of the cloth, the increase in reflectance (ΔR) is 17.4 units and the loss of peroxide in the solution not in contact with the cloth is no greater than when no copper is used.

The best results are usually obtained, with copper and cobalt salts. Cobalt salts have special advantages because of their outstanding effectiveness in improving peroxide bleaching when used with a chelating agent in accordance with the process. The chelating agent chosen should preferably be one which forms a chelate complex with the heavy metal or mixture of heavy metal ions which chelate is soluble in the bleaching solution to the extent of at least $5 \times 10^{-5}$ mols per liter of solution. There are special advantages in using as the chelating agent a particular subgroup of amino carboxylic acids having not more than 2 carbon atoms separating the carboxyl group from the amino nitrogen atom. These are the pyridine-2-carboxylic acids such as pyridine-2-carboxylic acid itself and pyridine-2,6-dicarboxylic acid and the like, which meet the foregoing requirements as to complexing power.

Another type of aminocarboxylic acid chelating agents which can be used in bleaching materials which have a stronger adsorption power for heavy metal ions than does cellulose are the compounds which contain at least one N,N-dicarboxyalkylamino group, $N—(R·COOX)_2$ wherein R is an alkylene radical of up to two carbon atoms and X is hydrogen or a salt-forming cation such, for instance, as an alkali metal or alkaline earth metal or ammonium ion, the two indicated X's being the same or different.

The process can be applied to bleaching baths such as are used for treating textiles, wood pulp and the like, to wash liquors, such as are used in commercial laundering and to solid bleaching compositions. Solid bleaching compositions prepared according to the process, preferably contain the peroxy bleaching agent, source of heavy metal ions and organic chelating agent therefor in the ratios previously indicated as desirable in the bleaching bath. In addition, inert salts and any of the conventional adjuncts may be used with or without detergents or other auxiliary agents.

Example 1: Cotton was bleached using a bath containing 10 grams of sodium perborate, 0.25 mmol. of pyridine-2,6-dicarboxylic acid and 0.012 mmol. of copper sulfate per liter. After 15 minutes bleaching at 60°C. the reflectance of the cotton had increased 10.4 units. The loss of peroxide through decomposition which did not result in bleaching was 16%. In another test in which 0.005 mmol./l. of copper sulfate and 0.042 mmol. of pyridine-2,6-dicarboxylic acid were used with the perborate in the same way the increase in reflectance was 13.3 units.

Example 2: Tests to show the amount of heavy metal adsorbed by the cotton cloth in the bleaching were carried out using 15 minutes bleaching time at 60°C. with a bath containing 10 grams of sodium perborate per liter to which was added different amounts of cobalt sulfate and pyridine-2-carboxylic acid. After bleaching and rinsing the samples of cloth, reflectance measurements were made in the usual way and the amount of cobalt adsorbed was determined by neutron activation. The following results were obtained:

| Cobalt Sulfate added (millimoles per liter) | Pyridine-2-carboxylic acid added (millimoles per liter) | Cobalt Adsorbed on Cloth (millimoles per 1,000 grams) | Bleach (increase in reflectance ΔR units) |
|---|---|---|---|
| 0.050 | 1.00 | 0.12 | 12.4 |
| 0.050 | 0.25 | 0.92 | 17.0 |

## Acylated Phosphonic Acid and Ester Activators for Persalt Bleach

K. Dithmar and P. Koblischek; U.S. Patent 3,379,493; April 23, 1968; assigned to Deutsche Gold- und Silber-Scheideanstalt vormals Roessler, Germany describe a process to bleach natural and synthetic fibers with sodium perborate and other persalts, at lower temperatures, by the use of acylated phosphonic acid esters or acylated phosphinic acids. These activators are compounds which contain organic groups bound directly to the phosphorus atom through a carbon atom and not through an intermediate oxygen atom. Examples of such acylated phosphonic acid esters or acylated phosphinic acids are as follows:

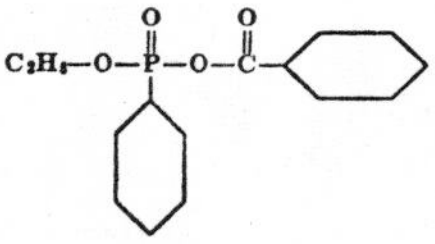

or

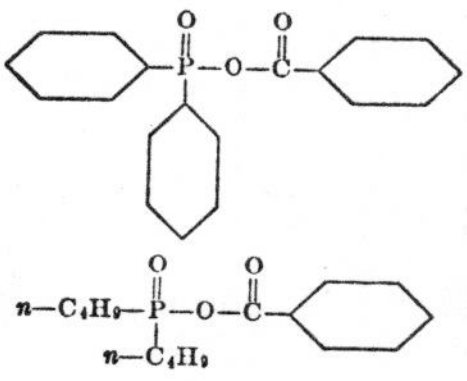

Additionally, the organic groups directly connected to the phosphorus atom through a carbon atom such as alkyl or aryl groups may carry substituents. These activators depending upon their composition, are liquid or solid and easily or difficultly soluble in water. The quantities of such activators required are relatively small and as a rule 0.1 to 1 mol of the activator per mol of active oxygen in the bleaching bath suffices. Usual stabilizers, buffers, wetting agents, detergents, protective agents for fibers and/or optical brighteners may also be added to the activated bleaching baths according to the process.

The best action according to the process is obtained in neutral weakly acidic or weakly alkaline baths or, in other words, at a pH between 3 and 11. The active oxygen content of such baths preferably is between 0.05 and 20 grams per liter. The presence of the activators in the bleaching baths causes a bleaching action to take place at relatively low temperatures. For example, bleaching actions which can only be achieved at temperatures between 70° and 100°C. in the absence of activators can be attained in the presence of the activators at temperatures between 30° and 60°C.

Example: A rayon staple fiber woven fabric was boiled in red wine and dried. The white content thereof when measured with a Zeiss-Elrepho-whiteness degree measurer against white standard 565 with filter 6 was 31%. The strongly soiled fabric was washed 3 times for 15 minutes at 60°C. in a Launder-O-Meter (Atlas) with an aqueous bath which per liter contained the ingredients shown on the following page.

| | Grams |
|---|---|
| Lauryl sulfate | 1.250 |
| Sodium pyrophosphate | 0.625 |
| Sodium tripolyphosphate | 0.625 |
| Dry water glass | 0.350 |
| Sodium perborate | 0.700 |
| O-benzoyl-phenyl phosphonic acid ethyl ester | 1.000 |

The pH of the bath was 7 to 8. After such wash the fabric had a 64.3% white content. In comparison, when the soiled fabric was given a washing treatment under the same conditions with the same bath composition except for the omission of the O-benzyol-phenyl phosphonic acid ethyl ester its white content was only raised to 44.5%.

Cobalt Catalyst for Low Temperature Persalt Bleach

B. Das and K.G. van Senden; U.S. Patent 3,398,096; August 20, 1968; assigned to Lever Brothers Company describe stable, dry catalysts for enhancing the bleaching action of inorganic persalts at temperatures of 20° to 50°C. The catalysts comprise a metal ion of a transition element like cobalt, adsorbed on a water-insoluble carrier like zinc silicate. The bleaching compositions containing the catalyst are effective in the removal of persistent stains, e.g., tea stains, from hard surfaces, such as dishes, tiles and the like, and are also useful for bleaching stains on fibrous materials, such as finished and unfinished textile fabrics. These compositions are also useful for dishwashing and scouring.

For use as the metal ion of the catalyst, it has been found that the metal ions of the transition elements in the periodic system especially those having an unpaired electron spin, such as $Co^{2+}$, $Mn^{2+}$, $Ni^{2+}$, $Cr^{3+}$, $Mo^{2+}$, or $Cu^{2+}$ are particularly effective. The carriers which can be advantageously used may be any water-insoluble or slightly soluble compound of Zn, Cd, Ca, Mg, Al, Sn, Be, Ti, Sb, Bi or $SiO_2$.

Of these possible combinations the most active catalysts are found when $Co^{2+}$ is adsorbed on: zinc silicate; titanium dioxide or hydroxide; tin dioxide; antimony trioxide; aluminum oxide, hydroxide or silicate; alkaline zinc carbonate; alkaline cadmium carbonate; cadmium carbonate, silicate or oxide. The metal ion content may be from 1 to 100 mg. or more per gram catalyst, and preferably from 20 to 60 mg./g. It goes without saying that the catalysts according to the process are also effective to aid the bleaching action of inorganic percompounds on colored solutions at low temperature. The catalysts may be prepared in the following ways:

(1) By dry mixing a finely divided powdered salt, e.g., $CoCl_2$, containing the metal ion, with the finely divided carrier in such a way that a homogeneous mixture is obtained, in which both substances are in very good contact.

(2) By adding an aqueous solution containing the metal ion, e.g., $CoCl_2$ solution, to the required amount of finely powdered carrier. Water is then added and the whole mixture is stirred until a paste-like substance is obtained. The product is then dried at temperatures up to 100°C., vacuum may or may not be applied, preferably at temperatures of 40° to 60°C. in a vacuum of 1 to 15 mm. Hg. Much higher temperatures

may also be applied, depending on the type of catalyst, although it is not advisable since it would affect the catalysts' activity.

(3) By freeze-drying the paste-like substances as described in (2), at temperatures of -50° to 0°C. in a vacuum of 0.01 to 1 mm. Hg.

The catalysts obtained by these methods assist bleaching at low temperatures (20°C. or even lower), as well as at higher temperatures, with sodium perborate and/or with other inorganic percompounds including percompounds, such as Du Pont's "Oxone" (a triple salt consisting of potassium permonosulfate, potassium hydrogen sulfate and potassium sulfate in the approximate molecular ratio of 2:1:1). It is desirable that the bleaching compositions should give an alkaline solution, preferably one having a pH value of between 7 and 11.

It has been found that catalysts prepared according to this process can be stored for some time without losing their activity. They can be dry mixed with inorganic percompounds to form low temperature bleaching compositions, preferably at weight ratios of between 1:1 and 1:8. Such mixtures when incorporated in normal detergent compositions enable the latter to bleach effectively at relatively low temperatures. In order to avoid unwanted coloring of the bleaching solution and of the object to be bleached, which may be caused by the catalyzing metal ion, sequestering agents may be incorporated in the bleaching composition, e.g., sodium hexametaphosphate. Other condensed phosphates, particularly sodium tripolyphosphate, appear to reduce the catalysts' activity.

Example 1: Catalysts having a cobalt content of approximately 4% (40 mg./g. catalyst) were prepared in the following ways:

(1) 100 grams finely powdered zinc silicate (100 mesh, British Standard Sieve) and 19 grams $CoCl_2 \cdot 6H_2O$ were mixed together in a mortar so that a homogeneous mixture was obtained. The zinc silicate was previously dried in an oven at 80°C. under vacuum for one hour.

(2) To 200 grams finely powdered zinc silicate (100 mesh, British Standard Sieve) a solution of 38 grams $CoCl_2 \cdot 6H_2O$ was added in 180 ml. water and the mixture was stirred so that a homogeneous paste was obtained. Part of the paste was then dried in a vacuum drier at 40°C. and 10 mm. Hg for 5 hours. The product obtained by powdering the granules in a mortar had a free moisture content of approximately 1.5%.

(3) The other part of the paste was dried at a temperature of less than -20°C. and at a pressure of 0.05 mm. Hg. The moisture content of the product was approximately 1.5%.

Example 2: A number of scouring powder compositions were prepared containing sodium perborate and catalyst prepared according to Example 1. For purposes of comparison two compositions were also prepared, one without catalyst and the other containing the metal ion without a carrier.

| | Parts by Weight | | | |
|---|---|---|---|---|
| | 1 | 2 | 3 | 4 |
| Sodium dodecylbenzenesulfonate | 9.0 | 9.0 | 9.0 | 9.0 |
| Silica | 72.0 | 65.5 | 82.5 | 82.5 |

(continued)

| | Parts by Weight | | | |
|---|---|---|---|---|
| | 1 | 2 | 3 | 4 |
| $NaBO_3 \cdot 4H_2O$ | 8.5 | 15.0 | 8.5 | 8.5 |
| Catalyst, containing 4% $Co^{2+}$ | 4.0 | 4.0 | -- | * |
| Urea | 3.0 | 3.0 | -- | -- |
| Sodium hexametaphosphate | 3.5 | 3.5 | -- | -- |

*0.16% $Co^{2+}$ as $CoCl_2$ without a carrier.

These compositions were tested at room temperature for the bleaching of tea stains on unglazed tiles (Royal Sphinx Maastricht, 15 x 15 cm.) by appyling them as a paste-like substance to the tile surface. The following test method was used. The reflectance of a tea stained tile was measured. 10 grams of powder were mixed with 7 ml. distilled water and stirred for 30 seconds. The mixture was poured on the tile surface. After a contact time of 1 or 2 minutes the paste was rinsed from the tile surface with running tap water for at least 10 seconds. After standing in the air for 24 hours the reflectance of the tile was measured again. In the following table the increase in reflectance of the tile surface after treatment is given. From this table it can be seen that the type of carrier used is an important factor.

| Composition | Carrier | Increase in Reflectance % After Contact Time of | |
|---|---|---|---|
| | | 1 min. | 2 min. |
| 1 | Zinc silicate | 27 | 42 |
| | Cadmium carbonate | 27 | 29 |
| | Stannic oxide | 20 | 24 |
| | Antimony oxide | 27 | 30 |
| | Zinc silicate* | 26 | 30 |
| 2 | Zinc silicate | 27 | 42 |
| 3 | No catalyst | 3 | 11 |
| 4 | No carrier | 0 | 5 |

*Catalyst prepared by dry mixing. The other catalysts were prepared by process (2).

## Diacyl Methylene Diformamide Activator for Persalts

K. Dithmar and P. Koblischek; U.S. Patent 3,425,786; February 4, 1969; assigned to Deutsche Gold- und Silber-Scheideanstalt vormals Roessler, Germany found that derivatives of methylene diformamide can be used to activate sodium perborate containing bleaching and washing compositions in aqueous solutions to be effective at temperatures below 100°C. The methylene diformamide derivatives are of the following formula:

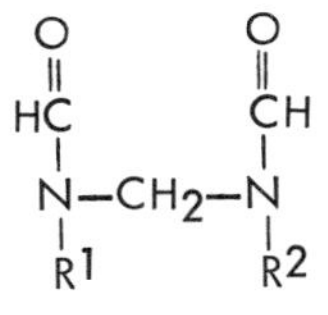

In the formula, $R^1$ and $R^2$ are carboxylic acid acyl groups containing 1 to 12 carbon atoms. N,N'-dibenzoyl methylene diformamide, N,N'-dipropionyl methylene diformamide and N,N'-diacetyl methylene diformamide come into question. It was found that such methylene diformamide derivatives react with hydrogen peroxide or other active oxygen carriers, such as, sodium perborate, sodium percarbonate and the like, when dissolved in water at moderately raised temperatures, such as, for example, 60° to 70°C., with simultaneous increase in the bleaching action of such active oxygen carriers. They, on the other hand, do not have the tendency to split off the acids of the acyl groups upon storage. The high activating action as well as the better stability are both still evident even when longer chained aliphatic acyl groups are present on the nitrogen atoms in addition to the formyl groups. N,N'-diacetyl methylene diformamide of the formula:

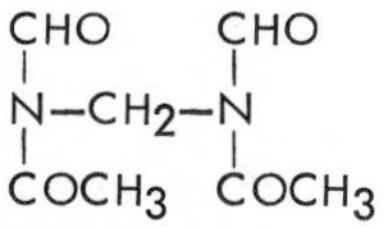

has proved particularly satisfactory as it is very reactive but, on the other hand, also is stable and does not split off acetic acid upon storage in contact with the atmosphere. N,N'-dipropionyl methylene diformamide is just as effective. It was found that the methylene diformamide derivatives in question which contain two nitogen atoms and an additional acyl group besides the formyl group on each nitrogen atom possess a stability on storage and activating action on peroxidic bleaching agents which is not possessed when such additional acyl group is absent or when a compound containing only one nitrogen group is concerned.

The activator employed can be mixed in with the washing and bleaching composition during the production of such mixture or later. Such activator also may be added directly to the washing and bleaching bath. The quantities of the activators required amounts to 0.3 to 2 mols of the activator per mol of active oxygen in the washing and bleaching composition. The activating action occurs in neutral, weakly acidic or weakly alkaline baths or, in other words, at a pH between 3 and 11. The active oxygen content of the baths preferably is between 0.05 and 20 grams per liter. The usual stabilizers, buffers, wetting agents, detergent protective agents for textile fibers, and/or optical brighteners employed in washing and bleaching compositions or baths according to the process.

Example 1: The production of N,N'-diacetyl methylene diformamide is as follows. 102 grams of methylene diformamide (1.0 mol) were boiled under reflux with an excess of acetic acid anhydride for 5 hours. Thereafter the acetic acid formed and the unreacted acetic acid anhydride were distilled off under vacuum and the oily residue immediately mixed with 100 ml. of ethanol. Upon cooling the N,N'-diacetyl methylene diformamide crystallized out and was filtered off on a suction filter and washed with ethanol and ether and dried. The product was white in color and had a melting point of 89° to 91°C. The yield was 82% of theory. Analysis calculated for $C_7H_{10}N_2O_4$: C, 45.16%; H, 5.42%; N, 15.05%. Found: C, 45.09%; H, 5.41%; N, 15.31%. If the ethanol is not added immediately to the residue remaining after the acetic acid and the acetic acid anhydride has been distilled off, only a highly viscous brown oil is obtained and it is practically impossible to recover pure N,N'-diacetyl methylene formamide therefrom.

Example 2: A cotton fabric which had been boiled in red wine for 1 hour and dried was washed three times for 15 minutes at 60°C. in an aqueous bath which per liter contained the following:

| | Grams |
|---|---|
| Lauryl sulfate | 1.250 |
| Sodium pyrophosphate | 0.625 |
| Sodium tripolyphosphate | 0.625 |
| Dry water glass | 0.350 |
| Sodium perborate | 0.600 |
| N,N'-diacetyl methylene diformamide | 0.725 |

As a comparison the bleaching effect attained with the same washing composition under the same conditions except for the omission of the N,N'-diacetyl methylene diformamide was also investigated. The white content of the red wine stained fabric before washing was 20% when measured with a Zeiss-Elrepho-whiteness measurer against white standard 565 with filter 6. After washing in the unactivated bath the whiteness content was 30% whereas after washing in the activated bath according to the process the whiteness content was 48.6%. The increase in whiteness effected by the activation according to the process therefore was 18.6%. N,N'-dibenzyol methylene diformamide has a melting point of 137° to 139°C. and is very slightly soluble in water and gives an increase of whiteness of 10%; N,N'-dipropionyl methylene diformamide is a oily substance and easily soluble in water and gives an increase of whiteness of 20 to 21%.

Chloroformate Activators for Perborate

L.T. Murray; U.S. Patent 3,589,857; June 29, 1971; assigned to Colgate-Palmolive Company describes the use of chloroformate in sodium perborate containing household detergent to improve the bleaching effect at washing temperatures below 85°C. The chloroformates useful in this process have the following structural formula:

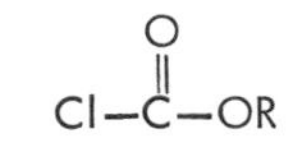


wherein R represents alkyl and preferably lower alkyl of from 1 to 4 carbon atoms, e.g., methyl, ethyl, propyl, isobutyl, etc. and aryl, e.g., phenyl. It will be understood that other substituent groups may be present as integral components of the activator molecule, the primary requirement with respect thereto being that such substituents be of an innocuous nature, i.e., devoid of any tendency to impair or otherwise deleteriously affect fabric materials or alternatively, to retard or otherwise interfere with the desired activator-bleaching agent interaction leading to the in situ generation of bleaching species. Suitable substituents in this regard include, for example, halogen, e.g., chloro, bromo, etc. Particularly beneficial results are obtained with the use of compounds of the above formula wherein R represents methyl, ethyl, and phenyl respectively, i.e., methyl chloroformate, ethylchloroformate, and phenylchloroformate.

The relative proportions of bleaching agent and chloroformate activator employed may vary over a relatively wide range depending somewhat upon the nature of the composition being formulated. In general, beneficial results are readily obtained by the use of the activator in amounts sufficient to yield a chloroformate peroxide mol ratio within the range of 0.1 to 2.0. Thus, in the case of a simple bleach composition, the involved ingredients will comprise, essentially, the chloroformate activator and peroxide. When formulating detergent compositions, the peroxide compound will usually be utilized in amounts sufficient to yield a concentration within the range of from 1 to 50%, weight basis, of total composition, with other ingredients including detergent, brightener, perfume, etc.

The limits are not critical per se but serve only to define those values found to yield optimum results for the broad spectrum of operations to which such compositions may be applied. The chloroformate activators described herein can likewise be employed to outstanding advantage in combination with one or more of the conventional activator compounds currently available commercially. Again, the choice of particular systems as well as concentrations lies largely within the discretion of the manufacturer. In any event, it is preferred that the chloroformate compound be used in major proportions in those instances wherein activator mixtures are employed.

The activator/bleaching agent system may be formulated together in a built detergent composition or alternatively as a separate bleach product. When provided in the latter form, the activator and bleach may be either intimately mixed or included in separate compartments of a water-soluble film packet. Any of the usual methods for providing the normally liquid activator compound in powder or other suitable solid form may be resorted to for such purposes, e.g., encapsulation. The following examples are given for purposes of illustration. In each of the examples, the following procedure is observed. A series of washing compositions is prepared in Terg-o-Tometer buckets by dissolving in 1,000 ml. of water 2 grams of a detergent of the following composition:

| | Percent |
|---|---|
| Linear tridecyl benzene sulfonate sodium salt | 21 |
| Sodium sulfate | 26.4 |
| Phosphates, sodium tripolyphosphate, trisodium orthophosphate, sodium pyrophosphate | 35 |
| Sodium silicate | 7 |
| Carboxymethylcellulose | 0.4 |
| Antioxidant, perfume, etc. | Balance |

To each of the Terg-o-Tometer buckets is added 2 mmol. each of sodium perborate and chloroformate activator specified so as to yield a concentration of $2 \times 10^{-3}$ M. Control samples are similarly prepared but omitting the activator. The wash samples comprise grape-stained cotton swatches. All washes are conducted at a temperature of 120°F. for 10 minutes follwed by rinsing and air drying. The average reflectance unit reading (Rd) is determined both before and after washing and the difference between such measurements, represented in the table by ΔRd, for each of the systems set forth. The results obtained are itemized in the following table:

| Example | Activator | ΔRd |
|---|---|---|
| 1 | Perborate (control) | 38.0 |
| 2 | Sodium perborate plus methylchloroformate | 49.6 |
| 3 | Sodium perborate plus ethylchloroformate | 51.3 |
| 4 | Sodium perborate plus phenylchloroformate | 47.7 |

As the above examples make manifestly clear, the significant increase in ΔRd for those cotton samples treated with the alkyl chloroformate activator-perborate systems indicates that a highly efficient and active bleaching function obtained when compared to the control sample subjected to the bleaching action of the perborate alone absent the activator additive. As will be appreciated, the reflectance for a given material is a direct function of its degree of whiteness; thus, greater difference in the ΔRd value are indicative of correspondingly higher bleaching activity. The chloroformate activators may be used in detergent compositions with anionic, cationic or nonionic surfactants and the conventional builders.

Sodium Alkyl Benzene Sulfonate in Perborate Detergent Bleach

J. Boldingh, L. Heslinga, E. Schmidl, and R. Syrovataka; U.S. Patent 3,686,127; August 22, 1972; assigned to Lever Brothers Company found that compounds of the formula

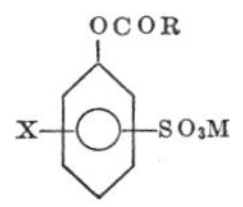

where X is $C_6$ to $C_{17}$, preferably $C_8$ to $C_{14}$, alkyl or acyl; R is H or $C_1$ to $C_7$ alkyl; M is alkali metal, ammonium or substituted ammonium undergo perhydrolysis with aqueous hydrogen peroxide to give percarboxylic acids which are good low temperature bleaching agents. As well as being bleach precursors the compounds and their perhydrolysis products have detergent properties. The compounds are therefore suited for incorporation in washing products. During the perhydrolysis reaction the compound according to the process forms percarboxylic acid and the respective phenol sulfonate at relatively low temperature, of which products the former as distinct from sodium perborate or $H_2O_2$ has efficient bleaching activity at low temperature and the latter has detergent properties, e.g.,

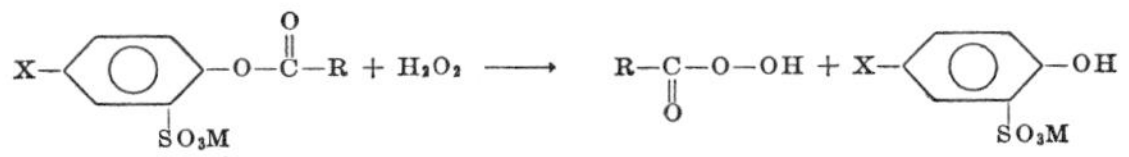

It has also been found that the solubility and hence the precursor activity of the compounds decreases with increasing chain length of the alkyl or acyl radical X. On the other hand the washing efficiency increases with increasing chain length of the alkyl or acyl group X. Suitable compounds giving satisfactory performance are those in which the alkyl or acyl group X contains 6 to 17 carbon atoms. Compounds in which $X > C_{17}H_{35}$ or $C_{16}H_{33}CO$ are almost or completely insoluble in water; compounds in which $X < C_5H_{13}$ or $C_5H_{11}CO$ have poor washing efficiency. The preferred compounds are those in which X is a branched or straight chain alkyl or acyl radical containing 8 to 14 carbon atoms.

It has further been found that these compounds in which the X radical in the benzene nucleus is in the para position to the —OCOR group are particularly effective. Evidently, in compounds in which the X radical is in the ortho position to the —OCOR group, the perhydrolysis of the acyl group is hampered by the X radical in the ortho position, probably because of the steric hindrance. This is also implied by the fact that the corresponding o-alkyl-phenol sulfonates are always more difficult to acetylate than the corresponding para isomers. Although it is known that percarboxylic acids in general are efficient bleaching agents at lower temperatures, it was found that the best results are only obtained from those compounds which on perhydrolysis release performic to percaprylic acid, i.e., from those compounds in which R is H or an alkyl radical having 1 to 7 carbon atoms. The preferred compounds are those in which R is methyl, i.e., acetoxy alkyl benzene sulfonate or acetoxy acyl benzene sulfonates, since they are easy to prepare. They may be prepared by sulfonation of an alkyl or acylphenol with either oleum, $SO_3$ or concentrated sulfuric acid and acetylating the sulfonate with acetyl chloride or acetic anhydride. The alkylphenol may be a commercially available alkylphenol, e.g., nonylphenol, octylphenol, etc. or may be prepared from phenol through acylation and Fries' rearrangement to acylphenol followed by reduction. Acetoxy acyl benzene sulfonates may be obtained by leaving out the reduction step.

These compounds when combined with an inorganic persalt, such as sodium perborate, form detergent bleaching compositions which are suitable for bleaching and cleansing at relatively low temperatures, e.g., 40° to 60°C., as well as at higher temperatures.

Example: A spray dried detergent powder base containing the following materials in parts by weight was prepared and 63 parts used as described below.

| | |
|---|---|
| Sodium dodecylbenzene sulfonate | 17.5 |
| Alkylphenol polyglycol ether | 5.8 |
| Lauric monoethanolamide | 1.7 |
| Sodium carboxymethylcellulose | 1.2 |
| Pentasodium triphosphate | 30.0 |
| Tetrasodium pyrophosphate | 11.0 |
| Alkaline sodium silicate anhydrous | 8.0 |
| Sodium sulfate | 12.8 |
| Water | 12.0 |

These materials were added to 11 parts of sodium perborate tetrahydrate and 26 parts of sodium 2-acetoxy-5-nonylbenzene sulfonate. The resulting composition was thoroughly mixed. Test swatches stained with "immedial black" were washed three times with a solution containing 5.75 g./l. of the above composition during 10 min. at a temperature of 60°C. with a heating up time of 20 min. As compared with the results obtained with the above composition without sodium 2-acetoxy-5-nonylbenzene sulfonate there was an increase of 5 in whiteness degree.

The sodium 2-acetoxy-5-nonylbenzene sulfonate was prepared as follows. A mixture of 94.7 grams (0.43 mol) commercial nonylphenol and 48.5 grams (0.475 mol) 96% sulfuric acid was stirred for 5 hours at 60° to 65°C. After neutralization with 10% sodium hydroxide solution, extraction with ether of the nonconverted nonylphenol and evaporation to dryness of the sulfonate solution, the sulfonate was isolated by extracting the evaporation residue with boiling ethanol.

A mixture of 40 grams crystallized sulfonate and 160 ml. acetic anhydride was stirred at 125°C. for 2 hours and excess acetic anhydride was then evaporated under vacuum, followed by sequential extractive evaporation with ether and carbon tetrachloride.

## PERBORATES WITH CARBONATE OR CARBOXYL ACTIVATORS

### Perborates with Organic Carbonates

E.A. Matzner; U.S. Patent 3,256,198; June 14, 1966; assigned to Monsanto Co. found that the bleaching activity of perborate containing compositions is improved by the presence of organic carbonates. In aqueous solutions, combinations of persalts, such as sodium perborate and an organic carbonate such as diphenyl carbonate are effective at temperatures as low as 50° to 70°C. The formula of the organic carbonates is the following:

$$R-O-\overset{\overset{\displaystyle O}{\|}}{C}-O-R$$

where R is selected from like or dissimilar organic radicals, at least one of such radicals being characterized in that its corresponding alcohol (ROH) has a $pK_a$, below 11.7. In the formula, R may be any of a wide variety of organic radicals provided that one or both of the organic radicals has or forms a corresponding alcohol, e.g., ROH which alcohol is characterized in having a $pK_a$ of below 11.7. Thus, in the above formula, R may represent organic radicals whose corresponding alcohols are characterized in having a $pK_a$ below 11.7 and such alcohols usually have a $pK_a$ between 11.7 and 5.0. On the other hand, R may represent dissimilar organic radicals, where the corresponding alcohol of only one radical is characterized in having a $pK_a$ between 11.7 and 5.0.

The term $pK_a$ is well-known, and the $pK_a$ of a compound is defined as the negative logarithm of the dissociation constant of the compound. Organic radicals whose corresponding alcohols are characterized in having a $pK_a$ below 11.7 include substituted or unsubstituted branched chain aliphatic radicals, and substituted or unsubstituted aromatic radicals. Organic carbonates falling within the scope of the formula in which R represents at least one of such organic radicals have been found to be particularly advantageous. Such organic carbonates have limited but effective water-solubility, e.g., from 0.01 to 5.0 grams per 100 grams of water and are somewhat more soluble in alkaline wash solutions. The carbonates are stable, and effectively promote the bleaching activity of compositions containing oxygen-releasing compounds when such compositions are dissolved in water.

In organic carbonates falling within the scope of the formula, the organic radical R may represent any of a wide variety of substituted or unsubstituted branched chain aliphatic groups or radicals. Examples of unsubstituted branched chain aliphatic groups or radicals include isopropyl, isobutyl, sec-butyl, t-butyl, isoamyl, isohexyl, isononyl, isodecyl, 2-ethylhexyl, 2-propylamyl, 2-butylamyl, etc., groups or radicals. The substituents of the substituted branched chain aliphatic groups or radicals may include, halogen atoms sulfo-, nitro-, carboxy-, methoxy-, carbethoxy-, amino-, etc. substituted branched chain aliphatic groups or radicals. The branched chain aliphatic radicals preferably contain from 3 to 10 carbon atoms in the aliphatic group. Although such radicals may contain more than 10 carbon atoms, organic carbonates containing them often have limited solubility in water.

Examples of unsubstituted aromatic radicals which may be represented by R in the formula may be phenyl, pyridyl, benzyl, alpha- and beta-naphthyl, quinoyl, anthryl, phenanthryl, benzquinolyl and the like. The substituents of substituted aromatic radicals include, for example, halo-, nitro-, sulfo- and alkyl-subsituted groups or radicals and the alkyl-substituted groups or radicals may contain from 1 to 20 carbon atoms in the alkyl group.

In addition to perborates other organic or inorganic oxygen-releasing compounds may be used. Examples, of organic oxygen-releasing compounds include organic peroxides, such as urea peroxide, benzoyl peroxide, methyl ethyl ketone peroxide and the like. Examples of inorganic oxygen-releasing compounds include inorganic peroxides, such as alkaline earth metal peroxides, for example, calcium, magnesium, zinc and barium peroxides. Other suitable inorganic peroxides include alkali metal carbonate peroxides, such as sodium carbonate peroxide and alkali metal pyrophosphate peroxides, such as sodium pyrophosphate peroxide. Particularly suitable inorganic oxygen-releasing compounds include inorganic persalts, such as metal and ammonium persulfates, perchlorates, and perborates. Conventional phosphate or silicate soap builders and surfactants may be included in the composition.

Example: Dry mixed compositions containing the following ingredients in the percentages given in Table 1 were prepared by homogeneously blending the ingredients:

TABLE 1

| Ingredient | Composition Number | | | | | | | | | |
|---|---|---|---|---|---|---|---|---|---|---|
| | 1 | 2 | 3 | 4 | 5 | 6 | 7 | 8 | 9 | 10 |
| Sodium perborate | 8.0 | 4.0 | 4.0 | 8.0 | 4.0 | 6.0 | 6.0 | 6.0 | 6.0 | 6.0 |
| Di-ortho-tolyl carbonate | 14.0 | -- | -- | -- | -- | 8.5 | -- | -- | -- | -- |
| Di-para-tolyl carbonate | -- | 7.0 | -- | -- | -- | -- | 8.5 | -- | -- | -- |
| Diphenyl carbonate | -- | -- | 6.0 | -- | -- | -- | -- | 8.0 | -- | -- |
| Bis(ortho-methoxyphenyl) carbonate | -- | -- | -- | 12.0 | -- | -- | -- | -- | 8.0 | -- |
| Diisobutyl carbonate | -- | -- | -- | -- | 8.0 | -- | -- | -- | -- | 12.0 |
| Sodium tripolyphosphate | 40.0 | 40.0 | 30.0 | 40.0 | 40.0 | 40.0 | 25.0 | 40.0 | 40.0 | 40.0 |
| Sodium sulfate | 38.0 | 34.0 | 30.0 | -- | 30.0 | 33.0 | 24.0 | 36.0 | -- | 20.0 |
| Sodium carbonate | -- | -- | 30.0 | 30.0 | 18.0 | -- | 12.0 | -- | 36.0 | 12.0 |
| Sodium silicate | -- | -- | 0 | -- | -- | -- | 12.0 | -- | -- | 10.0 |
| Sodium dodecylbenzene sulfonate | 0 | 15.0 | 0 | 10.0 | 0 | 12.5 | 12.5 | 10.0 | 10.0 | 0 |

The bleaching activity of compositions 1 through 10 was determined by dissolving 0.35 gram of each composition in 1 liter of water in separate cylindrical receptacles. The receptacles were provided with a mechanical agitator and the solutions therein were agitated and maintained at a temperature of 60°C. Solutions of compositions 2, 3 and 5 had an available oxygen concentration of 10 ppm; solutions of compositions 1 and 4 had an available oxygen content of 20 ppm and solutions 6 through 10 had an available oxygen content of 15 ppm. The solutions of the compositions contained a mol ratio of sodium perborate to organic carbonate of approximately 1:1.

Twenty 5" x 5" swatches of unbleached, naturally yellowed muslin were analyzed for reflectance (Rd) and (a) and (b) color values on a Gardner Automatic Color Difference Meter (a tristimulus colorimeter). Two swatches were then placed in each of 10 receptacles containing one of the dissolved compositions and washed for 10 min. After this period, the swatches were dried, pressed, and again analyzed on the Gardner Automatic Color Difference Meter. The reflectance $\Delta$Rd (brightening) and bleaching efficiency $\Delta$(a) and $\Delta$(b) were calculated by subtracting the readings before and after the washing operation. The loss or consumption of available oxygen was also determined for each solution. The results are summarized in Table 2 shown on the following page.

TABLE 2

| Composition Number | Loss of Available Oxygen, %* | ΔRd** | Δ(a)*** | Δ(b)*** |
|---|---|---|---|---|
| 1 | 84 | +9.0 | -0.4 | -3.1 |
| 2 | 84 | +7.6 | -0.2 | -1.9 |
| 3 | 82 | +8.4 | -0.3 | -2.8 |
| 4 | 76 | +6.4 | -0.2 | -2.2 |
| 5 | 72 | +6.1 | -0.2 | -1.6 |
| 6 | 84 | +8.4 | -0.4 | -3.0 |
| 7 | 82 | +8.0 | -0.3 | -2.2 |
| 8 | 84 | +8.2 | -0.3 | -3.0 |
| 9 | 74 | 6.8 | -0.3 | -2.4 |
| 10 | 72 | 6.2 | -0.2 | -1.9 |

*Determined by iodometric titration of spent wash solutions.
**Positive values indicate degree of increase in reflectance or brightening.
***Negative values indicate the degree of color disappearance or bleaching.

## Persalt and Organic Carbonate Bleach and Detergent

B.H. Chase; U.S. Patent 3,272,750; September 13, 1966; assigned to Lever Brothers Company describes a process to improve the performance of sodium perborate or other persalts containing bleach and detergent compositions by the addition of esters of carbonic or pyrocarbonic acids. The performance of hydrogen peroxide in bleaching solutions is also improved by the organic carbonates. The compositions are suitable for use in both commercial laundries or in the textile industry.

To evaluate the organic carbonates capability to improve the performance of the persalts the following test was developed. To a solution at 60°C. containing the following in 500 ml. distilled water: 1.25 grams sodium pyrophosphate decahydrate ($Na_4P_2O_7 \cdot 10H_2O$) and 0.154 gram sodium perborate ($NaBO_2 \cdot H_2O_2 \cdot 3H_2O$) (10.4% available oxygen) is added an amount of ester, in equimolecular ratio to the available oxygen. Water-soluble esters are added directly to the aqueous solution; other esters should be dissolved in 10 ml. ethyl alcohol before addition, the volume of distilled water being reduced in such cases to 490 ml. Acidic derivatives should be neutralized before addition.

The mixture is mechanically stirred by means of a 3/4" glass stirrer at 600 revolutions per minute and maintained at 60°C. After 1 minute and after 5 minutes 100 ml. portions are withdrawn and immediately pipetted onto a mixture of 250 grams of cracked ice and 15 ml. of glacial acetic acid. 0.4 gram of potassium iodide is then added. The liberated iodine is immediately titrated with 0.1 N sodium thiosulfate, using starch as indicator, until the first disappearance of the blue color.

Esters which may be used in bleaching compositions and processes of this method are those which give a titre of 1.0 ml. or more in this test after 1 minute or after 5 minutes, or both. According to this process there is provided a bleaching process in which is used an aqueous solution of hydrogen peroxide and an ester of carbonic acid or pyrocarbonic acid which gives a titre of not less than 1.0 ml. of 0.1 N sodium thiosulfate in the test. This process further provides a bleaching composition which contains an inorganic persalt together with an ester of carbonic acid or pyrocarbonic acid which gives a titre of not less than 1.0 ml. of 0.1 N sodium thiosulfate in the test. Esters which give a titre of not less than 1.0 ml. of 0.1 N sodium thiosulfate and hence may be used include compounds within the class $R_1O \cdot CO \cdot OR_2$ in which $R_1$ exerts an electron attracting effect and $R_2$ is an alkyl, aryl, or alicyclic radical or a substituted alkyl, aryl or alicyclic radical. The esters should not yield easily oxidizable hydrolysis products such as polyhydric phenols. Examples of esters which give a titre of at least 1.0 ml. of 0.1 N sodium thiosulfate in the test, are: sodium p-sulfophenyl ethyl carbonate, diethyl pyrocarbonate, p-carboxyphenyl methyl carbonate, and benzyl p-carboxyphenyl carbonate, etc.

Hydrogen peroxide cannot, of course, be included in a solid composition, and bleaching solutions prepared from hydrogen peroxide should be prepared as required for use. The hydrogen peroxide may be added to the solution as such, or may be liberated in situ from a persalt. By "inorganic persalt" is meant a salt which will give rise to hydrogen peroxide in aqueous solution. Suitable compounds are alkali metal perborates, percarbonates, perpyrophosphates and persilicates. These are believed not to be true persalts in the strict chemical sense but to contain hydrogen peroxide of crystallization, which is liberated in aqueous solution.

When the process is applied to bleaching or wash liquors, alkali sufficient to give an initial pH of 9 to 11 is preferably present in the bleaching or wash liquor before addition of the ester. Compositions according to the process may contain any of the conventional adjuncts present in detergent compositions. There may be mentioned, supplementary builders, inert and organic materials such as alkali metal sulfates, chlorides, carboxymethylcellulose and fluorescent agents. Compositions must not contain water in an amount sufficient to permit appreciable chemical reaction between the components prior to use.

Example: Bleaching solutions were prepared containing:

| | Percent |
|---|---|
| Sodium dodecylbenzene sulfonate | 0.054 |
| Coconut monoethanolamide | 0.018 |
| Sodium tripolyphosphate | 0.115 |
| Sodium sulfate | 0.045 |
| Anhydrous alkaline sodium silcate | 0.029 |
| Sodium carboxymethylcellulose | 0.004 |
| Sodium perborate tetrahydrate | 0.032 |

and an ester in concentration as given in the table below. The carboxylic acids were converted to their sodium salts immediately before use. A length of cotton cloth was stained by immersion in boiling tea extract for one hour. It was then thoroughly rinsed, dried and cut into pieces whose percent reflectances were measured in a Hunter reflectometer using the blue filter. The percent reflectance of the test pieces was measured again after bleaching. The bleach obtained was expressed as the difference in the two percent readings on each test piece. A piece of stained cloth was immersed in each of the freshly prepared bleaching solutions at 60°C. for 10 minutes with stirring. The cloths were then removed from the solutions, rinsed three times in distilled water, ironed and their reflectance measured.

| Ester | Concentration of Ester in Solution (%) | Increase in % Reflectance | Increase in % Reflectance in Absence of Ester (Control) |
|---|---|---|---|
| Sodium p-sulfophenyl ethyl carbonate | 0.054 | 19.1 | 12.6 |
| p-Carboxyphenyl ethyl carbonate | 0.042 | 17.9 | 12.6 |
| p-Carboxyphenyl methyl carbonate | 0.039 | 15.1 | 11.3 |
| p-Carboxyphenyl phenyl carbonate | 0.052 | 19.0 | 11.9 |
| o-Carboxyphenyl ethyl carbonate | 0.042 | 13.6 | 11.3 |
| o-Carboxyphenyl methyl carbonate | 0.039 | 13.6 | 11.3 |
| Diethyl pyrocarbonate | 0.032 | 22.5 | 15.9 |
| p-Carboxyphenyl n-propyl carbonate | 0.045 | 20.7 | 15.9 |
| p-Carboxyphenyl n-butyl carbonate | 0.048 | 16.3 | 10.6 |
| Benzyl p-carboxyphenyl carbonate | 0.054 | 16.4 | 10.6 |
| Sodium p-sulfophenyl n-propyl carbonate | 0.056 | 21.6 | 15.9 |
| Sodium p-sulfophenyl n-butyl carbonate | 0.060 | 18.4 | 10.6 |
| Sodium p-sulfophenyl benzyl carbonate | 0.066 | 17.5 | 10.6 |

The esters other than p-carboxyphenyl phenyl carbonate and diethyl pyrocarbonate can be prepared by reacting an alkyl chloroformate and a phenol.

## Stable Perborate Bleach with Vinyl Polymer

K. Lindner and E. Eichler; U.S. Patent 3,658,712; April 25, 1972; assigned to Henkel & Cie GmbH, Germany found that the stability of an aqueous sodium perborate suspension is improved by the addition of a nonoxidizable vinyl polymer containing carboxyl groups. The polymer is characterized in that a 1% aqueous solution of the polymer in the form of its sodium salt has a viscosity of at least 5,000 cp. at a pH of 7 and a temperature of 20°C.

The polymers containing carboxyl groups can be obtained by the polymerization of α, β-unsaturated monocarboxylic acids having 3 to 5 carbon atoms. Acrylic acid, methacrylic acid, α-chloroacrylic acid and α-cyanacrylic acid are preferably employed as monomers. The polymerization of the monomeric compounds takes place in the presence of small amounts of cross-linking agents, i.e., of polymerizable substances having at least two terminal olefin groups. These include many different hydrocarbons, esters, ethers or amides, such as divinyl benzene, divinyl naphthalene, polybutadiene, ethylene glycol diacrylate, methylene-bis-acrylamide, allyl acrylate, alkenyl ethers of sugars or sugar alcohols, acid anhydrides, etc. The polymerization can be carried out under conditions in which the free carboxyl groups form anhydrides. These anhydride groups are again split in the manufacture of the suspensions.

In the polymerization process, the cross-linking produces a considerable increase in the molecular weight of the polymerization products. The cross-linking, however, may progress only to the point where the polymers are completely water-soluble in the form of their alkali salts and the viscosity of the 1% aqueous solution of the sodium salts falls within the indicated range. The viscosities apply to solutions of the polymers in distilled water, i.e., without the addition of other soluble substances, as otherwise the viscosities of the solutions might be effected. The measurements are carried out in a Brookfield viscosimeter. Water-soluble carboxyl group containing vinyl polymers suitable for use in the process are disclosed in U.S. Patent 2,798,053. A commercially available embodiment of the polymer is Carbopol 934, which is described in "Chemical Products," December 1959, pages 459 to 460. The suspensions can contain, in addition to the perborate, further materials which are capable of supporting the action of the perborate in the particular oxidizing, bleaching, washing, cleaning or disinfecting process involved. Such materials may be inorganic or substantially nonoxidizable organic, dissolved, emulsified or suspended materials.

The inorganic substances which may be present partially in the undissolved state are preferably the acid, neutral or alkalinely reacting sodium salts of the acids of phosphorus, which are often used in detergents and cleaning agents, such as the salts of orthophosphoric acid, and particularly the sodium salts of condensed phosphoric acids, such as of pyrophosphoric acid or tripolyphosphoric acid. However, the sodium salts of other condensed phosphoric acids can also be present, such as, for example, the sodium salts of metaphosphoric acid or the still water-soluble sodium salts of the higher, medium or long chain phosphoric acids. Illustrative of other adjuvants which can be present are sodium sulfate, sodium metasilicate and sodium polysilicate.

The stability of the active oxygen is also dependent upon the pH of the aqueous phase, which may range from 6 to 11 and preferably from 7 to 9 in the case of concentrates intended for washing and bleaching in the household and in certain industries. The acid, neutral or alkalinely reacting condensed phosphates, sodium carbonates, sodium silicates, trisodium orthophosphate, etc., which have been mentioned above can be used for the adjustment of the pH to the desired range, although the danger of oxygen losses increases with the alkalinity.

The suspensions, i.e., dispersions, can be prepared by bringing together components in any desired order. It is advantageous first to mix together the solid components including the vinyl polymer containing carboxyl groups, and to make a paste of this mixture with water. However, an aqueous solution of a salt of the polymer can be prepared first, and the other components can be then incorporated into this solution. It is less advantageous to add the solid polymer as the last component to the solid components of the suspension after they have been already mixed with water.

It has been found advantageous to add known percompound stabilizers. These include the known stabilizers of the group of organic acids, such as tannic acid, citric acid, barbituric acid, ascorbic acid, or of the group of neutrally reacting compounds, such as acetophenetidine-acetanilide or 8-oxyquinoline. Polymeric phosphates, such as the previously mentioned tetrasodium pyrophosphate; or sodium polysilicates, have a stabilizing effect. Solid percompound stabilizers, such as magnesium silicate and components that can be used for the formation of preferably colloidal magnesium silicate can be incorporated. Lastly, certain surfactants are also characterized by a pronounced ability to stabilize the active oxygen of the sodium perborate.

A slightly cross-linked, nonoxidizable vinyl polymer containing carboxyl groups and available in the form of the free acid, having an acid number of 754, was used in the preparation of the suspensions described in the examples. In order to determine the viscosity of a 1% aqueous solution of the sodium salt of the polymer, a corresponding amount of the free acid was steeped in distilled water, thereafter somewhat less than the amount of dilute caustic soda solution required for neutralization was added and the mixture obtained was stirred occasionally until the polymer had dissolved. The polymer solution was adjusted to a pH of 7 by the addition of additional caustic soda solution. (In preparing the polymer solutions it is advantageous to check the pH after 30 minutes of standing; if it has decreased, lye is added and the pH determination repeated after 30 minutes.) The aqueous solution which is thusly obtained had a viscosity of 50,000 cp. measured at 20°C. in the Brookfield "RVT" rotation viscosimeter using a No. 6 spindle.

Distilled water was used for the production of the thickly liquid suspensions described in the examples. The other components were available in standard technical grade. The term perborate as used herein is to be taken to mean a commercial, finely crystalline sodium perborate of the approximate composition $NaBO_2 \cdot H_2O_2 \cdot 3H_2O$. The other inorganic salts named in the examples are used in the calcined state. The quantity data given for the surfactants refer to 100% active substance.

Unless otherwise expressly stated, the indicated amount of water was gradually added to the mixture of the solid components, whereupon the polymer, after first passing through a swelling state, passes into solution and the suspension is formed after a period of stirring. In other cases a portion of the components were dissolved in water and then mixed with the rest of the components which were in the form of a solid mixture. pH values in paretheses are of a 1% solution of the suspension in distilled water.

Example 1:

| Component | Percent |
|---|---|
| Perborate | 30 |
| Polymer containing carboxy groups (see above description) | 1 |
| Water | 69 |

The pH of the suspension is 8.36. Active oxygen loss after 3 weeks is 3.4%.

Example 2:

| Component | Percent |
|---|---|
| Perborate | 20 |
| $Na_4P_2O_7$ | 10 |
| Polymer as in Example 1 | 1 |
| Water | 69 |

The pH of the suspension is 8.1 (9.9). Active oxygen loss after 8 weeks is 2.9%.

Example 3:

| Component | Percent |
|---|---|
| Perborate | 20 |
| $Na_5P_3O_{10}$ | 10 |

(continued)

| Component | Percent |
|---|---|
| Polymer as in Example 1 | 1 |
| Water | 69 |

The pH of the suspension is 7.9 (9.6). Active oxygen loss after 8 weeks is 3.9%.

Solid Persalt Bleach with Carboxymethylcellulose

L.L. Maddox; U.S. Patent 3,697,217; October 10, 1972; assigned to The Clorox Company found that a solid bleaching mixture of a peroxygen compound such as sodium perborate tetrahydrate, sodium perborate monohydrate, or sodium monopersulfate, and an alkalinity booster such as sodium carbonate, sodium silicate or trisodium phosphate, effectively removes stains from household laundry when the mixture is added to the wash water in an amount which provides between 50 and 150 ppm available oxygen and a pH between 10 and 12. When between 50 and 500 ppm of a surfactant, and between 5 and 25 ppm of an antideposition agent such as carboxymethylcellulose or polyvinyl pyrrolidone are added to the wash water either together with the bleaching mixture or separately, the laundering is particularly effective. A sequestering agent for softening water, such as sodium tripolyphosphate may be included.

A granular, free-flowing solid bleaching mixture of sodium carbonate, sodium perborate tetrahydrate and sodium tripolyphosphate was prepared by mixing the following components in a commercial mixing unit for solids:

| | Weight Ratio |
|---|---|
| $Na_2CO_3$ | 53.55 |
| $NaBO_3 \cdot 4H_2O$ | 30.00 |
| $Na_5P_3O_{10}$ | 16.00 |

Small amounts of ultramarine blue coloring agent, perfume antidust agents and commercial optical brighteners totaling 0.45 part by weight were also added to the solid mixture to bring the total to 100 parts by weight. One-half cup (about 120 g.) of the solid bleaching mixture was added to a household washing machine containing 17 gallons or 64,000 ml. water. This provided 36 grams of sodium perborate to the wash water resulting in 56.5 ppm available oxygen, and a pH of 10.2 measured after addition of the detergent. The detergent mixture added to the wash water provided the following additional ingredients to the water:

| | Parts per Million |
|---|---|
| Alkylbenzene sulfonate | 320 |
| Sodium tripolyphosphate | 800 |
| Sodium metasilicate | 95 |
| Sodium sulfate | 375 |
| Carboxymethylcellulose | 10 |

Clothes stained with ink, blood and mustard were placed in the wash water. The clothes were washed for 10 minutes at a water temperature of 120°F., and dried in a tumbler drier for 20 to 25 minutes. The stain removal was excellent. Also, tests were conducted at varying available oxygen contents and the bleaching effectiveness of the bleaching mixtures hereof was measured quantitatively. Representative stains were padded onto commercially available unbleached cotton muslin test cloth. The stained cloth was cut into swatches, each swatch was numbered and Hunter D25 Color Difference meter readings were made on the test fabrics. They were then attached to carrier towels and the assembly was washed, dried, pressed wrinkle-free and reread on the color meter. The total change in the color values between unwashed and washed test swatches indicated the amount of stain removed during the test and is indicated as $\Delta E$.

The water washing-bleaching operation was carried out in a commercial automatic laundry unit having a normal cleaning cycle of 10 minutes and employing a cleaning solution volume of 16.5 gallons (120 ppm water hardness) at 120 ± 3°F. One hundred grams of a representative detergent, a linear alkylbenzene sulfonate composite:

| | Parts by Weight |
|---|---|
| Linear alkylbenzene sulfonate | 20 |
| Sodium tripolyphosphate | 50 |
| Sodium metasilicate | 6 |
| Sodium sulfate | 23.5 |
| Sodium carboxymethylcellulose | 0.5 |
| | 100.0 |

plus the carrier towels and ballast towels sufficient to provide a 5 lb. load, were charged to the washer followed by a solid granular mixture of sodium perborate tetrahydrate, sodium carbonate and sodium tripolyphosphate proportioned to provide the pH and available oxygen values set forth in the table below. Drying, ironing and reading under standardized conditions completed the test. Permanent blue-black ink is a representative stain encountered in household laundry and was, for that reason, selected for the test with the following results.

| Stain | pH | Available $O_2$ | E | Stain removal |
|---|---|---|---|---|
| Permanent blue-black ink. | 9 | 0 | 51.1 | Poor. |
| | | 20 | 56.1 | Do. |
| | | 40 | 58.0 | Do. |
| | | 60 | 58.0 | Do. |
| | | 80 | 54.6 | Do. |
| | | 100 | 54.6 | Do. |
| | 10.0 | 0 | 52.3 | Poor. |
| | | 20 | 53.7 | Do. |
| | | 40 | 56.8 | Do. |
| | | 60 | 65.5 | Fair. |
| | | 80 | 74.9 | Excellent. |
| | | 100 | 76.5 | Do. |
| | 10.5 | 0 | 50.6 | Poor. |
| | | 20 | 56.7 | Do. |
| | | 40 | 64.6 | Fair. |
| | | 60 | 67.1 | Good. |
| | | 80 | 76.8 | Excellent. |
| | | 100 | 76.7 | Do. |
| | 11.0 | 0 | 51.4 | Poor. |
| | | 20 | 60.1 | Do. |
| | | 40 | 68.3 | Fair. |
| | | 60 | 74.5 | Excellent. |
| | | 80 | 77.7 | Do. |
| | | 100 | 76.1 | Do. |

## PERBORATES WITH ENZYME ADDITIVES

### Proteolytic Enzymes and Peroxy Compounds

K.F. Blomeyer and F.J. Cracco; U.S. Patent 3,519,379; July 7, 1970; assigned to The Procter & Gamble Company describe a process for improved stain removal in household laundry by the combined use of proteolytic enzymes and sodium perborate in detergent compositions. Proteolytic enzymes have become available commercially which can be used successfully in solid detergent compositions. These enzymes can successfully attack protein stains, as for example egg and aged blood stains which are difficult to remove with oxygen bleaches and thus increase the detergent performance of compositions containing them. Since these enzymes have normally no bleaching activity, evidently the combined action of an oxygen bleach compound and the proteolytic enzyme would be desirable. However, investigations carried out seemed to indicate that proteolytic enzymes and oxygen bleaches do not work together. The enzymatic action is strongly inhibited by the presence of active oxygen produced by oxygen bleach compounds.

It has been found that this inhibiting action is only a temporary one and disappears practically completely after a certain time, after which the enzyme activity is again apparent. The advantages of the combined action is of course evident, particularly with the presence of colored protein stains. The enzyme prepares the way for the oxygen bleach. The soaking and laundering process is carried out by:

(a) Dissolving a solid detergent composition, which, besides an organic detergent and an alkaline builder salt contains 0.5 to 4.5% available oxygen in the form of a solid peroxy compound and a proteolytic enzyme in an amount which provides from 80 to 64,000 units of proteolytic enzymatic activity (per gram of the detergent composition), and which, when dissolved in water, gives a solution with a pH of 6.5 to 10.5 and by

(b) soaking and laundering soiled laundry in this solution to act for a period of at least a quarter of an hour and preferably at least half an hour at a temperature from 1° to 60°C.

The enzymatic proteolytic activity is expressed as the Löhlein Volhard unit which corresponds to the amount of enzyme required to hydrolize 1.725 mg. casein. Enzymes suitable for use in this process are those active in a pH range of from 4 to 12 and, preferably, are active in the pH range of from 7 to 11 and at a temperature in the range of from 50° to 185°F. preferably from 70° to 170°F.

Proteolytic enzymes catalyze the addition or removal of water and degrade soil, especially of a protein type. They include hydrolyzing enzymes (hydrolases), which cleave ester linkages (carboxylic ester hydrolases, phosphoric monoester hydrolases, phosphoric diester hydrolases) or cleave glycosides (glycosidases) or cleave peptide linkages ($\alpha$-aminopeptide amine acid hydrolases, $\alpha$-carboxypeptide amino acid hydrolases). They also include hydrating enzymes (hydrases). (Hydrating enzymes can also be classed as oxiodoreductases.)

The preferred hydrolases catalyze the addition of water to the substrate, i.e., the substance such as soil with which they interact, and thus, generally, cause a breakdown or degradation of such a substrate. This breakdown of the substrate is particularly valuable in the ordinary washing procedures, as the substrate and the soil adhering to the substrate is loosened and thus more easily removed. For this reason, the hydrolases are the most important and most preferred class of enzymes for use in cleaning applications. Particularly preferred hydrolases are the proteases, esterases, carbohydrases and nucleases, with the proteases having the broadest range of soil degradation capability.

The enzymes are generally utilized in a dry, powdered form. It is desirable that the enzymes be used in a dry form prior to use since degradation of the enzymes is minimized. The enzymes per se have molecular diameters of from 30 to several 1000 A. However, the particle diameters of the enzyme powder as utilized herein are normally much larger due to agglomeration of individual enzyme molecules or addition of inert vehicles such as starch, organic clays, sodium or calcium sulfate or sodium chloride, during enzyme manufacture. Enzymes are grown in solution.

Such vehicles are added after filtration of such solution to precipitate the enzyme in fine form which is then dried; calcium salts also stabilize enzymes. The combination of enzyme and inert vehicle usually comprises from 2 to 80% active enzyme. The enzyme powders of this process, including the examples, mostly are fine enough to pass through a Tyler Standard 20 mesh screen (0.85 mm.) although larger agglomerates are often found. Some particles of commercially available enzyme powders are fine enough to pass through a Tyler Standard 100 mesh screen. Generally a major amount of particles will remain on a 150 mesh screen. Thus, the powdered enzymes utilized herein usually range in size from 1 mm. to 1 micron, most generally from 0.1 to 0.01 mm. The enzyme powders of the examples have particle size distribution in these ranges.

Specific examples of commercial enzyme products include: Alcalase, Maxatase, Protease B-4000 and Protease AP, CRD-Protease, Viokase, Pronase-P, Pronase-AS and Pronase-AF, Rapidase P-2000, Takamine, Bromelain 1:10, HT proteolytic enzyme 200, Enzyme-L-W (derived from fungi rather than bacterial), Rhozyme P-11 concentrate, Pectinol, Lipase B, Rhozyme PF, Rhozyme J-25 (Rhozyme PF and J-25 have salt and corn starch vehicles and are proteases having diastase activity), Amprozyme 200.

Peroxy compounds tend to inhibit the activity of enzymes. The inhibiting action of the peroxy compounds is more pronounced in the case of those enzymes which contain sulfhydryl groups or disulfide bonds, e.g., pepsin, tripsin, papain, lipase, diastase and urease. Therefore, the preferred class of proteases for use in the detergent composition of the process of this method are those which are free from sulfhydryl groups or disulfide bonds. These are exemplified by the subtilisin family of enzymes. Specific examples of these enzymes include: Alcalase, Bakterie, Proteinase, Maxatase, and the alkaline protease portion of CRD-Protease.

In this process, the amount of active enzyme in a detergent composition is from 0.005 to 4.0% by weight of the composition. When the preferred enzyme, Alcalase, is so utilized, the final detergent composition preferably contains from 0.006 to 0.12% active enzyme by weight. These figures correspond to a range of from 0.1 to 2.0% by weight of the final detergent composition of Alcalase. The range of 80 to 64,000 Löhlein Volhard (LV) units corresponds approximately to a detergent composition containing from 0.054 to 43.5% Alcalase which in turn is 6% active enzyme.

The range of 0.5 to 4.5% available oxygen corresponds approximately to a detergent composition containing from 5 to 45% sodium perborate. (Sodium perborate per se contains 10.4% available oxygen.) Sodium perborate is the preferred compound and is used preferably in an amount ranging from 8 to 25% of the composition. Other useful peroxy compounds are sodium persulfate and percarbonate. In addition to the enzyme and the peroxy compound, the detergent compositions of this process contain 10 to 95% of the usual mixtures of organic detergent and alkaline builder salts in a ratio in the range of 2:1 to 1:10. The builder salts and organic detergent compounds are more fully described hereinafter.

Example: The following detergent composition is prepared.

| Composition | Percent |
|---|---|
| Sodium linear dodecylbenzene sulfonate | 11.0 |
| Sodium tallow/coconut (80/20) soap | 2.5 |
| Coconut fatty alcohol-ethylene oxide condensate with 6 mols of ethylene oxide | 2.0 |
| Sodium tripolyphosphate | 36.0 |
| Sodium silicate | 7.0 |
| Sodium carboxymethylcellulose | 1.0 |
| Sodium perborate | 0 to 12.5 |
| Proteolytic enzyme (content corresponding to 960 LV units is 0.6% Alcalase) | 0 to 0.6 |
| $H_2O$, $Na_2SO_4$, brightener, and perfumes | Balance |

Performance tests are carried out under the following conditions.

Prewash and Main Wash — A 4 kg. drum washing machine is used for a one lye process. The water used has a hardness of 18 grains/U.S. gallon. The washing machine is loaded with 4 kg. of the soiled household clothes to be tested which are first prewashed (rinsed) for 6 minutes with cold water without product. Then 180 g. of the detergent product is added and the main wash is started by increasing gradually the temperature from 15° to 60°C. over a 30 minute period and from 60° to 100°C. over a further 30 minute period. The testing data are:

| | (a) | (b) | (c) |
|---|---|---|---|
| Detergent used | As indicated | | |
| Perborate, percent | 0 | 12.5 | 12.5 |
| Enzyme (LV units) | 960 | 0 | 960 |
| Terry towels | | | |
| Cleaning | —58 | —19 | 0 |
| Whiteness | —69 | —24 | 0 |
| Kitchen towels | | | |
| Cleaning | —47 | —16 | 0 |
| Whiteness | —70 | —7 | 0 |
| Shirts, cleaning | | | |
| Collars | —9 | —22 | 0 |
| Cuffs | —5 | —31 | 0 |
| Whiteness | —24 | —7 | 0 |

The results represent the sum of the gradings given by a panel of 4 independent judges, who are experts in cleaning evaluations, by grading the pairs (ac) and (bc). Each pair of towels is graded on 9 replicates and each pair of shirts on 6 replicates. A standard scale of -3, -2, -1, 0, 1, 2 and 3 is used wherein 0 means that the cloths are equal, 1 means there is a slight difference, 2 means there is a moderate difference and 3 means there is a large difference. The minus (plus) value indicates that the chosen standard is better (worse) than the test swatch. The results show the surprising and uniform cleaning power of the composition containing both enzymes and available oxygen over what is achieved with compositions containing only one of these.

Persalts, Enzymes and Inhibitor

M.R.R. Gobert and G. Mouret; U.S. Patent 3,606,990; September 21, 1971; assigned to Colgate-Palmolive Company describe a process to prevent catalytic decomposition of persalts such as perborate during the washing process by enzymes contained in the stains of soiled cloth by the addition of an inhibitor. Inhibitor compounds which may be effectively employed encompass a relatively wide range of materials of diverse chemical nomenclature. Special representatives include, hydroxylamine salts like the neutral sulfate, hydrochloride, etc., hydrazine and phenylhydrazine and their salts, e.g., sulfates; substituted phenols and polyphenols, including mono- and polysubstituted, the substituents including at least one $-NH_2$, $SO_2NH_2$, $-Cl$, $-Br$, $-NO_2$, alkyl, etc., for example, aminophenols like o-amino-p-chlorophenol, aminotriazoles and derivatives thereof like 3-amino-1,2,4-triazole-3-ureido-5-triazone, etc.; alkali metal chlorates; sodium nitride; alkali metal cyanurates, etc., as well as mixtures comprising two or more of the foregoing.

The inhibitors may be contacted with the stained cloth, during a soaking or prewashing stage prior to contact with the peroxide bleaching agent or, alternatively, during the washing step simultaneous to contact with such bleaching agent. They may also be added directly to the bath as an individual component or more preferably, incorporated in a detergent composition further containing a detergent of conventional type, the latter procedure being preferred when the composition is added during the soaking step. Alternatively, the inhibitor may be included a component of a composition further containing peroxide bleaching agent, and particularly in those instances wherein the composition is contemplated for use, i.e., is to be added, during the washing step.

The compositions of the process may be defined as containing as essential ingredients on a weight basis, from 0.1 to 15% of an inhibitor compound capable of inhibiting or retarding enzyme-induced decomposition of peroxide bleaching agent; and a substance selected from the group consisting of (a) from 2 to 85% of a water-soluble peroxide bleaching agent, (b) from 5.0 to 99% of a water-soluble organic detergent and (c) mixtures of (a) and (b) with the provision that, in the case of mixtures, at least 10% and about 10% respectively of bleaching agent and detergent be present.

The peroxide bleaching agents prescribed for use in the process are preferably those of the water-soluble inorganic type including, for example, alkali metal perborates such as sodium perborate monohydrate and/or tetrahydrate; alkali metal perphosphates such as sodium and potassium perphosphate; alkali metal persilicates such as sodium and potassium persilicate; alkali metal percarbonate such as sodium and potassium percarbonate; peroxides such as sodium peroxide, hydrogen peroxide, etc., including mixtures of two or more of the foregoing. In any event, the peroxide bleaching agent is employed in amounts ranging from 2 to 90% by weight of total composition with a range of 5 to 20% being preferred.

Anionic, cationic, nonionic surfactants, conventional builder salts and chelating agents can be used in the compositions containing the inhibitors; also bluing agents, anticorrosive agents and perfumes. Soaking and washing steps in the presence of the inhibitor composition are carried out in the usual way and preferably under the usual temperature conditions (80° to 100°C. for example) preferably for the usual time, typically one-half hour.

Certain enzyme inhibitors may in some instances be found to be somewhat toxic to or exhibit suboptimum compatibility with hydrogen peroxide. It is preferable to use such inhibitors in the process by adding them at the soaking or prewashing step of the process prior to contact with the peroxide bleaching agent. Such inhibitors are more effectively provided as a component of detergent compositions which omit peroxide bleaching agent. Among the inhibitors found to be particularly effective for soaking are reducing agents like formaldehyde; strong oxidizers like alkali metal hypochlorites, alkali metal salts of chlorocyanuric acids, potassium salts of monopersulfuric acid, etc.; acid solutions which lower the pH of the bath below 4; basic solutions which raise the pH of the bath above 12; heavy metal salts, e.g., mercuric salts, like mercuric chloride; certain solvents like acetone, etc., which precipitate proteins.

Example 1: This example illustrates the rate of sodium perborate decomposition when the goods are soaked by conventional methods. Three grams of commercial sodium perborate are dissolved at ordinary temperature in 500 $cm.^3$ tap water and the increase in the volume of liberated gas measured over an interval at ordinary temperature by standard methods. There is no volume of gas 10 minutes after perborate is added to the water.

The preceding test is repeated, except that tap water is replaced by the same volume of the aqueous suspension of soil (this suspension is obtained by soaking 1 kilo of normally soiled cloth in 4 liters water, treated with 20 grams of sodium dodecylbenzene sulfonate, for 30 minutes at room temperature, and the soaking bath drained off). An intense liberation of gas is observed a few seconds after perborate is added to the soaking bath; this gaseous liberation is 140 cc after 5 minutes and 217 cc after 10 minutes. The liberated gas is identified as oxygen. Calculation shows that the volume of gas liberated after 10 minutes corresponds to the complete decomposition of 3 grams sodium perborate. The above test results in complete decomposition of the perborate in a soaking bath from heavily soiled cloth in less than 2 minutes. Example 2 illustrates the process wherein the inhibitor compound is added during the soaking step.

Example 2: Two loads of soiled cloth, as identical as possible, are prepared and soaked overnight at ordinary temperature in a 5 g./l. solution of sodium dodecylbenzene sulfonate without bleaching agent but 2 $cm.^3$/l. of a commercial 13% solution of sodium hypochlorite is added to one of the loads. A sodium dodecylbenzene sulfonate detergent composition containing 10% sodium perborate is then added and the washing of the soiled cloth continued at 80°C. The decomposition of perborate is measured after soaking as described in Example 1.

| | % Perborate Decomposed After 5 minutes* |
|---|---|
| Soaking without hypochlorite | 90 |
| Soaking with hypochlorite | 0 |

*Average of 12 tests.

Each load is then washed in a boiler tub (rug boiler) for 30 minutes at boiling, with the same 5 g./l. detergent composition, and bleaching effectiveness measured during washing, as described in French Patent 1,338,856.

| | Increase in Brilliancy of Dyed Cloths After One Washing* |
|---|---|
| Load soaked without hypochlorite | 6.3 |
| Load soaked with hypochlorite | 21 |

*Average of 12 tests.

The pieces washed during the 12 tests, on the other hand, were visually examined and numerically recorded. The statistical interpretation of the results illustrate that the washing method, which employs soaking in the presence of hypochlorite, yields significantly better results than the test washing without hypochlorite.

## Stain Removal with Enzyme, Perborate and Activator

F.W. Gray; U.S. Patent 3,637,339; January 25, 1972 describes a stain removing detergent composition containing a proteolytic enzyme, perborate, an activator for the persalt and the conventional detergent components. The enzymes used in this composition are Alcalase, Maxatase, Protease AP, Protease ATP40 and Rapidase. Metalloproteases which contain divalent ions such as calcium, magnesium and zinc bound to their protein chain are of particular interest. Among suitable activators are the following:

N-acetyl phthalimide
Triacetyl cyanurate
N-benzoyl succinimide
N-p-chlorobenzoyl-5,5-dimethyl hydantoin
N-o-chlorobenzoyl succinimide
N-methoxycarbonyl-5,5-dimethyl hydantoin
1,3-Di(N-methoxycarbonyl) hydantoin
1,3-Di(N-methoxycarbonyl)-5,5-dimethyl hydantoin
N-m-chlorobenzoyl-5,5-dimethyl hydantoin
N-m-chlorobenzoyl succinimide
N-benzenesulfonyl phthalimide
N-benzenesulfonyl succinimide
Methylchloroformate
Phenylchloroformate

Since individual activators vary in structure and molecular weight as well as performance, it is convenient to relate the quantity of activator to be employed to the desired available oxygen present in the particular percompound being used. For reactive aromatic monoacyl compounds such as m-chlorobenzoyl dimethyl hydantoin and m-chlorobenzoyl succinimide, strong bleaching is obtained when approximately equimolecular quantities of activator and peroxygen are present.

Bleaching is enhanced with increase in the concentration of activator and maintenance of a 1:1 mol ratio of activator and the peroxygen present in the percompound. By increase of the mol ratio of available oxygen to activator, milder bleaching is obtained particularly when the ratio is greater than 2:1. For reactive aliphatic polyacylated compounds such as tetraacetyl ethylenediamine, tetraacetyl hydrazine, triacetyl cyanurate, the mol ratio of available oxygen to activator is preferably 2:1, although higher (e.g., 6:1) or lower (e.g., 1:1 or less) mol ratios may be employed.

The enzyme concentration can be varied widely. Typically the enzyme is present in the range of 0.1 to 0.5% of the total detergent formulation. The optimum proportion of enzyme to be used in a detergent composition containing percompound and activator will of course depend upon the effective enzyme content of the enzyme preparation. As with percompound content, the quantity of enzyme to be used for stains susceptible to enzyme action will be dependent upon a number of factors, particularly time, temperature, and proportions of percompound and activator. For the enzyme sold as Alcalase (having an activity of 1.5 Anson units per gram) a preferred range of proportions is one which gives 1 to 40 ppm, more preferably 2 to 8 ppm, of the Alcalase in the wash water. This 2 to 8 ppm concentration corresponds to 0.003 to 0.012 Anson units per liter of wash water or, in a detergent formulation designed for use at a concentration of 1.5 gram per liter of wash water, 0.002 to 0.008 Anson units per gram of detergent formulation. The enzyme, perborate and activator may be used together, as in the water used for a prerinse of the soiled clothes, without any surface-active detergent being present. It is preferable, however, to mix these ingredients into a surface-active detergent composition, such as a heavy duty built granular detergent composition.

Example: (A) A detergent composition contains the following ingredients: 10% nonionic detergent consisting of a primary alkanol of an average of 14 to 15 carbon atoms ethoxylated with an average of 11 mols of ethylene oxide per mol of alkanol (Shell "Neodol 45-11"); 30% anhydrous pentasodium tripolyphosphate (designated TPP below); 5% trisodium nitrilotriacetate monohydrate (designated NTA below); 0.5% sodium carboxymethylcellulose (designated CMC below); 0.8% of the commercial proteolytic enzyme preparation known as Alcalase; 16% of sodium perborate ($NaBO_3 \cdot 4H_2O$); 24% of m-chlorobenzoyl dimethyl hydantoin (an activator for the perborate); the balance sodium sulfate. In this composition the mol ratio of perborate to activator is 1:1 (specifically it is 1.1:1).

The composition is used for the washing of standard cocoa stained fabric and standard coffee/tea stained fabrics at 120°F. for a period of 10 minutes, using 1 gram of the composition per liter of water in a Terg-O-Tometer. The effectiveness of the composition is determined by reflectance readings ($R_d$) on the fabrics before and after washing, using a Gardner Color Difference Meter for the measurement. The difference in reflectance before and after the washing is reported as $\Delta R_d$. Cocoa is a proteinaceous substance and cocoa stained fabric is a commonly used material for testing the effectiveness of detergents for removing protein stains.

For comparison the same washing tests are made on otherwise identical compostions, (B) containing no enzyme, perborate or activator; (C) containing enzyme, but no perborate or activator; and (D) containing the enzyme plus a commercial bleach yielding hypochlorous ion (specifically a mixture of 4/5 potassium dichloroisocyanurate (KDCC) and 1/5 trichloroisocyanuric acid (TCCA). The results are tabulated below:

| Ingredients, % | Composition (A) | (B) | (C) | (D) |
|---|---|---|---|---|
| Detergent | 10 | 10 | 10 | 10 |
| TPP | 30 | 30 | 30 | 30 |
| NTA | 5 | 5 | 5 | 5 |
| CMC | 0.5 | 0.5 | 0.5 | 0.5 |
| Alcalase | 0.8 | -- | 0.8 | 0.8 |
| Perborate | 16 | -- | -- | -- |
| Activator | 24.8 | -- | -- | -- |
| KDCC-TCCA | -- | -- | -- | 18.1 |
| $Na_2SO_4$ | q.s. | q.s. | q.s. | q.s. |

| Composition | ΔRd Cocoa Stained | Coffee/Tea Stained |
|---|---|---|
| A | +13.4 | +8.0 |
| B | +7.6 | -0.1 |
| C | +11.0 | +0.4 |
| D | +4.0 | +8.2 |

The foregoing results show that on addition of the perborate plus activator to the enzyme-containing composition, there is a marked improvement in detergency for the protein stain even in the short washing period of 10 minutes, at a moderate washing temperature. In contrast when another strong oxidizing agent (the chlorine bleach, KDCC-TCCA) is used with the enzyme the protein stain removal effectiveness drops to below even the level obtained for the enzyme-free composition. The activator-perborate-enzyme composition is also highly effective for removal of coffee/tea stain.

## MISCELLANEOUS PERBORATES

### Perborate-Polyphosphate Agglomerates

A.S. Roald; U.S. Patent 3,154,496; October 27, 1964; assigned to The Procter & Gamble Company describes a process for preparing quick dissolving agglomerates of finely divided particulate perborate for use in bleaching solutions which is accomplished in four stages.

The process comprises these stages. In the first stage a mixture is formed of finely divided particulate sodium perborate tetrahydrate and a particulate hydratable inorganic salt selected from the group consisting of disodium orthophosphate, trisodium orthophosphate, sodium carbonate, sodium pyrophosphate, the corresponding potassium salts, sodium tripolyphosphate and the mixtures thereof. The particulate mixture is formed into a falling curtain. A sodium silicate solution is sprayed onto the particles in a curtain as an agglomerating agent. This sodium silicate solution is in a finely divided form having an average by volume diameter of about 20 to 150 microns, the ratio of $SiO_2:Na_2O$ in the silicate ranging from 2.6:1 to 3.8:1 and the percentage of water in the solution being from 55 to 70% by weight.

The final step involves contacting with each other the particles which have received spray-on treatment in a tumbling bed whereby the perborate particles and hydratable inorganic salt particles are agglomerated when the hydratable inorganic salts dehydrate the silicate solution sufficiently to form solid silicate bonds between particles. The agglomerated particles of the perborate, hydratable inorganic salt, and silicate have an average by weight particle diameter of about 200 to 700 microns and having a bulk density in the range of 0.4 to 0.7 $g./cm.^3$. The particles in the curtain should fall with a vertical velocity less than that achieved in a free fall of 10 feet (preferably less than 6 feet) starting with zero initial vertical velocity and the thickness of the tumbling bed being less than 20 in. (preferably more than 2 in. and less than 12 in.) whereby undue compaction and break-up of the agglomerates are avoided.

It is essential in the process to have a falling curtain of perborate and hydratable inorganic salt particles. Spraying the silicate solution onto a tumbling bed of such fine particles will not give the desired agglomerate characteristics, i.e., quick-dissolving and nonsegregating. The agglomerates, having a larger particle size and a lower bulk density, tend to congregate at the top of a tumbling bed so that there is a greater tendency for further agglomeration of the already agglomerated particles than for agglomeration of the small particles. A falling curtain, on the other hand gives the smaller particles at least an equal chance of being contacted by the silicate solution droplets. This curtain can be prepared in a variety of ways. It is essential that the curtain be formed so as to avoid compaction and break-up of the finished agglomerates. This preferably involves forming the curtain with an initial downward component of velocity of zero and with a minimal horizontal component of velocity. It also involves having a minimum free fall of the particles.

The perborate is in the form of finely divided particulate matter having an average by weight diameter of from 50 to 200 microns. The perborate with the above average particle size range preferably has a particle size distribution in which from 0 to 20% is on 35 mesh, from 0 to 50% is on 48 mesh, from 0 to 63% is on 65 mesh, from 0 to 87% is on 100 mesh, and from 10 to 96% is on 200 mesh. All mesh figures herein refer to Tyler Standard Screen sizes. The perborate is present in the finished agglomerate in an amount from 40 to 90% by weight (preferably 55 to 70%).

The hydratable inorganic salt should have an average by weight diameter of from 100 to 300 microns. The salt with the above average particle size range preferably has a particle size distribution in which from 0 to 1% is on 20 mesh, from 0 to 35% is on 35 mesh, from 0 to 95% is on 65 mesh, from 1 to 99% is on 100 mesh, from 11 to 100% is on 200 mesh, from 15 to 100% is on 270 mesh. Since a function of the hydratable inorganic salt is dehydration of the silicate solution to form a solid bond which holds the agglomerate together, there should be from 10 to 30%, by weight of the hydratable inorganic salt on an anhydrous basis in the final agglomerates.

Example: A mixture of 7.5 lbs. of sodium perborate tetrahydrate and 2.1 lbs. of anhydrous sodium tripolyphosphate (STP) with the varied particle sizes were agglomerated in a cement mixer of 2.5 $ft.^3$ volume. The mixer, which had only one speed, turned at 28 rpm and had three baffles equally spaced. This apparatus in operation with its particulate charge gave a falling curtain of 1 ft. in height and a tumbling bed depth of 4 in. Sodium silicate solution in the amount of 3.2 lbs. with a 3.2:1 $SiO_2$:$Na_2O$ ratio and a water content of 62.5% was sprayed at a rate of 150 cc/min. onto the falling curtain using a single two fluid nozzle. Air at a pressure of 50 pounds per square inch was used as the atomizing fluid, giving an average by volume droplet diameter of 50$\mu$. An agglomerate of coarse particle size and light density was obtained.

The agglomerated product has improved segregation characteristics when mixed with light density synthetic detergent granules, compared to regular crystalline perborate, and has markedly quicker solubility than unagglomerated perborate of the same particle size, providing rapid and effective oxygen bleaching action for soiled and stained clothing and other fabrics. When stored for 4 weeks in closed containers at 120°F. the oxygen stability of the perborate was also improved by agglomeration. In the example, the particulate sodium tripolyphosphate can be replaced with particulate sodium carbonate or trisodium orthophosphate having similar characteristics with substantially equal results.

Fatty Amide Encapsulated Persalts

J. Schiefer and M. Dohr; U.S. Patent 3,441,507; April 29, 1969; assigned to Henkel & Cie GmbH, Germany describe a process to coat granules of persalts, such as perborate with fatty amides, which are water-insoluble at room temperature, but melt in an aqueous medium at temperatures above 50°C. In addition to the amides, glycerides or nitrogenous fatty acid esters may be used as coating or encapsulating materials. The consistancy of the coating materials can be measured in a known manner by determing the penetrometric value [DFG-Einheitsmethoden, Method C-4 10 (53)].

The ultimate bending tension as determined by DIN standards 51,030 is suitable for specifying the hardness or brittleness. The glycerides serving as coatings advantageously have a penetrometric value at 25°C. of no more than 4 mm., preferably of 0.3 to 3.5 mm., and especially 0.5 to 2.5 mm. The hardness or brittleness, expressed by the ultimate bending tension, is best not lower than 120 g./mm.$^2$; advantageously it ranges from 130 to 350, and especially from 140 to 250 g./mm.$^2$.

The encapsulated persalts are used in detergent bleaching compositions in combination with hypochlorites or chlorinated cyanurates. At low washing temperature the active chlorine part of the composition does the bleaching. As the temperature is increased, the residual chlorine is eliminated by the persalt to prevent damage to the fabric and the persalt continues the bleaching action at the higher temperatures. For optimum results 90 to 98% of the persalt should be encapsulated.

The amount of coating substance depends to some extent on the grain size of the per compounds. The grain size of the coated per compounds should best be about the same as the average grain size of the other bleaching and detergent components in which they are contained. This size ranges from 0.2 to 3.2 mm., and preferably from 0.3 to 2.0 mm., and there should be practically no dust-like particles of a grain size below 0.1 mm., and no large particles of a grain size above 3.5 mm. In this range of grain sizes, the amount of coating substance necessary for the achievement of a satisfactory coated perborate can range from 15 to 50 weight percent, and preferably from 25 to 40 weight percent, based on the oxygen-yielding components, e.g., perborate, the products of finer granularity generally requiring more coating substance than coarser ones.

The coating percentage of coated per compounds is determined by the following standard. About 10 grams of the coated per compound is extracted with chloroform in a Soxhlet-type extraction apparatus. After the extraction has ended and the chloroform has been evaporated, the coating substance remains as a residue. The percentage "H" of coating material is calculated from the amount of residue and the initial weight.

To determine the quantity of uncoated or incompletely coated perborate, 5.00 grams of the coated perborate is suspended in 500 ml. of water; the suspension is let stand for one hour at room temperature with frequent gentle agitation. The residue is removed by filtration, the filter is washed and the filtrate is made up to 1 liter. Fifty milliliters thereof are titrated with 0.1 N $KMnO_4$ solution. From the amount consumed (a = ml. 0.1 N $KMnO_4$ solution) it is possible to compute the percentage A of uncoated or incompletely coated perborate on the basis of the formula shown on the following page.

$$A = \frac{10,000a}{31.9(100 - H)}$$

The difference of 100 - A is considered as the degree of coating, and represents the percentage of completely coated perborate. In the following examples, unless otherwise expressly specified, such amounts of coated perborate are used that oxidation-equivalent quantities of active chlorine and active oxygen are present, i.e., the quantities of the two oxidizing agents are such that they would completely annihilate one another if simultaneously dissolved in water. In each of the detergents or washing adjuvants described in the examples, each of the following coated sodium perborates is incorporated.

(A) Coating material: Lauric acid isopropanolamide, melting point 62°C., penetrometric value 1.1 mm., and ultimate bending tension 125 g./mm.$^2$.

| | Weight Percent |
|---|---|
| Degree of coating | 90 |
| Perborate content | 69 |
| Active oxygen content | 7.1 |

(B) Coating material: One weight part of lauric acid isopropanolamide as in (A) above as the inner layer, and 1 weight part of a triglyceride with 1 mol of hardened tallow fatty acid and 2 mols of acetic acid per mol of glycerin, a melting point of 54°C., a penetrometric value of 2.9 mm., and an ultimate bending tension of 150 g./mm.$^2$, as the outside coating.

| | Weight Percent |
|---|---|
| Degree of coating | 96 |
| Perborate content | 67 |
| Active oxygen content | 6.5 |

(C) Coating material: Coconut fatty acid monoethanolamide with melting point of 74°C., a penetrometric value of 2.7 mm. and an ultimate bending tension of 250 g./mm.$^2$.

| | Weight Percent |
|---|---|
| Degree of coating | 70 |
| Perborate content | 64 |
| Active oxygen content | 6.4 |

(D) Coating material: Triethanolamide-behenic acid diester, obtained by esterifying 2 mols of behenic acid with 1 mol of triethanolamine, melting point 58°C., penetrometric value 1.8 mm. and ultimate bending tension 205 g./mm.$^2$.

| | Weight Percent |
|---|---|
| Degree of coating | 85 |
| Perborate content | 62 |
| Active oxygen content | 6.2 |

Example 1: To prepare a washing adjuvant, coated sodium perborate is mixed with a mixture made of 1 part by weight of technical calcium hypochlorite containing sodium chloride (36 weight percent active chlorine), and 4 parts of sodium polyphosphate.

Example 2: To prepare a bleaching detergent designed for use in washing machines and having the composition stated below, a powder containing all components except the potassium dichloroisocyanurate and the coated perborate are mixed with these two bleaching agents. The salts present in the detergent, unless otherwise specified, are sodium salts.

The composition consists of 5 weight percent alkylbenzenesulfonate, alkyl $C_{10}$ to $C_{15}$, average chain length $C_{12}$; 5 weight percent tallow fatty alcohol sulfate; 5 weight percent of a water-soluble addition product (molecular weight 8,000) of ethylene oxide on a polypropylene oxide with a molecular weight of 1,600; 5 weight percent soap made from a fatty acid mixture described later; 45 weight percent tripolyphosphate; 10 weight percent water glass; 7.8 weight percent potassium dichloroisocyanurate (59 weight percent active chlorine); and the remainder, coated perborate, sulfate and water. The fatty acid mixture had the following composition:

| | Weight Percent |
|---|---|
| Myristic acid | 8 |
| Palmitic and stearic acid | 47 |
| Arachic acid and behenic acid | 43 |
| Unsaturated fatty acids | 2 |

## Fatty Acid Glyceride Encapsulated Persalts

J. Schiefer and M. Dohr; U.S. Patent 3,459,665; August 5, 1969; assigned to Henkel & Cie GmbH, Germany describe the use of glycerides of fatty acids as encapsulant for persalt for the same purpose as the fatty acid amides described above in U.S. Patent 3,441,507. The glycerides used soften between 45° to 65°C. in an aqueous phase and release the encapsulated persalt.

## Granular Perborates with Additives

Y. Nakagawa and I. Maruta; U.S. Patent 3,522,184; July 28, 1970; assigned to Kao Soap Company, Limited, Japan describe a process to granulate sodium perborate containing bleaching and washing composition by adding an inorganic salt or sugar to assist granulation, by stirring the composition and heating it to 64° to 95°C. The following compounds are suitable to assist granulation: beryllium sulfate, magnesium sulfate, potassium chloride, potassium nitrate, sugar, and sodium borate (borax).

Example 1:

| | A, parts[1] | B, parts |
|---|---|---|
| Sodium perborate $NaBO_3.4H_2O$ | 30 | 30 |
| Anhydrous sodium tripolyphosphate | 30 | 30 |
| Anhydrous sodium sulfate | 35 | 40 |
| Magnesium sulfate (7-hydrate) | 5 | ---------- |

[1] The parts are by weight here and also in the following examples.

First of all, mixtures of compositions A and B in the above table were prepared. Both mixtures passed through a sieve of 60 mesh size. (This is also the case in the following examples.) When a container was filled with such mixtures, a considerable amount of the powder was seen to fly and disperse. When each mixture was left at 30°C. under a relative humidity of 80%, it agglomerated in 7 days. On the other hand, a container made of glass was filled to 1/3 its capacity with each mixture, was rotated quietly at 80°C. for 2 minutes and was then left at room temperature. The composition A became granulated and its granularity distribution was as listed in the following table.

| Granularity | Yield in % |
|---|---|
| On a sieve of 24 mesh | 38 |
| 24 to 32 mesh | 45 |
| 32 to 60 mesh | 13 |
| Below a sieve of 60 mesh | 4 |

On the other hand, the composition B in which no magnesium sulfate had been mixed did not granulate as did the composition A and it all passed through a sieve of 60 mesh, the same as the initial raw material mixture. Further, when the composition A as granulated was left at 30°C. under a relative humidity of 80% for 7 days, no agglomerating phenomenon occurred. The oxidation ability of the composition A after the granulation was substantially exactly the same as before the granulation.

Example 2: A mixture of the composition listed in the table below was prepared.

| | Parts |
|---|---|
| Sodium perborate | 35 |
| Sodium bicarbonate | 20 |
| Sodium sulfate | 32.8 |
| Sodium alkylbenzene sulfonate (powder product containing 40% sodium sulfate) | 6 |
| Whitex BO Conc (fluorescent bleaching agent) | 0.2 |
| Beryllium sulfate | 5 |

Twenty kilograms of this mixture were placed in a cylindrical rotary blender of a capacity of 100 liters and were rotated at 30 rpm at 68°C.. The mixture was taken out in 20 minutes and was left to cool at room temperature. Its granularity distribution was as follows:

| | Percent |
|---|---|
| Remaining on a sieve of 24 mesh size | 63 |
| 24 to 60 mesh size | 34 |
| Passing through a sieve of 60 mesh size | 3 |

## Perborate Bleaching and Washing Composition

G. Mouret and M. Gobert; U.S. Patent 3,525,695; August 25, 1970; assigned to Colgate-Palmolive Company found that the efficiency of sodium perborate bleaching and washing composition is improved if the cloth is given a presoak in water or in a detergent solution below 60°C. and then adding the perborate containing detergent at 80° to 100°C. This process eliminates the loss of oxygen from the perborate while the water is heated up to above 60°C. An alternate process uses fatty acid coated granules of the sodium perborate to prevent loss of oxygen by decomposition of the perborate until the melting fatty acid releases the perborate granules.

Example 1: This examples illustrates the rate of sodium perborate decomposition when the goods are soaked by conventional methods. Three grams of commercial sodium perborate are dissolved at ordinary temperature in 500 $cm.^3$ tap water and the increase in the volume of liberated gas measured over an interval at ordinary temperature by standard methods. There is no volume of gas 10 minutes after perborate is added to the water.

The preceding test is repeated, except that tap water is replaced by the same volume of an aqueous suspension of soil (this suspension is obtained by soaking one kilo of normally soiled cloth in 4 liters water treated with 20 grams ordinary detergent composition for 30 minutes at room temperature, and the soaking bath drained off). An intense liberation of gas is observed a few seconds after perborate is added to the soaking bath; this gaseous liberation is 140 cc after 5 minutes and 217 cc after 10 minutes. The liberated gas was identified as oxygen. Calculation shows that the volume of gas liberated after 10 minutes corresponds to the complete decomposition of 3 grams sodium perborate. The above test resulted in complete decomposition of the perborate in a soaking bath of heavily soiled cloth in less than 2 minutes. The following Examples 2 and 3 illustrate the practice of the process.

Example 2: Soiled cloth is machine washed with commercial laundry soap powder, especially designed for machines, containing 10% sodium perborate, according to a cycle which provides for a prewash and washing for 5 minutes at boiling. A cloth dyed with pyrogene black is mixed with the soiled goods to measure bleaching effectiveness during washing. The dyed cloth is transferred from one washing to another 5 times. The decomposition rate of perborate is also measured after prewashing by the test described in Example 1. Three different tests are conducted at prewashing temperature. The following results are obtained:

| | Amount of perborate decomposed after five minutes | Increase in brilliancy of dyed cloth after five washings |
|---|---|---|
| Pre-wash temperature, ° C.: | | |
| 40 | 80 | 6.9 |
| 60 | 50 | 24.2 |
| 100 | 0 | 35.7 |

It is apparent that by first contacting the soiled cloth with perborate at elevated temperatures vastly increased the effectiveness of the bleaching.

Example 3: Soiled cloth is washed with a detergent composition without perborate 30 minutes at the boil by a technique similar to that described in Example 2, except that a boiler tub (rag boiler) is used. Bleaching effectiveness is measured as described above. The two methods for introducing sodium perborate (10% by weight based on detergent composition) are compared.

| | Increase of brilliancy of dyed cloths after one washing (average of ten tests) |
|---|---|
| Perborate added with cold detergent solution | 6 |
| Perborate added as boiling begins | 23.7 |

It is apparent that first contacting the soiled cloth with the perborate at elevated temperature vastly increased the effectiveness of bleaching.

## Stable Perborate Bleaching Concentrates

K. Lindner and E. Eichler; U.S. Patent 3,553,140; January 5, 1971; assigned to Henkel & Cie GmbH, Germany describe a process to prepare perborate concentrates suitable in washing and bleaching textiles in the household and industry also in cosmetics, in the manufacture of building materials and for use as a catalyst in the polymerization of organic compounds.

It has been found that solid perborate containing compositions characterized by prolonged shelf life which can be used to provide aqueous concentrates containing dissolved perborate in an amount substantially in excess of the solubility of perborate in water and which can be used as bleaching agents and oxidants are provided by concentrates of the following composition: (a) 3 to 40% by weight and preferably 5 to 30% by weight of perborate (as $NaBO_2 \cdot H_2O_2 \cdot 3H_2O$), and (b) at least 0.3 and preferably 0.5 to 10 parts by weight of potassium hydroxide and/or nonoxidizable potassium salt of an inorganic polybasic acid, for each part by weight of perborate (as $NaBO_2 \cdot H_2O_2 \cdot 3H_2O$).

The total amount of potassium present in the concentrates in the form of the above named compounds is at least half as great as the total amount of sodium present in the concentrate in the form of its compounds, and amounts preferably to from 1 to 16 times, and most preferably to from 2 to 10 times this amount, a potassium-to-sodium ratio of 20:1 usually not being exceeded.

In addition to potassium hydroxide, there can be used, for example, the potassium salts of the following polybasic, nonoxidizable inorganic acids: sulfuric acid, ortho-, pyro- or polyphosphoric acid, carbonic acid or silicic acid. Where acid potassium salts of these acids exist, they can be used also. Sodium-potassium double salts can also be used as long as they do not change the ratio of potassium to sodium to an undesirable one.

Of these salts, dipotassium dihydrogen pyrophosphate, tetrapotassium pyrophosphate and pentapotassium tripolyphosphate have a particular practical importance to the extent that the concentrates manufactured with these salts are especially stable against oxygen loss. The pyrophosphates are superior to the tripolyphosphates in this respect. If the presence of tripolyphosphate is desirable for a special reason, as for example, the fact that the concentrates are going to be used as detergents or bleaches for textiles, the tripolyphosphate should be present advantageously in a quantity that does not exceed two-thirds of the total amount of pyro- and tripolyphosphate.

The preparations may contain, in addition, other substances which are capable of supporting the action of the perborate in the oxidation and bleaching process involved. These may be inorganic or organic dissolved, emulsified liquid or suspended solid substances. They include preferably the known stabilizers for per compounds. However, bactericides or fungicides, anticorrosives, optical brighteners, perfumes and dyes can also be incorporated into the concentrates. Whenever the oxidizers and bleaches contain emulsified or suspended components, it is recommended that thickeners or other emulsion or suspension stabilizers, such as the water-soluble salts of polyacrylic acid or polymethacrylic acid be additionally incorporated into the concentrates. It has been found that the water-soluble salts of those carboxyl-group-containing vinyl polymers are most advantageously used, the 1% aqueous solutions, of which while being free of other dissolved substances, especially of other electrolytes, have a viscosity of at least 5,000 and preferably from 7,000 to 100,000 cp. at a pH of 7 and a temperature of 20°C.

These carboxyl-group-containing polymers are obtained by the polymerization of α,β-unsaturated monocarboxylic acids having 3 to 5 carbon atoms. Acrylic acid, methacrylic acid, α-chloracrylic acid and α-cyanacrylic acid are instances of preferable monomers. The polymerization of these monomers is carried out in the presence of slight amounts of cross-linking agents, i.e., polymerizable substances containing at least two terminal olefin groups. These include many different hydrocarbons, esters, ethers or amides, such as for example divinylbenzene, divinylnaphthalene, polybutadiene, ethylene glycol diacrylate, methylene-bis-acrylamide, allylacrylate, alkenyl ethers of sugars or sugar alcohols, acid anhydrides, etc. The polymerization can be conducted under conditions in which the free carboxyl groups form anhydrides. These anhydride groups are split again in the manufacture of the suspensions according to the process.

In the polymerization process, the cross-linking results in a considerable increase in the molecular weight of the products. However, it may proceed only to the point that the polymers are perfectly soluble in the form of their alkali salts and the viscosity of the 1% aqueous solution of the sodium salts is within the indicated range. The viscosities are based on solutions in distilled water, i.e., without the addition of other soluble substances, inasmuch as the latter would affect the viscosities and are determined using a Brookfield Viscosimeter.

These polymers are not only excellent suspension stabilizers and thickeners, but they also stabilize the per compounds. The concentrates can also contain surfactants, particularly nonoxidizable surfactants, i.e., those that are free of olefinic double bonds or other oxidizable groups. In the examples, the amounts are given as parts by weight and percentages by weight. The caustic potash solutions and the potassium salt solutions used were aqueous solutions. Unless otherwise specified, the preparations were manufactured and stored at a temperature of 20°C.

Example 1: Five parts of commercial sodium perborate were dissolved with stirring in 80 parts of a 10% potassium sulfate solution. The clear aqueous solution thus prepared had a pH of 10. It released its active oxygen slowly and uniformly. After 24 hours, approximately 77% of the original amount of active oxygen was still present, and after 48 hours about 49%.

Example 2: Ten parts of sodium perborate were dissolved with stirring in 80 parts of an approximately 25% aqueous solution of potassium bicarbonate. The clear solution thus prepared had a pH of 8.8. It released all of its active oxygen within 24 hours, the release of the oxygen proceeded uniformly throughout this period.

Example 3: Fifteen parts of sodium perborate were dissolved with stirring in 80 parts of a 40% potassium carbonate solution. The clear solution thus obtained had a pH of 9.7. The active oxygen was released from the solution rapidly. Only 28% of the active oxygen was still present after 24 hours and only 10% after 48 hours. It was possible to produce the same type of solution using the same quantities as above set out and by dissolving a mixture of the same salts in water.

Example 4: Twenty parts of sodium perborate were dissolved with stirring in 80 parts of a 50% solution of tetrapotassium pyrophosphate. The clear solution had a pH of 10.3. In spite of this relatively high pH, the solution released its active oxygen more slowly than the solutions of Examples 1 through 3. After 24 hours, about 98% of the original amount of active oxygen was still present, after 48 hours 97% and after 72 hours 92%.

Spray Dried Compositions

H. Pistor; U.S. Patent 3,627,684; December 14, 1971; assigned to Deutsche Gold- und Silber-Scheideanstalt vormals Roessler, Germany describes a process to spray dry from a slurry at high temperature a washing composition, containing persalts without decomposition and a nearly completely homogeneous mixture of the components.

Example 1: To produce a slurry the following constituents were introduced into 2,600 parts of water at 80°C.: 200 parts of Marlon A 375 (a 75% solution in water of sodium alkylbenzene sulfonate containing 12 carbon atoms in the alkyl group), 600 parts of neutral sodium pyrophosphate, 60 parts magnesium silicate, 40 parts carboxymethylcellulose, 232 parts sodium metaborate ($NaBO_2 \cdot 4H_2O$), and 400 parts anhydrous sodium sulfate. The slurry obtained in this manner was treated in a spray tower in the course of 1 hour. Immediately before introduction of the slurry into the spray nozzle there was uniformly added 85 parts of hydrogen peroxide (70%). The temperature of the drying at the entrance to the tower was 183°C. and at the exit 93°C. The wash powder mixture obtained had a content of 2.11% active oxygen material and 1.43% boron which corresponds to a sodium perborate tetrahydrate ($NaBO_2 \cdot 4H_2O$) content of 20.3%.

Example 2: To produce a slurry there were introduced to 2,628 parts of water at 80°C. the following constituents: 100 parts of Marlon A 375, 300 parts of neutral sodium pyrophosphate, 15 parts of magnesium silicate, 36 parts of magnesium chloride hexahydrate, 20 parts of carboxymethylcellulose, and 128 parts of anhydrous sodium sulfate. The slurry thus produced was introduced into a spray tower in the course of an hour. Immediately before introduction of the slurry into the spray nozzle there was uniformly added the following mixture.

Two hundred parts of sodium tetraborate decahydrate dissolved in 180 parts of hydrogen peroxide (35%) to which previously there had been added 2 parts of sodium pyrophosphate and 0.03 part of the sodium salt of ethylenediaminetetraacetic acid. The temperature of the drying air amounted to 165°C. at the entrance and 98°C. at the exit. The wash powder obtained had a content of 1.38% active oxygen and 1.25% boron, which corresponds to 8.73% perborax ($Na_2B_4O_7 \cdot 3H_2O_2$).

Conventional spray drying temperatures can be used. The entrance temperature is usually between 100° and 600°C. and the exit temperature usually between 50° and 300°C. The slurry which is spray dried generally contains 15 to 95% solids, the balance being water. In addition to hydrogen peroxide as the active oxygen carrier there can be used sodium peroxide or potassium peroxide. There can be employed any conventional detergents, builders and other additives used in washing compositions.

## Peroxygen-Anhydride Absorbed on Expanded Perlite

J.R. Moyer and W.G. Moore; U.S. Patent 3,639,248; February 1, 1972; assigned to The Dow Chemical Company describe a process to prepare a dry bleaching composition of a peroxygen compound and an organic anhydride by first absorbing the anhydride on an inert clay and then mixing the granulated anhydride clay combination with granular peroxygen compound.

Usually the inert absorbent material and the organic anhydride are combined in amounts to provide a granulated product containing from 30 to 65% by weight of the inert absorbent material and 35 to 70% by weight of the anhydride. The exact quantity of inert absorbent required will depend upon the oil capacity of the particular inert absorbent employed. A sufficient amount of the inert material should be employed, however, so as to provide a free-flowing product at temperatures above 108°F. Usually 35 to 50% by weight of inert absorbent in the granulated product is preferred.

The technique employed for providing the absorbed anhydride-inert material in granulated form is not ordinarily critical to improving the stability of the final product. However, the particle size does govern the resultant stability and larger sizes are preferred, since they reduce the surface to mass ratio of the active ingredients in the granulated compositions and thus impart a high temperature stability to the bleaching compositions. The size of the particles should be as large as possible while still sufficiently small such that a substantially homogeneous mixture of the granulated components of the bleaching compositions may be made. Usually the particles should range in size from 5 to 100 mesh (U.S. Standard Sieve). Particles ranging in size from -8 to 20 mesh have been found to be particularly useful in providing a product which can be readily admixed with the other components of dry solid bleach compositions and are stable for sustained periods of time.

Inert absorbent materials which have found particular utility include, for example, diatomaceous earths; synthetic silica-alumina compounds; certain metal salts, such as sodium sulfate, magnesium sulfate, calcium sulfate; and certain substantially inert clays such as, for example, kaolinite-type clays, montmorillonite-type clays, vermiculite, attapulgus clay, fuller's earth, and the like. An especially useful absorbent consists of an expanded perlite material. As employed in the process the inert absorbent material should be soluble in aqueous solutions or have a particle size which is small enough to pass through woven fabrics.

As indicated, the process concerns an improvement in dry solid bleach compositions which yield organic monoperacids or salts when dissolved in an aqueous media. These dry bleach compositions usually consist of a dry stable particulate mixture of at least one solid inorganic peroxygen compound and at least one solid organic acid anhydride. Examples of inorganic peroxygen compounds include, the inorganic perborates, peroxides, percarbonates, perphosphates and the like. Sodium perborate monohydrate has been widely used because of its low cost and availability. Potassium perborate and ammonium perborate are especially desirable because of their stability for sustained periods of time. Also included in the commonly used dry bleach compositions is a solid organic acid anhydride. Examples of suitable anhydrides include succinic anhydride, maleic anhydride, phthalic anhydride, glutaric anhydride and benzoic anhydride.

Example: Molten benzoic anhydride was absorbed onto several different particulate inert absorbent materials including a commercially obtained expanded perlite material. The absorbed materials were allowed to cool. The cooled materials were then compacted and ground to provide a granulated product. The various granulated materials were screened to separate particles ranging in size from -10 to 20 mesh. These particles were dry blended with about half their weight of sodium perborate monohydrate. The different samples were maintained in open containers at a temperature of 50°C. They were analyzed by infrared techniques after 4 and 10 days to determine the percent of unreacted benzoic anhydride remaining in each sample. The percent by weight of absorbent employed and results are tabulated in the following table. The natural diatomaceous earths used as absorbents are obtained from various commercial sources.

| Absorbent Material | Weight percent of Absorbent | Unreacted Benzoic Anhydride, % | |
|---|---|---|---|
| | | 4 days | 10 days |
| Natural diatomaceous earth | | | |
| (1) | 47 | 78 | 64 |
| (2) | 41 | 81 | 68 |
| (3) | 38 | 82 | 71 |
| (4) | 38 | 82 | 77 |
| Calcined diatomaceous earth | 44 | 81 | 76 |
| Flux calcined diatomaceous earth (1) | 41 | 81 | 71 |
| Flux calcined diatomaceous earth (2) | 44 | 85 | 81 |
| Expanded perlite | 38 | 97 | 94 |

All the samples maintained a granulated form over the test period whereas a bleaching composition in which an inert absorbent was not employed fused into a solid mass. In a similar manner, an organic anhydride can be dissolved in an inert volatile solvent, such as methylene chloride, and slurried with from 100 to 200 parts by weight of an inert absorbent material, such as, for example, expanded perlite. Upon evaporation of the solvent a free-flowing powder containing the anhydride absorbed on the inert material is formed. The powder can be granulated to prepare particles of a desired size and these be employed in the production of dry bleach compositions. Also, a molten anhydride may be sprayed onto an agitated bed of an inert absorbent material. The resulting powder can be compacted and granulated to the desired particle size range.

## Perborate Suspensions in Trichloroethane with Benzoic Anhydride

R.G. La Barge and D. K. Bradley; U.S. Patent 3,660,295; May 2, 1972; assigned to The Dow Chemical Company describe a process to prepare viscous liquid bleaching compositions by dispersing an inorganic peroxygen compound such as sodium perborate in a chlorinated carrier solvent such as 1,1,1-trichloroethane. The carrier liquid contains a dissolved carboxylic acid anhydride. Suitable inorganic peroxygen compounds are, for example, the inorganic perborates and peroxides; hydrated sodium perborate (e.g., the monohydrate) is frequently employed. Potassium perborate and ammonium perborate are also desirable because of their stability for extended periods of time. Other peroxygen compounds which may be employed are zinc peroxide, magnesium peroxide and calcium peroxide. The peroxygen compounds employed may be either anhydrous (e.g., $KBO_3 \cdot 1/2H_2O$) or in the hydrated form (e.g., $NaBO_2 \cdot H_2O_2 \cdot 3H_2O$) as long as they are sufficiently free of uncombined water so as to be unreactive toward the organic acid anhydride. The solvent for the organic anhydride is an organic hydrocarbon containing from 1 to 12 carbon atoms which is essentially nonreactive with the anhydride or with the inorganic peroxygen compound. Suitably nonreactive compounds are also nonacidic in the sense that they do not contain reactive nucleophilic sites such as are present in amines, acids, alcohols, or mercaptans. In addition, the solvents have melting points below 5°C. and boil at temperatures in excess of 60°C.

Suitable solvents having the characteristics set forth above include polyhalogenated hydrocarbons having from 1 to 7 carbon atoms; aliphatic hydrocarbons having from 6 to 10 carbon atoms; aromatic hydrocarbons and alkyl-substituted aromatic hydrocarbons containing from 6 to 12 carbon atoms; nitro-substituted aliphatic hydrocarbons having 1 to 5 carbon atoms; and nitro-substituted aromatic hydrocarbons having from 6 to 10 carbon atoms. Suitable solvents include, for example, nitroethane, nitropropanes, benzene, toluene, xylenes, heptanes, octanes, and nitrobenzene. Preferred polyhalogenated solvents are the polychlorinated hydrocarbons having from 1 to 6 carbon atoms. Specific examples of these compounds include 1,1,1-trichloroethane, carbon tetrachloride, chloroform, o-dichlorobenzene, dichloroethyl ether, 1,2-dichloroethylene, methyl chloroform, perchloroethylene, propylene dichloride and chlorobenzene.

Preferably, the acid anhydride is benzoic anhydride, the inorganic peroxygen compound is potassium perborate, and the solvent is 1,1,1-trichloroethane. The bleaching compositions of the process are prepared by dissolving the anhydride in the solvent and admixing the particulate peroxygen compound into the solution. The bleaching formulations are used dissolved in water. Detergents, water softeners, etc., can be added to the aqueous solution of bleach.

When laundering clothes, good results will be obtained if the temperature of the aqueous bleach bath is from 100° to 160°F., with temperatures of from 120° to 140°F., being quite feasible. The latter temperature range can easily be provided by commonly-employed hot water heaters. The pH of the aqueous bleach bath should be regulated at from 7 to 10, with optimum results being obtained at pH levels of from 8.0 to 8.5. Regulation of pH is easily accomplished by incorporating a buffering composition into the bleach formulation. Suitable buffer compositions include, for example, trisodium phosphate, ammonium sulfite, sodium sulfate anhydride and sodium tripolyphosphate.

Bleaching baths prepared with these formulations are generally characterized as containing from 5 to 150 ppm of active oxygen. For purposes of the process, active oxygen is determined by admixing 10 ml. of the bleach solution (heated sufficiently to conform with the normal use procedures described above) to be measured, with 50 ml. of deionized water and 25 ml. of 20% sulfuric acid. About one gram of potassium iodide crystals are added to the mixture along with several drops of a 5% (by weight) solution of ammonium molybdate, as catalyst. The resulting mixture is titrated with 0.01 normal sodium thiosulfate solution. The active oxygen content as weight percent of the solution is computed by the formula shown on the following page.

$$\text{Active oxygen} = \frac{\text{N x ml. x } O_2 \text{ factor}}{\text{Amount of sample}}$$

In the formula, N is the normality of the sodium thiosulfite; ml. is the volume of sodium thiosulfite employed; and $O_2$ factor is a constant equal to 0.08. The active oxygen content in weight percent can be converted to ppm is desired. With liquid bleaching compositions such as those described above, it is possible to blend various particulate thickening agents into the composition without causing detrimental reactions between the components thereof. Such thickening agents are generally employed where it is desired to reduce settling out of components of the bleaching composition. An example is a pourable thickened bleaching composition comprising by weight from 0.1 to 5.0% of a thickening agent. Preferably, from 2.0 to 3.0% by weight of the composition is a thickening agent. The other components of the thickened composition and the amounts thereof are the same as for the nonthickened systems described herein above.

Thickening agents employed are particulate substances having a generally uniform particle size of from 0.01μ to 10μ, and preferably from 0.015μ to 2μ. The viscosity of the thickened bleach is in excess of 500,000 cp. and is usually from 1,000,000 to 1,500,000 cp. Suitable particulate thickening agents are, for example, fumed silicas, expanded clays (e.g., attapulgites), polyacrylates, and substituted mineral clays.

The bleaching compositions described above are generally prepared as a suspension of solid particulate peroxygen compound dispersed in a liquid carrier. The resulting composition is packaged in a single container. It may be desirable, however, to prepare and separately package the components of the bleach with mixing occurring only when the bleach bath is formed. An example is a multichambered package of bleaching composition comprising a first chamber containing a solution of organic carboxylic acid anhydride. A second chamber of the package contains the inorganic peroxygen compound, either in solid form or as an aqueous solution. In use, the separately-housed components are poured from the package and mixed in the presence of sufficient water to form a bleaching bath of the desired strength.

Where the inorganic peroxygen compound is present in the container as a solid compound, a reaction will take place in the washing medium to produce hydrogen peroxide. A sufficient amount of inorganic peroxygen compound is employed so that the concentration of $H_2O_2$ is from 0.1 to 12% by weight. If desired, the solution of inorganic peroxygen compound can also be replaced with an aqueous solution of $H_2O_2$ (0.1 to 8% by weight).

Example: A liquid bleaching composition is prepared by dissolving 31 grams of benzoic anhydride in 12 grams of 1,1,1-trichloroethane as solvent. 13.2 grams of sodium perborate monohydrate are blended into the anhydride solution to form a liquid bleaching composition of the process.

The composition is stored in a sealed polyethylene container for 4 weeks at a temperature of 70°F. Subsequently, 2 grams of the composition are dissolved in 1,000 grams of water. The temperature of the water is 160°F. The resulting aqueous bleaching bath is used to bleach swatches of muslin cloth measuring 4 inches on a side. As a comparison, similarly sized swatches of the cloth are bleached with an aqueous solution of sodium hypochlorite (5.25% NaOCl by weight) corresponding closely to chlorine bleaches commercially available. The test procedure involves agitating the swatches in the aqueous oxygen type bleaching bath and the chlorine bleach bath. The swatches are allowed to remain in the bleach baths for 10 minutes during which time the temperature of the baths are maintained at 160°F. At the end of 10 minutes time, the swatches are removed from the bleach bath and are rinsed with fresh water for 5 minutes. The temperature of the rinse water is 160°F. Upon completion of the rinse cycle, the swatches are ironed dry and whiteness readings are taken.

The whiteness of the cloth is determined by measuring blue and luminous (green) reflectance on a D40 type reflectometer. The whiteness reading for the muslin swatches subjected to the oxygen bleaching composition of the process is from 62 to 74 units and compares favorably with the chlorine bleached swatches which exhibit a whiteness of 65 to 75 units.

## PERACIDS

### Heat Treatment plus Peracetic Bleach for Cellulose Acetate

W.A.P. Schoeneberg and F. Fortess; U.S. Patent 3,077,371; February 12, 1963; assigned to Celanese Corporation of America describe a process to heat treat knitted cellulose acetate fabrics, during the bleaching process, to insure dimensional stability. The bleaching agents may be aqueous solutions of hydrogen peroxide, peractic acid, oxalic acid, sodium hypochlorite or sodium chlorite. In accordance with this process, a knitted fabric having a basis of fibers of a cellulose ester of low hydroxyl content is treated by a procedure which involves bleaching of the fabric in aqueous medium and, before completion of the bleaching, i.e., prior to or during the bleaching, subjecting the fabric to an aqueous liquid under superatmospheric pressure and at a temperature of at least 250°F.

The cellulose esters of low hydroxyl content employed in the process contain not more than 0.29, preferably 0.0 to 0.12, alcoholic hydroxyl groups per anhydroglucose unit in the cellulose molecules thereof. Best results are obtained by the use of cellulose acetate of low hydroxyl content and of correspondingly high acetyl value, e.g., an acetyl value of at least 59%, preferably 61 to 62.5%, calculated as combined acetic acid. However, other lower aliphatic acid esters of cellulose of low hydroxyl content may be employed. Examples of such esters are cellulose propionate, cellulose butyrate, cellulose acetate-propionate, cellulose acetate-butyrate and cellulose acetate-formate.

The knitted fabric is scoured before the bleaching treatment, in order to remove any foreign materials such as sizes and yarn lubricants. For best results the scouring is carried out in an aqueous liquid at a relatively high temperature of at least 250°F. and at a superatmospheric pressure sufficiently high to maintain the scouring medium in the liquid state at this temperature. Thus, for example, temperatures of 250° to 280°F. and pressures of 12 to 30 psig may be used. This high temperature treatment under pressure has the added effect of stabilizing the knitted fabric so that it will not tend to shrink or change its shape substantially on subsequent washing. Preferably during the high temperature scouring the fabric is maintained at substantially constant dimensions by suitable mechanical means. To effect a high degree of stabilization the high temperature scouring should be continued for at least 60 minutes when the scouring temperature is 250°F. With higher scouring temperatures less time is needed to attain the same stabilization. The aqueous liquid used in the scouring treatment may contain any of the usual scouring agents, such as soaps or synthetic detergents. A solution containing soap and a sequestering agent, such as sodium hexametaphosphate, yields excellent results. Other useful scouring agents are the sodium salt of oleyl taurate and combinations of sulfonated oils and nonionic detergents. The scouring solution generally contains at least 95% of water. When the fabric has previously been given a scouring treatment at a sufficiently high temperature and for a sufficient length of time to stabilize the fabric dimensions and configuration, as described above, the bleaching treatment may be conducted at a lower temperature, i.e., a temperature below that used for scouring. Thus, in this case bleaching temperatures of, for example, 230° to 240°F. are suitable.

Where there has been no preliminary scouring treatment or when the scouring treatment has not been such as to stabilize the fabric dimensions and configuration, the bleaching solution should be at a high temperature of at least 250°F. and under a superatmospheric pressure sufficiently high to maintain the bleaching solution in its liquid state, e.g., a temperature of 250° to 280°F. and a pressure of 12 to 30 psig. This high temperature bleaching results in stabilization of the size and shape of the fabric, and it is therefore desirable, during such bleaching, to maintain the fabric at constant dimensions by suitable mechanical means. The high temperature bleaching should be continued for a sufficient period of time to impart the desired dimensional stability to the fabric. Thus, about 30 to 60 minutes of bleaching at a temperature of 250° to 260°F. has been found to give very good results.

It will be understood, of course, that the fabric may be partially stabilized by a relatively short high temperature scour and that the stabilization may be completed during a high temperature bleaching treatment, which in this case need not be of such long duration as if there had been no previous partial stabilization of the fabric.

During the high temperature treatment a crystalline structure is developed in the cellulose ester of low hydroxyl content. Thus, in the case of cellulose acetate of low hydroxyl content the "crystalline order index" (as determined by study of the x-ray diffraction pattern, as described below) is increased to 1.4 or higher. In addition, the safe ironing point of the cellulose acetate of low hydroxyl content is raised to 250°C. (about 480°F.). The treatment improves the dimensional stability of the knitted fabric to such an extent that even after five washes in a conventional household washing machine at 140°F. the fabric shrinks a total of less than 7% in area. It also improves the resistance of the fabric to wrinkling on washing.

The crystalline order index referred to above is obtained by calculation from curves based on x-ray diffractometer studies of samples of the treated fabric. These curves are plots of intensity of the diffracted beam against the angle of the beam. Thus, the curve for cellulose triacetate has four pronounced peaks at angles of 8, 10, 12.6 and 16°, the angle being that which the diffracted x-ray beam makes with respect to the incoming x-ray beam. The degree of heat treatment will affect the sharpness of these peaks. The crystalline order index, which is an indication of the sharpness of the peaks, is the average of the ratios of each peak height to its width at half height for the four peaks mentioned above. For producing the diffractometer curves used for measuring the crystalline order index there is employed a North American Philips Geiger counter diffractometer operated under the following conditions.

Radiation: Ni filtered Cu Kα radiation 35 kv., 14 ma., using both voltage and ma. stabilization. Diffractometer constants: 3° take-off, 1° divergence slit; 0.003" receiving slit (0.025°), 1° scatter slit; 2° 2θ/min. scanning speed. Recorder constants: ratemeter, scaler, 8; multiplier, 0.6 (full scale 240 counts per second). Time constant: 16. Chart speed: 0.5"/min.

As stated, it is desirable to maintain the fabric at constant dimensions during treatment. In one suitable arrangement the fabric is wound, in flat unfolded condition, on a perforated tubular metal beam, and the resulting roll is placed into an appropriate pressure-tight apparatus where treating fluid under pressure is circulated through the roll of fabric passing through the perforations in the beam. In order to prevent widthwise shrinkage of the fabric during treatment, the selvedges of the fabric are held in place by taping them or by introducing a suitable binding cord onto the selvedges as the fabric is wound onto the beam. To further restrain movement of the fabric the entire roll is wrapped with cotton fabric and, finally, with a girdle of steel mesh before the roll is placed in the treating apparatus. In order to avoid stressing the fabric unduly after the treatment and thus introducing strains which may have an effect on the dimensional stability of the knitted fabric, the treated material should be maintained in a relaxed condition during subsequent drying and finishing operations.

Example 1: A tricot fabric warp-knitted of yarns composed of continuous filaments of cellulose acetate of an acetyl value of 61.7% calculated as combined acetic acid, and having 76 courses per inch and 28 wales per inch, is scoured for 30 min. at a temperature of 250°F. and under a pressure of 15 psig with an aqueous solution containing 1 g./l. of "Calgon" (sodium hexametaphosphate) and 1 g./l. of soap (sodium oleate). Thereafter, the scoured fabric is rinsed twice with water and then bleached for one hour with an aqueous solution having a pH of 6.5 and containing 2.3 g./l. of peracetic acid, 0.85 g./l. of caustic soda and 0.5 g./l. of Calgon at a temperature of 260°F. and a pressure of 20 psig. During the entire process the fabric is held at constant dimensions. The resulting fabric has excellent dimensional stability and a high degree of whiteness, and shows very little tendency to yellow when exposed to sunlight or nitrogen oxides. It has little tendency to become wrinkled when washed.

Example 2: A fabric as described in Example 1 is scoured for one hour at a tempera-of 270°F. and under a pressure of 27 psig with an aqueous solution containing 2% of "Calsolene oil HS" (a highly sulfonated oil), 1% of "Triton X-100" (an alkyl aryl polyoxyethylene ether alcohol) and 0.5% of soda ash, and having a pH of 8.5. The fabric is then rinsed and thereafter bleached at a temperature of 230°F. for 1 hour at a pressure of 6 psig with an aqueous solution containing 4.7 g./gal. of peracetic acid, 1.6 g./gal. of soda ash, 1.4 g./gal. of Calgon and 0.2 g./gal. of "Igepon T-77" (sodium salt of oleyl taurate), and having a pH of 6.5. Thereafter the fabric is rinsed thoroughly. During the entire process the fabric is held at constant dimensions. The resulting fabric has excellent dimensional stability, a high degree of whiteness and shows very little tendency to yellow on exposure to sunlight or nitrogen oxides. It has little tendency to become wrinkled on washing.

Peroxybenzoic Acid Bleach and Sterilizer

J.H. Blumberg; U.S. Patent 3,248,336; April 26, 1966; assigned to FMC Corp. describes a process to prepare stable aqueous bleaching solutions of peroxybenzoic acid by the use of tertiary butyl alcohol as a mutual solvent. The tert-butyl alcohol can be used to make up the aqueous-based solutions of the peroxybenzoic acids up to 12% by weight when the alcohol is present in a 1:1 weight ratio with water. Lower amounts of the tert-butyl alcohol can be used to make up more dilute peroxy acid solutions. A dilute solution of 2% by weight peroxy acid requires 10% by weight of the tertiary alcohols to impart stability and homogeneity to the aqueous solution.

The tertiary alcohols used in making up the aqueous bleach solution should be free of impurities that can cause rapid decomposition of the peroxybenzoic acids or which can be oxidized by these acids, thus wastefully consuming them. The most common offending impurities are heavy metals. Heavy metals initiate catalytic decomposition of the peroxybenzoic acids, and therefore, are highly undesirable, even in small amounts.

Two methods can be employed to eliminate decomposition by heavy metals. The first of these is to remove the heavy metals by distilling the tertiary alcohols in glass or glass-lined equipment. Another is to render the heavy metals inactive by adding a small amount, on the order of 100 ppm, of a metal chelating agent. A compound such as dipicolinic acid is an ideal chelating agent and effectively ties up the heavy metal in a complex organic structure, thereby making it unavailable for initiating decomposition of the peroxybenzoic acids.

The peroxy acid which has been found most suitable in the present bleaching solution is peroxybenzoic acid because of its high solubility in the tert-butyl alcohol and because it has proportionately more active oxygen per molecular weight than do the substituted peroxybenzoic acids. Aqueous solutions having up to 12% by weight of peroxybenzoic acid can be made up readily when tert-butyl alcohol is employed as the tertiary alcohol. In making up such solutions, they should contain 44% by weight water, 12% by weight peroxybenzoic acid, and from 44% by weight tert-butyl alcohol. A preferred bleaching mixture contains from 52% by weight water, 9% peroxybenzoic acid, and 39% by weight of tert-butyl alcohol.

In addition to peroxybenzoic acid, substituted peroxybenzoic acids can be employed which do not contain groups that can be oxidized by the peroxybenzoic acids. These substituted peroxybenzoic acids further must be sufficiently soluble in the tertiary alcohol-water mixture to reach the peroxy acid content desired. For example, nitro groups are nonoxidizable, but they are unsuitable in this process because the nitro group so lowers the solubility of the resultant peroxybenzoic acid that it cannot be used in making up the bleach solutions.

Substituted peroxybenzoic acids which contain chloro, lower saturated aliphatic, methoxy, and other such groups have been found suitable. Compounds such as ortho-chloroperoxybenzoic acid, para-chloroperoxybenzoic acid, meta-chloroperoxybenzoic acid, para-tertiary-butyl peroxybenzoic acid, and para-methoxy peroxybenzoic acid, have been found suitable. Examples of substituents on the peroxybenzoic acid which are oxidizable, and therefore unsuitable, include primary and secondary alcoholic groups, hydroxyl, ketonic and aldehydic groups.

Example: Peroxybenzoic acid was dissolved in tert-butyl alcohol and this solution was mixed with various proportions of distilled water. To these solutions were added small amounts of dipicolinic acid to act as sequestering agents for the removal of heavy metal ion impurities. The samples were stored at room temperature (25°C.) for up to 6 months and were analyzed periodically for their peroxy acid and hydrogen peroxide content by the method described in "Analytical Chemistry," 20 (1948), p. 1061. The concentration of the ingredients in the solution and the results of the analysis are listed in Table 1 below:

TABLE 1

| Sample No. | Sample Composition | Peracid and Hydrogen Peroxide Content | | | |
|---|---|---|---|---|---|
| | | At Start | After 1 Month Storage | After 3 Months Storage | After 6 Months Storage |
| 1 | Peroxybenzoic Acid, percent | 11.40 | 11.27 | 11.16 | 11.04 |
| | Hydrogen Peroxide, percent | 0.00 | Trace | Trace | Trace |
| | Tert-butanol, percent | 44.3 | | | |
| | Distilled Water, percent | 44.3 | | | |
| | Dipicolinic Acid, p.p.m. | 100 | | | |
| 2 | Peroxybenzoic Acid, percent | 7.22 | 7.10 | 7.05 | 6.92 |
| | Hydrogen Peroxide, percent | 0.00 | 0.03 | 0.03 | 0.02 |
| | Tert-butanol, percent | 32.7 | | | |
| | Distilled Water, percent | 60.0 | | | |
| | Dipicolinic Acid, p.p.m. | 100 | | | |
| 3 | Peroxybenzoic Acid, percent | 5.00 | 4.92 | 4.90 | 4.77 |
| | Hydrogen Peroxide, percent | 0.00 | Trace | 0.02 | 0.03 |
| | Tert-butanol, percent | 30.3 | | | |
| | Distilled Water, percent | 64.7 | | | |
| | Dipicolinic Acid, p.p.m. | 100 | | | |
| 4 | Peroxybenzoic Acid, percent | 1.40 | 1.39 | 1.31 | 1.22 |
| | Hydrogen Peroxide, percent | 0.00 | Trace | Trace | 0.02 |
| | Tert-butanol, percent | 9.5 | | | |
| | Distilled Water, percent | 89.1 | | | |
| | Dipicolinic Acid, p.p.m. | 100 | | | |

In this example, the amount of hydrogen peroxide in the sample indicates the degree of hydrolysis of the peroxybenzoic acid in accordance with the formula set forth below:

$$R{-}\overset{\displaystyle O}{\overset{\|}{C}}{-}OOH + H_2O \rightleftharpoons R{-}\overset{\displaystyle O}{\overset{\|}{C}}{-}OH + H_2O_2$$

The greater the amount of hydrogen peroxide in the solutions, the greater the hydrolysis and the less effective is the bleaching and germicidal activity of the peroxybenzoic acid solution. Sample 1 after storage for 6 months at room temperature was tested for germicidal activity against Staphylococcus aureus by the standard procedure for phenol coefficient determination as described in "Official Methods of Analyses of the AOAC," AOAC (1960), pp. 63 to 65. The phenol coefficient was found to be 400, calculated on the basis of the active ingredient. An acceptable, active germicide should have a phenol coefficient of 200.

The same sample was also tested for bleaching effectiveness. The procedure used follows. Thirty-two cotton swatches (5" x 5" desized cotton Indianhead fabric, uniform in weave and thread count) were stained with tea, coffee and wine in the following way. Five tea bags were placed in a liter of water and boiled for 5 min. The swatches were immersed in the tea and the boiling continued for 5 minutes. Thirty-two additional swatches of the same cloth were coffee stained by boiling 50 grams of coffee in a liter of water, immersing the swatches in the coffee solution and boiling for an additional 5 minutes. The wine stains were created by soaking swatches of the same cloth in a red wine at room temperature. The stained swatches were then squeezed to remove excess fluid, dried, rinsed in cold water and dried. Three of the stained cotton swatches were then added to each of a series of stainless steel Terg-O-Tometer vessels containing 1,000 ml. of a 0.2% standard detergent solution at a temperature of 120°F. Measured amounts of the bleach solution were then added to each vessel sufficient to correspond to an active oxygen content of 20, 40, 60 and 80 ppm respectively. The pH of the solutions were adjusted to 9.5, using soda ash. Cut-up pieces of white terry cloth toweling were then added to provide a typical household wash water/cloth ratio of 20:1. The Terg-O-Tometer was then operated at 72 cycles per minute for 15 minutes at a temperature of 120°F. At the end of the wash cycle, the swatches were removed, rinsed under cold tap water and dried in a Procter-Schwartz skein dryer. The tests were run in triplicate and included detergent blanks. Reflectance readings of the swatches were then taken before and after the wash cycle with a Reflectometer and the readings were averaged. The percent stain removal was obtained in accordance with the following formula:

$$\text{Total percent stain removal} = \frac{\text{Reflectance after bleaching} - \text{reflectance before bleaching}}{\text{Reflectance before staining} - \text{reflectance stained}} \times 100$$

The results are reported in Table 2 below:

TABLE 2

| Sample Tested (Sample 1 from Table I) | Contents, percent | Active Oxygen Content, p.p.m. | Percent Stain Removal | | |
|---|---|---|---|---|---|
| | | | Tea | Coffee | Wine |
| Peroxybenzoic Acid | 11.04 | 20 | 74.2 | 92.7 | 92.6 |
| Hydrogen peroxide | Trace | 40 | 79.3 | 97.0 | 96.2 |
| Tert-butanol | 44.3 | 60 | 80.0 | 96.4 | 96.8 |
| Water | 44.3 | 80 | 82.2 | 95.7 | 94.5 |
| Dipicolinic Acid | [1] 100 | | | | |
| Detergent Blank | | | 26.9 | 45.0 | 61.0 |

[1] Parts per million.

## Polytetrafluoroethylene Fiber Bleach Using Perchloric Acid

N.C. Jeckel; U.S. Patent 3,318,656; May 9, 1967; assigned to United States Catheter & Instrument Corporation describes a process to bleach polytetrafluoroethylene fibers with perchloric acid, nitric acid and sulfuric acid mixtures. Medical prosthetic devices made with polytetrafluoroethylene fibers can also be bleached by the same process. Teflon in its solid form is white but acquires a small percentage of impurities when made into yarn or fiber due to high temperature involved which makes it dark brown. These brown fibers have less tissue reactivity than Dacron or nylon but somewhat more than purified white Teflon. Therefore, from the aspect of reactivity and also for appearance, it is preferable to have purified (bleached) white Teflon as the material from which artificial blood vessels, sutures, catheters, or other surgical devices are made. The various advantages of Teflon and the white purified Teflon for these purposes have been set forth by Harrison in American Journal of Surgery 3 (1958).

As pointed out, purified Teflon grafts were boiled for 24 hours in each of concentrated $HNO_3$, concentrated $H_2SO_4$, 50% NaOH, and aqua regia. This treatment has been shortened by others to boiling for 10 to 20 hours in a mixture of concentrated $HNO_3$ and $H_2SO_4$. However, this is still an expensive and time-consuming process, particularly when dealing with hot concentrated acids. It has been found that brown Teflon filaments alone or in a fabric can be bleached to the desirable purified white Teflon by boiling for a relatively short period in a mixture of sulfuric, nitric and perchloric acid (lower temperatures take longer time).

As pointed out brown Teflon may be used in surgical applications but it is apparent that purified or white Teflon is the preferred material to be used in producing blood vessel grafts, sutures, etc. After forming a Teflon tube as a blood vessel graft by knitting (preferably), weaving, or braiding, the brown tube is treated in a solution of 1 part concentrated $HNO_3$, 1 part concentrated $H_2SO_4$, and 2 parts $HClO_4$ (70%) for one-half hour, during which the solution is brought to a boil at 285°F. and heating continued to raise the boiling point to 375°F. At this point 1 part $HClO_4$ is again added to the boiling solution and the treatment continued for another half hour. This second addition of $HClO_4$ is not necessary but has been found to quicken the bleaching process.

At the end of this one hour treatment, the Teflon is generally white. If it is still tan, the treatment is continued until the Teflon is white, the time depending somewhat on the size of the filaments, tightness in the tube, etc. If the treatment continues past an hour, more $HClO_4$ may be added to aid the process.

Various relative amounts of the three acids may be used as long as all are present but the 1:1:2 mixture has been found to give the quickest results. If less concentrated acids are used, they will become concentrated during the boiling treatment. For safety purposes, the conventional 60 or 70% $HClO_4$ is used. The primary value of the $H_2SO_4$ is believed to be its high boiling point which thus permits the $HNO_3$ and $HClO_4$ to react with the Teflon at higher temperatures. Therefore, an essential aspect of the process is treatment of the Teflon at a temperature range of above 285°F. For practical reasons, it does not generally go above 375°F. In fact, bleaching will occur if a mixture of $HNO_3$ and $HClO_4$ is added dropwise to $H_2SO_4$. Any relative quantities of the three acids cause bleaching but it is convenient to have substantial amounts of each. When the mixture is being heated, it is obvious, of course, that $HNO_3$ and $HClO_4$ will boil away, leaving $H_2SO_4$. Thus the former two must be replenished to maintain a given ratio.

## Organic Peroxy Acid Bleaching Systems

J.R. Moyer; U.S. Patent 3,384,596; May 21, 1968; assigned to The Dow Chemical Company found that the bleaching activity of peroxy acids, such as metachloroperoxybenzoic acid, monoperoxyphthalic acid and peroxyacetic acid is increased by the presence of a water-soluble alkaline earth metal ion. The process introduces a water-soluble source of an alkaline earth metal ion, either magnesium or calcium, into an aqueous alkaline solution of metachloroperoxybenzoic acid, monoperoxyphthalic acid, or peroxyacetic acid. The aqueous alkaline solution has an apparent pH of about 10, and a peroxy acid concentration expressed as active oxygen of about 10 to 150 ppm.

The aqueous alkaline bleaching solution is characterized in that when metachloroperoxybenzoic acid is selected, either calcium or magnesium ion is introduced in a metal ion to peroxy acid ratio of about 0.5 to 1; when monoperoxyphthalic acid is selected, magnesium ion is introduced in an ion to peroxy acid ratio of about 0.5 to 2.5; and when peroxyacetic acid is selected, calcium ion is introduced in a calcium ion to peroxy acid ratio of about 0.5 to 1.

The calcium or magnesium ions are normally introduced into the peroxy acid solution as soluble inorganic salts. In bleaching utilizing this process, a temperature within the range of from 100° to 150°F. is ordinarily employed, but a temperature range generally limited only by the 100°F. as a minimum temperature and the boiling point of the respective bleaching solution may operably be employed. Preferably a temperature within the range of from 120° to 150°F. is used during bleaching.

The brightness increase obtained is directly related to the active oxygen concentration, i.e., greater increases are obtained at the higher active oxygen concentrations. Generally an active oxygen concentration of from 10 to 150 ppm is employed in bleaching with peroxy acids, preferably from 50 to 100 ppm. It is noteworthy that all metal ions and, in fact, even all alkaline earth metal ions do not operate to activate or increase the bleaching potential of the aforesaid peroxy acids. For example, the transition metal ions, iron, nickel, copper, mercury and cobalt actually serve to decrease the brightness of articles bleached by peroxy acids. Similarly, the alkali metal ions or the alkaline earth metal ions other than calcium and magnesium are not operable to any significant extent in the process.

The presence of the calcium or magnesium metal ions in the indicated ratio with the peroxy acid in the solution very significantly enhances and increases the bleaching activity of the particular peroxy acid specified providing a powerful oxidizing solution, thereby obtaining unusually high brightness increases in fabric articles bleached therein. The amount of water employed in the composition is that amount necessary to provide the aforesaid concentration of active oxygen in the solution. The pH of the bleaching solutions of the process are adjusted to an alkaline pH by, e.g., the addition of caustic, etc.

The process distinguishes from methods concerning the stabilization of bleaching compositions. For example, though magnesium sulfate in bleach baths containing hydrogen peroxide serves to stabilize the baths by, e.g., minimizing decomposition, neither calcium nor magnesium ions enhance the bleaching activity of hydrogen peroxide-containing solutions. In addition, as is shown in the following examples, neither magnesium nor calcium ion is effective to increase the bleaching activity of all peroxy acids, contrary to what might be expected if the ion were acting or being employed as a stabilizer.

For purposes of the following examples and in order to make relative quantitative brightness determinations, a brightness (reflectance) value of zero was assigned to a piece of unbleached muslin which had been washed in a 2.0 gram solution of a household detergent called "Tide." All brightness readings in the following examples were obtained using a Hunter Photovolt Reflectometer with a green tristimulus filter.

Example 1: Five by five (5" x 5") inch squares of unbleached muslin (brightness of 0) were bleached in a solution at 130°F. and pH of 9.8 containing 2 grams of Tide detergent and 0.254 gram of 85% pure metachloroperoxybenzoic acid, corresponding to 20 ppm active oxygen. The reflectance of the muslin bleached in the metachloroperoxybenzoic acid was determined to be 65. Two identical solutions to that above were prepared. Calcium nitrate was added to one and magnesium sulfate to the other in a molar amount equal to that of the metachloroperoxybenzoic acid. Unbleached muslin squares were then exposed to each solution at 130°F. for about 10 minutes, rinsed, dried and then tested for brightness. Brightness increases of 11 and 6 points respectively on the scale defined above, were obtained over the brightness of muslin bleached in the detergent-metachloroperoxybenzoic acid solution not containing the calcium or magnesium values.

Example 2: Four bleaching baths were prepared. Each contained one liter of tap water and 2.0 grams of Tide detergent. Three contained sufficient metachloroperoxybenzoic acid to give 50 ppm active oxygen. To one of these three baths, no additional chemicals were added. To a second was added a solution of magnesium sulfate sufficient to give one mol of magnesium ion peracid group in the metachloroperoxybenzoic acid. To the third bath was added a solution of calcium nitrate, in the same molar proportion with the peroxy acid. Each bath was maintained at 130°F. and adjusted to pH 9.8 by addition of NaOH solution.

A five inch square of unbleached cotton muslin was agitated in each bath for 10 min. and then rinsed and dried. The reflectance of the cloth was measured with the Hunter Photovolt Reflectometer. The reflectance of the unbleached muslin was set at zero and that of the muslin bleached in 50 ppm active oxygen as metachloroperoxybenzoic acid alone was established at 39. The bleaching results obtained by introduction of the aforesaid calcium and magnesium ions showed that the magnesium and calcium ion additives provide outstanding brightness increases over the brightness obtained with metachloroperoxybenzoic acid alone.

## Cotton Bleach Using Peracetic Acid

L. Peloquin; U.S. Patent 3,457,023; July 22, 1969; assigned to Union Carbide Corporation describes a process of bleaching cotton fabrics with peracetic acid by first impregnating the fabrics with an aqueous solution of peracetic acid at a pH of 3 followed by exposure of the impregnated fabric to ammonia vapors. The bleached fabrics were evaluated by the following techniques:

(1) Reflectance determined with a Hunter Multipurpose Reflectometer employing green, blue and amber filters, with the reflectometer calibrated to read 100% reflectance from a magnesium oxide block. Green and blue reflectance are reported as percents of the reflectance from a magnesium oxide block.

(2) Whiteness was calculated by the equation:

$$\text{Whiteness} = (4 \times \text{blue reflectance}) - (3 \times \text{green reflectance})$$

(3) Yellowness index was calculated by the equation:

$$\text{Yellowness} = \frac{\text{amber reflectance} - \text{blue reflectance}}{\text{green reflectance}}$$

Example 1: Scoured cotton printcloth was padded to 100% wet pick-up with an aqueous solution having a pH of 3 and containing 2.4 weight percent peracetic acid, 0.5 weight percent sodium lauryl sulfate as a wetting agent and 0.2 weight percent sodium hexametaphosphate as a stabilizer. The impregnated fabric was then treated with ammonia vapors for a period of one hour, washed and dried. A second piece of printcloth was impregnated with the same solution, air dried for one hour and then washed and dried. The analyses of these two fabrics are summarized in Table 1, together with the data from the untreated fabric for comparison.

TABLE 1

| Property | Ammonia treatment | Air dry | Untreated |
|---|---|---|---|
| Blue reflectance, percent | 84.2 | 74.6 | 65.2 |
| Green reflectance, percent | 89.0 | 82.2 | 72.1 |
| Whiteness, percent | 69.8 | 52.0 | 44.5 |
| Yellowness index | 0.064 | 0.118 | 0.12 |

Example 2: The experiment of Example 1 was repeated, except that the aqueous bath also contained 1.25 weight percent sodium hydroxide to adjust the pH to 5.5. The results for this experiment are summarized in Table 2.

TABLE 2

| Property | Ammonia treatment | Air dry | Untreated |
|---|---|---|---|
| Blue reflectance, percent | 84.4 | 80.0 | 65.2 |
| Green reflectance, percent | 89.0 | 86.2 | 72.1 |
| Whiteness, percent | 70.6 | 61.2 | 44.5 |
| Yellowness index | 0.060 | 0.089 | 0.12 |

A comparison of the results of Examples 1 and 2 indicated that the pH of the pad bath has little effect on the extent of bleaching by the ammonia treatment of this process, although a significant effect is observed with the air dry.

## Peracetic Acid plus Hexamethylenetetramine

B.K. Easton; U.S. Patent 3,521,992; July 28, 1970; assigned to FMC Corporation found that brown stains which form on polyester fibers when they are bleached with peracetic acid can be eliminated by the addition of hexamethylenetetramine or trioxane. In the makeup of the bleach solution the peracetic acid, whether in situ or preformed, is diluted with water so that the solution contains 0.25 to 1.25% of peracetic acid. The pH of the bleach solution should be on the order of 5.5 to 7.0, and preferably 5.5 to 6.0. The pH of the bleach solution is adjusted by adding ammonium hydroxide and acetic acid, as required, to reach the desired pH level.

It is also desirable to add phosphates, particularly the water-soluble polyphosphates, such as sodium tripolyphosphate, tetrasodium pyrophosphate, sodium hexametaphosphate and long chained polyphosphates normally termed, polyphosphate glasses. These glasses are produced by condensing molecules of sodium orthophosphate to form long chains of molecules having P—O—P bonds, and should have a chain length sufficiently small to allow sufficient solubility in the solution. Typical polyphosphates include "Sodaphos," having a chain length of about 6 and a $P_2O_5$ content of 63.8% by weight, "Hexaphos," having a chain length of about 13 and a $P_2O_5$ content of 67.5% by weight and other such glasses having chain lengths up to 20. The phosphates are added in amounts sufficient to supply 0.1 to 0.5% of the bleach solution. To the above solution is then added at least 0.025% (and preferably 0.05 to 0.5%) of hexamethylenetetramine or trioxane. Larger amounts of either hexamethylenetetramine or trioxane can be employed but do not improve the stain-inhibiting effect.

The polyester-containing fibers or goods which are to be bleached are placed in a saturator containing the above solution and removed with 50 to 150% by weight pickup of the bleach solution. The exact bleach solution pickup is not critical, but about 100% has been found effective in obtaining the desired bleach without removing excessive bleach solution from the saturation. The temperature of bleach solution being applied is not critical, so long as it is below that at which peracetic acid becomes unstable, and ambient temperatures are most conveniently used. After saturating the polyester fibers or goods, they are heated by contact with steam or steam-air mixture at temperatures of from 150° to 212°F. The time of treatment will vary depending upon the degree of bleach desired and the original condition of the fibers and/or cloth. The time of bleaching may vary from 30 minutes up to 4 hours; however, most bleaching is conducted within a period of about 40 minutes to 2 hours. After steaming of the bleach-impregnated goods, the goods may be washed, dried and packaged. In some cases, washing of the cloth is not necessary, and it is directly fed to subsequent stages, i.e., dyeing, etc. for further processing.

## Peracetic Acid Bleach Formed on the Fabric

B.C. Lawes; U.S. Patent 3,528,115; September 15, 1970; assigned to E.I. du Pont de Nemours and Company describes a process to bleach cotton and mixed textiles like cotton-nylon with peracetic acid formed on the fabric, in situ. Textile fabrics which are amenable to bleaching with peracetic acid are saturated with an alkaline hydrogen peroxide bleaching solution, the saturated fabric is contacted with acetic anhydride vapor to effect absorption of the vapor on the fabric and reaction of the absorbed acetic anhydride with the hydrogen peroxide thereby producing peracetic acid in situ on the fabric, and the fabric is bleached with the peracetic acid so produced. Alternatively, the fabric can be partially bleached with the alkaline hydrogen peroxide solution with which it is initially saturated, following which, peracetic acid is formed in situ on the fabric as indicated above and bleaching is completed with the peracetic acid so formed.

The alkaline hydrogen peroxide solution with which the fabric to be bleached is saturated and which serves as the precursor of the peracetic acid solution formed in situ will generally contain from around 0.1 to 5%, preferably 0.4 to 2%, $H_2O_2$ by weight and an alkali which will be present in an amount sufficient to render the solution distinctly alkaline, e.g., have a pH of at least 9.0. Preferably, the solution will have a pH of at least 10.5, e.g., 10.5 to 13, and will contain sufficient alkali so that when it is reacted on the fabric with the acetic anhydride vapor to produce the peracetic acid, the alkalinity of the solution will be reduced to a pH not lower than 4.5, most preferably not lower than pH 5.5.

## Peracetic Acid for Bleaching and Shrinkproofing Wool

W.E. Helmick; U.S. Patent 3,634,020; January 11, 1972; assigned to PPG Industries, Incorporated describes a process to shrinkproof and at the same time bleach wool by exposure to a bath of peracetic acid dissolved in methylchloroform. Other chlorinated hydrocarbon, dry cleaning solvents may also be used. Although peracetic acid is ideal and readily available other organic peracids (peroxycarboxylic acids) are useful in lieu of or in combination with peracetic acid. Useful peracids conform to the structure $R(COOOH)_x$, x being a whole integer notably 1 or 2. When x is 1, the acids are monoperacids and when x is 2 the acids are diperacids. R is usually an alkyl, cycloalkyl, aryl or haloalkyl group but may be hydrogen when x is 1 or an alkaryl or haloalkaryl group when x is 2. That is, R of these peroxycarboxylic acids is principally composed of a carbon chain or carbocyclic ring and hydrogen atoms. Ether linkages, —C—O—C—, may be present in the chain or ring (e.g., heterocylcic rather than carbocyclic rings). R may include a carboxylic group, as when the peracid is a monoperacid of a dicarboxylic acid, e.g., monopersuccinic acid. Although olefinic and acetylenic unsaturation may be present in the R group, saturated peracids are to be preferred. Peracids of acetic, formic, propionic, n-butyric, isobutyric, n-valeric, caproic, monochloroacetic, dichloroacetic, trichloroacetic, monobromacetic, $\alpha$-chloropropionic, $\beta$-chloropropionic, glycolic and lactic are among the aliphatic monocarboxylic acids. Others include perbenzoic, mono- and diperacids of maleic, succinic acids, the perphthalic acids such as diperisophthalic acid, diperterephthalic acid, monoperphthalic acid, and other dicarboxylic acids.

Typically, woolen fiber (unwoven or as woven fabric) is immersed in the liquid shrinkproofing bath made up of an organic solution of percarboxylic acid which is free of any separate aqueous phase for a period adequate to effect shrinkproofing, usually of at least 10 or 15 minutes up to 45 minutes. Rarely need the period of contact exceed about 90 minutes, although it may be more extended. Because it is possible to achieve significant shrinkproofing in these shorter contact periods (i.e., up to 10 to 15 minutes), this process is particularly suited to performance continuously.

Peracid concentration in the treating bath initially at the time of initial immersion should be sufficient to establish an active oxygen content of at least 0.04% by weight of the bath up to 0.7 weight percent active oxygen. This is provided for example, by including from 0.2% peracetic acid by weight of the bath (peracid and solvent), to 1.5% peracetic acid. Although baths with higher active oxygen concentrations effect shrinkproofing, economics usually dictate use of the lesser concentrations.

Example: A solution of peracetic acid in methylchloroform is prepared by mixing vigorously for several minutes an aqueous peracetic acid solution (40 weight percent peracetic acid) with methylchloroform in a separatory funnel, following which the lower organic phase is separated. Several further amounts methylchloroform are added to the aqueous phase remaining after each removal of the organic phase. An organic solution containing 4% peracetic acid and free of any water phase is thus prepared.

A 24" x 24" piece of 50% woolen, 50% worsted fabric is immersed for 15 minutes in a bath at 84°F. of the methylchloroform solution of peracetic acid (diluted with an appropriate quantity of water-free methylchloroform to give the particular peracetic acid concentration) at 10:1 liquor to goods ratio. After being withdrawn from the solution, the piece is rinsed with warm water, air dried and its shrink resistance is measured. Following this procedure a plurality of runs were performed, the table below lists the results.

| Run | Peracid, wt. % | Shrinkage, % Total | Relaxation | Felting |
|---|---|---|---|---|
| 1 | Control | 56.92 | 5.65 | 51.27 |
| 2 | 0.3 | 22.75 | 2.37 | 20.38 |
| 3 | 0.6 | 12.82 | 5.22 | 7.60 |
| 4 | 0.9 | 4.03 | 5.22 | +1.19 |
| 5 | 1.5 | +4.35 | +1.05 | +3.30 |

(+ indicates stretch or elongation)

Perphthalic Acid Stabilized with Alkali Metal Salt

D.R. Nielsen; U.S. Patent 3,639,285; February 1, 1972; assigned to PPG Industries, Incorporated describes a process to stabilize perphthalic acid bleaches by alkali or alkali earth metal salts of an acid with an ionization constant at 25°C. of at least $1 \times 10^{-3}$ for the first hydrogen. Anhydrous alkali metal or alkaline earth metal sulfates, such as sodium sulfate or magnesium sulfate, are especially effective for this purpose. Other sulfates which can be used include potassium sulfate ($K_2SO_4$), ammonium sulfate, lithium sulfate, and like sulfates of alkali metals. Bisulfates also may be used. Thus, sodium bisulfate or potassium bisulfate or like alkali metal bisulfates stabilize perphthalic acids although not as sodium sulfate or like neutral sulfates.

The process is directed to the stabilization of perphthalic acids, such as the peracid of phthalic, terephthalic or isophthalic acids, and the halogenated (particularly the fluorinated or chlorinated) perphthalic acids, and their preparation and use as bleaching agents especially for the bleaching of textiles. These acids are capable of liberating active oxygen for bleaching and various oxidizing reactions. The amount of stabilizing salt used normally is large, particularly when water-soluble salts are used. As a general rule, amounts at least equal to one-half part by weight of alkali metal or alkaline earth metal salt per part by weight of the peracid are found satisfactory, and amounts in the range of 1 to 10 parts of stabilizing salt per part of the peracid are found useful.

As a general rule, it is undesirable to incorporate other organic materials which can contaminate the product or promote instability of the perphthalic acid. Of course, other inert materials can be added. However, the incorporation of synthetic organic detergents, such as alkyl sulfates and sulfonates and the like, may be found to be disadvantageous to the stability of the ultimate product. This difficulty may be avoided, however, by either coating or encapsulating the mixture of perphthalic acid and sodium sulfate or the detergent with a suitable inert protective coating, such as paraffin wax, polyvinyl alcohol or the solid polymeric glycols or the like, which are water-soluble or water-dispersible, in order to segregate the detergent, alkaline agent or like material from the perphthalic-stabilizing salt mixtures. The mixture thus protected may be mixed with dry detergents or other materials, whether or not they attack unprotected perphthalic acid.

Example: Samples of diperisophthalic acid of approximately 85% assay were intimately mixed with equal weights of finely powdered sodium sulfate (technical grade), potassium sulfate and a 3:1 by weight mixture of sodium carbonate and sodium sulfate. These samples and a portion of the undiluted peracid were allowed to stand at 60°C. under an air atmosphere of 52% relative humidity for 45 days. Each sample was analyzed for active oxygen by dissolving in dilute sulfuric acid solution containing potassium iodide and titrating the liberated iodine with thiosulfate. The results are shown in the following table.

| Additive | Average percent Decomposition/Day* | Improvement in Stability, % |
|---|---|---|
| None | 8.6 | -- |
| $Na_2SO_4$ | 0.2 | 4,300 |

(continued)

| Additive | Average percent Decomposition/Day* | Improvement in Stability, % |
|---|---|---|
| $K_2SO_4$ | 0.7 | 1,130 |
| $Na_2CO_3$, $Na_2SO_4$ | 0.8 | 980 |

*Calculated as percent of the total active oxygen lost per day.

Bleaching Fabrics and Wood Pulp with Peracetic Acid

L.H. Dierdorff, Jr. and J.H. Kosciolek; U.S. Patent 3,655,698; April 11, 1972; assigned to FMC Corporation describe a method of increasing the oxygen content of oxidizable organic materials by contacting the oxidizable organic material with a vaporous effluent containing 5% by weight of peracetic acid coming directly from a peracetic acid generator. The method is useful in bleaching liquids, textiles and wood pulp. Two processes are given for the production of peracetic acid. Figure 5.3 illustrates a flow diagram in which the peracetic acid vaporous effluent is generated by the peroxidation of acetic acid.

The peracetic acid generator, the oxidation zone, their components such as stirrers and packing, and all other surfaces which come in contact with the peracetic acid should be constructed of inert materials which do not catalyze the decomposition of peracetic acid. Suitable materials of construction include aluminum, certain types of stainless steel, tantalum, zirconium, tin, glass, quartz, ceramics, and many types of plastics. However, when peracetic acid is generated by the peroxidation of acetic acid, the generator should not be constructed of aluminum since this material is attacked by the strong acid used therein. All exposed surfaces should be thoroughly cleaned and free of contaminating materials prior to use. Certain materials, even in small quantities, are highly catalytic towards peracetic acid, causing it to decompose into acetic acid and oxygen, and thus must be specifically excluded. Among these materials are metals such as iron, cobalt, copper, zinc, silver, and lead.

Referring to Figure 5.3, acetic acid, hydrogen peroxide, and water are continuously charged to reactor (53) via lines (50), (51) and (52), respectively, at a safe and useful ratio. Sulfuric acid is initially charged to reactor (53) via line (54) when starting up the generator. With a temperature in the range of 20° to 80°C. and a pressure in the range of 15 to 350 mm. of mercury, distillation conditions are maintained in reactor (53).

The vapors produced therein are passed through rectifying column (55), where any entrained sulfuric acid is separated from the peracetic acid, acetic acid, and water vapors and returned to reactor (53). When the acetic acid content of the equilibrium mixture in reactor (53) is sufficiently high that no sulfuric acid is entrained in the vaporous effluent, rectifying column (55) can be eliminated. A vaporous stream containing the desired amount of peracetic acid, acetic acid, and water passes through line (56) to oxidizer (57). The organic material to be oxidized is introduced into oxidizer (57) via line (58). Oxidizer (57) is preferably a packed fractionating column. When the oxidation process involves the bleaching of a solid material such as cloth, or wood, it is generally desirable to have two or more oxidizers which can be used alternately, thus providing an off cycle for loading and unloading the oxidizers. Uncondensed vapors from oxidizer (57) pass through line (59) to condenser (60) wherein all condensable materials are recovered. The vacuum necessary

FIGURE 5.3: PERACETIC ACID FROM PEROXIDATION OF ACETIC ACID

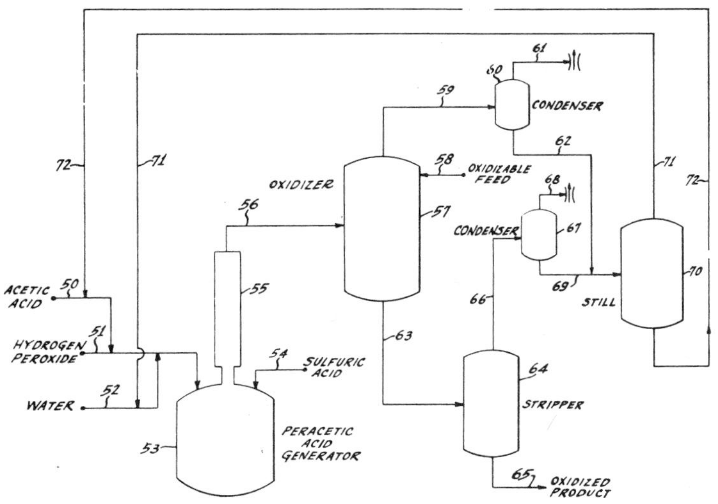


Source: L.H. Dierdorff, Jr. and J.H. Kosciolek; U.S. Patent 3,655,698; April 11, 1972

to provide reduced pressure in the generator and oxidizer system is provided via line (61). Condenser material, primarily aqueous acetic acid, is removed from condenser (60) via line (62). Oxidized product, along with any aqueous acetic acid, is removed from oxidizer (57) and passed via line (63) to stripper (64). Aqueous acetic acid is removed from the oxidized product in stripper (64) and passed overhead to condenser (67) via line (66). Vacuum for the stripping operation is provided via line (68). Substantially pure oxidized product is recovered from stripper (64) via line (65). Aqueous acetic acid vapors are condensed in condenser (67) and passed to still (70) via line (69), along with the aqueous acetic acid stream coming from condenser (60) via line (62). Water and acetic acid are separated in still (70), the water being passed overhead and recycled for reuse in the process via line (71). The acetic acid remaining in still (70) is recycled to the process via line (72).

Example 1: Using a peracetic acid generator attached to a vertical glass column a 50% peracetic acid effluent containing 130 parts per hour of peracetic acid was obtained. Eleven parts of dry, unbleached cotton cloth was saturated with water and placed in the vertical glass column. After 43 minutes, bleaching was stopped and the cloth was water washed. An acceptable reflectance reading of 84.3 was obtained.

Example 2: By using the peracetic acid generator of Example 1, a 50% peracetic acid vaporous effluent containing 100 parts per hour of peracetic acid was obtained. Twenty parts of unbleached bisulfite treated softwood pulp was placed in the vertical glass column. The peracetic acid vapors coming directly from the generator were passed through the pulp for 0.5 hour. After this treatment, the sample was washed with water until free of acid, thickened on a Büchner funnel, and pressed into paper. This example was repeated using different bleaching temperatures and samples of the same pulp which had been pretreated in different ways. The results are set forth in the following table.

Peracetic Acid Vapor Bleaching of Wood Pulp

| Sample | Pulp pretreatment | Bleaching Temp., °C. | Bleaching Time, min. | Brightness |
|---|---|---|---|---|
| A | None (control) | | | 55.0 |
| B | None | 35 | 30 | 58.9 |
| C | do | 45 | 30 | 60.9 |
| D | Versenex 80 | 45 | 30 | 67.0 |
| E | Na pyrophosphate | 45 | 30 | 69.2 |

## Bleaching Wool with Performic Acid

O. Schmidt and K. Heinz; U.S. Patent 3,374,177; March 19, 1968; assigned to Badische Anilin- & Soda-Fabrik AG, Germany describe a process to prepare bleaching solutions containing performic acid from hydrogen peroxide and a substance yielding formaldehyde or a low molecular weight formic ester or formamide. The substances are mixed in an acid to neutral aqueous solution at a temperature up to 60°C. Stable solutions containing performic acid are obtained which may be diluted and adjusted to a pH of 2 to 8.5. Substances yielding formaldehyde are defined as substances which contain at least once a group having the general formula:

$$Y{-}CH_2{-}X \quad (1)$$

in which Y denotes a nitrogen or oxygen atom, X denotes a nitrogen or oxygen atom or the group $-SO_3M$, M being a proton, an ammonium ion, an alkali metal ion or half an alkaline earth metal ion, and the valencies of the nitrogen and/or oxygen atoms which are not attached to the central carbon atom being, in the absence of a $SO_3M$ group, at the most partly saturated by hydrogen.

Examples of substances which contain at least once a group having the Formula (1) are: formaldehyde polymers, such as trioxymethylene, tetraoxymethylene and paraformaldehyde; formals, such as dimethylformal and diethylformal; hydroxymethanesulfonic acid; aminomethanesulfonic acid; iminobismethanesulfonic acid and N-alkyl and N-hydroxyalkyl derivatives of these two acids; nitrilotrismethanesulfonic acid, ammonium salts, alkali metal salts and alkaline earth metal salts of these acids, such as preferably the sodium, potassium and calcium salts; N-methylol compounds of carbamides, such as particularly urea, imidazolidones, tetrahydrotriazinones, hexahydropyrimidones and urones; of melamines, of dicyanodiamide; and also ethers of such N-methylol compounds with alcohols having one to five carbon atoms. Examples of low molecular weight formic esters and formamides are substances having the general formula:

$$QHC{=}O \quad (2)$$

in which Q denotes a radical $-OR$ or $-NR^1R^2$, R being an alkyl radical having one to five carbon atoms, $R^1$ being a hydrogen atom or an alkyl radical having one to five carbon atoms and $R^2$ being a hydrogen atom or an alkyl radical having one to five carbon atoms. Examples of substances having the Formula (2) are: methyl formate, ethyl formate, propyl formate, butyl formate and diethylformamide. The drop in the pH value during bleaching may be prevented or minimized by employing or additionally employing substances containing at least one group having the Formula (1) or substances having the Formula (2) which form amines or preferably ammonia by reaction with hydrogen peroxide. A particularly suitable and therefore preferred substance of this type is hexamethylene tetramine. Formamide, dimethylformamide, aminomethanesulfonic acid, iminobismethanesulfonic acid, their alkyl derivatives, particularly their methyl derivatives, and their salts however act in this way.

To form performic acid with hydrogen peroxide, the following substances are preferred because they are solid at room temperature, stable and readily available: formaldehyde polymers, such as trioxymethylene, tetraoxymethylene and paraformaldehyde; hydroxymethanesulfonic acid and its sodium, potassium and calcium salts; the reaction products of this acid with ammonia, i.e., aminomethanesulfonic acid, aminobismethanesulfonic acid and nitrilotrismethanesulfonic acid and the sodium, potassium and calcium salts of these acids and hexamethylene tetramine.

Substances of the abovementioned type are reacted with hydrogen peroxide in aqueous medium. The speed of this reaction is dependent on the hydrogen ion concentration of the medium. An acid or neutral range is used to achieve rapidly a bleaching solution which is ready for use. The acid reaction may be produced by adding inorganic or organic acids or acid salts. The acids may be chosen at will. Strong acids are very suitable, such as sulfuric acid and sulfonic acids, and acid salts such as sodium bisulfite and sodium bisulfate.

The amount of acid substance required depends on its acidity. Amounts of 0.5 to 10% of strong acids, with reference to the weight of hydrogen peroxide to be reacted, are adequate, whereas 10 to 50% of weaker acids or acid salts may be necessary. It seems as though even in acid solution, the reaction to form performic acid does not immediately proceed to the point of completely exhausting one reaction component; rather there is apparently set up an equilibrium condition which continues during bleaching and consequently during consumption of performic acid, as long as all the reactants are still present.

The ratio between substances which contain at least one group having the Formula (1) or substances having the Formula (2) and hydrogen peroxide is preferably chosen so that for each mol of groups having the Formula (1) or of substances having the Formula (2), one to twelve mols of hydrogen peroxide are available; particularly successful results are achieved with a molar ratio of 1:3 to 1:10. The amount of hydrogen peroxide depends on whether it is desired to bleach in a long liquor or with a padding liquor. Amounts of 35% hydrogen peroxide of from 10 to 30 cc/l. have proved to be particularly useful for long liquors and from 40 to 130 cc/l. for padding liquors.

The performic acid bleaching liquor may be prepared by adding the ingredients direct to the liquor in the amounts desired for bleaching. Alternatively a concentrated stock bleaching solution may first be prepared which is diluted to the concentration for use immediately prior to use. Stock bleaching solutions may be prepared in concentrations which are equivalent to 100 to 500 cc/l. of 35% hydrogen peroxide. When preparing stock bleaching solutions it is recommendable that the temperature should not exceed 40°C.; bleaching liquors to which the ingredients are added direct may however be prepared at higher temperatures up to 60°C. The liquors may be used for example for bleaching textile materials of polyamide, polyester, polyacrylonitrile, cellulose, cellulose ester and regenerated cellulose fibers. They are particularly suitable for bleaching textile material which contains or consists of protein fibers, for example, wool, silk and animal hairs, and also feathers, bristles and the like.

Example: An agent for the production of a bleaching liquor containing performic acid has the following composition: 55 parts of sodium hydroxymethanesulfonate, 22 parts of hexamethylene tetramine, 15 parts of sodium tripolyphosphate, 6 parts of ethylenediaminetetraacetic acid and 2 parts of sodium bisulfite. Thirty liters of 35% hydrogen peroxide and then 7.5 kg. of the agent described above are added to 2,000 liters of water in a wood, stoneware or alloyed steel winch dyeing machine having a capacity of 3,000 liters. A bleaching bath is immediately obtained which is ready for use and whose wetting out property may be improved by adding 0.6 kg. of octylphenol heptaglycol ether.

The winch dyeing machine is charged with 50 kg. of unbleached piece goods of wool and polyacrylonitrile (70:30) for bleaching. The whole is left for 5 minutes at room temperature, the pH value of the liquor is controlled and if necessary adjusted to 5.5 with ammonia or formic acid. The temperature is then raised in about 15 minutes to 80°C. and kept steady for 30 minutes. The pH value remains practically constant during this bleaching. The material is then thoroughly rinsed hot and cold and further treated in any desired way. The material has a high degree of whiteness after this bleach.

## PERSULFATES AND PERPHOSPHATES

### Potassium Monopersulfate and Sodium Chloride

D.B. Lake and P.T.B. Shaffer; U.S. Patent 3,048,546; August 7, 1962; assigned to E.I. du Pont de Nemours and Company found that alkali metal monopersulfate and alkali chloride mixtures are comparatively stable solid bleaching and cleaning compositions. These solid compositions are easily prepared by simply mixing the chloride salt in the desired proportions. Preferred compositions are comprised of sodium chloride and potassium monopersulfate. The proportions of these compositions can vary widely and increased bleaching activity can be readily detected when the weight ratio of sodium chloride to potassium monopersulfate is about 0.1 to 40/1, respectively. Highly active compositions have a weight ratio of sodium chloride to potassium monopersulfate of about 1 to 10/1, respectively. These ranges also embrace other alkali metal chlorides including potassium chloride and sodium monopersulfate which approximated the preferred compounds in effectiveness.

The preferred active oxygen compound, potassium monopersulfate, can be conveniently prepared with potassium bisulfate and potassium sulfate in the form of a triple salt $KHSO_5{:}KHSO_4{:}K_2SO_4$ in the mol ratio of about 2:1:1. The triple salt is very stable and contains about 50% by weight of the active oxygen compound; when added to water, a pH of 2 to 3 is developed. Sodium and potassium monopersulfates can be used without the bisulfate or sulfate salts; however, a convenient method of preparing potassium or sodium monopersulfate is by reacting a mixture of monopersulfuric and sulfuric acid with the alkali metal carbonate or hydroxide and drying the resulting mixture. The dried mixture corresponds to the triple salt in the above ratio. A description of the preparation of these mixtures is disclosed in U.S. Patent 2,802,722. The active oxygen content of the alkali metal monopersulfate is about 10% by weight. Bleaching solutions will usually contain sufficient monopersulfate to yield about 10 to 700 ppm active oxygen, depending on the particular use intended. Corresponding chloride ion content is within the range of 5 to 140,000 ppm. When used in scouring, the solid composition is used in the form of a slurry and the concentrations of active oxygen and chloride are much higher.

Example 1: The following compositions were prepared by intimately mixing together in the proportions indicated, sodium chloride and a composition containing 47.4% $KHSO_5$, the balance being $KHSO_4$ and $K_2SO_4$ (inert) in about equimolar proportions to each other.

| Components | Composition in wt. % A | B | C |
|---|---|---|---|
| $KHSO_5$ | 45 | 47 | 24 |
| NaCl | 5 | 1 | 50 |
| Inert | 50 | 52 | 26 |

The above anhydrous compositions are stable when kept under anhydrous conditions. When dissolved in water to give solutions containing about 50 to 700 ppm of active oxygen, the resulting solutions are highly effective for bleaching. Solutions containing less than 50 ppm active oxygen, about 10 to 30 ppm, are employed as preventive bleaching solutions in detergent formulations.

Example 2: A cleanser formulation containing the following ingredients is particularly adapted to erase hard-to-remove stains from porcelain sinks and bowls:

| | Percent by Weight |
|---|---|
| $KHSO_5$ | 0.7 |
| NaCl | 20 |
| Detergent (alkyl aryl sulfonate) | 3 |
| Ground silica and inert | 75.8 |

The large amount of chloride salt in this formulation augments the germicidal properties of the formulation and a slurry of this formulation has a pH about 9. It has been found that the bleaching action of monopersulfate solutions on heat-discolored nylon and other textiles is promoted significantly and beneficially by the presence of chloride ion in the monopersulfate treating solution.

The concentration of the monopersulfate in treating solutions used to bleach fabrics generally should correspond to an active oxygen concentration of at least 50 ppm but not greater than about 700 ppm, since higher concentrations can damage the fabric. Active oxygen concentrations of 100 to 400 ppm are preferred. The treating solutions will generally be used at temperatures ranging from 120°F. to the boiling temperature, temperatures of from 180° to 200°F. being preferred. The time required for effective bleaching is inversely related to the temperature used. Thus, at 140°F., effective bleaching may require 1 hour as compared with 15 minutes at 200° to 212°F. While temperatures below 120°F., e.g., room temperature, can be used, bleaching at such temperatures is usually too slow for most purposes. The bleaching treatment can be carried out in any desired manner which will maintain contact of the fabric with the treating solution for a time sufficient to cause effective bleaching. The fabric can simply be immersed for the required time in the treating solution maintained at the chosen temperatures, or the fabric can be passed back and forth through the treating solution. The preferred compositions containing about 1 to 10/1 alkali metal salt to alkali metal monopersulfate are particularly active when tea-coffee stained cotton swatches were bleached with a detergent formulation containing sufficient $K_2CO_3$ and $Na_4P_2O_7$ to yield a pH about 9. The active oxygen content was about 25 ppm.

As the ratio $NaCl/KHSO_5$ is increased the increase in brightness becomes smaller. In the bleaching of fabrics the preferred ratio of chloride salt to monopersulfate salt is 1 to 10/1; higher ratios do not greatly enhance the bleaching effect. Higher ratios of $NaCl/KHSO_5$ in cleansers impart superior activity until the ratio reaches about 30 to 40/1. Increasing the amount of chloride in these formulations above 40/1 has little effect in promoting the activity of the monopersulfate salt. Less than 0.1/1 $NaCl/KHSO_5$ will enhance the bleaching effect to some extent, but at least 0.1/1 $NaCl/KHSO_5$ is necessary to increase the bleaching action of the composition as a practical matter.

Peroxy Monosulfate-Diethylenetriamine Pentaacetic Acid

G.I. Jenkins and E.V. Kring; U.S. Patent 3,049,495; August 14, 1962; assigned to E.I. du Pont de Nemours and Company found that peroxymonosulfate compounds can be used as effective bleaching agents for cotton fabrics by the addition of a sequestering agent even if the water used contains manganese ions. It is well-known that peroxymonosulfate compounds containing $HSO_5^-$ such as $NaHSO_5$ and $KSO_5$ are highly effective bleaching compounds, being stronger oxidizing agents than hydrogen peroxide and many other commonly used per compounds.

However, the use of such strong oxidizing agents in common tap water presents a problem because some tap water supplies contain manganese in amounts of 0.1 ppm or more. Analysis has shown that ordinary tap water rarely contains more than 0.2 ppm manganese. Under the influence of peroxymonosulfate the manganese becomes oxidized to a high valence state (+3 or +4) and imparts a yellow discoloration to the object being treated. The problem is especially acute when white cloth is laundered or white porcelain is cleansed. Tests have shown that as little as 0.1 ppm manganese in the water will noticeably discolor white cloth or porcelain when manganese is oxidized to a higher valence state.

It has been found that manganese ions can be effectively sequestered from the oxidizing action of peroxymonosulfate with diethylenetriamine pentaacetic acid (DTPA) or the alkali metal salts thereof and that these sequestering agents are not attacked by peroxymonosulfate to an appreciable extent. It has also been found that DTPA does not decompose peroxymonosulfate to any appreciable extent, and that the monopersulfate compound and DTPA may be formulated in solid bleaching and cleansing compositions which can be added to common tap water containing substantial amounts of manganese with the result that the latter is effectively sequestered so that it will not discolor the substrate being bleached.

## Quaternary Ammonium Peroxysulfates

L.H. Diamond and J.H. Blumbergs; U.S. Patent 3,353,902; November 21, 1967; assigned to FMC Corporation found that quaternary ammonium peroxysulfates are useful bleaching and softening compounds for textiles. These compounds are quaternary ammonium peroxysulfates having the formula:

$$\left[R_2-\underset{\underset{R_3}{|}}{\overset{\overset{R_1}{|}}{N}}-R_4\right]HSO_5$$

and diquaternary ammonium peroxydisulfates having the formula:

$$\left[R_2-\underset{\underset{R_3}{|}}{\overset{\overset{R_1}{|}}{N}}-R_4\right]_2S_2O_8$$

in which $R_1$, $R_2$, $R_3$ and $R_4$ may be saturated aliphatic, saturated cycloaliphatic or aromatic groups having up to 18 carbon atoms which are not oxidized by peroxysulfates and which are attached to the nitrogen atoms through a carbon atom. The quaternary ammonium peroxymonosulfates and the diquaternary ammonium peroxydisulfates are produced by reaction of a quaternary ammonium salt (preferably one in which the anion of the salt is a halogen element) and an inorganic mono- or dipersulfate. Typical reactions proceed according to the following equations:

$$(1)\quad \left[R_2-\underset{\underset{R_3}{|}}{\overset{\overset{R_1}{|}}{N}}-R_4\right]Cl + KHSO_5 \longrightarrow \left[R_2-\underset{\underset{R_3}{|}}{\overset{\overset{R_1}{|}}{N}}-R_4\right]HSO_5 + KCl$$

$$(2)\quad 2\left[R_2-\underset{\underset{R_3}{|}}{\overset{\overset{R_1}{|}}{N}}-R_4\right]Cl + K_2S_2O_8 \longrightarrow \left[R_2-\underset{\underset{R_3}{|}}{\overset{\overset{R_1}{|}}{N}}-R_4\right]_2S_2O_8 + 2KCl$$

where $R_1$, $R_2$, $R_3$ and $R_4$ are saturated aliphatic, saturated cycloaliphatic or aromatic groups having up to about 18 carbon atoms which are not oxidized by peroxysulfates and which are attached to the nitrogen atom through a carbon atom. An alternate method for preparing monopersulfates is carried out by reacting quaternary ammonium hydroxides with Caro's acid solutions.

The particular quaternary ammonium salt precursor which is chosen will depend upon the requirements of the final quaternary ammonium persulfate. For example, if the final product is to be used for both softening and bleaching of textiles or in detergent applications, a quaternary ammonium salt having all alkyl groups, at least one of which is a fatty alkyl group, is preferred. Examples of these are dodecyl trimethyl ammonium chloride, dihydrogenated tallow dimethyl quaternary ammonium chloride, and others. If the final product is to be used only for bleaching, lower molecular weight quaternary ammonium salt precursors can also be used with good effect such as tetramethyl ammonium chloride or tetraethyl ammonium chloride.

The peroxysulfates may be used in laundering solutions in amounts sufficient to supply up to 30 ppm of active oxygen based on the weight of the bleaching solution. Larger quantities may be used, but for economic considerations, larger amounts are undesirable; moreover, the bleaching effect is sufficient at this concentration to satisfy home laundering requirements. The precise quantities used depends upon the particular peroxysulfate chosen. For example, if a lower alkyl quaternary ammonium peroxysulfate is selected which is intended primarily as a bleaching agent without any fabric softening effect, it is added in amounts sufficient to supply about 30 ppm of active oxygen.

On the other hand, if a long chain quaternary ammonium peroxysulfate is employed to obtain a fabric softening effect, it can be added in amounts of 0.1 to 0.15% on the weight of the fabric. When using low concentrations of these compounds to obtain fabric softening, the bleaching effect will not be as substantial as when higher concentrations of the quaternary ammonium persulfates are employed. However, even at low concentrations, the bleaching effect aids in maintaining the original whiteness of the fabric while giving a softening effect; this is important because subsequent yellowing of fabrics treated with chemical softeners often occurs due to chemical changes in the softener per se.

Example 1: A charge of 7.56 grams of tetramethyl ammonium chloride dissolved in 75 grams of methanol was poured into a 250 ml. beaker. To the solution was added, with stirring, 6.85 grams of ammonium persulfate dissolved in 25 grams of dimethyl formamide. The reaction mixture was cooled to approximately 2° to 5°C. with an ice-water bath. An insoluble product crystallized from the reaction mixture and was separated from the mother liquor by filtration. The crude product was washed three times with 25 ml. each of cold methanol and was dried in a Rinco Evaporator under reduced pressure. A white crystal material was obtained weighing 9.5 grams. It had an active oxygen content of 4.73% by weight as determined by iodometric titration. The crude product was purified by recrystallization from methanol and was identified as tetramethyl ammonium diperoxysulfate.

Example 2: A charge of 44 grams of Arquad 2HT was dissolved in a 150 ml. of methanol and poured into a 400 ml. beaker. Arquad 2HT is a proprietary composition containing 75% by weight of dimethyl dihydrogenated tallow ammonium chloride. The fatty alkyl groups in this proprietary composition contain approximately 65% $C_{18}$, 30% $C_{16}$ and 5% of $C_{14}$ alkyl chains. The Arquad 2HT solution was cooled to 5°C.

A solution made up of 15 grams of potassium monopersulfate (potassium caroate) assaying 88% dissolved in 150 ml. of distilled water was then added to the Arquad 2HT solution. The reaction mixture was stirred and further cooled to 0°C. A white solid product crystallized from the reaction mixture and was separated by filtration from the mother liquor. The resulting solids were washed twice with 100 ml. of a cold (-5°C.) methanol-water mixture and dried in a Rinco Evaporator under reduced pressure. The resultant white solid product weighed 36 grams and had an active oxygen content of 2.20% by weight. A sample of the product was purified by recrystallization from methylene chloride and was identified as dimethyl dihydrogenated tallow ammonium monoperoxysulfate.

Example 3: The bleaching effect of dimethyl dihydrogenated tallow ammonium monoperoxysulfate (DDTAM), produced by the method set forth in Example 2, was compared with a standard bleaching agent, potassium monopersulfate, using the following procedure. Thirty-two cotton swatches (5" x 5" desized cotton Indianhead fabric, uniform in weave and thread count) were stained with tea. The staining was accomplished by placing 5 tea bags in a liter of water and boiling for 5 minutes, and thereafter immersing the swatches in the tea and continuing the boiling for an additional 5 minutes. The stained swatches were then squeezed to remove excess fluid, dried, rinsed in cold water and dried.

Three of the stained cotton swatches were added to each of a series of stainless steel Terg-O-Tometer vessels containing 1,000 ml. of a 0.2% standard detergent solution at a temperature of 120°F. Measured amounts of each of the bleaches were then added to separate vessels sufficient to correspond to predetermined active oxygen contents. The pH of the solutions in the vessels were adjusted to 9.5 using soda ash. Cut-up pieces of white terry cloth toweling were added to provide a typical household wash water/cloth ratio of 20:1. The Terg-O-Tometer was then operated at 72 cycles per minute for 15 minutes at a temperature of 120°F.

At the end of the wash cycle, the swatches were removed, rinsed under cold tap water and dried in a Procter-Schwartz skein dryer. The tests were run in triplicate and include detergent blanks. Reflectance readings of the swatches were taken before and after the wash cycle with a Hunter Model D-40 Reflectometer using the blue filter. Each swatch was read twice (warp and fill) on either side, with a backing of 5 similarly soiled swatches. Fluorescent effect was excluded in all readings. The resultant reflectance increase over blank samples at various active oxygen concentrations is given in Figure 5.4 for both the DDTAM and the potassium peroxymonosulfate.

FIGURE 5.4: REFLECTANCE INCREASE VERSUS ACTIVE OXYGEN CONCENTRATION

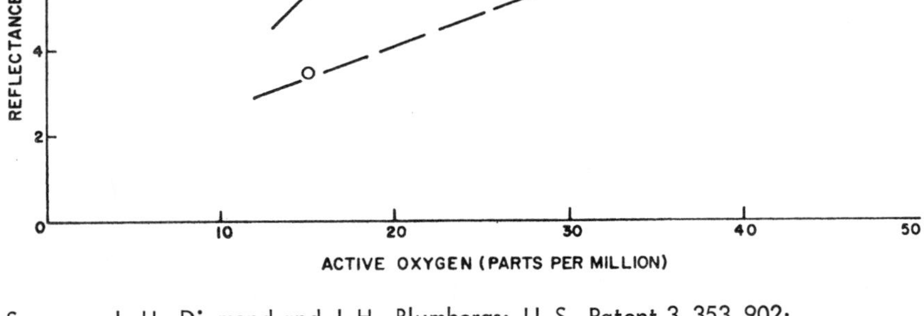

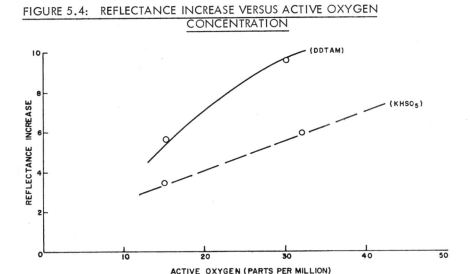


Source: L.H. Diamond and J.H. Blumbergs; U.S. Patent 3,353,902; November 21, 1967

Peroxymonosulfate for Sensitive Dyestuffs

N.J. Stalter; U.S. Patent 3,556,710; January 19, 1971; assigned to E.I. du Pont de Nemours and Company describes a process to bleach dyed textiles without bleeding or "mark off" by the use of peroxymonosulfate bleaching solution. The preferred peroxymonosulfate is the potassium peroxymonosulfate triple salt compound of the formula: $KHSO_4 \cdot K_2SO_4 \cdot 2KHSO_5$. A product comprising such triple salt compound and containing about 4.7% active oxygen by weight is available commercially. The peroxymonosulfate component of the solution can also be formed in situ in the solution by the reaction of peroxymonosulfuric acid, also called Caro's acid, $H_2SO_5$, and an alkali such as caustic soda, caustic potash, ammonium hydroxide or an alkaline salt such as sodium or potassium carbonate, or the like. Preferably, the concentration in the solution of active oxygen derived from the peroxymonosulfate will range from 0.005 to 0.02% for immersion bleaching, and from 0.03 to 0.2% for saturation bleaching.

To be effective for the process purpose, the peroxymonosulfate bleaching solution should be buffered to a pH within the range 5 to 8.5, preferably 5.5 to 7. If the pH is lower than about pH 5, the solution may be sufficiently acidic to damage cellulosic fabrics. On the other hand, if the solution has a pH substantially higher than about 8.5, the solution becomes less stable.

Although bleaching may be effected at ordinary temperatures, it generally will be preferred to employ somewhat elevated temperatures in order to decrease the bleaching time required. Thus, bleaching will generally be effected at a temperature of at least 100°F., e.g., temperatures from 100°F. up to as high as 212°F. At such temperatures, bleaching times ranging from 10 minutes to 5 hours may be employed, the higher temperatures generally requiring shorter times than lower temperatures.

Bleaching may be effected by immersing the fabric in the bleaching solution maintained at the desired temperature, but preferably the fabric will be impregnated or saturated with an amount of the bleaching solution equal to from 80 to 120%, of the fabric weight, and the saturated fabric will then be heated at the desired temperature, e.g., by contacting it with live steam or a mixture of steam and air, depending upon the temperature desired. Most preferably, a continuous length of the fabric to be bleached will be continuously saturated with the bleaching solution, following which the saturated fabric will be continuously heated by contact with steam or a mixture of steam and air and the heated fabric then stored, e.g., in a J-box or other equipment, from which it can be continuously withdrawn after a residence therein equal to the desired bleaching time.

The method is applicable to the bleaching of fabrics containing portions dyed with naphthol dyes, disperse dyes, vat dyes and copper phthalocyanine dyes. It is also applicable to fabrics containing one or more of the above types and, thus, avoids the necessity of segregating differently dyed fabrics during bleaching. The process is suitable for bleaching the fabric in open width or rope form in the widely used continuous bleach J-box ranges with no significant modification of the equipment being required. Although continuous operation is most generally preferred, the method can be practiced in batch operations when desired. The following abbreviations are used in some of the examples to designate certain of the constituents of the bleaching solutions employed: DTPA designates a commercial diethylenetriamine pentaacetic acid sequestrant; TSPP designates tetrasodium pyrophosphate decahydrate;

and PMS designates the commercially available potassium peroxymonosulfate product comprising the triple salt $KHSO_4 \cdot K_2SO_4 \cdot 2KHSO_5$ having an active oxygen content of about 4.7%. Bleaching solutions containing peracetic acid were formulated using the commercially available 40% peracetic acid solution.

Example: Samples of a cotton-polyester (ethylene glycol terephthalate) shirt fabric containing red stripes formed of yarns dyed with red disperse and naphthol red dyes, and black stripes formed of yarns dyed with black disperse and black vat dyes, woven into the fabric were bleached using the bleach solutions indicated in the tabulation below. Before bleaching, the fabric samples had been scoured by saturating them with an aqueous solution containing 0.83% TSPP and 0.5% soda ash, and then heating the saturated fabric at 150°F. for 60 minutes. The results obtained were as follows:

| Run | Bleach solution | Bleach conditions | Bleeding or mark-off |
|---|---|---|---|
| A | 0.70% $H_2O_2$<br>1.25% sodium silicate<br>0.25% $NaHCO_3$<br>0.25% sulfonated castor oil<br>0.20% DTPA<br>0.02% Epsom salt<br>pH 10.2 | 1 hour at 212° F | Considerable. |
| B | 0.80% peracetic acid<br>0.40% TSPP<br>pH 5.8 (NaOH) | 30 minutes at 210° F | Visible but less than in A. |
| C | 1.0% PMS<br>4.0% TSPP<br>0.7% acetic acid<br>pH 7.0 (NaOH) | 15 minutes at 212° F | None visible. |

Only a short bleaching time was required in Run C because essentially all of the active oxygen of the PMS salt was used up in that time. All of the samples were well bleached, but the data show that only in the case of those bleached with the PMS bleaching solution did bleaching occur without causing the dye to bleed or mark off onto the undyed portions of the fabric.

When the general procedure of the above example was followed in bleaching samples of a 50:50 cotton/polyester fabric having a red checked pattern (1/2" checks) containing disperse and naphthol red dyes, and the dye bleeding that occurred was assessed by observing the relatively pink coloration of white threads pulled from the bleached samples, it was noted that white threads from the samples bleached with alkaline peroxide solutions and peracetic acid solutions had visibly more pink coloration than did the white threads pulled from samples bleached with PMS solutions. Viewing the whole fabric samples, those bleached with PMS solution had distinctly whiter looking background areas than did the samples that were bleached with alkaline peroxide or peracetic solutions.

## Activated Peroxymonosulfate

N.J. Stalter; U.S. Patent 3,556,711; January 19, 1971; assigned to E.I. du Pont de Nemours and Company describes a process to bleach textiles dyed in portions with sensitive dyestuff by means of peroxymonosulfate activated with water-soluble salts of monocarboxylic acids such as sodium acetate. The two components are mixed in particulate form, in anhydrous state and kept dry by a desiccating agent such as magnesium sulfate. By the addition of detergents, alkalis, scouring powders, household cleaning and stain removing compositions can be formulated. For use in this process potassium peroxymonosulfate monohydrate, $KHSO_5 \cdot H_2O$ and the triple salt of potassium peroxymonosulfate of the formula $2KHSO_4 \cdot K_2SO_4 \cdot 2KHSO_5$ are preferred.

The commercially available form of this triple salt having an active oxygen content of about 4.7 weight percent is the preferred peroxymonosulfate for use. For stability reasons, the particulate compositions should be essentially dry, i.e., they should be essentially devoid of free water and of water loosely held in the form of water of hydration. Preferably, all components will be in their essentially anhydrous forms and the presence of a desiccant material in particulate form to protect against the presence of free water due to moisture picked up from the atmosphere will be desirable.

When formulating the solution compositions any of the abovementioned peroxymonosulfates or peroxymonosulfuric acid itself may be employed as the source of the peroxymonosulfate ion, $HSO_5^-$. Similarly, any of the abovementioned carboxylic acids, or the salts thereof, may be employed as the source of the acylate ion promoter. However, the aqueous solution containing both peroxymonosulfate and acylate ions should have a pH in the range of 5 to 10, since no significant promotional effect by the above acylate ions upon the oxidizing action of the peroxymonosulfate ion has been observed to be exerted at higher or lower pH values.

Solutions having a pH in the range of 5.5 to 8 are preferred. Adjustment of the pH of the solution so as to bring it within the pH range of 5 to 10, preferably 5.5 to 8, may be accomplished by the addition of any of the common soluble alkaline materials, when an adjustment of the pH upward is necessary or desired; or by the addition of any of the common soluble acidic materials, when an adjustment downward is necessary or desired.

The particulate compositions may be entirely free of components other than the peroxymonosulfate and carboxylic acid salt. However, they may also contain a source of alkali and/or a buffering agent such as tetrasodium pyrophosphate in such an amount that when the composition is dissolved in water to give a 3% solution thereof, the resulting solution will have a pH within the desired range of 5 to 10. Still other materials such as stabilizers, wetting agents, fluorescent brighteners and the like may also be incorporated, generally in minor amounts. The particulate compositions may be used in formulating dry bleaching compositions, e.g., for home laundry use, stain removal compositions, and as a component in home laundry detergent formulations and cleanser compositions. They may also be employed for sanitizing the waters of swimming pools and as components of denture cleaner compositions.

A dry home laundry bleach composition can be formulated, e.g., with a 1:1 weight mixture of sodium acetate and the commercially available potassium peroxymonosulfate triple salt product. A bleaching composition containing 50% of such a mixture with the balance consisting of light granular soda ash can be used effectively for home laundry bleaching purposes when the composition is added to the laundry water to provide an oxygen concentration of 20 to 50 ppm. The incorporation of a small amount, e.g., 0.1% of a fluorescent brightener is generally advantageous.

Mixtures of sodium acetate and the above potassium peroxymonosulfate triple salt product, e.g., at a 1:1 weight ratio, are also usable in formulating home laundry detergent compositions. Thus, such a detergent composition may contain 5 to 20% of such a mixture, 20% of an alkyl aryl sulfonate, 35 to 50% of sodium tripolyphosphate, 10% of sodium sulfate, and 10% of sodium metasilicate. The incorporation of small amounts of a soil-release agent such as sodium carboxymethylcellulose and an optical brightener is generally advantageous. Powdered cleanser compositions for removing stains from the surfaces of porcelain, aluminum ware and stainless steel ware may also be formulated using a 1:1 mixture of sodium acetate and the potassium peroxymonosulfate triple salt product. Such a cleanser composition may be formulated using 5% of such a mixture, 5% of an alkyl aryl sulfonate, 5% of anhydrous tetrasodium pyrophosphate, 5% sodium carbonate and 80% powdered silica.

When the particulate composition of the process is intended for use in preparing solutions for bleaching textiles, it is preferred that it contain a peroxymonosulfate and a salt of a 1 to 4 carbon saturated aliphatic monocarboxylic acid together with a buffering agent in such an amount that when the composition is dissolved in water to give a bleach solution having the desired active oxygen content, e.g., 0.01 to 0.2%, the solution will also have the desired pH, 5.5 to 8, without requiring the separate addition of any pH adjusting agent.

Buffering agents suitable for use in formulating such compositions include sodium tripolyphosphate ($Na_5P_3O_{10}$), sodium metasilicate ($Na_2SiO_3$), sodium tetraborate ($Na_2B_4O_7$), disodium phosphate ($Na_2HPO_4$), tetrasodium pyrophosphate ($Na_4P_2O_7$), soda ash ($Na_2CO_3$) and trisodium phosphate ($Na_3PO_4$). Of such agents, soda ash, tetrasodium pyrophosphate (TSPP), trisodium phosphate (TSP) and disodium phosphate (DSP) are generally preferred. These preferred agents are highly effective and behave generally similarly, particularly in solutions for bleaching textiles, e.g., at temperatures from room temperature to 212°F., preferably 150° to 190°F.

Example: A standard tea-coffee stained cotton fabric was prepared. The whiteness value of the stained fabric was then determined using a Hunter Reflectometer. In such a determination, the whiteness values are reported as percent reflectance representing the percent light reflected from the sample as measured using the Hunter Reflectometer with a blue filter for which magnesium oxide gives a reflectance of 100%. Samples of the stained fabric were then employed in home laundry bleaching trials using a standard Launder-O-Meter.

In one series of trials the laundering liquid contained a commercial alkyl benzene sulfonate-based laundry detergent at a concentration of 0.1% with no other additive. In a parallel series of trials the laundering liquid contained in addition to the 0.1% detergent an amount of the abovementioned commercial potassium peroxymonosulfate triple salt product (active oxygen content 4.7%) to provide 25 ppm of active oxygen. In a third series of trials the laundering liquid contained in addition to the detergent and the potassium peroxymonosulfate product at the above concentrations, sodium formate, or sodium acetate, or sodium butyrate at a concentration of 1.25%. In each of the series of trials, the laundering operation was repeated 8 times for 10 minutes each at 130°F. in the Launder-O-Meter, following which the fabric samples were washed and dried and the whiteness values determined.

It was found that the addition of either sodium formate, sodium acetate, or sodium butyrate to the laundering liquid containing the peroxymonosulfate increased significantly the final whiteness of the fabric. These salts clearly actively promoted the bleaching activity of the peroxymonosulfate. The promotional effect of sodium acetate was particularly outstanding in that the gain in reflectance points over the base plus the detergent was 6.7 as compared with only 3.9 when only the peroxymonosulfate was used with the detergent. The increase of 2.8 points due to the promotional effect of the sodium acetate represents a 72% increase in whiteness over that obtained when the peroxymonosulfate was used alone with the detergent.

## Desizing and Bleaching with Peroxymonosulfate

M.H. Rowe; U.S. Patent 3,619,111; November 9, 1971; assigned to E.I. du Pont de Nemours and Company describes a continuous process to desize, wash and bleach cotton-polyester fabrics, dyed in part with sensitive dyestuffs. The process consists of the following steps: (1) singeing the dye greige fabric with a flame; (2) quenching or extinguishing flame on the fabric by exposing the fabric to steam or to a water mist controlled to avoid a water pickup by the fabric greater than 30% of the fabric weight; (3) desizing and scouring the fabric and removing loosely held dyestuff therefrom by treating in a conventional multidip rope washer with an alkaline water solution of a surface active agent at a temperature of 160° to 210°F. and at a solution; fabric weight ratio of 20 to 40:1, then washing the fabric thoroughly with water; and, (4) saturating the fabric with a solution of a peroxymonosulfate buffered at a pH of 100° to 212°F. to effect bleaching thereof.

The process is illustrated by the following examples. In the examples all proportions or ratios and all compositions or concentrations expressed as percentages are by weight. In the examples, the abbreviation PMS is used to designate the commercially available potassium peroxymonosulfate product ($KHSO_4 \cdot K_2SO_4 \cdot 2KHSO_5$) and having an active oxygen content of 4.7%.

Example 1: A greige gingham fabric which was a 50:50 blend of cotton and polyester (ethylene glycol terephthalate) fibers containing naphthol red dyed areas in check and stripe patterns containing an excess of loosely held dye in the fibers in the dyed areas was singed in conventional manner by passing the dry fabric in open width form between opposing flame jets. Flame on the fabric was immediately quenched by next passing the open width fabric between opposing jets of steam. The fabric was then roped through poteyes and passed in rope form at a rate of 200 yards per minute through a multidip (28 dips) rope washer to which was fed continuously a water solution containing 0.3% of a commercial sodium dodecylbenzene sulfonate and 0.5% trisodium orthophosphate and having a pH of 11.

The rate of feed of the solution to the washer and the overflow of solution therefrom was at a rate of about 1.5 times the rate of feed of the fabric through the washer, on a weight bases. The fabric passed through the washer in serpentine fashion so as to make about 28 dips into and out of the solution. The solution in the washer was maintained at 180°F. and the residence time of the fabric in the solution was 28 seconds. The fabric exiting from the above washer was fed through a second similar washer through which was similarly passed water maintained at 150°F. It was then similarly washed in a third similar washer through which was similarly passed water at room temperature (76°F.).

The washed fabric was then saturated with an amount of a bleach solution (about equal to the dry weight of the fabric) by immersing it in the bleach solution at room temperature, then expressing excess solution from the fabric. The bleach solution contained 1.6% PMS, 0.25% trisodium phosphate, and 1.1% sodium acetate. It had a pH of 6.4. The saturated fabric was then bleached by passing it continuously through a J-box maintained at 170°F. with a mixture of steam and air. The rate of travel of the fabric through the J-box was such as to give a residence time therein of 90 minutes. The bleached fabric withdrawn from the J-box was washed thoroughly with water at 150°F. The resulting fabric was well bleached and showed no evidence of color bleeding or markoff onto the white background portions, even though the starting fabric contained excess loosely held dye on the fibers in the colored areas.

Example 2: A greige fabric which was about a 50:50 blend of cotton and polyester (ethylene glycol terephthalate) fibers and contained blue stripes on a white background with the blue stripes consisting of cotton fibers dyed with a vat blue dye and polyester fibers dyed with a disperse blue dye, was singed as described in Example 1 then all surfaces of the fabric were exposed to a fine mist of water to quench or extinguish surface flames. The time of exposure to the water mist was such that the water pickup was about 10% of the dry weight of the fabric.

The fabric was roped through poteyes, then passed through a multidip rope washer as described in Example 1 where it was treated at 190°F. with a solution containing 0.5% sodium lauryl sulfate and sufficient soda ash to give a pH of 11. The rate of feed of the solution to the washer and the overflow of the solution therefrom was at a rate of 1.2 times the rate of feed of the fabric through the washer, on a weight basis. The residence time of the fabric in the washer was about 45 seconds. The fabric exiting from the washer was washed with hot water, then with cold water and then bleached, all as described in Example 1. The resulting fabric was well bleached and showed no evidence of color bleeding or markoff onto the white background portions, even though the colored stripes of the starting fabric contained excess loosely held dyes on the fibers.

## Peroxymonophosphate by Enzymatic Hydrolysis

H.M. Castrantas; U.S. Patent 3,666,399; May 30, 1972; assigned to FMC Corp. describes a process to generate peroxymonophosphate by enzymatic hydrolysis of peroxydiphosphate. The controlled generation of the active peracid, peroxymonophosphate has substantial advantages over preformed peroxymonophosphate in bleaching and germicidal processes. The controlled enzymatic hydrolysis of peroxydiphosphates takes place in an aqueous environment under proper conditions of temperature, peroxydiphosphate concentration, phosphatase/peroxydiphosphate ratio, pH, and time. Previously, it was not known how to control the hydrolysis of peroxydiphosphates except by acidification at a strongly acidic pH of about 3.

The reactants consist of a peroxydiphosphate, a phosphatase enzyme or phosphatase enzyme-containing substance and water. The operating temperature is between 0° and 50°C., although ambient temperatures, 20° to 30°C., is preferred. Any of the peroxydiphosphates or their corresponding acid salts that are water-soluble to the extent of 0.001 weight percent can be used in the composition of this process. Examples of these are potassium peroxydiphosphate ($K_4P_2O_8$), lithium peroxydiphosphate ($Li_4P_2O_8$) sodium peroxydiphosphate ($Na_4P_2O_8$), ammonium peroxydiphosphate $[NH_4)_4P_2O_8]$, tripotassium monosodium peroxydiphosphate ($K_3NaP_2O_8$), monoammonium tripotassium peroxydiphosphate $[K_3(NH_4)P_2O_8]$, etc.

The peroxymonophosphate which is produced by the hydrolysis gradually hydrolyzes to produce hydrogen peroxide, which supplements the activity of the peroxymonophosphate in such application as bleaching and germicidal applications. All the analyses given herein report the total amount of active-peroxygen products of the hydrolysis and includes the small amount of hydrogen peroxide which results from this hydrolysis, that is from essentially none to as much as 15 to 20%. Phosphatase enzymes are divided into two classifications, the acid and the alkaline. Both of these are active in hydrolyzing peroxydiphosphate, and therefore both can be used in practicing this process. Phosphatase enzymes that have been derived from a phosphatase enzyme-containing substance, or the phosphatase enzyme-containing substances themselves without any purification, can be used in this process as the source of the phosphatase enzyme. However, the degree of purity of the phosphatase enzyme affects the quantity needed to obtain a specific hydrolysis rate. Acid phosphatase-containing substances that can be used include wheat flour, wheat germ, rice, barley, potatoes and regular household all-purpose bleached flour. Alkaline phosphatase-containing substances that can be used include intestinal mucosa and liver cortex.

The hydrolysis rate and the yield of peroxymonophosphate are regulated by adjusting the peroxydiphosphate concentration, the phosphatase/peroxydiphosphate ratio, the purity of the phosphatase and the temperature. By the proper regulation of these conditions, slow, controlled release of peroxymonophosphate over several hours or weeks can be obtained readily. The concentration of the peroxydiphosphate and the phosphatase/peroxydiphosphate ratio can be varied greatly depending upon the application. The minimum amount of phosphatase enzyme needed to activate the hydrolysis of peroxydiphosphates is 0.0004 part of phosphatase enzyme by weight per part of peroxydiphosphate, for an enzyme having an activity of 6 units per milligram. (A phosphatase activity unit is defined as that amount of enzyme required to release one micromol of phenol from p-nitro phenyl phosphate at 37°C.) There is no maximum limit on the proportion of enzyme in the combination of this process, other than a practical economic one. Use of more enzyme generally increases the rate of hydrolysis.

The minimum concentration of peroxydiphosphate in solution depends upon the application. In germicidal applications, there should be at least 0.1 weight percent of peroxydiphosphate in the water-phosphatase solution. In bleaching operations, there should be at least 0.02 weight percent of peroxydiphosphate in the water-phosphatase solution. Hydrolysis of the peroxydiphosphate takes place at concentrations as low as 0.001 weight percent peroxydiphosphate in solution, which is sufficient for some applications.

When bleaching is the application, the preferred concentration of peroxydiphosphate is about 1 to 5 weight percent of the total solution and the preferred weight ratio of phosphatase enzyme to peroxydiphosphate is between 1/100 and 1/300. For germicidal processes, the preferred concentration of peroxydiphosphate is 0.1 to 1 weight percent of the total solution and the preferred weight ratio of phosphatase enzyme to peroxydiphosphate is between 1/100 and 1/300. The optimum pH for the hydrolysis depends upon the specific phosphatase enzyme used. In general, acid phosphatase are most active between pH 4 and 7, and alkaline phosphatases between pH 8 and 10. However, both are quite active over much wider pH ranges of 3.5 to 10 for acid phosphatase and 7 to 11 for alkaline phosphatase.

In bleaching processes, the controlled generation of peroxymonophosphate permits the optimum concentration of peroxymonophosphate to be maintained throughout the process even when prolonged bleaching is desired. This can be accomplished by adjusting the rate at which peroxymonophosphate is generated to approximately equal the rate at which peroxymonophosphate is dissipated during bleaching, thereby maintaining the peroxymonophosphate concentration consistently at the optimum value. The addition of the total quantity of peroxymonophosphate at the start of the bleaching process as practiced heretofore, resulted in a deleterious environment for many materials. The prior art acidification method of generating peroxymonophosphate also resulted in a deleterious environment because of the strongly acidic solution required for reasonable hydrolysis rates. Figures 5.5a and 5.5b show the effect of the variables upon the hydrolysis rate. Figure 5.5a shows the effect of an alkaline phosphatase enzyme derived from calf intestine, on the hydrolysis of potassium peroxydiphosphate at pH 9.1. The data shown in Figure 5.5a was obtained as described in Example 1. Figure 5.5b shows the effect of an acid phosphatase enzyme derived from wheat germ on the hydrolysis of potassium peroxydiphosphate at different phosphatase/peroxydiphosphate ratios. The data shown in Figure 5.5b was obtained as described in Example 2.

FIGURE 5.5: ENZYMATIC PEROXYDIPHOSPHATE HYDROLYSIS

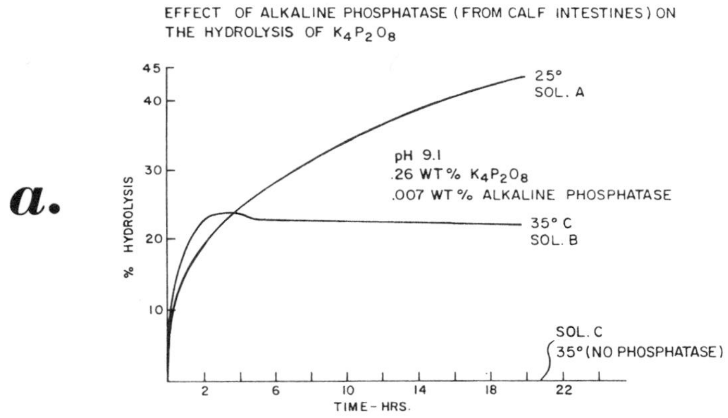


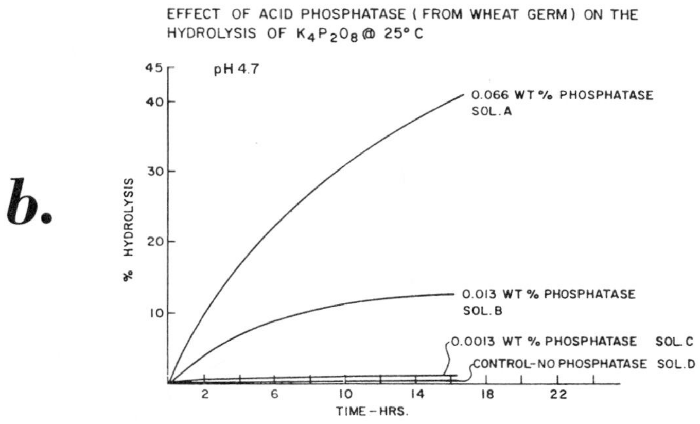


Source: H.M. Castrantas; U.S. Patent 3,666,399; May 30, 1972

Example 1: Three solutions (A, B and C) were prepared, each containing 0.26 grams of potassium peroxydiphosphate ($K_4P_2O_8$) and 100 grams of distilled water. The pH value of the solution was 9.1. To each of solutions A and B was added 0.007 gram of an alkaline phosphatase enzyme derived from calf intestines. The alkaline phosphatase was omitted from comparative solution C. Solution A was maintained at 25°C. and solutions B and C were maintained at 35°C.

Periodically aliquots of each solution were titrated for peroxymonophosphate anion following the procedure of M.M. Crutchfield, "The Acidity, Complexes, Hydrolysis and Decomposition of Peroxydiphosphate and Its Salts," June 1960 (Thesis in Brown University Library). The results are shown in Figure 5.5a, and demonstrate that in the absence of the phosphatase enzyme, no peroxymonophosphate was formed, while with alkaline phosphatase enzyme present, peroxymonophosphate formed at different rates, depending upon temperature.

Example 2: Four solutions (A, B, C and D) were prepared, each containing 0.5 gram of potassium peroxydiphosphate ($K_4P_2O_8$) and 100 grams of distilled water. The pH values of the solutions were adjusted to 4.7 with a sodium acetate, acetic acid buffer. An acid phosphatase enzyme derived from wheat germ was added to solutions A, B, and C, but not to comparative solution D.

The amount of enzyme added to solution A equaled 0.066 weight percent of the solution. The amount of enzyme added to solution B was one-fifth of the amount added to solution A (0.013 weight percent). The amount of enzyme added to solution C was one-fiftieth of the amount added to A (0.0013 weight percent). Aliquots from each solution were periodically withdrawn and analyzed for peroxymonophosphate anion following the procedure used in Example 1. The temperature was maintained at 25°C. The results are shown in Figure 5.5b and show that increasing the phosphatase ratio increases the rate of formation of peroxymonophosphate.

# CHLORINE RELEASING ORGANIC COMPOSITIONS

## STABILIZED BLEACHING COMPOSITIONS

### Trichlorocyanuric Acid Stabilized with Sulfonamides

W.F. Symes; U.S. Patent 3,002,931; October 3, 1961; assigned to Monsanto Chemical Company found that the stability of trichlorocyanurate bleach, containing alkaline metal salts is improved, in the presence of moisture, by the addition of aryl sulfonamides, such as benzene sulfonamide. The process is applicable to compositions containing in excess of 0.5% by weight of moisture, but is particularly applicable to solid and free-flowing compositions containing in excess of 1% by weight of moisture, as well as aqueous slurries or solutions of such compositions. The trichlorocyanuric acid employed in the compositions has a theoretical available chlorine content of 91.5%. The commercially available product containing between about 84 and 91.5% available chlorine can also be used.

A variety of alkaline, water-soluble, alkali metal salts can be used in the compositions, including, preferably, those inorganic salts which are employed as detergent builders. As examples of such salts may be mentioned trialkali metal phosphates such as trisodium phosphate and tripotassium phosphate; dialkali metal hydrogen phosphates such as disodium hydrogen phosphate and dipotassium hydrogen phosphate; the alkaline, water-soluble molecularly dehydrated alkali metal phosphate salts such as the alkali metal pyrophosphates, for example, tetrasodium pyrophosphate, trisodium hydrogen pyrophosphate and tetrapotassium pyrophosphate.

Further examples include the alkali metal tripolyphosphates such as sodium tripolyphosphate ($Na_5P_3O_{10}$) and potassium tripolyphosphate; the alkaline, water-soluble alkali metal metaphosphates such as sodium hexametaphosphate; the water-soluble alkali metal silicates such as sodium silicates having an $Na_2O$ to $SiO_2$ mol ratio of 1.5:1 to 1:3.6, preferably 1:1 to 1:3.5, and the corresponding potassium silicates; the water-soluble alkali metal borates such as calcined sodium tetraborate or borax, and the water-soluble alkali metal carbonates or bicarbonates such as sodium or potassium carbonates.

In addition the composition may contain anionic wetting agents and inert diluent salts. The various ingredients can be used in the compositions in various proportions depending on whether the composition is to be used as a bleaching composition, a disinfecting composition, a dishwashing composition, a detergent composition, etc.

As an example, the compositions comprise, on a solids basis, from about 3 to 15% by weight of trichlorocyanuric acid, about 0.3 to 1.2 mol of benzene sulfonamide per mol of trichlorocyanuric acid, about 10 to 60% by weight of sodium tripolyphosphate or a mixture of such phosphate and sodium silicate and the remainder consisting substantially of sodium sulfate. Such compositions are useful as commercial laundry bleaches and dishwashing compositions. Also, the compositions may comprise, on a solids basis, about 3 to 15% by weight of trichlorocyanuric acid, about 0.3 to 1.2 mols of benzene sulfonamide per mol of trichlorocyanuric acid, about 10 to 60% by weight of sodium tripolyphosphate, about 0.1 to 5% by weight of anionic wetting agent, preferably sodium dodecyl benzene sulfonate, and the remainder consisting substantially of sodium sulfate. Such compositions are useful as household bleaches and sanitizers.

In another variation, the compositions comprise from about 3 to 15% by weight of trichlorocyanuric acid, from about 0.4 to 1.4 mols of benzene sulfonamide per mol of trichlorocyanuric acid, and the remainder consisting substantially of sodium carbonate or mixtures with sodium tripolyphosphate. These composition are useful as sanitizer detergents, particularly in cleaning and sanitizing food processing equipment and containers. Comparisons indicate that moisture tends to decompose compositions containing trichlorocyanuric acid and alkaline salts, whereas this decomposition is suppressed or inhibited or retarded when benzene sulfonamide is present in the composition with the most effective results being obtained by using from 0.18 to 1.1 mols of benzene sulfonamide per mol of trichlorocyanuric acid.

## Stable Dichlorocyanurate Complex Salts

R.W. Marek; U.S. Patent 3,055,889; September 25, 1962; assigned to Olin Mathieson Chemical Corporation describes the reaction of alkali or alkaline earth metal salts of dichlorocyanuric acid with salts of cadmium, nickel or copper. The resulting complexes contain available chlorine and are more stable than the dichlorocyanurates used in their preparation. Thus, the compounds are useful for general bleaching of textiles and fabrics and disinfecting; as in the prevention of growth of fungi and algae in water cooling towers and swimming pools.

These compounds are prepared by reacting an alkali metal or alkaline earth metal dichlorocyanurate with a nickel, copper or cadmium salt. The reaction is started at a pH of about 4 to 7 in a solvent for the reactants such as water or aqueous acetone from which the product precipitates. The temperature and reactant concentrations are not critical except that enough of the reactants be used to exceed the solubility of the product to facilitate its recovery by filtration. Room temperature is most convenient. The pH should be on the acid side to prevent precipitation of the copper, cadmium or nickel oxides or hydroxides. After mixing the reactants, a brief stirring period, such as a few minutes to a half hour, facilitates precipitation of the complex compound.

The structure of all of these compounds is believed to be of the form: $A_n(MZ_4)\cdot xH_2O$ wherein A is the alkali or alkaline earth metal, n is 2 for the alkali metals and 1 for the alkaline earth metals, M is copper, cadmium or nickel and x is 0 to about 6 depending upon the nature of the various complexes. In the formula, Z is the dichlorocyanurate radical which has the formula $(C_3N_3O_3Cl_2)^-$ or, structurally:

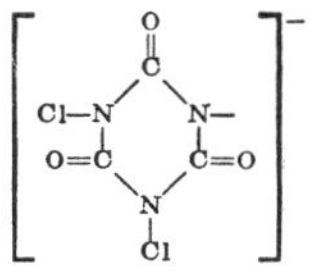

The salts of dichlorocyanuric acid are readily prepared by dissolving or suspending dichlorocyanuric acid or trichlorocyanuric acid or a mixture of the two in water or aqueous acetone and adding thereto a base such as sodium or potassium hydroxide. Weaker bases such as sodium acetate are also useful. At a pH of about 7 all of the dichlorocyanuric acid has been converted to the salt. The pH of the reaction mixture should not be allowed to exceed that at which the insoluble oxides or hydroxides of copper, cadmium or nickel will precipitate. A pH range of 4 to 7 is satisfactory for most reactions.

Example 1: 1 mol of sodium dichlorocyanurate was dissolved in 1,200 g. of water and a solution containing 1 mol of cupric sulfate in 890 g. of water was added to it. The mixture was allowed to stand about 5 minutes and the precipitate was filtered off. It was washed with 100 g. of water and dried to constant weight at 55°C. The color of the precipitate was Periwinkle (A Dictionary of Color, Maerz and Rea Paul, McGraw-Hill and Co., Inc., 1950). The yield was 206 g. or 86% of material having the composition $Na_2[Cu(C_3N_3O_3Cl_2)_4]$.

Example 2: 1/2 mol of dichlorocyanuric acid was dissolved in 500 ml. of acetone. To it was added a solution containing 2 mols of sodium acetate in 400 g. of water and then 0.25 mol of cupric sulfate in 323 g. of water. After 5 minutes the precipitate which formed was filtered off. It was washed with 50 ml. of water and then with 50 ml. of acetone. This product was dried at 55°C. to a constant weight of 95 g. The available chlorine was 58.9% corresponding to the formula $Na_2[Cu(C_3N_3O_3Cl_2)_4]$.

Example 3: 1/10 of a mol of trichlorocyanuric acid was dissolved in 100 ml. of acetone. To this was added 0.4 mol of sodium acetate in 80 ml. of water and 0.04 mol of cupric sulfate in 100 ml. of water. The precipitate formed immediately. It was filtered off, washed with 50 ml. of water and dried. The available chlorine was 61.2%.

Example 4: The following experiment demonstrates the bleaching ability of several of these complex compounds. The performance of the simple calcium and sodium dichlorocyanurates were included for comparison. Each of the compounds listed below were dissolved in water to give a solution containing 200 parts per million of available chlorine. To each solution was added 2,500 parts per million of a commercially available laundering detergent whose active ingredients were alkyl aryl sulfonate and tallow alcohol sulfate and 400 parts per million of sodium metasilicate as buffer. Sections of Indian Head cotton cloth which had been stained with aqueous tea solutions were placed in each bath for 8 minutes at 140°F. After this, they were rinsed in water and then in dilute acetic acid and the brightness, or light reflectancy, was measured by means of a Photovolt Brightness Meter. Two blanks are included to show the effect of water alone and water with detergent:

| Composition | Brightness |
|---|---|
| Water alone | 63 |
| Aqueous detergent | 65.5 |
| Aqueous detergent and sodium dichloro-cyanurate | 86.5 |
| Aqueous detergent and calcium dichloro-cyanurate | 85.0 |
| Aqueous detergent and cupric ion complex of: | |
| Sodium dichlorocyanurate | 84.5 |
| Potassium dichlorocyanurate | 86.0 |
| Calcium dichlorocyanurate | 87.5 |

The initial reflectancy of the tea-stained cloth was 55. Thus the complexes of this process bleach as well as the simple dichlorocyanurates.

Dichloroisocyanuric Acid and Tripolyphosphate

P.J. Schauer and R.N. Weston; U.S. Patent 3,096,291; July 2, 1963; assigned to Monsanto Chemical Company describe a process to prepare dry, stable and granular mixtures of dichloroisocyanuric acid and tripolyphosphate hexahydrate. The product dissolves readily in water and forms clear solutions with strong bleaching, disinfecting and germicidal action. The granular product is obtained by mixing 5 to 50 parts dichloroisocyanuric acid, anhydrous powder, 40 to 85 parts sodium tripolyphosphate, anhydrous granular, 10 to 17 parts water, the components being so selected that they add up to 100 parts by weight.

The dichloroisocyanuric acid component used is anhydrous (i.e., contain less than about 1% by weight of water) and powdery, i.e., 90% or more will pass through 100 mesh screen. The sodium tripolyphosphate component used is anhydrous (i.e., contain less than about 1% by weight of water) and granular, i.e., 98% or more will pass through 20 mesh screen and 98% or more will be retained on 100 mesh screen. This salt exists in two forms designated as $Na_5P_3O_{10}$ I and $Na_5P_3O_{10}$ II as described in volume 63, Journal of the American Chemical Society, pp. 461 to 462. In this process either form can be used or any mixture thereof.

The respective materials, i.e., anhydrous powdered dichloroisocyanuric acid and granular anhydrous sodium tripolyphosphate are brought together and intimately mixed in any desired fashion and while mixing same the water is applied in the form of droplets as for example by spraying water on the mixing mass. It is preferred that the mixing of the three components take place at about room temperature, however, temperatures in the range of 10° to 50°C. can be employed. The agglomerated mass so obtained is then heated at about 50° to 90°C. to reduce the water content to that in the range of 1 to 10% by weight of the total mass. It is preferred that the agglomerated solids be aged for at least 6 hours at about room temperature prior to heating the solids to reduce the water content.

During the mixing and heating operation a reaction takes place producing sodium dichloroisocyanurate and a sodium acid tripolyphosphate having the empirical formula

$$Na_{5-n}H_nP_3O_{10}\cdot XH_2O$$

wherein n is a whole number from 1 to 4, usually 1 or 2, and wherein X is an integer from 0 to 2. In this reaction the hydrates of sodium dichloroisocyanurate form preferentially, the dihydrate of sodium dichloroisocyanurate forming preferentially to the monohydrate. In this reaction the amount of sodium acid tripolyphosphate as sodium will be equal, in other words, for each sodium atom given up by anhydrous granular sodium tripolyphosphate one molecular proportion of dichloroisocyanuric acid is converted to the sodium salt thereof.

Example: To a rotating drum at room temperature is added and intimately mixed 41 parts by weight of anhydrous powdery dichloroisocyanuric acid (100% passed through 100 mesh screen) and 45 parts by weight of anhydrous granular sodium tripolyphosphate (100% passed through 20 mesh and 100% was retained on 100 mesh). After about 5 minutes of mixing and while continuing the mixing 14 parts of water is sprayed onto the mixture. Upon completion of the water addition the mass is mixed for an additional 5 minutes. The granules or agglomerates are then removed from the drum and placed in a hot air circulating oven and heated at 70°C. solids temperature for about 30 minutes and then permitted to cool to room temperature. This nondusty granular product contains the following percentage by weight.

35% dichloroisocyanuric acid
10% sodium dichloroisocyanurate monohydrate
3% sodium dichloroisocyanurate dihydrate
19% sodium acid tripolyphosphate, $Na_4HP_3O_{10}$
33% sodium tripolyphosphate hexahydrate

The available chlorine content of the granular product is 27.1%. This product is very stable under normal storage conditions.

Chlorinated Acetone-Urea

M. Kokorudz; U.S. Patent 3,104,260; September 17, 1963; assigned to Wyandotte Chemicals Corporation describes a process to use chlorinated acetone-urea condensation product as an organic, active chlorine bleaching composition for cotton fabrics. The product is stable at room temperature and can be used in a formulated bleaching composition by mixing it in the substantially dry state with alkaline salts, such as sodium carbonate, sodium borate, sodium silicate, trisodium phosphate, tetrasodium pyrophosphate, sodium triphosphate or mixtures of these. In addition, wetting agents, synthetic detergents generally, soaps, fillers, abrasives and water softening agents of the inorganic or organic type may be incorporated in bleach compositions containing the organic active chlorine compound of the process in order to impart special properties.

The organic active chlorine bleach compound is readily prepared by suspending an acetone-urea condensation product in an aqueous solution having a pH of less than about 12.5, cooling the resulting mixture to below about 14°C. and passing chlorine gas into the cooled mixture while maintaining the temperature at no more than about 30°C. The acetone-urea condensation product is prepared by reacting a mixture of acetone and urea with a strong mineral acid dehydrator such as hydrochloric acid and sulfuric acid. Urea is suspended in a molar excess of acetone sufficient to serve as the reaction media as well as one of the reactants. In practice it has been found that a ratio of about 3.5:1 mols of acetone to mols of urea is sufficient to obtain high yields. The mineral acid must be present in sufficient quantity to remove the water formed by the reaction. Successful reactions have been carried out where the mineral acid was present in the ratio of 1:1 mols of acid to mols of urea. The reaction is carried out at a temperature up to and including about 60°C. and until the crystals of the acetone-urea condensation product have been formed. The crystals are separated from the reaction mixture by filtration and air dried. The condensation product produced has been analyzed and found to contain carbon, hydrogen and nitrogen in a weight ratio of about 47 parts to 8.6 parts to 20.3 parts, respectively. This analysis corresponds to an approximate empirical formula of $C_{11}H_{20}N_4O_2$.

The aqueous solution in which the chlorination is carried out can be alkaline, neutral or acidic before the chlorination is commenced. The solution may be rendered alkaline or acidic by the use of any material which does not have an adverse effect resulting in the decomposition of the acetone-urea condensation product. This problem can generally be avoided if the pH is kept below about 12.5. Examples of suitable materials are sodium or potassium carbonate or bicarbonate and hydrochloric acid. However, ammonium salts should not be used because of the danger of forming nitrogen trichloride.

Sodium hydroxide and potassium hydroxide present in concentrations of about 10% decomposed in the acetone-urea condensation product and therefore should be avoided. Good yields of the chlorinated acetone-urea condensation product have been obtained when the acetone-urea condensation product was suspended in water and higher yields were obtained when the condensation product was suspended in an aqueous alkaline solution having 1.4 mol equivalents of sodium carbonate per 100 grams of condensation product. The temperature of the mixture while the chlorine gas is passed into the mixture is desirably about 0° to 30°C. and preferably should not be substantially above about 15°C.

Chlorine gas is passed in over a period of time in the range of about 0.5 to about 6 hours until the acetone-urea condensation product is chlorinated. In the case where chlorination is carried out in an alkaline solution containing about 1.4 mol equivalents of alkaline compound per 100 grams of acetone-urea condensation product, chlorine gas is passed into the cooled mixture until the pH of the reaction mixture is in the range of 7 to 5, desirably about 6. It should be apparent that the chlorination is completed at a higher pH when higher concentrations of alkaline compound are used.

In the case of mixtures initially neutral, acidic, or slightly alkaline (containing less than 1.4 mol equivalents of alkaline compound per 100 g. of acetone-urea condensation product), the chlorine addition is stopped when the solution becomes noticeably yellow due to the temination of the acetone-urea condensation product chlorination. When the addition of chlorine gas is completed the white solid product is easily separated from the reaction mixture by filtration and air dried. The compound is very stable under ordinary conditions but undergoes rapid decomposition at about 90°C.

Example: The acetone-urea condensation product was prepared by suspending 120 grams of urea in 400 ml. of acetone in a 1-liter 3-necked flask equipped with a stirrer, thermometer, condenser, and addition funnel for introducing the sulfuric acid. 200 grams of anhydrous sulfuric acid were introduced into the mixture continuously over a period of about 1 hour. The mixture was initially at room temperature and when the sulfuric acid was introduced the temperature increased because of the exothermic reaction and was held within the range of 50° to 56°C. After the addition of the sulfuric acid was completed, the mixture was stirred for about 7 hours at a temperature of 58° to 60°C. and allowed to stand at room temperature for about 15 hours. The mixture was filtered and the crude product washed with methanol. After washing, the product was suspended in 750 ml. of ice water and 200 ml. of 25% NaOH solution were added to neutralize the free acid present. The white solid product was filtered and dried. The weight of the product after drying was 214 g. The active chlorine bleach compound was prepared by the chlorination of the acetone-urea condensation product prepared by the procedure described. An alkaline solution in which the chlorination was carried out was prepared by dissolving 75 g. (0.7 mol) of $Na_2CO_3$ in 1,000 ml. of water. The aqueous sodium carbonate solution was cooled to 10°C. and 100 g. of acetone-urea condensation product were added. The pH after addition of the acetone-urea condensation product was 11.5.

Chlorine gas was introduced to the aqueous alkaline solution of the acetone-urea condensation product while cooling the mixture so that the temperature was in the range of 3° to 6°C. Chlorine gas was passed into the alkaline solution over a period of about 5 hours until the pH was 6.0. The white solid product which was produced was filtered and air dried at room temperature and weighed. A yield of 125 g. of the chlorinated acetone-urea condensation product was obtained. Analysis showed the compound contained 33.8% active chlorine. The test of the active chlorine compound prepared in this example at 200 parts per million provided a bleached swatch which had a 65.8% reflectance at pH of 10. These results are generally equal to or superior to the bleaching effectiveness of 1,3-dichloro-5,5-dimethylhydantoin which gave a bleached swatch under comparable conditions having a 65.6% reflectance.

Halo-Glycoluril Compositions

I. Rosen and F.B. Slezak; U.S. Patent 3,019,075; January 30, 1962; and U.S. Patent 3,019,160; January 30, 1962; both assigned to Diamond Alkali Company describe in two related patents the use of chlorinated glycolurils for the bleaching of cotton fabrics and as fungicides. The chlorinated glycolurils have the following general formula: 2,4,6,8-tetrachloro-2,4,6,8-tetrazabicyclo[3.3.0]octa-3,7-dione

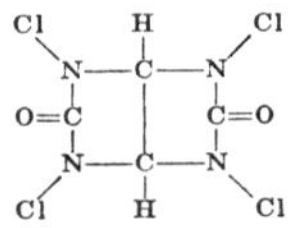

1,5-dimethyl-2,4,6,8-tetrachloro-2,4,6,8-tetrazabicyclo[3.3.1]nona-3,7-dione

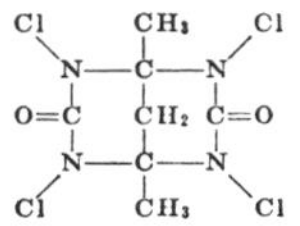

dichloro-2,4,6,8-tetrazabicyclo[3.3.0]octa-3,7-dione

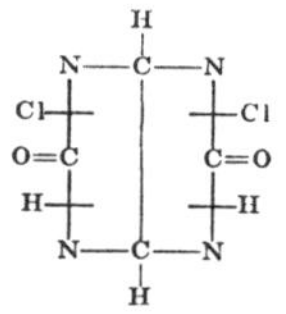

In addition to the chlorinated glycolurils, the composition may contain major or minor proportions of alkali soap builders, such as carbonates, silicates and phosphates; wetting agents such as sulfonates and polyether alcohols; also soaps and water softening agents. Examples 1 and 2 describe the preparation of the chlorinated glycolurils and Example 3, their use.

Example 1: Preparation of Dichloro-2,4,6,8-Tetrazabicyclo[3.3.0]Octa-3,7-Dione — There is suspended in 800 ml. of water 14.2 g. (0.1 mol) of glycoluril and the introduction of chlorine is begun, a 6N solution of NaOH being added portionwise to maintain the pH within the range from 7 to 8. A total of 89.2% of the theoretical amount of chlorine (based on stoichiometric amount required to form trichloroglycoluril) and 45 ml. of 6N NaOH introduced. The resultant material is filtered and the filtrate evaporated to dryness to yield 34.8 g. of a white solid. This solid is washed twice with water and dried under suction. There results 17.1 g. of a white solid. Chemical analysis indicates preparation of the desired $C_4H_4Cl_2N_4O_2$.

Example 2: Preparation of 1,5-Dimethyl-2,4,6,8-Tetrachloro-2,4,6,8-Tetrazabicyclo[3.3.1]Nona-3,7-Dione — Into 3 liters of water is introduced 56 g. (0.3 mol) of 1,5-dimethyl-2,4,6,8-tetrazobicyclo[3.3.1]nona-3,7-dione [Rec. trav. chim. 27, 162-91 (1908)]. Chlorine is then gradually introduced into the stirred solution simultaneously with the addition of 6N sodium hydroxide at a rate to maintain the pH of the reaction mixture within the range from approximately 5 to 8. A total of 130% of the theoretical amount of chlorine (110 g.) and 125% of the theoretical amount of 6N sodium hydroxide (250.2 ml.) are added. The resultant reaction mixture is filtered and the pasty residue washed with 400 to 600 ml. of water and filtered again. The thus-obtained solid is allowed to dry to yield a white powder weighing 86.8 g. Chemical analysis indicates that it contains 78.3% available chlorine (theoretical available chlorine 88%). Chemical analysis indicates preparation of the desired $C_7H_8Cl_4N_4O_2$.

Example 3: Part A — To illustrate the effectiveness of compounds of this process as bleaching agents, tests are conducted whereby tea-stained, unbleached muslin is bleached in solutions containing such compounds. More particularly, the procedure employed is as follows: Unbleached muslin is scoured for 6 hours at the boil in 9 liters of a 1% NaOH solution containing 20 g. of Nacconal NR (alkyl aryl sulfonate) and 2.0 g. of Rapidase-Z (starch and size-removing bacteria). The thus-scoured muslin (465 g.) is then stained by immersion in a solution consisting of 9 liters of water containing 140 ml. of a stock tea solution prepared by leaching 15 conventional tea bags in 1 liter of water for 35 minutes at 97°C. The thus-stained cloth is then rinsed in cold water and dried.

Bleach baths are prepared by adding sufficient of a saturated aqueous solution of the compound being investigated to provide a bath 300 ml. in volume and containing 100 ppm of available chlorine, using a 5% aqueous solution of sodium tripolyphosphate to buffer the bath to a pH of about 9.4. The available chlorine content is checked just prior to conducting the tests of titration against sodium thiosulfate using starch as an indicator. Using such bleach solutions in glass jars, one strip (9 x 7 inches) of unbleached, scoured, tea-stained muslin is immersed in each bath and the jar placed in the water bath of a Launder-Ometer (Atlas Electric Devices Company) (Model B-5, Type LHD-EF) for 20 minutes at a predetermined test temperature. The muslin is then washed with cold tap water, dried, and the bleach effectiveness determined by measuring the reflectance of the bleached samples with a reflectometer (Hunter

Multipurpose Reflectometer). Using this procedure, tests are conducted at temperatures of 80°, 100°, 120°, 140° and 160°F. The resultant data, presented as percent whiteness increase, are as follows:

| | Whiteness Increase (Percent) | | | | |
|---|---|---|---|---|---|
| Bleach Mixture | 80°F. | 100°F. | 120°F. | 140°F. | 160°F. |
| Dichloro-2,4,6,8-tetrazabicyclo[3.3.0]octa-3,7-dione | 19.6 | 22.8 | 24.2 | 25.7 | 27.8 |
| 2,4,6,8-tetrachloro-2,4,6,8-tetrazabicyclo[3.3.0]octa-3,7-dione | 25.4 | 28.6 | 30.3 | 29.5 | 32.1 |

Part B — To illustrate the synergistic effect obtained by combining two compounds of this process, experiments are carried out wherein a mixture of dichloro-2,4,6,8-tetrazabicyclo[3.3.0]octa-3,7-dione and 2,4,6,8-tetrachloro-2,4,6,8-tetrazabicyclo[3.3.0]octa-3,7-dione is employed in bleaching tests as described hereinbefore. The results of such tests indicate that more effective bleaching is obtained with this mixture than with either compound alone.

## Trichlorocyanuric Acid Deodorized by Spray Dried Silicate Salt

R.C. Ferris; U.S. Patent 3,093,590; June 11, 1963; assigned to Purex Corporation, Ltd. describes a process to sterilize and deodorize trichlorocyanuric acid bleach by the admixture of spray dried base containing sodium silicate, sodium phosphate and sodium sulfate. Examples 1 through 5 give the composition of the type of base used. The ingredients are slurried and spray dried in form of beads. Mixed with 10% by weight of trichlorocyanuric acid, the mixtures are useful as bleaches and disinfectants.

Example 1:

| | |
|---|---|
| "N" grade sodium silicate (anhydrous basis)* | 15.0% |
| Sodium tripolyphosphate | 10.0% |
| Sodium sulfate | 75.0% |

*"N" grade silicate has an approximate $Na_2O:SiO_2$ ratio of 1:3.22. A syrupy liquid containing 62.4% water.

This base is essentially of inorganic nature, but may contain small amounts of optical dyes, dedusting and surface active agents, ordinarily known to the trade.

Example 2:

| | |
|---|---|
| Metso 99* (anhydrous basis) | 9.0% |
| Sodium sulfate | 88.0% |

*Metso 99 is a hydrous sodium silicate powder which has an $Na_2O:SiO_2$ ratio of 3:2. The water content is 38.7%.

This base is essentially inorganic, but preferably contains up to 4% of a surface active agent.

Example 3:

| | |
|---|---|
| "D" grade sodium silicate* (anhydrous basis) | 10.0% |
| Sodium sulfate | ---- |

*"D" grade silicate is a syrupy solution which has an $Na_2O:SiO_2$ ratio of 1:2. The water content is 56.5%.

Example 4:

| | |
|---|---|
| Metso 99 (anhydrous basis) | 6.0% |
| Sodium tripolyphosphate | 10.0% |
| Sodium sulfate | 81.0% |

Example 5:

| | |
|---|---|
| "D" grade silicate (anhydrous basis) | 3.0% |
| Sodium tripolyphosphate | 17.0% |
| Sodium sulfate | 77.0% |

These bases (Examples 3,4, and 5) are essentially inorganic and may contain up to 4% surface active agent.

Example 6: A base resulting from spray drying an aqueous slurry of:

| | Parts |
|---|---|
| Sodium silicate ("N" grade, anhydrous basis) | 20 |
| Trisodium phosphate | 40 |
| Sodium carbonate | 31 1/2 |
| was uniformly mixed with | |
| Trichlorocyanuric acid | 8 1/2 |

The resulting product is odor stable and commercially acceptable.

Example 7: A base resulting from spray drying an aqueous slurry of:

| | Parts |
|---|---|
| Sodium silicate ("N" grade anhydrous basis) | 15 |
| Trisodium polyphosphate | 40 |
| Sesqui-carbonate | 30 |
| Dodecylbenzenesulfonate | 6 1/2 |
| was uniformly mixed with | |
| Trichlorocyanuric acid | 8 1/2 |

The resulting product is odor stable and commercially acceptable. Formulations made in exact correspondence with Example 6 and 7 simply by dry mixing the corresponding anhydrous silicate, phosphate, carbonate and trichlorocyanuric acid, early after preparation develop and continue to evolve highly irritating odors. As long as the base material is essentially inorganic and contains silicate to achieve the desired physical and chemical stability, the balance of the composition may consist of water-soluble inorganic salt. Small amounts of additive materials such as surfactants, fluorescent dyes, perfumes, etc. may be used.

Lead-Copper Salt of Dichlorocyanuric Acid

R.W. Marek; U.S. Patent 3,115,493; December 24, 1963; assigned to Olin

Mathieson Chemical Corporation found that a stable bleaching composition is obtained by reacting the lead salt of dichlorocyanuric acid with a copper salt. The structure of this lead-copper complex of dicyanuric acid is believed to be in the form:

$$PbCu(Z)_4$$

wherein Z is the dichlorocyanurate radical consisting of the group $(C_3N_3O_3Cl_2)^-$ or structurally

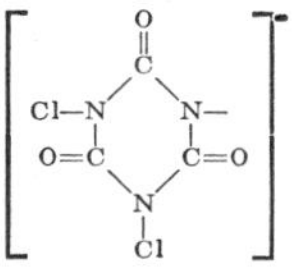

The reaction of the lead dichlorocyanurate with the cupric salt is preferably performed in water or acetone since the product readily precipitates from these solvents. In any event, solvents must be employed which are not susceptible to chlorination or oxidation. Reaction is preferably carried out at a pH of about 5.5 although a pH range of 4 to 7 may be utilized. Care must be taken to prevent reaction from proceeding in basic medium in view of the insolubility of cupric oxide and hydroxide in basic medium. The desired reaction proceeds well at room temperature although higher temperatures have been used. There are no critical reactant concentrations, but obviously sufficient reactants must be present in the reaction medium to ensure that the product precipitates from solution.

The reactant ratios are not critical since the product will form until the reactant present in lesser stoichiometric amount is consumed. However, best results have been obtained when about 0.5 to 20 equivalent weights of the lead dichlorocyanurate are reacted with an equivalent of cupric salt. Formation of the precipitate proceeds rapidly when the reactants are mixed together, but a brief stirring period ensures that reaction is complete before filtration of the product is completed.

Example 1: A solution of lead acetate was prepared by dissolving 0.25 mol of the salt in 500 cc of water. This solution was added at room temperature with stirring to a solution of 1 mol of potassium dichlorocyanurate in 1,500 cc of water. A white precipitate formed almost immediately upon mixing of the solutions. Precipitation of the lead dichlorocyanurate was completed by cooling in an ice bath. The filtered and dried lead dichlorocyanurate contained 43.1% available chlorine and the yield was 79%.

Example 2: To a solution of 0.05 mol of lead dichlorocyanurate in 1,200 cc of water was added 0.025 mol of cupric chloride dihydrate. A precipitate formed over a 15 minute period and after an additional stirring period, the precipitate was collected by filtration. After drying, 11.0 g. of a solid was obtained which closely approximated the color of Hortense V (A Dictionary of Color, Maerz and Rea Paul, McGraw-Hill and Co., Inc., 1950). The product analyses agree with the product formula:

$$PbCu(C_3N_3O_3Cl_2)_4$$

| | Theoretical, percent | Found, percent |
|---|---|---|
| Lead | 19.6 | 20.4 |
| Copper | 6.0 | 5.7 |
| Available chlorine | 53.6 | 51.0 |

In this composition, the molar ratio of Pb:Cu:Cl is 1:1:8. The product obtained in Example 2 was soluble in water to the extent of 0.50 g. per 100 cc of solution. The thermal stability of the product was determined by slowly heating a small sample of the compound to a temperature of about 250°C., cooling and recording the available chlorine before and after the heat treatment. For comparative purposes, dichlorocyanuric acid and the sodium salt thereof were subjected to the same treatment.

The thermal stability test is a severe one, and it is noted that PbCu(Z) exhibits a much higher degree of stability than the reference compounds. This high degree of stability is of special value as it ensures a prolonged shelf life for the product without excessive deterioration of the compound through loss of available chlorine. This composition has commercial utility as a bleaching agent, and this is so especially in those applications where an improved thermal stability is a prime requisite, e.g., laundry bleaches, bleaching agents in cleanser, tropical bleaches and war gas decontaminants.

## Chlorinated Piperazines

M. Kokorudz; U.S. Patent 3,142,530; July 28, 1964; assigned to Wyandotte Chemicals Corporation and U.S. Patent 3,158,436; November 24, 1964; assigned to Wyandotte Chemicals Corporation found that chlorinated piperazines can be mixed in dry state with sodium carbonate, borates, or phosphates and used as bleaching or disinfecting compositions. The active chlorine piperazines have the following general formula:

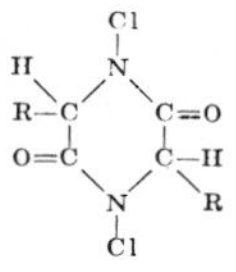

In the formula R is hydrogen or a methyl or ethyl radical. R may be the same or different in each occurrence in the formula. Thus, these compounds include N,N'-dichloro-2,5-diketopiperazine, N,N'-dichloro-2,5-diketo-3-methylpiperazine, N,N'-dichloro-2,5-diketo-3,6-dimethylpiperazine, N,N'-dichloro-2,5-diketo-3-ethylpiperazine, N,N'-dichloro-2,5-diketo-3,6-diethylpiperazine and N,N'-dichloro-2,5-diketo-3-methyl-6-ethylpiperazine.

These compounds are readily prepared in high yields as white solids which require little or no purification by the direct chlorination of the corresponding 2,5-diketopiperazine. These compounds can be used in formulated bleaching or germicidal composition by mixing in the substantially dry state with alkaline salts, such as sodium carbonate, sodium borate, sodium silicate, trisodium phosphate, tetrasodium pyrophosphate, sodium triphosphate or mixtures of these. In addition, wetting agents, synthetic detergents generally, soaps, fillers, abrasives and water softening agents of the inorganic or organic type may be incorporated in the bleach and germicidal compositions.

These compounds are characterized by a high degree of stability when dry and may be stored for long periods of time and transported over considerable distances without substantial decomposition. When these compounds are dissolved in water, the ingredients of the mixture apparently react to yield hypochlorite chlorine which is responsible for the efficient oxidizing, bleaching, disinfecting and sterilizing action. The germicidal and bleach compounds are readily prepared by suspending a 2,5-diketopiperazine corresponding to the formula:

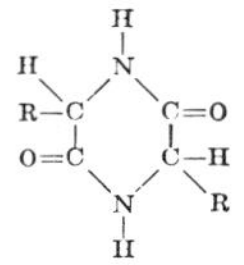

where R is hydrogen or a methyl or ethyl radical and may be the same or different in each occurrence in an aqueous solution having a pH of less than about 12.5, cooling the resulting mixture to below about 14°C. and passing chlorine gas into the cooled mixture, while maintaining the temperature at no more than about 30°C. The 2,5-diketopiperazines corresponding to the aforesaid formula are prepared from the condensation reaction of 2 mols of ethyl glycinate hydrochloride, alanine ethylester hydrochloride, or alpha-amino butyric acid ethylester hydrochloride or from the condensation of 2 mols of any combination of the aforesaid compounds as set forth in E. Fischer; Ber. 39, 2930 (1906).

The aqueous solution in which the chlorination is carried out can be alkaline, neutral, or acidic before the chlorination is commenced. The solution may be rendered alkaline or acidic by the use of any material which does not have an adverse effect resulting in the breaking of the ring structure of the starting material. This problem can generally be avoided if the pH is kept below about 12.5. Examples of suitable materials are sodium or potassium carbonate or bicarbonate and hydrochloric acid. However, ammonium salts should not be used because of the danger of forming nitrogen trichloride.

Good yields of the chlorinated diketopiperazine have been obtained when the diketopiperazine was suspended in water or in an aqueous alkaline solution having 2 mol equivalents of sodium carbonate per mol of diketopiperazine. The temperature of the mixture while the chlorine gas is passed into the mixture is desirably about 0° to 30°C. and preferably should not be substantially above about 15°C.

Chlorine gas is passed in over a period of time in the range of about 0.5 to about 6 hours until the diketopiperazine is chlorinated. In the case where chlorination is carried out in an alkaline solution containing about 2 mol equivalents of alkaline compound per mol of diketopiperazine, chlorine gas is passed into the cooled mixture until the pH of the reaction mixture is in the range of 7 to 5, desirably about 6. It should be apparent that the chlorination is completed at a higher pH when higher concentrations of alkaline compound are employed. In the case of mixtures initially neutral, acidic, or slightly alkaline (containing less than 2 mol equivalents of alkaline compound per mol of diketopiperazine), the chlorine addition is stopped when the solution becomes noticeably yellow due to the termination of the diketopiperazine chlorination. When the addition of chlorine gas is completed the white solid product is easily separated from the reaction mixture by filtration and air dried.

Example: 2,5-diketopiperazine was prepared by the condensation reaction of ethyl glycinate hydrochloride as set forth in E. Fischer; Ber. 39, 2930 (1906). N,N'-dichloro-2,5-diketopiperazine was prepared by the chlorination of 2,5-diketopiperazine in an alkaline solution. The alkaline solution in which the chlorination was carried out was prepared by dissolving 70 g. (0.66 mol) of $Na_2CO_3$ in 950 ml. of water. The aqueous sodium carbonate solution was cooled to 12°C. and 57 g. (0.5 mol) of 2,5-diketopiperazine were added. Chlorine gas was introduced to the aqueous alkaline solution of the 2,5-diketopiperazine while cooling the mixture so that the temperature was in the range of 12° to 14°C. Chlorine gas was passed into the alkaline solution over a period of about 5 hours until the pH was 6.0. The white solid product which was produced was filtered, air dried at room temperature and weighed. A yield of 56 g. which was 62% of the theoretical amount of N,N'-dichloro-2,5-diketopiperazine, was obtained. Analysis of the product showed the product to have an active chlorine content of 34.7%, theoretical being 38.8%.

The chlorine compound prepared was tested for bleaching activity by dissolving a sufficient quantity of the product in a liter of water so that the concentration of the product was sufficient to provide 200 parts per million of available chlorine. Deionized water was used to prepare the solution. The test procedure used was that described in U.S. Patent 2,957,915, column 3, lines 16 to 40, using a Hunter Multipurpose Reflectometer. The test of the active chlorine compound prepared in this example at 200 parts per million provided a bleached swatch which had a 65.0% reflectance at pH 4, 63.7% reflectance at pH 7, and 60.9% reflectance at pH 10. The results when compared to a reflectance of 50% for the unbleached swatch illustrate the bleaching characteristics of this active chlorine compound.

## Stable Sodium or Potassium Dichloroisocyanurates

S.J. Kovalsky and R.A. Olson; U.S. Patent 3,299,060; January 17, 1967; assigned to FMC Corporation describe a process to manufacture thermally stable sodium or potassium dichloroisocyanurates, using either a batch or a continuous process. It has been observed that the start-up of plant runs of these salts have good thermal stability, essentially like the salts obtained in laboratory operations. However, after the plant has been in operation intermittently the thermal stability of the resultant product commences to decrease and finally reaches a low. It has been found that the poor thermal stability of dichloroisocyanuric acid salts is due to the presence of insoluble impurities in the mother liquor which is used in making up the reaction slurry; the mother liquor normally is an effluent stream recovered from a previous reaction slurry.

It has also been found that dichloroisocyanuric acid salts having a uniformly high thermal stability can be produced by polish-filtering the recycled mother liquor to remove all insoluble impurities larger than 15 microns prior to using the mother liquor as an aqueous reaction medium in the make-up of additional slurries of dichloroisocyanuric acid and an alkali metal hydroxide.

In the above reaction, the recycled mother liquor is polish-filtered to remove all insolubles larger than 15 microns. A paper filter equivalent to a Whatman No. 42 was found effective in removing such particles. These insolubles are believed to be used as the reaction medium in forming the salts of dichloroisocyanuric acid. In this reaction, the pH of the reaction mixture should not be allowed to rise above about 7 because of the possible formation of nitrogen trichloride, which is undesirable since it is detonable, even in small quantities. The pH of the reaction slurry should also not be permitted to fall below 6.0 in order to ensure complete conversion of the dichloroisocyanuric acid to the desired product. At lowered pH values, the final product is often contaminated by unreacted agents such as dichloroisocyanuric acid. Within the pH range from about 6 to 7, desirably high yields of pure salts of dichloroisocyanuric acid are obtained with a minimum of undesired side products. After completion of the above reaction, a slurry is removed from the reactor, cooled to about 20°C., and treated to separate the salts of dichloroisocyanuric acid from the mother liquor. The mother liquor is recycled back to the reactor along with make-up dichloroisocyanuric acid and additional potassium or sodium hydroxide, while the separated salts of dichloroisocyanuric acid are passed to a dryer. There they are contacted with a heated gas stream to remove residual water, either in combined or uncombined form, so that the final salts of dichloroisocyanuric acid contain less than about 0.2% by weight of water.

Example: Run A — A continuous preparation of potassium dichloroisocyanurate was carried out by feeding a 20% by weight dichloroisocyanuric acid aqueous slurry into a reactor. Simultaneously, a 50% potassium hydroxide solution was fed to the reactor at a rate sufficient to maintain the pH at 6.8. The temperature of the reaction mixture was controlled between 20° and 25°C. by cooling the reaction mixture in a water cooled heat exchanger. The aqueous liquor used in the make-up of the dichloroisocyanuric acid slurry was mother liquor from a prior bath centrifuging operation in the production of potassium dichloroisocyanurate, which had been stored for over 72 hours at 35°C. The resultant potassium dichloroisocyanurate salt precipitate was continually removed as a slurry and centrifuged to separate the salt from its mother liquor. The mother liquor was returned directly to the reactor as make-up reaction medium without any intermediate treatment.

The wet salt product was dried by contact with hot air to less than 0.2% water. The resultant dried salt product was found to be composed of fine crystals having a cross-sectional area of below 50 square microns. The resultant dry potassium dichloroisocyanurate crystals were tested for thermal stability by placing a 5 to 10 g. sample of the dry salt in a test tube and inserting the tube into a molten salt bath at a temperature of 275°C. The time required for initiating decomposition was recorded as the stability time of the product. The resultant crystals were found to have a stability time of only 30 seconds when tested as set forth above.

Run B — The process set forth in Run A was repeated except that the recycled mother liquor was filtered through a fine filter paper equivalent to a Whatman No. 42 and the resultant clear liquor was used to prepare a new batch of potassium dichloroisocyanurate. The conditions of reaction were identical with those carried out in Run A. The resultant potassium dichloroisocyanurate crystals were separated from the mother liquor and were found to be in the form of large rhomboid crystals having a cross-sectional area of about 1,500 square microns. This potassium salt product had a stability time of 150 seconds when tested as set forth in Run A.

## Potassium Salts of Chlorinated Cyanuric Acids

E.A. Casey and R.L. Liss; U.S. Patent 3,364,146; January 16, 1968; assigned to Monsanto Company describe the use of potassium salts of chlorinated cyanuric acids, such as monopotassium dichlorocyanurate, $KCl_2C_3N_3O_3$, in bleaching, sterilizing and disinfecting compositions. Other salts are: dipotassium monochlorocyanurate, $K_2ClC_3N_3O_3$, and monopotassium monochloro hydrogen cyanurate, $KClHC_3N_3O_3$. The most important of these salts is the monopotassium dichlorocyanurate. Depending on the reaction temperature employed it is obtained in one or two anhydrous forms. Form 1 is a white crystalline solid whose internal and external symmetry is monoclinic. Form 2 is also a white, crystalline solid whose internal symmetry is monoclinic but whose external symmetry is triclinic. In the preparation of dipotassium monochlorocyanurate, two molecular proportions of tripotassium cyanurate are reacted with one molecular proportion of trichlorocyanuric acid in an aqueous medium at about room temperature. Similarly, monopotassium monochloro hydrogen cyanurate is prepared by reacting equimolecular proportions of dipotassium hydrogen cyanurate and dichloro hydrogen cyanuric acid ($Cl_2HC_3N_3O_3$) in an aqueous medium at about room temperature.

Tripotassium cyanurate was prepared by adding 500 g. of 50% potassium hydroxide to 334 g. of wet cyanuric acid (42.8% moisture) to form a thick slurry which was immediately diluted with water until the solids had completely dissolved. The resulting solution contained 12% by weight of tripotassium cyanurate. 464 g. of dry trichlorocyanuric acid were then added to 2,028 g. of the 12% solution of tripotassium cyanurate, the addition being sufficiently slow to keep the temperature of the mixture below about 50°C. The mixture was cooled to about 5°C. and allowed to stand for about 30 minutes during which time the monohydrate of potassium dichlorocyanurate precipitated from solution and settled to the bottom of the reaction mixture.

The solids which contained 7% water of hydration were filtered and dried in an oven at 105°C. to remove the water of hydration. The dried product was a crystalline white solid which to the naked eye appeared to be of a hexagonal crystal structure, however, the crystals are in fact characterized by having a monoclinic internal symmetry and a triclinic external symmetry. This dried crystalline product has the same size and shape of the precipitated monohydrate. This anhydrous salt is termed Form 2 potassium dichlorocyanurate. Examples of bleaching, sanitizing and cleaning compositions follow.

Typical Household Laundry Bleach

| | Weight Percent |
|---|---|
| Sodium tripolyphosphate | 40 |
| Sodium sulfate | 24 |
| Sodium metasilicate | 20 |
| Sodium dodecylbenzenesulfonate | 5 |
| Potassium silicate | 1 |
| Potassium dichlorocyanurate (Form 1 or Form 2 or mixture thereof) | 10 |

Typical Scouring Powder

| | Weight Percent |
|---|---|
| Silica | 90 |
| Sodium tripolyphosphate | 5 |
| Soda ash | 2.5 |
| Sodium lauryl sulfate | 2.2 |
| Potassium dichlorocyanurate (Form 2) | 0.3 |

## Bleach Containing Zinc Di(Dichlorocyanurate)

E.A. Matzner; U.S. Patent 3,456,054; July 15, 1969; assigned to Monsanto Co. describes a process for the production of zinc di(dichlorocyanurate) and its use in bleaching and disinfecting compounds in combination with sodium tripolyphosphate, surfactants and inorganic carriers. Zinc di(dichloroisocyanurate) and hydrates can be represented structurally as:

$$\begin{bmatrix} Cl-N-\overset{O}{\underset{\|}{C}}-N- \\ \phantom{Cl-N}| \phantom{-C-}| \\ O=C-N-C=O \\ \phantom{O=C-}| \\ \phantom{O=C}Cl \end{bmatrix}_2 Zn \cdot mH_2O$$

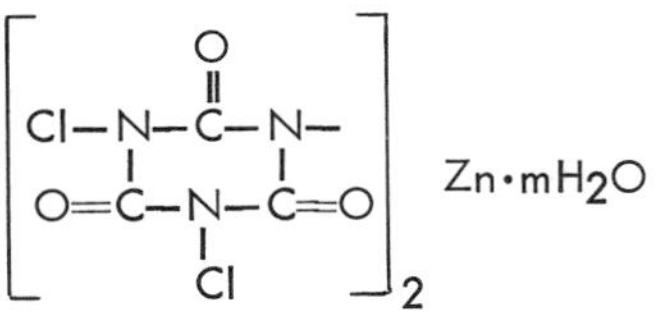

wherein m is an integer in the range of from 0 to 10, inclusive, preferably 0 to 6 inclusive, are useful sources of available chlorine in bleaching, oxidizing, disinfecting, sterilizing and/or detergent formulations. All of these salts are white solids exhibiting moderate water-solubility. Examples of these salts include anhydrous, amorphous zinc di(dichloroisocyanurate) trihydrate, tetrahydrate and the hexahydrate. Although the formula for the above compounds is represented structurally as being in the keto or iso form compounds may also exist in the enol form or as mixtures of the enol and keto or iso forms.

The compounds can be prepared by reacting an aqueous solution containing from about 3.0 to 80% by weight of a water-soluble zinc salt such as, for example, the chloride, bromide, sulfate, and acetate salts of zinc, preferably zinc chloride, and an aqueous solution containing at least 5% by weight of sodium, lithium or cesium dichlorocyanurate in a reaction zone at a temperature in the range of from about 5° to about 60°C. The zinc salt and such dichlorocyanurate are preferably employed in a molar ratio of about 1:2. The product, which precipitates from the aqueous reaction system, is a hydrate of zinc di(dichloroisocyanurate) which can be dehydrated to lower hydrates or to the amorphous anhydrous salt by the well-known means for removing water of hydration from hydrates.

Example 1: To a suitable reaction vessel equipped with a thermometer and agitator was added 440 g. of sodium dichloroisocyanurate in the form of a saturated aqueous solution thereof. To the solution over a period of about 15 minutes was added a solution of 150 g. of zinc chloride in 50 g. of water while maintaining the temperature of about 20°C. Upon completion of the zinc chloride addition the reaction mass was agitated for about 1 hour, cooled to about 10°C. and thereafter filtered. (The pH of the filtrate was 5.8). The filter cake was then dried for several hours at room temperature. The dried product was a white solid containing 10.5% by weight of hydration and identified as the trihydrate of zinc di(dichloroisocyanurate). This solid is characterized by an available chlorine content of 53.9%.

The trihydrate of zinc di(dichloroisocyanurate) on drying in an air-circulating oven at 120°C. for 2 to 3 hours yielded anhydrous white solid zinc di(dichloroisocyanurate) which was characterized by an available chlorine content of 59.2% and a zinc content determined by ignition, of 13.8%. Unlike the trihydrate which is a crystalline compound, the anhydrous zinc di(dichlorocyanurate) is an amorphous compound.

Example 2: A formulation containing 45% by weight of anhydrous sodium tripolyphosphate, 33% by weight sodium sulfate, 20% by weight sodium carbonate and 2% by weight zinc di(dichloroisocyanurate) trihydrate was prepared by mechanically admixing the ingredients in a mechanical mixer. Similar formulations were prepared in one of which the 2% zinc di(dichlorocyanurate) trihydrate was replaced by 2% by weight of calcium di(dichlorocyanurate) anhydrous; in another 2% by weight of magnesium di(dichlorocyanurate) hexahydrate, replaced the zinc salt and 2% by weight of sodium dichlorocyanurate replaced the zinc salt in still another formulation. The four formulations so prepared were analyzed for available chlorine content, were placed in open vials and stored in a humidity cabinet maintained at a temperature of 32°C. and a relative humidity of 85% for 25 days. The formulations were then reanalyzed for available chlorine. Losses in available chlorine content of the four chlorocyanurate containing compositions are shown in the following table.

| Formulation | Loss of Available Chlorine After 25 Days (%) |
|---|---|
| Zinc di(dichlorocyanurate) trihydrate | 9.0 |
| Calcium di(dichlorocyanurate) anhydrous | 20.0 |
| Magnesium di(dichlorocyanurate) hexahydrate | 32.0 |
| Sodium dichlorocyanurate | 32.0 |

The tests clearly show the zinc di(dichlorocyanurate) either as the amorphous anhydrous salt or as the crystalline trihydrate salt is unexpectedly more stable toward loss of available chlorine under severe conditions of humidity and long storage periods than a variety of other chlorocyanurate salts in a wide range of formulations. The anhydrous zinc di(dichlorocyanurate) and hydrates thereof are also surprisingly effective as bacteriostatic and fungistatic agents.

N, N', N''-Trichlorosuccinimidine

C.L. Coon; U.S. Patent 3,642,824; February 15, 1972; assigned to Stanford Research Institute describes a process to prepare N, N', N''-trichlorosuccinimidine by reacting succinimidine with an excess of a hypochlorite compound in an aqueous medium under acid conditions, the reaction proceeding rapidly at room temperatures with the formation of the desired chlorinated product. This succinimidine reactant is conveniently formed by dissolving succinamidine in water, and this reaction, together with that which occurs when the succinimidine is chlorinated can be illustrated as follows:

$$NH_2\overset{NH}{\overset{\|}{C}}CH_2CH_2\overset{NH}{\overset{\|}{C}}NH_2 \cdot 2HCl \xrightarrow{H_2O} \begin{array}{c} CH_2-C{\overset{NH}{\diagup}} \\ | \quad\quad\quad NH \cdot HCl \\ CH_2-C{\diagdown}_{NH} \end{array} + NH_4Cl$$

succinamidine

succinimidine

$$\begin{array}{c} CH_2-C{\overset{NH}{\diagup}} \\ | \quad\quad\quad NH \cdot HCl \\ CH_2-C{\diagdown}_{NH} \end{array} + 3NaOCl + 2HCl \longrightarrow \begin{array}{c} CH_2-C{\overset{NCl}{\diagup}} \\ | \quad\quad\quad NCl \\ CH_2-C{\diagdown}_{NCl} \end{array} + 3NaCl + 3H_2O$$

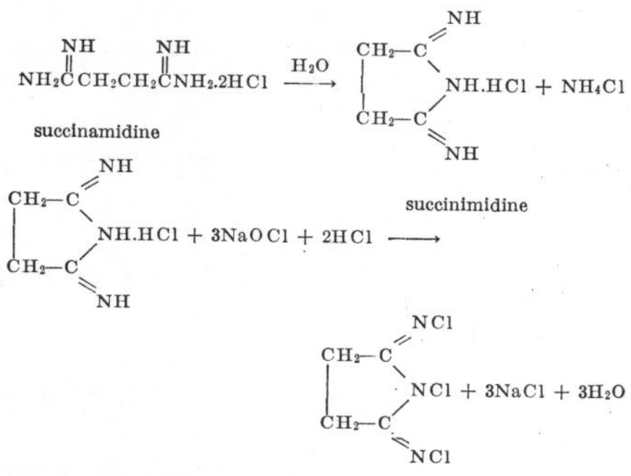

N, N', N''-trichlorosuccinimidine is a white crystalline solid having a melting point of 211° to 212°C. which is substantially insoluble in water and of good solubility in chloroform, acetone, benzene and other organic solvents. It has a high content of available chlorine and is useful as a chlorinating agent, bleach and disinfectant.

The hypochlorite reactant can be added in the form of a sodium, potassium or other alkali metal hypochlorite salt. This reactant is employed in excess over the amount stoichiometrically required to replace all the nitrogen-attached hydrogen atoms in the succinimidine reactant with chlorine. Preferably this excess is at least 2 X over the theoretically required amount. The reaction is conducted in the presence of a strong acid such as hydrochloric or sulfuric acid, for example, the acid normally being employed on essentially an equimolar basis with respect to the hypochlorite. A moderate excess of either hypochlorite or acid does not interfere with the reaction. The succinimidine reactant is normally employed in the form of a water-soluble salt such, for example, as succinimide hydrochloride. This compound can readily be formed from succinamidine hydrochloride merely by adding the latter to water.

Example 1: N, N', N"-Trichlorosuccinimidine – A solution was prepared by admixing 60.0 ml. of an aqueous solution of 5.25% sodium hypochlorite (0.042 mol) and 50 ml. of dilute HCl (0.042 mol). This solution was then stirred over 25 ml. of $CH_2Cl_2$ as 1 g. of succinamidine hydrochloride (0.0054 mol) was added in increments over a 10 minute period, the reaction mixture being at approximately 25°C. After stirring for an additional 5 minutes, the $CH_2Cl_2$ phase was separated and the aqueous phase was twice extracted with $CH_2Cl_2$. The $CH_2Cl_2$ phase then was combined with the extracts.

After drying this combined material with $MgSO_4$ and separating off the latter salt, the $CH_2Cl_2$ was distilled off under reduced pressure leaving a white crystalline solid which was then purified by recrystallization from hot $CHCl_3$. There was recovered 0.92 g. of a white, plate-like material having a MP of 211° to 212°C. This product, which was insoluble in water and of good solubility in various organic solvents, was identified as N, N', N"-trichlorosuccinimidine by infrared and elemental analysis.

Example 2: The process of Example 1 is repeated, but with the succinamidine hydrochloride being combined with the dilute acid solution and the resulting solution then being slowly added to the stirred solution of sodium hypochlorite over $CH_2Cl_2$. Yields of from 80 to 90% can be obtained. N, N', N"-trichlorosuccinimidine is characterized by a high content of available chlorine and has utility as a chlorinating agent. Thus, chlorine is released as a solution of the compound is subjected to ultraviolet light. The compound also has utility as a bactericide, among other applications. Thus, acetone solutions of N, N', N"-trichlorosuccinimidine, as well as acetone-water solutions containing as much as 88% by volume of water, give effective control of E. coli and S. aureus at concentrations containing as little as 0.0005% available chlorine.

## DETERGENT LAUNDRY BLEACHES

### Stabilized Trichlorocyanuric Acid

H.E. Wixon; U.S. Patent 3,108,078; October 22, 1963; assigned to Colgate-Palmolive Company found that trichlorocyanuric acid is stabilized by olefin and specifically by polyunsaturated terpenes, such as limonene and myrcene. From 5 to 20% of the terpenes are used to stabilize the trichlorocyanuric acid. The stabilized trichlorocyanuric acid may be employed in any substantially dry composition in which trichlorocyanuric acid is otherwise suitable for use, such as washing, bleaching, sterilizing and disinfecting compositions. Thus it may be used in admixture with "inert diluents" including surface active agents and synthetic detergents stable in the presence of trichlorocyanuric acid. Such detergents include a wide variety of anionic detergent salts such as the water-soluble higher fatty acid alkali metal soaps, e.g., sodium myristate and sodium palmitate; water-soluble sulfated and sulfonated anionic foaming alkali metal and alkaline earth metal detergent salts containing a hydrophobic higher alkyl moiety (typically containing from about 8 to 22 carbon atoms) such as salts of higher alkyl mono- or polynuclear aryl sulfonates having from about 10 to 16 carbon atoms in the alkyl group (e.g., sodium dodecylbenzene sulfonate, magnesium tridecylbenzene sulfonate, lithium or potassium pentapropylene benzene sulfonate); alkali metal salts of higher alkyl naphthalene sulfonic acids; sulfated higher fatty acid monoglycerides such as the sodium salt of the sulfated monoglyceride of coconut oil fatty acids and the potassium salt of the sulfated monoglyceride of tallow fatty acids.

Also, alkali metal salts of sulfated fatty alcohols containing from about 10 to 18 carbon atoms (e.g., sodium lauryl sulfate and sodium stearyl sulfate); alkali metal salts of higher fatty acid esters of low molecular weight alkylol sulfonic acids, e.g., fatty acid esters of the sodium salt of isethionic acid; the fatty ethanolamide sulfates; the fatty acid amides of amino alkyl sulfonic acids, e.g., lauric acid amide of taurine; as well as numerous other ionic organic surface active agents such as sodium toluenesulfonate, sodium xylenesulfonate, sodium naphthalene sulfonate; and mixtures thereof can be used. In general these detergents are employed in the form of their alkali metal or alkaline earth metal salts as these salts possess the requisite stability, solubility, and low cost essential to practical utility.

Example: A water-soluble washing composition having substantial bleaching power and suitable for use on heavily soiled fabrics is prepared by spray drying an aqueous slurry to form a granular product having a particle size such that 95% of the granules pass through a 20-mesh sieve (sieve opening 0.84 mm.). The granular product has the following composition:

| | Parts by Weight |
|---|---|
| Sodium dodecyl benzene sulfonate | 35.5 |
| Pentasodium tripolyphosphate | 39.0 |
| Moisture | 3.0 |
| Sodium carboxymethylcellulose | 0.8 |
| Bleach resistant optical brightener, rancidity retardant, and antitarnishing agent | 0.6 |
| Sodium sulfate, balance to 98.57 parts | |

These granules are tumbled in a rotating drum and sprayed with 0.14 part by weight (based on the final composition) of limonene. Thereafter 1.29 parts by weight of trichlorocyanuric acid are added slowly and are thoroughly admixed therewith. The trichlorocyanuric acid employed has a particle size such that at least 80% passes through a 100-mesh sieve (sieve openings 0.149 mm.). The product, which by analysis is found to contain 1.00% available chlorine, is then packaged in aluminum foil-covered paper board containers. After standing under room conditions for 8 months, there is substantially no diminution in the available chlorine content of the product. A control, prepared in the same way except that the limonene is omitted, loses 2/3 of its initial content of available chlorine within 6 months' aging under the same conditions. In use, approximately 9 lbs. of soiled color-fast household laundry including cottons and nylons stained by ink, tea, coffee, and grape juice is placed in a conventional household automatic washing machine, the machine is filled with water to its customary working level and the water-soluble composition of this example is then introduced added to the water in an amount such as to form a 0.2% solution having a pH of 9.0. The composition dissolves readily and liberates hypochlorite chlorine in a highly effective manner and at a desirable rate. The laundry is washed and rinsed by the machine in the conventional manner. On inspection of the fabrics at the end of the cycle, it is found that they are evenly and effectively bleached and cleaned, the stains are removed, and there is virtually no deterioration of the fibers either locally or generally.

A detergent having greater bleaching power than that of the foregoing example may be prepared by increasing the proportions of limonene and trichlorocyanuric acid therein to 0.5% and 14.2% respectively by weight of the final product. If desired, camphene may be employed in place of limonene, in which case the camphene may be dissolved in ethanol to obtain a liquid form suitable for spraying on the spray dried granules.

## Chlorinated Hydantoin

R.B. Wearn, P.T. Vitale and G.F. Marion; U.S. Patent 3,257,324; June 21, 1966; assigned to Colgate-Palmolive Company describe detergent bleach compositions based on 1,3-dichloro-5,5-dimethylhydantoin, together with a surfactant, tetrasodium pyrophosphate and other sodium salts. Among the outstanding properties of such compositions are their ability to remove stains such as ink, coffee and like stains on fabrics, food stains on enamel sinks, tarnish stains on metals such as copper, stains which form on porcelain toilet bowls, and the like; their ability to preserve or retain whiteness of cotton on repeated bleaching treatment; their ease of storage and use; and their versatility for such multiple uses as a complete washing composition for clothes, particularly in automatic type washing machines, a bleaching rinse, a metal cleaner, a toilet bowl cleaner, a kitchen sink cleaner, and other like uses, and as a growth inhibitor for algae in water tanks, and the like.

Tests demonstrate that the combination of properties possessed by the compositions of this process depends critically upon the presence in the composition of at least about 10% of at least 1 ingredient of the group of water-soluble molecularly dehydrated phosphate salts, alkali metal silicates and fully neutralized alkali metal carbonates. Thus, compositions consisting of about 10% dichloro-dimethylhydantoin, 2.2% sodium dodecylbenzene sulfonate, 4.5% sodium toluene sulfonate, 53.3% of sodium sulfate and 30% of various inorganic sodium salts are used to wash a plurality of cotton swatches for 10 cycles.

Each cycle consists of a 20 minute wash at 110°F. in a Tergometer in tap water (about 50 ppm hardness) at 2.3 g. of each composition per liter, a rinse in warm tap water for 3 minutes and drying by ironing. After 10 cycles the cotton swatches are evaluated visually for preference and also on a Hunter color difference meter. Compositions containing tetrasodium pyrophosphate, sodium tripolyphosphate, sodium hexametaphosphate, sodium carbonate and sodium silicate give significantly higher whiteness (visual preference and higher percent white by the meter) and lower yellowness than compositions containing other alkaline salts such as sodium bicarbonate, trisodium phosphate or borax as the sodium salt content.

Example 1: A composition is made up having the ingredients shown on the next page.

| | Percent |
|---|---|
| 1,3-dichloro-5,5-dimethylhydantoin | 14.7 |
| Wetting agent (an alkyl aryl sulfonate that achieves a quicker and better dispersion and dissolution of the particles of the bleach composition; more particularly, a mixture of mono-butyl and dibutyl naphthalene sodium sulfonates) | 5.9 |
| Sodium sulfate | 29.4 |
| Sodium chloride | 23.5 |
| Tetrasodium pyrophosphate | 23.5 |
| Foaming and emulsifying agent (an alkyl aryl sulfonate which adds foaming and emulsifying properties to the composition; more particularly, a roll-dried sodium dodecyl benzene sulfonate wherein the source of the alkyl group is propylene tetramer) | 2.9 |
| Fluorescent dye (a bleach-stable commercially available optical dye or bleach) | 0.1 |
| | 100.0 |

The foregoing components are mechanically mixed in dry form and are then pulverized in a mill to a particle size varying from about 2 to 90 microns, with over 90% of the composition consisting of particles in the range between 5 and 30 microns. The finely divided composition thus obtained shows good solubility in water and an available chlorine content of about 9.2%. Tests show that this bleach composition is very effective in cleaning and removing stains.

Towel tests performed upon towels soiled by workmen employed in a railroad maintenance shop show that this composition is extremely effective in respect of its total washing power, its low soil redeposition qualities, and the whiteness or bleaching effect of the optical dye contained therein, while at the same time having no adverse effect upon the tensile strength of the fabric. The towel tests are run using 0.25% by weight of a commercially available sodium alkyl aromatic sulfonate synthetic detergent dispersed in tap water, having about 30 to 50 parts per million hardness as $CaCO_3$, the solid bleach composition of the process being added during the wash cycle. The foregoing bleach composition is also found to be very stable on standing, accelerated aging tests under conditions of 90% relative humidity and 90°F. revealing only relatively minor (i.e., from 15 to 30%) losses of available chlorine content after 4 weeks exposure to same. The pH of the aqueous solution obtained with the composition of this example is in excess of 9.

Example 2:

| | Percent |
|---|---|
| 1,3-dichloro-5,5-dimethylhydantoin | 14.7 |
| Wetting agent (sodium isopropyl naphthalene sulfonate) | 5.9 |
| Tetrasodium pyrophosphate (anhydrous) | 23.5 |
| Sodium dodecyl benzene sulfonate | 2.9 |
| Borax (anhydrous) | 16.0 |
| Sodium sulfate | 37.0 |
| | 100.0 |

## Stabilized Chlorinated Cyanuric Acid

S.C. Bright and A. Alsbury; U.S. Patent 3,278,443; October 11, 1966; assigned to Lever Brothers Company describe a process to stabilize trichloro- and dichlorocyanuric acids by phenols and phenyl derivatives for use in bleaching compositions. The chlorinated cyanuric acids are stabilized by compounds of the general formula: R·CH:CH·X, where at least one of R and X is an electron-releasing group.

Preferably at least one of the groups R and X is a phenyl group, a substituted phenyl group, a benzyl ($C_6H_5 \cdot CH_2$—) group or a benzyl group substituted in the aromatic nucleus. One of the groups R and X may be a hydrogen atom, a methyl group, an hydroxymethyl ($HO \cdot CH_2$—) group or an ester or ether of an hydroxymethyl group. The groups R and X may be joined together to form a cyclic system. Especially preferred stabilizing compounds have the group —CH:CH— as part of a $C_3$ side-chain of a benzene ring which is substituted by at least one ether group. Compounds chosen to stabilize chlorinated cyanuric acids should not be readily polymerizable. Styrene, beta-methyl styrene, isopropyl styrene and allylbenzene are not suitable for the stabilization of chlorinated cyanuric acids.

Compounds chosen to stabilize chlorinated cyanuric acids should react with hypochlorous acid but should not have a very low vapor pressure nor yet be extremely volatile. Compounds of boiling point at atmospheric pressure outside the range 150° to 360°C. are not contemplated as being normally useful within the scope of this process, although compounds boiling below 150°C. may be employed where sealed containers or volatility depressants such as mineral oil are employed. Compounds of boiling point at atmospheric pressure within the range 225° to 270°C. are preferred. 4,4'-dimethoxystilbene, 4,4'-diaminostilbene and cinnamyl stearate, which boil above 360°C., are not suitable for the stabilization of chlorinated cyanuric acids.

Compounds suitable for the stabilization of chlorinated cyanuric acids include cinnamyl alcohol, cinnamyl cinnamate, anethole, stilbene, indene, methyl eugenol and methyl isoeugenol. Safrole, isosafrole, eugenol and isoeugenol are preferred. The amount of stabilizing compound to be used depends upon a number of factors, such as the physical form of the chlorinated cyanuric acid and the nature of the stabilizing compound to be used, the nature and amount of the other ingredients of the composition and the conditions under which it may be stored. It is preferred to use a granulated chlorinated cyanuric acid. Appreciable increases in stability may be obtained with as little as 1%, by weight of the chlorinated cyanuric acid, of stabilizing compound, while amounts of up to 50% are generally more than adequate to give substantially complete stability for normal compositions and times and conditions of storage. It is preferred to use from 2 to 20% of stabilizing compound by weight of the chlorinated cyanuric acid.

In preparing compositions of this process, care must be taken that the chlorinated cyanuric acid and the stabilizing compound are not brought into contact with one another before at least one of them is, and preferably both are, diluted with or adsorbed on some of the other ingredients of the composition. A slight loss of available chlorine is generally found to have taken place during the process of mixing a chlorinated cyanuric acid with the other ingredients of a bleaching composition such as a bleaching scouring powder, and this mixing loss is often increased by the presence of organic compounds added for the purpose of stabilizing the chlorinated cyanuric acid on storage. It is a further advantage of many of the compositions herein described that this mixing loss is found to be relatively small. Many of the stabilizing compositions are soluble or miscible with the perfume normally used in commercial bleaching or detergent compositions. A preferred method of incorporating the stabilizing compounds in such cases is to form a perfume-stabilizer mixture which is then sprayed onto the remainder of the bleaching composition.

## Dichlorocyanurate, Sodium Tripolyphosphate and Sodium Sulfate

A.G. Brown, W.W. Lee and K.M. Sancier; U.S. Patent 3,293,188; December 20, 1966; assigned to Procter & Gamble Company found that dichlorocyanurate base, dry bleaching, sterilizing and disinfecting compositions are more stable and more effective than compounds based on trichlorocyanuric acid. Dichlorocyanuric acid is obtained by chlorinating cyanuric acid or by reacting trichlorocyanuric acid with cyanuric acid in the presence of water. When cyanuric acid is chlorinated in acid solution, hydrogen dihalocyanurate, or in other words dihalocyanuric acid, is the product; and in basic solution the product is composed of the dihalocyanurate anion and the cation of the base.

Both dihalocyanuric acid and its salts are referred to as dihalocyanurates, since they contain in the dihalocyanurate anion which provides the available halogen that gives the compound its properties. The dihalocyanurate is dried for storage and packaging, and then placed in the presence of water when it is to be used as a bleaching, sterilizing and disinfecting agent. Although the dihalocyanuric acid or its salts may be employed alone, for most purposes they are more advantageously utilized in a mixture with a nonhygroscopic synergistic carrier agent which is soluble in water, inert to the dihalocyanuric acid, and which increases and controls the pH of the solution of dihalocyanuric acid up to about 11. The dihalocyanurate is most advantageously utilized in a solution having a pH range of from about 6.5 to 9.5, and the carrier agent is preferably a buffer which controls the pH within this optimum activity range.

The direct chlorination to the dichlorocyanurate may be conducted in acidic or basic solution. However, it is most advantageous to increase the pH of the solution by adding a base, because more cyanuric acid dissolves as the pH is increased and the chlorination reaction proceeds more rapidly. It has been found most convenient to conduct the chlorination by dissolving in water a ratio of about 1 mol of cyanuric acid with sufficient base to provide a molar equivalent of 2 mols of hydroxide ion, and then bubbling in about 2 mols of chlorine. In this procedure, 2 of the 3 hydrogen atoms on the cyanuric acid are neutralized and then replaced by chlorine to form a dichlorocyanuric acid precipitate.

The structure of dichlorocyanuric acid varies between a number of forms because of tautomerism in the molecule. For example, the three oxygens can be attached to the three carbons by double bonds, and the two chlorines attached to two nitrogens with the other nitrogen being occupied by a hydrogen. The empirical formula of the dichlorocyanuric acid is in all cases $C_3N_3O_3Cl_2H$. If more than 2 mols of base is employed so that the sodium salt of dichlorocyanuric acid is produced, acid may be conveniently added in order to recover dichlorocyanuric acid as a precipitate.

Another method of chlorinating cyanuric acid to form dichlorocyanuric acid is by mixing trichlorocyanuric acid with cyanuric acid in the presence of water or a suitable solvent for both compounds. A ratio of 1 mol of cyanuric acid reacts with 2 mols of trichlorocyanuric acid to form dichlorocyanuric acid. Any excess of trichlorocyanuric acid above the 2 mols remains in the mixture and tends to decompose. However, as long as less than about 2 mols of trichlorocyanuric acid is employed per mol of cyanuric acid, the trichlorocyanuric acid is converted into the lower chlorinated compounds of cyanuric acid which are relatively stable compared to the trichlorocyanuric acid and which are excellent bleaching and disinfecting agents. Consequently, mixing cyanuric acid with trichlorocyanuric acid provides a method of stabilizing trichlorocyanuric acid so that it may be spray dried without a large loss of available chlorine.

Also, when dichlorocyanuric acid is prepared by mixing cyanuric acid and trichlorocyanuric acid, there is no chloride formed as a by-product of the reaction. This result is desirable because the dichlorocyanuric acid becomes somewhat unstable in the presence of the chloride ion. Furthermore, this method of forming dichlorocyanuric acid from cyanuric acid and trichlorocyanuric acid requires very little equipment, and the solid reactants can advantageously be mixed in the presence of a small amount of water in order to form the dichlorocyanuric acid. The amount of water employed for conducting the reaction between cyanuric acid and trichlorocyanuric acid is not particularly critical. The reaction may be carried out with sufficient water to provide a solution of the reactants, a slurry, a plastic mass, or merely sufficient water to moisten the reactants.

Although the dihalocyanurate anion, in the form of dihalocyanuric acid or its salts, may be utilized alone for bleaching, disinfecting or deodorizing, the active compound is most advantageously mixed with a synergistic carrier agent when it is to be used for such purposes. Incorporation of the synergistic agent enables the final product to be more readily dispensed, it prevents caking by keeping apart the organic ingredients, provides a product that is stable on storage, and provides the pH in the desired range for greatest activity of the dihalocyanuric acid. The synergistic agent is selected from inexpensive nonhygroscopic compounds that are stable, nonreactive with dihalocyanuric acid, have good flow properties, and which readily dissolve in water. When used for household purposes, the carrier agent should yield a clear solution when it is dissolved.

In addition a carrier which is also buffering agent that increases and controls the pH up to about 11 and preferably in a range between about 6.5 and 9.5 with or without the presence of organic detergents is most advantageously employed. The stability of the dichlorocyanuric acid is somewhat reduced as the pH is increased above 9.5, and at a pH below about 6.5 there is a tendency for volatile hypochlorous acid to be formed in the solution. Since the pH of pure dichlorocyanuric acid dissolved in water in a concentration of 200 ppm is 3.6, the importance of the synergistic carrier agent in increasing and maintaining the increased pH is clearly evident. Suitable carrier agents having the foregoing properties are readily selected from the alkali metal (e.g., sodium potassium) phosphates, polyphosphates, tripolyphosphates, silicates, borates, and carbonates, as well as neutral salts, such as sodium sulfate.

The halides are generally unsatisfactory since they tend to cause decomposition of the dichlorocyanuric acid. Also, ammonium salts cannot be used as carrier agents because they cause serious degradation of the active ingredient as well as reactions which produce dangerous explosive combinations. Sodium sulfate and sodium tripolyphosphate are extremely useful as diluents, and mixtures of such compounds are of the greatest value since they provide a buffer solution having a pH in the preferred range of 6.5 to 9.5. Some household products, such as scouring cleaners and dishwashing compositions, which advantageously include a dichlorocyanurate, usually contain carrier agents which are more alkaline than sodium tripolyphosphate, e.g., sodium carbonate and sodium phosphate. The pH of such products can range up to about 11. The pH of solutions made up from dichlorocyanuric acid in combination with various ratios of sodium tripolyphosphate and sodium sulfate are given below. These solutions were prepared to contain 100 ppm available chlorine from a solid formulation containing 10% by weight available chlorine.

| Wt. Ratio, Sodium Tripolyphosphate to Sodium Sulfate | pH |
|---|---|
| 1:0 | 9.03 |
| 3:1 | 8.84 |
| 1:1 | 7.43 |
| 1:3 | 6.88 |
| 0:1 | 3.70 |

A particularly effective method of obtaining an intimate mixture of the carrier agent and dihalocyanurate is to employ sodium sulfate decahydrate in the synergistic carrier agent. When sodium sulfate decahydrate is heated to a temperature above about 32°C. which is conveniently just above usual room temperature, water is liberated from the hydrate. The hydrate again solidifies when the temperature falls below 32°C., and it forms a homogeneous mass with the intermixed dihalocyanurate. The amount of hydrate in the carrier agent may be readily regulated to provide either a plastic mass or an agglomeration of the mixture of dihalocyanurate and carrier agent by empolying some of the anhydrous carrier agent as well as other carrier agents. The plastic mass is advantageously extruded in sheet or string form while it is still plastic to provide a large surface area for efficient drying. Any convenient means of drying may be employed, and the solidified form can readily be broken up into a powder and screened.

In order to test the relative bleaching power of dichlorocyanurate anion and illustrate its effectiveness compared to the commonly used bleaching compounds, a test was conducted using solutions containing 100 parts per million of available active chlorine. One cup of Tide, a detergent of Procter and Gamble was inserted in 16 gallon solutions of each of the bleaching agents tested. Eight pounds of clothes were washed twice in each of the different solutions containing the various bleaching compounds using successive 15 minute washing cycles with fresh solution in each cycle, and employing a water temperature of 52°C. Measurements of the clothes were made with a reflectometer before the clothes were washed and after the two 15 minute washings. The following reflectance readings clearly illustrate the effectiveness of dichlorocyanurate as a bleach:

| | |
|---|---|
| Control | 72.0 |
| A commercial solution of sodium hypochlorite | 79.4 |
| A preparation of 10% by weight dichlorocyanuric acid mixed with 90% by weight carrier agent* | 81.1 |
| A commercial preparation of trichlorocyanuric acid | 79.9 |
| A commercial preparation of 1,3-dichloro-5,5-dimethylhydantoin | 76.2 |

*The carrier agent was composed of 1 part by weight of sodium tripolyphosphate and 3 parts by weight sodium sulfate.

In addition to its other advantages, the sterilizing and disinfecting properties of dichlorocyanurate render it extremely effective as a bactericide.

Complex Metal Halocyanurates

E.A. Matzner; U.S. Patent 3,294,690; December 27, 1966; assigned to Monsanto Company found that complex chlorinated cyanurate compounds such as monozinc dipotassium tetra(dichlorocyanurate) have excellent stability with reference to available chlorine and can be used in bleaching, cleaning and scouring compositions. The complex metal halocyanurate compounds have the general formula:

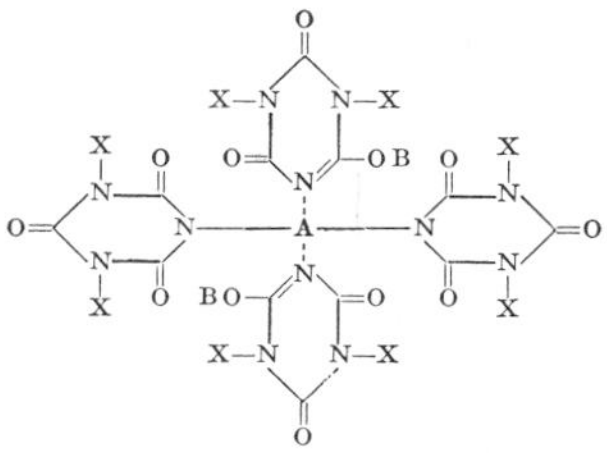

where A is magnesium or zinc, B is potassium or rubidium and X is a halogen atom such as bromine, chlorine, fluorine and iodine or mixtures thereof, but is preferably bromine or chlorine, and more preferably is chlorine. Such compounds exist in the hydrate or anhydrous forms and when in the hydrated state are crystalline solids. When in the anhydrous state these compounds are usually noncrystalline amorphous solids. Generally the complex halocyanurate compounds such as bromo-, fluoro- and iodocyanurates tend to be less stable than the corresponding complex chlorocyanurate compounds.

Preferred compounds are anhydrous or hydrated monomagnesium dipotassium tetra(dichlorocyanurate), monozinc dipotassium tetra(dichlorocyanurate), monomagnesium dirubidium tetra(dichlorocyanurate) and monozinc dirubidium tetra(dichlorocyanurate). Anhydrous monomagnesium dipotassium tetra(dichlorocyanurate) is characterized in being soluble to an extent of about 4.2 g. per 100 ml. of water at 25°C. and in having 63.7% of available chlorine. Anhydrous monozinc dipotassium tetra(dichlorocyanurate) is characterized in being soluble to an extent of about 2.3 grams per 100 ml. of water at 25°C. and in having 61.0% of available chlorine.

These compounds may be prepared, in general by reacting in an aqueous medium a halocyanurate, potassium and/or rubidium ions, and magnesium and/or zinc ions. By so proceeding, a complex dimetal tetra(dihalocyanurate) forms in and usually precipitates from the aqueous medium. Examples of halocyanurates which may be employed in the above process include tri- and dihalocyanuric acids such as, for example, tribromo-, trichloro-, triiodo- and trifluorocyanuric acids and the corresponding dihalocyanuric acids and metal salts of the dihalocyanuric acids. Examples of such metal salts include, for example, lithium, sodium, potassium, calcium, barium, strontium, magnesium, zinc, etc. dihalocyanurates. As will be hereinafter evident the above described potassium and/or rubidium and magnesium and/or zinc ions employed in the processes of this method are usually provided in the aqueous medium by water-soluble metal salts of such metals.

In the preparation of bleaching, laundering, cleaning and disinfecting compositions from the complex halocyanurates, conventional soap builders such as sodium phosphates, carbonates, and silicates may be used. The compositions may contain organic sequestering or chelating agents, essential oils, anionic or nonionic detergents.

Example 1: 336 g. of powdered dichlorocyanuric acid and 72 g. of powdered magnesium carbonate were added with agitation to 10 liters of water at 25°C. in a reaction vessel until complete solution occurred. The pH of the resulting solution was 6.7. To this solution was added, while agitation was continued, 110 g. of crystalline rubidium chloride dissolved in 100 ml. of water. The aqueous medium was thus provided with magnesium and rubidium ions. During the addition of the rubidium chloride, a crystalline precipitate formed in the reaction vessel. After the addition of the rubidium chloride was completed, stirring was discontinued and the aqueous liquid in the reaction vessel was permitted to stand for 6 hours during which time additional precipitate formed in and settled to the bottom of the reaction vessel.

Thereafter the liquid was removed from the reaction vessel by decantation, the precipitated solids were washed with cold (5°C.) water and dried in an oven at 90°C. A yield of 50 g. of a white crystalline material was obtained. Analysis of the liquid for available chlorine, magnesium and rubidium showed it to consist of an aqueous solution of a further 400 g. of product of which 190 g. were recovered by concentrating the solution at a temperature of 20°C. A portion of the crystals were further dried at 150°C. for 6 hours to constant weight during which time the crystals lost their crystalline character and became an amorphous powder. The amount of water lost during the last mentioned drying operation showed that the crystals contained 12.8% water.

Elemental analysis of this anhydrous material for Mg, Rb, C, N and available chlorine showed that the elemental content produced values which corresponded substantially to values for monomagnesium dirubidium tetra(dichlorocyanurate) having the general empirical formula $MgRb_2(Cl_2C_3N_3O_3)_4$ and the water content of the crystalline material initially obtained corresponded to that of an octahydrate of the above compound.

Example 2: Dry mixed compositions in the percentages given in Tables 1 and 2 were prepared:

TABLE 1

| Ingredient | Composition Number | | | | | |
|---|---|---|---|---|---|---|
| | 1 | 2 | 3 | 4 | 5 | 6 |
| $MgK_2(Cl_2C_3N_3O_3)_4$ | 8.0 | | | | | |
| $MgRb_2(Cl_2C_3N_3O_3)_4$ | | 4.0 | | | | |
| $ZnK_2(Cl_2C_3N_3O_3)_4$ | | | 6.0 | | | 1.0 |
| $ZnRb_2(Cl_2C_3N_8O_3)_4$ | | | | 10 | | |
| $MgK_2(Br_2C_3N_3O_3)_4$ | | | | | 4.0 | |
| Sodium tripolyphosphate | 40 | 23 | 50 | 45 | 30.0 | 30.0 |
| Sodium sulfate | 52 | 30 | | 40 | 20 | |
| Sodium carbonate | | 30 | 29.0 | | 22.5 | |
| Sodium silicate | | | 13.0 | | 10 | |
| Silica | | | | | | 64.5 |
| Sodium dodecylbenzene sulfonate | | 3 | 2.0 | 5 | 3.5 | 4.5 |

TABLE 2:

| Ingredient | Composition Number | | | | | |
|---|---|---|---|---|---|---|
| | 7 | 8 | 9 | 10 | 11 | 12 |
| $K(Cl_2C_3N_3O_4)$ | 8.0 | - | - | - | - | - |
| $Rb(Cl_2C_3N_3O_3)$ | - | 4.0 | - | - | - | - |
| $Zn(Cl_2C_3N_3O_3)_2$ | - | - | 6.0 | - | - | 1.0 |
| $Mg(Cl_2C_3N_3O_3)_2$ | - | - | - | 10.0 | - | - |
| $Na(Cl_2C_3N_3O_3)$ | - | - | - | - | 4.0 | - |
| Sodium tripolyphosphate | 40 | 23 | 50.0 | 45.0 | 30.0 | 30.0 |
| Sodium sulfate | 52 | 30 | - | 40.0 | 20.0 | - |
| Sodium carbonate | - | 30 | 29 | - | 22.5 | - |
| Sodium silicate | - | - | 13.0 | - | 10 | - |
| Silica | - | - | - | - | - | 64.5 |
| Sodium dodecylbenzene sulfonate | - | 3 | 2.0 | 5.0 | 3.5 | 4.5 |

Compositions 1,4,7 and 10 when dissolved in water effectively bleached stain from the textiles. Compositions 2,3,8 and 9 when dissolved in water were effective as cleaning and whitening compositions when employed in laundering operations. Compositions 8 and 11 were effective, when dissolved in water, and combined bleaching, laundering and sanitizing operations. Compositions 6 and 12 were effective bleaching and scouring powders.

## Alkylenediphosphonic Acid and Salts as Sequestering Agents

M.M. Crutchfield and R.R. Irani; U.S. Patent 3,297,578; January 10, 1967; assigned to Monsanto Company found that alkylenediphosphonic acids and their salts are stable sequestering agents in bleaching, cleaning and disinfecting compositions, containing organic or inorganic compounds which release chlorine during the bleaching or washing process. The alkylenediphosphonic acids have the following formula:

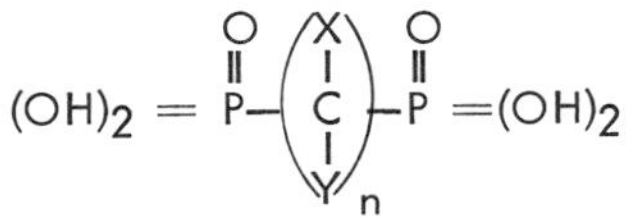

wherein n is an integer from 1 to 10, X represents hydrogen or lower alkyl (1 to 4 carbon atoms) and Y represents hydrogen, hydroxyl or lower alkyl (1 to 4 carbon atoms). Compounds illustrative of alkylenediphosphonic acids include the following:

(1) Methylenediphosphonic acid, $(OH)_2(O)PCH_2P(O)(OH)_2$

(2) Ethylidenediphosphonic acid, $(OH)_2(O)PCH(CH_3)P(O)(OH)_2$

(3) Isopropylidenediphosphonic acid, $(OH)_2(O)PC(CH_2CH_3)P(O)(OH)_2$

(4) 1-hydroxy, ethylidenediphosphonic acid, $(OH)_2(O)PC(OH)(CH_3)P(O)(OH)_2$

(5) Hexamethylenediphosphonic acid, $(OH)_2(O)PCH_2(CH_2)_4CH_2P(O)(OH)_2$

The free alkylenediphosphonic acids and their salts may be prepared by hydrolysis of the ester using strong mineral acids such as hydrochloric acid and the like. Although in general, any water-soluble salt of the alkylenediphosphonic acids may be employed, the alkali metal salts are preferred, and, in particular, the sodium salts such as the di-, tri- and tetrasodium salt; however, other alkali metal salts, such as potassium, lithium, and the like, as well as mixtures of the alkali metal salts, may be substituted therefor. In addition, any water-soluble salts, such as the ammonium salts and the amine salts, which exhibit the characteristics of the alkali metal salts may be used. In particular, amine salts prepared from low molecular weight amines, i.e., having a molecular weight below about 300, and more particularly the alkyl amines, alkylene amines and alkanol amines containing not more than 2 amine groups, such as, ethylamine, diethylamine, propylamine, propylenediamine, hexylamine, 2-ethylhexylamine, N-butylethanolamine, triethanolamine and the like are the preferred amine salts.

The alkylenediphosphonic acid and salts are hydrolytically stable. These compounds are not degraded or hydrolyzed under conditions prevailing in a bleaching solution.

Example 1: For household dry bleaching the following additives within the ranges specified when incorporated with the chlorine-releasing agent give an effective formulation.

| | |
|---|---|
| Chlorine-releasing agent (chlorinated trisodium phosphate, trichloroisocyanuric acid, dichloroisocyanuric acid, sodium dichloroisocyanurate, potassium dichloroisocyanurate or mixtures of these), (percent available chlorine per total weight of formulation) | 5 to 10 |
| Additives: | % by Wt. |
| Sequestering agent | 1 to 50 |
| Inorganic phosphate (sodium or potassium-tripolyphosphate, -pyrophosphate, -orthophosphate or mixtures of these) | 0 to 50 |
| Inert additive (sodium or potassium-carbonates, -borates, -silicates, -metasilicates, -sulfates, -chlorides or mixtures of these) | 30 to 75 |
| Organic anionic surfactant | 0 to 10 |

The following dry composition (parts by weight) is especially adapted for use as a household dry bleach in an aqueous system at a concentration of about 50 ppm to 100 ppm available chlorine for bleaching and stain removal.

| | |
|---|---|
| Potassium dichloroisocyanurate | 13.0 |
| Sodium tripolyphosphate | 25.0 |
| Tetrasodium methylenediphosphonate | 5.0 |
| Sodium sulfate | 55.0 |
| Sodium dodecyl benzene sulfonate | 2.0 |
| | 100.0 |

Example 2: For commercial laundry bleaches the following additives within the ranges specified when incorporated with the chlorine-releasing agent give an effective formulation. The additives are listed on the following page.

| | |
|---|---|
| Chlorine-releasing agent (chlorinated trisodium phosphate, trichloroisocyanuric acid, dichloroisocyanuric acid, sodium dichloroisocyanurate, potassium dichloroisocyanurate or mixtures of these) (percent available chlorine per total weight of formulation) | 5 to 20 |
| Additives: | % by Wt. |
| Sequestering agent | 1 to 50 |
| Inorganic phosphate (sodium or potassium-tripolyphosphate, -pyrophosphate, -orthophosphate or mixtures of these) | 0 to 50 |
| Inert additive (sodium or potassium-carbonates, -borates, -silicates, -meta-silicates, -sulfates, -chlorides or mixtures of these) | 30 to 50 |
| Organic anionic surfactant | 0 to 5 |

The following dry composition (parts by weight) is especially adapted for use as a commercial laundry dry bleach when used in an aqueous system at the rate of about 2 ounces per 100 pounds of clothes.

| | |
|---|---|
| Trichloroisocyanuric acid | 17.0 |
| Tetrasodium methylenediphosphonate | 10.0 |
| Tetrasodium pyrophosphate | 35.0 |
| Sodium sulfate | 35.0 |
| Sodium dodecylbenzene sulfonate | 3.0 |
| | 100.0 |

## Bleaching Wash and Wear Textiles with Sodium Dichlorocyanurate and Cyanuric Acid

T.B. Hilton and X. Kowalski; U.S. Patent 3,431,206; March 4, 1969; assigned to Monsanto Company describe a process to bleach treated, so-called "wash and wear" textiles with detergent compositions, containing, as the bleaching agent, dichlorocyanuric acid and cyanuric acid. The active chlorine, released during the bleaching step, reacts with the NH group of the cyanuric acid rather than with the NH group of the resins used for the wash and wear treatment. The bleaching agent of the process consists of (a) a polychlorocyanurate containing a metal cation; and (b) an organic nitrogen-containing compound having in one tautomeric form the structural formula:

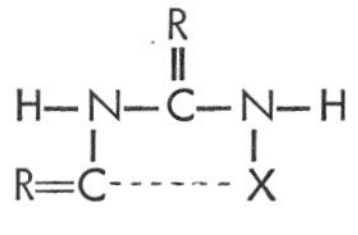

where X is selected from the group consisting of:

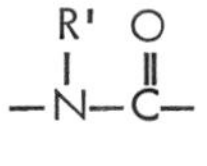

and

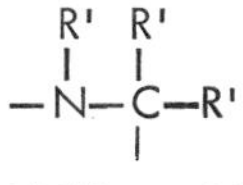

where R is selected from the group consisting of oxygen and NH, and R' is selected from the group consisting of hydrogen and alkyl radicals having from 1 to 10 carbon atoms, until the material is bleached and/or is characterized by having an improved color such as improved whiteness and brilliance. The process provides an effective way to protect any natural and/or synthetic fibrous base material which is sensitive to chlorination or oxidation due to presence in or on the surface thereof of an

$$-\overset{|}{N}H$$

group containing cured or water-insoluble resinous material for the purpose of crease proofing, or other effects. Typical fibrous base materials of which the textiles may be formed in whole or in part and which are subsequently treated with such resinous or polymeric materials are cotton, nylon, viscose rayon, Dacron, polyester, hemp, linen, jute, and blends thereof such as, for example, cotton-Dacron, cotton-Dacron-viscose rayon, cotton-nylon-viscose rayon, cotton-Dacron-nylon and cotton-nylon (all in various weight ratios). The bleaching composition comprises (a) a polychlorocyanurate, which has at least one metal cation in its molecular structure, selected from the group consisting of:

sodium dichloroisocyanurate and hydrates thereof;
potassium dichloroisocyanurate and hydrates thereof;
[(monotrichloro)(monopotassium dichloro)] diisocyanurate;
[(monotrichloro) tetra-(monopotassium dichloro)] pentaisocyanurate;
magnesium di(dichloroisocyanurate) and hydrates thereof;
calcium di(dichloroisocyanurate) and hydrates thereof;
monomagnesium, dipotassium tetra(dichlorocyanurate) and hydrates thereof;
monomagnesium, dirubidium tetra(dichlorocyanurate) and hydrates thereof;
monozinc, dirubidium tetra(dichlorocyanurate) and hydrates thereof;

and mixtures thereof; (b) cyanuric acid; and (c) a detergent builder salt selected from the group consisting of neutral and alkaline, alkali metal phosphates, alkali metal silicates having an alkali metal oxide to silica mol ratio of 1:1 to 1:3.6, alkali metal sulfates, alkali metal borates, alkali metal carbonates, alkali metal bicarbonates, and mixtures thereof. In addition to the various ingredients (a), (b), and (c) listed above or in place of the detergent builder salt, (c), there may be utilized a nonsoap synthetic organic detergent selected from the group consisting of nonsoap synthetic nonionic surface active agents and nonsoap synthetic anionic surface active agents.

Example: A series of experiments were conducted on a "treated" textile material in order to demonstrate the improved "brightening" or "whiteness" resulting from the utilization of the bleaching agents or compositions in the processes of this method. The bleaching agents set forth in the table were mixed together immediately prior to bleaching by weighing the individual solid ingredients separately and the mixing them together in a dry state. These dry mixed bleaching agents were individually dissolved in 1 liter of water contained in a Terg-O-Tometer (U.S. Testing Company, Inc.) which is a standard testing device used in the detergent art for conducting such bleaching tests. The amount of each bleaching agent used was sufficient to provide the chlorine or oxygen concentrations listed in the table.

In addition to the dry mixed compositions, there was added to each liter of water approximately 0.05% by weight, based on the weight of the compositions, of a surfactant such as sodium alkyl benzene sulfonate in which the alkyl group is a linear chain of about 10 to 15 carbon atoms and which is commercially available as

Santomerse. The pH of the aqueous bleaching solution was then adjusted with either NaOH or $HPO_3$ to the pH values listed in the table. To each of the aqueous bleaching agents or compositions prepared in the Terg-O-Tometer, there was then added 5" x 5" piece or swatch of unbleached cotton muslin which was not soiled but which had a natural grayish color and which had been previously treated with a urea-formaldehyde resin (cured to the water-insoluble state) in order to provide

$$-\overset{|}{N}H$$

groups on the surface of the muslin. (Specifically the muslin material was crease-proofed with a methylated urea-formaldehyde resin. The muslin was padded with an aqueous bath of 7.5% methylated urea-formaldehyde resin and 0.86% magnesium chloride hexahydrate. The resin was applied in a laboratory type padder, with one dip and one nip, at a 79% wet pick up. The muslin was then frame dried and cured in one operation at 305°F. for 12 minutes). In addition, there was also added to the bleaching solutions a sufficient amount of clean, white (unstained) fill cloth to give an aqueous solution:cotton cloth weight ratio of 20:1.

The Terg-O-Tometer was then set for a rate of agitation of 100 cycles per minute and and each cotton swatch was given a single 10-minute washing at 140°F. The cotton muslin swatch was then removed from the Terg-O-Tometer, air dried at about 70°C., pressed, and then subjected to a total light reflectance test in order to compute the difference in reflectance before and after washing, i.e., to determine the increase in brightening of the swatches due to bleaching. This reflectance test was carried out with the use of a Model D-1 Color-Eye, Instrument Development Laboratories, Inc., and the values obtained, designated at ΔRd, are set forth in the table. (The reflectance of both sides of each swatch was determined at least 5 times in order to attain a representative reading. The ΔRd, then, is the difference of these averages).

It will be noted that of the nitrogen compounds listed in the table, melamine, dimethyl hydantoin, and urea were of the prior art chlorine acceptors set forth and exemplified in U.S. Patent 3,099,625. When these acceptors are used with polychlorocyanurates, it has been found that the loss in tensile strength is 50% to as high as 400% greater than with cyanuric acid of the process. Since even relatively small losses in tensile strength are undesirable from a consumer viewpoint, it can readily be seen then that the use of cyanuric acid of this process is a significant improvement as contrasted to the prior art compositions. The use of sodium perborate as a bleach is exemplified in order to illustrate relative ΔRd obtained using an oxygen-based bleach and to show the more effective bleaching obtained by a chlorine-based bleach as contrasted to an oxygen-based bleach.

| Bleach [1] | Nitrogen Compound | Mol Ratio [2] | Chlorine or $O_2$ Concentration, p.p.m. | Washing Solution pH | ΔRd [3] of Textile [4] |
|---|---|---|---|---|---|
| Sodium Dichlorocyanurate | Cyanuric Acid | 0.5:1 | 200 | 7.5 | 8.71 |
| Do | do | 2:1 | 200 | 7.5 | 6.77 |
| Do | do | 4:1 | 200 | 7.5 | 5.76 |
| Do | do | 4:1 | 100 | 7.5 | 4.88 |
| Do | do | 4:1 | 50 | 7.5 | 4.14 |
| Do | Melamine [5] | 4:1 | 200 | 7.5 | 2.52 |
| Do | Dimethyl Hydantoin [5] | 4:1 | 200 | 7.5 | 3.74 |
| Do | Urea [5] | 4:1 | 200 | 7.5 | 2.21 |
| Do | | | 200 | 7.5 | 1.48 |
| Sodium Perborate | | | [6] 45.2 | 9.3 | 1.64 |
| Do | | | [6] 22.6 | 9.3 | 0.93 |
| Do | | | [6] 11.3 | 9.3 | 0.62 |
| Trichlorocyanuric Acid | Cyanuric Acid | 4:1 | 200 | 7.5 | 3.98 |

[1] Compound used to supply either available chlorine or oxygen for bleaching purposes.
[2] Mol ratio of nitrogen compound to bleach compound.
[3] Measurement of total light reflectance; the larger value indicates a more effective bleaching without "yellowing" of the fabric or deterioration thereof.
[4] Unbleached cotton sheeting (muslin) treated with urea-formaldehyde resin.
[5] Chlorine acceptors as exemplified in U.S. Patent 3,099,625.
[6] $O_2$ concentration.

The bleaching effect obtained by using a mixture of trichlorocyanuric acid and cyanuric acid is illustrated at the end of the table. It should be noted that trichlorocyanuric acid does not contain any metal cations in its molecule and, therefore, is not a polychlorocyanurate as defined here. The bleaching effect of trichlorocyanuric was included in the table in order to compare it with the bleaching effect obtained by the use of a polychlorocyanurate in accordance with the processes of this method. On a comparative testing basis (at the same level of available chlorine), the sodium dichlorocyanurate-cyanuric acid mixture subsequently yielded a processed muslin which has a ΔRd of 5.76 compared to the value of 3.98 which was obtained by processing another similar piece of muslin with the trichlorocyanuric acid-cyanuric acid mixture.

While cyanuric acid has in the past been suggested for use in compositions which contain trichlorocyanuric acid (note U.S. Patents 2,980,622 and 3,213,029), it can readily be seen, then, that there is a significant improvement and unexpected result over such trichlorocyanuric compositions when cyanuric acid is used with the herein defined polychlorocyanurate in the processes of this method.

Complex Chlorinated Cyanurates

W.F. Symes; U.S. Patent 3,350,317; October 31, 1967; assinned to Monsanto Co. describes a process for the use of potassium containing complex chlorinated isocyanates in bleaching, oxidizing, sterilizing and disinfecting compositions. One of the complex compounds is an anhydrous, crystalline solid having a distinct x-ray diffraction pattern and the general formula:

(I)

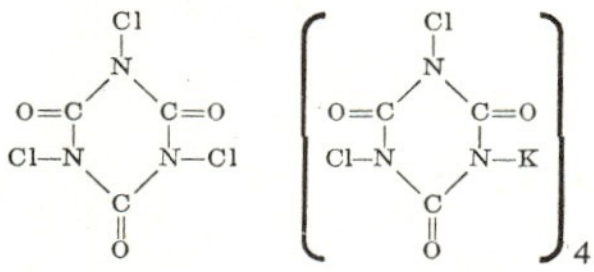

and is further characterized in that it has an available chlorine content of 66.4%, is soluble in distilled water at 25°C. to an extent of about 2.5% by weight; the pH of a saturated aqueous solution being about 4.3. This compound, as prepared, usually has an available chlorine content in the range of 66% and 67% decomposes without melting in the range of 260° to 275°C. The above described compound is designated either as Compound I or as [(monotrichloro), tetra(monopotassium dichloro)]-penta-isocyanurate.

Another cyanurate compound of this process is also an anhydrous, crystalline material but has the general formula:

(II)

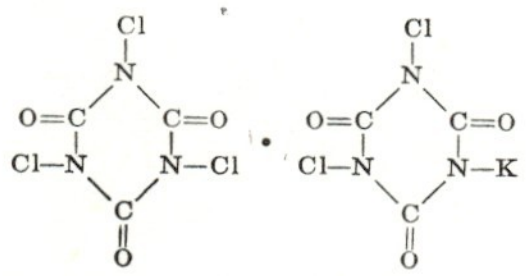

and is further characterized in that it has an available chlorine content of 75.8%, is soluble in water at 25°C. to an extent of about 1.0% by weight; the pH of an aqueous saturated solution thereof being about 4.1. This compound has an available chlorine content of from 75 to 77% and undergoes decrepitation when heated to 170° to 215°C. and decomposes without melting in the region of 260° to 275°C. This compound is designated as either Compound II or (trichloro) (monopotassium dichloro) diisocyanurate. These compounds may be prepared singly (in pure form) or in the form of mixtures by at least 2 different methods. One such method comprises bringing together and reacting monopotassium dichloroisocyanurate and trichloroisocyanurate acid in an inert liquid in a reaction zone and isolating the compound or mixture of compounds therefrom.

It is possible by varying and controlling the pH of the inert liquid and the molecular ratio between monopotassium dichloroisocyanurate and trichloroisocyanuric acid to prepare either Compound I or Compound II, in pure form, or mixtures of these compounds. This method comprises bringing together and reacting in a reaction zone, an aqueous solution containing from about 5 to 12% by weight of monopotassium dichloroisocyanurate and from about 15 to 36% by weight of trichloroisocyanuric acid dissolved in an inert organic solvent such as acetone, methyl or ethyl alcohol wherein 20 to 25% by weight of the solution is composed of trichloroisocyanuric acid.

When it is desired to prepare Compound I only, the amount and concentration of the solutions of monopotassium dichlorocyanurate and trichlorocyanuric acid in the reaction zone should be such that the molecular ratio of monopotassium dichloroisocyanurate to trichloroisocyanuric acid is in excess of between 6 to 1 and 8 to 1. Under these conditions essentially pure Compound I separates from the liquid phase of the reaction mixture as an insoluble precipitate which can be separated from the unreacted materials which remain in solution.

When it is desired to produce Compound II, it is generally preferred to bring together and react an aqueous solution of monopotassium dichloroisocyanurate containing 6 to 8% by weight of monopotassium dichloroisocyanurate and trichloroisocyanuric acid dissolved in an inert organic solvent such as acetone or methyl or ethyl alcohol wherein 30 to 35% by weight of the solution consists of trichloroisocyanuric acid, the amounts and concentrations of such solutions in the reaction zone being preferably adjusted so that the molecular ratio of monopotassium dichloroisocyanurate to trichloroisocyanuric acid is not more than between 1.15 to 1 and 1.25 to 1.

It is possible to prepare various mixtures of Compounds I and II by altering the quantities and/or concentrations of the above described solutions in such a manner as to provide a molecular ratio of monopotassium dichloroisocyanuric acid and trichloroisocyanuric acid in the range of from 3.95:1 to 1.35:1. However, it is preferred to make mixtures of Compounds I and II by bringing together an aqueous solution consisting of from 10 to 25% by weight of monopotassium dichloroisocyanurate and slurry containing from 15 to 60 parts by weight of trichloroisocyanuric acid and from 40 to 85 parts by weight of water; the quantities of the solution and slurry referred to being controlled to provide a molecular ratio within such range. The particular ratio used will depend on the relative amounts of Compounds I and II which may be desired in the mixture.

In all of the above reactions, which are preferably carried out at a temperature within the range of 5° to 50°C., a precipitate forms (usually as white, fine, particulate crystals) which can be separated from the bulk of the liquid phase of the reaction mixture by filtration, centrifugation and the like. The precipitate is then preferably dried, although it may be used directly in the wet state. This precipitate is, of course, Compound I, Compound II or mixtures thereof, depending on the ratio of reactants, etc., as described above.

Example 1: [(Monotrichloro), tetra-(monopotassium dichloro)], pentaisocyanurate or Compound I — 16 g. of monopotassium dichloroisocyanurate were dissolved in 194 g. of water and placed in a 300 ml. Pyrex beaker. To this solution was added 2.0 g. of trichloroisocyanuric acid dissolved in 8.0 ml. of acetone and the resulting mixture, which mixture had a pH of about 4.9, was stirred at about 300 rpm with a standard electric stirrer for about 10 minutes. A white precipitate, which formed almost immediately, was removed by filter paper filtration in a Büchner funnel. The precipitate, in the form of a filter cake, was successively washed 3 times with 3 ml. increments of water and was aspirated for 5 minutes to remove as much moisture as possible. The filtrate which consisted essentially of water, acetone, and monopotassium dichlorocyanurate was discarded. The filter cake was then dried to constant weight in an oven set at 100°C.

A white dry crystalline solid weighing 4.94 g. was obtained. X-ray diffraction patterns were obtained using the instrument and method described in Phillips Technical Reviews, vol. 10, p. 1, published in 1948. X-ray diffraction analysis of the above described white crystalline solid showed a diffraction pattern which was unique and distinct from diffraction patterns shown by either trichloroisocyanuric acid or monopotassium dichloroisocyanurate.

Example 2: (Monotrichloro), (monopotassium dichloro), diisocyanurate or Compound II — 12 8/10 g. of monopotassium dichloroisocyanurate was dissolved in 187.2 g. of water, placed in a 300 ml. beaker and the resulting product mixed with 10 g. of trichloroisocyanuric acid dissolved in 35 ml. of acetone. The resulting mixture which had a pH of 2.8 was stirred at 300 rpm with a standard electric stirrer for about 5 minutes. A white precipitate, which formed almost immediately, was removed by filter paper filtration in a Büchner funnel. The precipitate, in the form of a filter cake, was successively washed 3 times with 3 ml. increments of water, and was aspirated for 5 minutes to remove as much moisture as possible. The filtrate was discarded. The filter cake was dried to constant weight in an oven set at 100°C. A white dry crystalline solid weighing 14.96 g. was obtained.

As illustrations of useful bleaching, cleansing and sanitizing compositions containing the compounds of this process, the following typical specific formulations are shown.

Typical Household Laundry Bleach

| | Weight Percent |
|---|---|
| Sodium tripolyphosphate | 40 |
| Sodium sulfate | 24 |
| Sodium metasilicate | 20 |
| Sodium dodecylbenzene sulfonate | 5 |
| Potassium silicate | 1 |
| Compound I (may also be Compound II or mixtures of I and II) | 10 |

Typical Scouring Powder

| | Weight Percent |
|---|---|
| Silica | 90.0 |
| Sodium tripolyphosphate | 5.0 |
| Soda ash | 2.5 |
| Sodium lauryl sulfate | 2.2 |
| Compound I (may also be Compound II or mixtures of I and II) | 0.3 |

Typical Dishwashing Formulation (For Automatic Dishwasher)

| | Weight Percent |
|---|---|
| Sodium tripolyphosphate | 45 |
| Sodium sulfate | 2.3 |
| Sodium metasilicate | 2.3 |
| Soda ash | 8 |
| Compound I (may also be Compound II or mixtures of I and II) | 1 |

## Puffed Borax and Dichlorocyanurate

B. Weinstein and H.L. Marder; U.S. Patent 3,538,005; November 3, 1970; assigned to American Home Products Corporation describe a process to prepare dry bleach and detergent compositions containing the metal salts of dichlorocyanurate and puffed borax in the weight ratio of about 30:70. The isocyanurates useful in these compositions are available commercially. For example, potassium dichlorocyanurate is available as ACL-59 and [(monotrichloro)-tetra(monopotassium dichloro)]-pentaisocyanurate is available as ACL-66.

The puffed borax useful in the preparation of the compositions is also available commercially in various bulk densities and particle size distributions. It is a form of borax made by the rapid heating of hydrates of sodium tetraborate. It is characterized by versatility of bulk density, large surface area, rapid solubility rate, and high absorptive potential for many substances. For use in the compositions the puffed borax may be prepared by rapidly heating the pentahydrated form of borax in a hot air stream. Some of the water of hydration within the individual borax feed particles is flashed off, thereby resulting in a particle of greater surface area and lower bulk density.

By this method, most densities of from about 3 to about 40 pounds per cubic foot are attainable from an original stock feed having a bulk density of 60 to 65 pounds per cubic foot. These resulting lower bulk densities are a function of the rate of the feed of the stock borax, the temperature to which the stock borax is exposed, and the residence time of the particles in the heating area. The finished product particle size distribution is dependent upon the desired bulk density of the puffed borax and the particle size distribution of the starting borax feed. The selected isocyanurate and the puffed borax, each having the particle size distribution and bulk densities within the ranges set forth hereinabove, may be blended by suitable blending apparatus in weight ratios within the proportion ranges also set forth above, to result in the homogeneous products of the process, which exhibit uniform chlorine distribution as well as the other advantages referred to hereinbefore.

The successful blending of materials of substantially equal particle size distributions and bulk densities normally poses no real processing problems, per se. However, the compositions of the process demonstrate that the successful blending of the two components with different particle size distributions and particularly bulk densities, results surprisingly in useful, more stable, effective consumer products highly suitable for use as a powder bleach in washing machines, only when there is made a choice of weight proportions, particle size distributions, and bulk densities for the two components, as set forth hereinbefore.

Thus, preferably, the proportions by weight of the isocyanurate to the puffed borax ranges from about 15:85 to about 60:40; the major proportion of the particles of the isocyanurate and the puffed borax is of the size within the U.S. sieve range of from about 40 to 100; and the bulk density of the total isocyanurate present is within the range of from about 50 to about 65 pounds/ft.$^3$, and that of the total puffed borax present is within the range from about 3 to about 40 pounds/ft.$^3$. Although excellent results have been obtained with compositions in accordance with the process which consist essentially of the aforesaid two components; i.e., selected isocyanurate and puffed borax, various additives may be incorporated in minor proportions, as will appear to those skilled in this art. Thus, for example such additives as suitable detergent builders, sequestering agents, compatible surface active agents, mineral oil, perfume, optical brighteners and/or stable colors for identification purposes, and the like, may be incorporated in the compositions in amount up to about 15% by weight of total composition.

## Light Density Cyanurate Detergent Bleach

F.K. Rubin and C.J. Carmack; U.S. Patent 3,640,875; February 8, 1972; assigned to Lever Brothers Company describe a process to manufacture a light density, spray-dried bleaching composition containing uniformly dispersed potassium dichlorocyanurate in a conventional detergent composition. In the process, the spray-dried detergent base is formed first, then the bleaching agent is blended with the base. The three required components in the spray-dried base are a phosphate builder, a borax fluffing agent and a solubilizing agent. Suitable alkali metal phosphate builders, among others, are sodium tripolyphosphate, potassium tripolyphosphate, tetrasodium pyrophosphate, tetrapotassium pyrophosphate, sodium acid pyrophosphate, sodium trimetaphosphate, sodium tetraphosphate and sodium hexametaphosphate.

A borax fluffing agent is employed to promote a light density product. This includes, among others, borax pentahydrate and borax decahydrate or other inorganic compound containing water of hydration which will "pop" from the heat of spray-drying to form light weight puffed beads. Any solubilizing agent for the borax fluffing agent can be used in the process. Sodium toluene sulfonate and sodium xylene sulfonate, among others, are suitable. One or more optional components can be included in the spray-dried base, such as fillers, fluorescent dyes, cerium source, colorants and surfactants.

The components can be spray-dried by any known method. A suitable method used in the example is to form in a crutcher an aqueous slurry, ranging in concentration from 10 to 40%, with 25% as the preferred slurry concentration, by introducing with mixing a borax fluffing agent, a solubilizing agent, a filler, a fluorescent dye, a colorant and water. This mixture is heated, e.g., to about 160°F., and the phosphate builder is added with agitation. The resulting slurry at a temperature about 160° to 180°F. is pumped to a booster or holding tank where it is circulated through a Reitz mill to improve homogeneity. From this tank the slurry is passed through a high pressure pump and is routed to single liquid spray nozzles at the top of a spray tower where it is forced out in a spray pattern at 80 to 300 psig pressure. The slurry droplets fall by gravity through a countercurrent flow of heated air which has an inlet temperature of about 600° to 700°F. and an outlet temperature of about 200° to 300°F. to form hollow beads which is known as a spray-dried base. The moisture level, particle size and gravity of the spray-dried base can be varied by adjusting the tower air flow rate, the tower temperature, the slurry feed rate and the slurry moisture level. The final bleach composition generally has about 50 to 85% of spray-dried base.

A chlorinated agent is then blended with the spray-dried base by any acceptable procedure. For instance, the two components may be introduced into a rotating horizontal drum. As defined herein, a chlorinating agent is a stable, chlorine-releasing organic compound which is compatible with the spray-dried base and which liberates chlorine under conditions normally used for bleaching purposes. This includes the following, among others: potassium dichlorocyanurate, sodium dichlorocyanurate, [(monotrichloro)-tetra-(monopotassium dichloro)]penta-isocyanurate, 1,3-dichloro-5,5-dimethyl hydantoin, N,N'-dichlorobenzoyleneurea, paratoluene sulfodichloroamide, trichloromelamine, N-chloroammeline, N-chlorosuccinimide, N,N'-dichloroazo-dicarbonamidine, N-chloro acetyl urea, N,N'-dichlorobiuret, chlorinated dicyandiamide, chlorinated trisodium phosphate, and the sodium derivative of N-chloro-p-toluene sulfonamide. Generally about 6 to 42 parts of the chlorinating agent is blended with 100 parts of the spray-dried base. The chlorinating agent usually comprises about 5 to 25% of the final bleach composition of this process.

Example: A spray-dried base was formed from the components and at the conditions indicated below:

| Components: | Parts |
|---|---|
| Sodium tripolyphosphate | 20.00 |
| Sodium toluene sulfonate (active base) | 25.00 |
| $Na_2B_4O_7 \cdot 5H_2O$ | 20.00 |
| Calcofluor White 5B | 0.115 |
| Tinopal RBS | 0.104 |
| Ultramarine Blue | 0.50 |
| Sodium sulfate | 9.991 |
| Water | 4.00 |

| Conditions: | |
|---|---|
| Slurry temperature (°F.) | 160 |
| Pressure (psig) | 80-120 |
| Air inlet temperature (°F.) | 625-675 |
| Air outlet temperature (°F.) | 250-300 |
| Density* (g./cc) | 0.255-0.34 |
| Water content* (percent) | 8.5-15 |

*Samples taken at various intervals during spray-drying

After screening through a 10-mesh screen, the spray-dried base had a moisture content of 10% and a density of 0.285 g./cc. The spray-dried base (67 parts) was blended with potassium dichlorocyanurate (18 parts) by introducing the two components into a rotating horizontal drum. A 60% aqueous solution of tetrapotassium pyrophosphate was sprayed into the rotating blend (1 lb. per minute at room temperature) until 9 parts of tetrapotassium pyrophosphate solids were added. The resulting product was then aged by air blowing at 100° to 150°F. for 15 minutes to form low density bleach composition A.

Bleach composition B was provided by the same procedure except that no tetrapotassium pyrophosphate solution was added to the blend. The uniformity was determined by packaging bleach composition A in boxes about 6" x 2 1/8" x 8 1/2". Bleach composition B was packaged similarly. Some of the boxes of each composition were vibrated for 5 minutes on the pan of a Syntron Jogger, Model J-1A. The available chlorine values of standard cupfuls from both unvibrated and vibrated boxes were determined. The distribution of the chlorinating agent was found to be satisfactorily uniform.

## Dichlorocyanurate Sanitizing Presoak Composition

R.E. Keay and R.R. Keast; U.S. Patent 3,634,261; January 11, 1972; assigned to FMC Corporation describe a presoak composition containing sodium dichlorocyanurate to clean, sanitize and deodorize fabrics soiled with a combination of organic and bacterial soil, such as diapers. The compositions contain as essential ingredients an N-chloro compound which hydrolyzes to yield positive chlorine ions, and preferably a chloroisocyanuric compound in an amount to provide in aqueous solution about 50 to 250 ppm of available chlorine, sulfamic acid in an amount to provide an NH to $Cl^+$ ratio in solution of about 1 to 1, 25 to 35% by weight of a polyphsophate, preferably sodium tripolyphosphate, potassium tripolyphosphate, sodium pyrophosphate or potassium pyrophosphate, 25 to 35% by weight sodium tetraborate having an $Na_2O$ to $B_2O_3$ ratio of about 1 to 2.5 and having 1 to 5 mols of water of hydration, 25 to 35% by weight of an inorganic buffer salt, preferably an alkali metal bisulfate, and about 2.5 to 10% by weight of an anionic or nonionic surfactant.

These compositions are particularly effective when used in a presoak solution in which the soiled fabric article, preferably but not necessarily after a simple rinse in cold water, is soaked at room temperature in the solution for about 12 to 48 hours before being washed in any useful soap or detergent washing solution. Alternatively, the compositions may be used in ordinary washing cycles, for example, in a home automatic clothes washer, in commercial laundry equipment or by hand. These compositions are particularly effective when used at ambient temperature, although they are effective at a temperature as high as 140°F.

The organic N-chloro compound is included in the compositions as a source of available chlorine, which preferably should be present in the amount of about 100 to 120 parts per million (ppm) in the sanitizer bath for best germicidal effectiveness. Obviously, it is possible to depart from this available chlorine level to a reasonable extent while still providing satisfactory sanitization, and in this connection often the bath may contain about 50 to 250 ppm of available chlorine. However, in the event too little available chlorine (less than about 50 ppm) is employed insufficient sanitizing occurs, while if too much available chlorine (above about 250 ppm) is present, chlorine odor may become a problem both in the bath and on the rinsed product. Generally speaking the N-chloro compound when employed in the amount of about 1 to 15% by weight, depending on the available chlorine content of the particular compound, in these compositions provides the desired amount of available chlorine in the sanitizer bath. The chloroisocyanurates, including the sodium and potassium dichloroisocyanurates as well as trichlorocyanuric acid and complexes of potassium dichloroisocyanurate and trichloroisocyanuric acid, are the preferred N-chloro compounds.

Examples 1 and 2:

| Component | Examples, percent by weight | |
|---|---|---|
| | 1 | 2 |
| Sodium tetraborate. $5H_2O$ | 30.0 | 30.0 |
| Sodium tripolyphosphate | 27.8 | 28.95 |
| Sodium bisulfate | 30.0 | 30.0 |
| Sulfamic acid | 2.2 | 1.85 |
| Sodium dichloroisocyanurate | [1] 5.0 | [2] 4.2 |
| Anionic surfactant [3] | 5.0 | 5.0 |

[1] 120 p.p.m.
[2] 100 p.p.m.
[3] Sodium dodecyl benzenesulfonate (85% active).

The compositions of Examples 1 and 2 in the table were evaluated for performance as presoaks for soiled diapers. In a general test, these two compositions were dissolved in water (1/2 ounce per gallon), and soiled cotton diapers were soaked in the solution for 24 hours. They were then washed in a standard detergent wash. The presoak eliminated odors and stains and prepared the diapers well for washing. Diaper damage was also found to be negligible. In that test, after 24 and 48 hour soaks of clean cotton diapers in aqueous solution of the composition of Examples 1 and 2 (1/2 ounce per gallon in each case), the fabrics showed only slight fluidity increases. The test was run in accordance with AATCC Method 82-1961 and fluidity increases of only 1 Rhes units (24-hour soaked diaper) and 2 Rhes units (48-hour soaked diaper) were measured. These values are consistent with acceptable practice.

Halobenzoylimide Activators for Persalts

F.W. Gray; U.S. Patent 3,655,567; April 11, 1972; assigned to Colgate-Palmolive Company found that halogenzoylimides, such as N-m-chlorobenzoyl-succinimide and N-m-chlorobenzoyl-5,5-dimethylhydantoin are synergistic activators for persalts such as perborates used as the bleaching agent in detergent formulations. These halobenzoylimides have the following general formula:

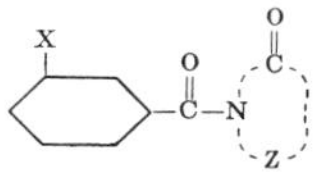


wherein X represents halogen, e.g., chloro, bromo, etc. and Z represents the atoms necessary to complete a heterocyclic nucleus selected from the group consisting of a hydantoin or succinimide.

Particularly beneficial results as regards the attainment of increased bleaching activity with lower concentrations of peroxide compound are obtained with compounds of the above formula wherein X represents chlorine, e.g., N-m-chlorobenzoyl-succinimide and N-m-chlorobenzoyl-5,5-dimethylhydantoin respectively. These particular compounds are singularly characterized in permitting the realization of manifold increases in bleaching activity despite their use in significantly reduced concentrations, i.e., concentrations constituting but a fraction of those heretofore required with analogous compounds promulgated in the art for such purposes. Thus, significant improvement in bleaching activity can be obtained with the use of peroxide quantities sufficient to yield a concentration of only 2 to 4 parts of available oxygen per million parts of wash solution. In general, peroxide quantities sufficient to yield available oxygen concentrations ranging from about 2 to 20 ppm in wash solutions are found to be eminently suitable for the vast majority of laundering operations.

The activator should be utilized in amounts sufficient to yield a mol ratio of perborate to activator with the range from about 0.5:1 to about 6:1. Within the foregoing range specific values may be selected in accordance with the bleaching problem to be negotiated; however, it is usually found that optimization of bleaching activity can be obtained by utilizing a perborate/activator mol ratio which approximates 1:1. In any event, and as will be explained in detail hereinafter, specific and unusual situations may arise dictating departures from use of the optimum ratio in order to suppress, for example, deleterious effects with regard to certain types of dyed fabrics. The bleaching compositions may be employed in various forms such as powders, tablets and the like with or without additional noninterfering ingredients such as inert fillers, e.g., sodium sulfate, sodium chloride, etc., conventional detergent composition components, e.g., synthetic detergent, soap, builder salts, soil suspending agents, brighteners, bacteriocides, bacteriostats, antioxidants, etc.

When used solely for bleaching purposes the composition may consist entirely or substantially (e.g., 50 to 100%) of the peroxygen compound and activator or alternatively there may be added inert filler or other noninterfering ingredients so that the peroxygen-activator combination comprises a minor fraction of the total composition (e.g., 1 to 50% by weight thereof). Where used as a component of a detergent formulation the peroxygen-activator combination may comprise from about 1 to 75% by weight of the total composition with 2 to 50% preferred and 5 to 40% being most preferred. The balance of such formulations comprises the usual ingredients, i.e., detergent, builder, etc., in the conventional concentrations, e.g., detergent 5 to 75% by weight, builder 5 to 75% by weight, fillers 0 to 75% by weight and minor amounts, e.g., 0.1 to 10% by weight of other adjuvants such as soil suspending agents, brighteners and antioxidants.

Examples 1 through 3: The superior bleaching capacity in terms of concentration efficiency will be made readily manifest by reference to the following examples wherein all percentages and parts are given by weight unless otherwise indicated. In each of the examples, testing is carried out at 120°F. in a tergotometer utilizing sodium perborate of formula weight 154, available oxygen 10% and activator in 1 liter of a 0.15% commercial detergent solution having the following composition:

| | Parts |
|---|---|
| Moisture | 10.5 |
| Linear tridecylbenzenesulfonate | 21.0 |
| Sodium sulfate | 25.8 |
| Pentasodium tripolyphosphate | 33.5 |
| Sodium silicate | 7.0 |
| Borax (as $Na_2B_4O_7 \cdot 10H_2O$) | 1.0 |
| Sodium carboxymethylcellulose | 0.4 |
| Polyvinyl alcohol | 0.2 |

with the remainder comprising antioxidant, brightener, etc.

In each instance, the test fabric comprises 3 x 6 inch coffee/tea-stained swatches as (cotton). The tergotometer receptacle containing the test swatches and activated perborate detergent solution, is agitated (75 rpm) for a period of 10 minutes. After a rinse, the load is dried and evaluated. Reflectance readings are taken both before and after completion of the immersion treatment, the numerical difference between readings being recorded as Δ Rd. The results obtained are itemized in the table below.

| | Detergent Composition | | | | | | |
|---|---|---|---|---|---|---|---|
| Ex. No. | Commercial deterg., percent | Na perborate, percent | N-m-cb[1] succinimide activator, percent | P.p.m. avail. $O_2$ | P.p.m. activator | Mole ratio perborate activator | Δ Rd |
| 1 | 93.2 | 2.5 | 4.3 | 4 | 70 | 1:1 | 4 |
| 2 | 88.3 | 4.1 | 7.6 | 7 | 130 | 1:1 | 7.8 |
| 3 | 84.3 | 8.4 | 7.3 | 15 | 130 | 2:1 | 6.2 |

[1] Chlorobenzoyl.

As the above data make manifestly clear, efficient bleaching activity is obtained despite an available oxygen concentration of only 4 ppm. Moreover, the reflectance values itemized as ΔRd are markedly superior to corresponding values obtained with analogous activator compounds typical of those described in the prior art when subjected to the identical testing procedure utilizing the identical compositions itemized in the table. The ΔRd value measurement of 4 obtained in Example 1 comprises a significantly high value in view of the minimal concentration of available oxygen as well as relatively mild temperature 120°F., at which the testing is carried out. Such value would indicate substantial bleaching activity and especially in view of the somewhat limited period of washing (10 minutes).

As the above data would further suggest, optimum Δ Rd values result with the use of a perborate/activator mol ratio of approximately 1:1. Mol ratios considerably in excess of the equimolar range apparently detract somewhat from the efficiency of the system. The reflectance values obtainable, however, are nevertheless of the first order of significance being markedly superior to many of the compounds heretofore provided. Thus, the 1:1 mol ratio value would be recommended for most home laundering operations; as a matter or pure economics, excess perborate should be avoided since correlative enhancement of bleaching activity for given increases in concentrations is not attainable thereby. Within the approximate 1:1 perborate/activator mol ratio range, increased bleaching activity attends the use of the activator system in increased quantities. This aspect is made clear in the above table wherein an approximate two-fold increase in Δ Rd is obtained when the relative proportions of activator and perborate and increased correspondingly while maintaining a 1:1 perborate/activator mol ratio.

Methods for the preparation of the activators are as follows: (a) Preparation of N-m-chlorobenzoylsuccinimide – To 20.0 g. of succinimide and 40 ml. of pyridine, initially at room temperature, was added in drops and with stirring 35.0 g. of m-chlorobenzoyl chloride in a period of about 0.5 hours. After stirring for an additional 0.5 hours the reaction mixture is allowed to stand for several hours at room temperature. The solidified mass is treated with 100 ml. of ethanol and the undissolved crude product removed by filtration. On recrystallization from hot ethanol, about 19.5 g. of N-m-chlorobenzoylsuccinimide, melting point 127.5° to 128°C., is obtained.

(b) Preparation of N-m-chlorobenzoyl dimethylhydantoin — To 102.4 g. of dimethylhydantoin and 160.0 g. of pyridine was added with stirring 140.0 g. of m-chlorobenzoyl chloride in a period of about 2.5 hours. Temperature of reaction is between 25° and 62°C. The reaction mixture is next added to 2 liters of water containing 400 ml. of concentrated hydrochloric acid. After allowing to stir for about 10 minutes, the white precipitate is removed by filtration and dried to constant weight. The crude product is recrystallized from ethanol to give N-m-chlorobenzoyldimethylhydantoin of melting point 144.5° to 145.5°C.

## PACKAGED OR TABLETTED BLEACHES

### Packaged Dry Cyanuric Acid Bleach

M.L. Dickey; U.S. Patent 3,061,549; October 30, 1962; assigned to Purex Corporation, Ltd. describes containers for dichloro- or trichlorocyanuric acid compositions with absorbents, so-called "deodorants", to eliminate objectionable volatiles which result from the partial decomposition of the cyanuric acid compositions. Water-soluble inorganic salts and mixtures thereof suitable for compounding with the dichlorocyanuric or trichlorocyanuric acid, include in weight percentages ranging, e.g., from 90 to 10%, alkali metal phosphates such as tripolyphosphate, tetrasodium pyrophosphates, orthophosphates, and hexametaphosphates, as well as such builders as sodium sulfate and sodium chloride.

For some purposes it may be desirable to compound with the acid-salt mixture, other additives such as small percentages of detergent, e.g., soap or lauryl sulfate or alkyl aryl sulfonate synthetics, and appropriate dyes or fluorescent compounds. The odor eliminating material referred to as the "deodorant" may consist of any of the following: mixtures of manganese dioxide and cupric oxide (hopcalite), activated alumina, activated carbon, zeolites, fuller's earth, bentonite, activated montmorillonite, silver phosphate, silver nitrate, mercuric nitrate, alkali metal silicates (e.g., Metso, anhydrous $Na_2O:SiO_2$ = 1:1), alkali metal hydroxides, potassium, rubidium or cesium carbonates, and alkaline earth oxides.

Figure 6.1a shows a package containing the dry bleach or disinfectant together with an odor stabilizing material contained in a bag; Figure 6.1b illustrates a package in which the odor stabilizing material is contained in a separate compartment of the package, and Figure 6.1c illustrates another form in which the deodorant is carried in a coating applied to an inside surface of the package.

Referring first to Figure 6.1a, the package generally indicated at (10) is shown to consist typically of a glass or paper board container (11) having top and bottom closures (12) and (13), and which may be adapted for opening in any suitable manner or by any suitable means, not shown, to dispense the package content (14) consisting essentially of dichlorocyanuric ro trichlorocyanuric acid which may be admixed with any appropriate water-soluble salt or mixture, such as alkali metal phosphate and sulfates. Conversion of noxious volatiles given off by the cyanuric acid to compounds having no objectionable odor is effected according to the showing of Figure 6.1a by inserting within the container a porous fabric bag (15) which may contain any of the deodorants listed above. By reason of the porosity of the bag, the deodorant is exposed to the package contents (14), but remains unmixed therewith.

In Figure 6.1b the deodorant (16) is shown to be accommodated within the perforated top compartment (17) in the container (18). Here the substance (16) remains exposed to the contents (19) and is contacted by an emanated volatiles, while remaining out of direct contact with the acid-salt mixture. The form shown in Figure 6.1c contemplates applying to the inside of a paper board package (20) containing the acid mixture (21) a coating (22) within which is incorporated any one or mixtures of the deodorant substances. Typically the coating (22) may consist of sodium silicate or any of the other deodorant substances admixed with an alkali metal silicate as a binder. Alternately, or in addition, the deodorant substance may be incorporated in the paper board itself, as by putting the deodorant into the pulp from which the paper board is made.

FIGURE 6.1: DRY BLEACH AND DISINFECTANT PACKAGES

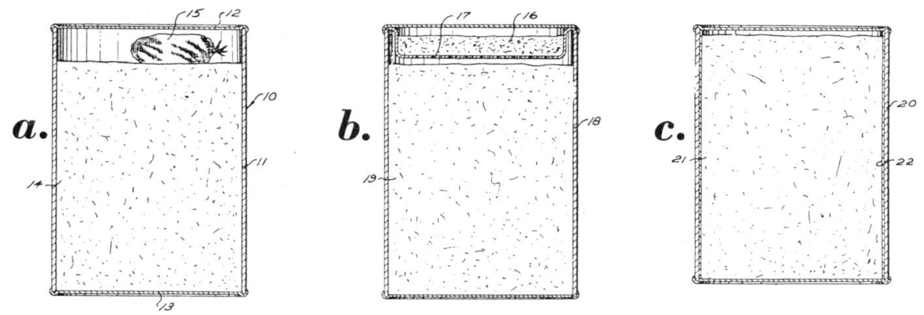


Source: M.L. Dickey; U.S. Patent 3,061,549; October 30, 1962

Example 1: 20 g. of a premix of 60% $MnO_2$ and 40% CuO (hopcalite) are placed in a small perforated bag of polyethylene plastic, the edges of which have been heat-sealed. This bag is suspended in a glass bottle which contains a formulation of 10% dichlorocyanuric acid, 30% sodium triphosphate, 3% sodium alkyl aryl sulfonate and 57% sodium sulfate, producing an excess of a strange, lachrymatory dichlorocyanuric acid odor. When checking the odor after 2 days, it was found that the offensive odor had disappeared and did not reappear after an extensive period of time.

Example 2: 20 g. of activated carbon (Norbit 20 x 60 mesh), were placed in a tea-bag (permeable paper construction) and suspended in the atmosphere of a glass container filled with commercial dichlorocyanuric acid. The characteristic odor disappeared after a few hours and did not reappear even after a period of many months.

Example 3: About 10 g. of a spray-dried base containing substantially 15% sodium silicate and 85% sodium sulfate was first poured into a 12 ounce glass bottle, which was then filled with an unstabilized dichlorocyanuric acid formulation. The odor had disappeared after a few hours, and the product remained free from odor. The dichlorocyanuric acid was exposed to but remained unmixed with the spray-dried base.

Dichlorocyanuric Acid-Sodium Carbonate Tablets

W.W. Lee and K.M. Sancier; U.S. Patent 3,120,378; February 4, 1964; assigned to The Procter & Gamble Company describe a process to produce readily water-soluble bleaching, sterilizing and disinfecting tablets containing dichlorocyanuric acid and alkali carbonates. A tablet is obtained by molding under pressure in the presence of sufficient water to bind the components together a water-soluble composition comprising an alkali metal carbonate, a hydrate forming compound, a dichlorocyanurate, and a solid acid. Water used for binding the components of the tablet combines with the hydrate forming compound to form a solid hydrate so that there is no free water in the resultant compressed tablet. The tablet may be quickly dissolved in excess water with accompanying effervescence as the acidic component reacts with the carbonate to liberate carbon dioxide, while the dissolved dichlorocyanurate releases available chlorine to the solution to effect the desired bleaching, disinfecting and germicidal action. Rapid solution of the tablet is caused by the effervescent release of carbon dioxide as a result of the reaction of solid acid and the alkali metal carbonate.

The solid acid is usually dichlorocyanuric acid since this one compound provides the acid for liberation of the carbon dioxide and also is the dichlorocyanurate which provides the available chlorine. In addition, the water employed during molding to provide a mechanically strong tablet is most advantageously provided from water contained within a hydrate compound. The alkali metal carbonates employed in the tablet are the normal and acid salts of alkali metals with carbonic acid. Examples of such carbonates are: $Na_2CO_3$, $Na_2CO_3 \cdot H_2O$, $Na_2CO_3 \cdot 10H_2O$, $NaHCO_3$, $NaHCO_3 \cdot Na_2CO_3 \cdot 2H_2O$, $K_2CO_3$, $K_2CO_3 \cdot 2H_2O$. These salts react with the acid to provide the source of carbon dioxide which causes effervescence and rapid solution of the tablet when it is placed in water.

The hydrate forming compound in the bleaching tablet is an alkali metal hydrate-forming salt. During formation of the tablet by compression of the composition while it is in contact with a small amount of binding water, the hydrate forming compound must be capable of accepting additional water in the hydrate structure. In other words, during this brief period of compression the hydrate forming compound should either be in the anhydrous form or in a lower hydrate form which is capable of accepting additional water of crystallization and becoming a higher hydrate. After compression the hydrate forming compound accepts and holds free water in the solid hydrate structure, and thus prevents premature effervescence or liberation of chlorine. Examples of alkali metal hydrate forming salts which may be employed as the hydrate forming compound are as follows:

$Na_2CO_3$ to $Na_2CO_3 \cdot H_2O$ or $Na_2CO_3 \cdot 10H_2O$
$Na_2SO_4$ to $Na_2SO_4 \cdot 10H_2O$
$Na_2B_4O_7$ to $Na_2B_4O_7 \cdot 10H_2O$
$Na_2HPO_4$ to $Na_2HPO_4 \cdot 7H_2O$ or $Na_2HPO_4 \cdot 12H_2O$
$Na_4P_2O_7$ to $Na_4P_2O_7 \cdot 10H_2O$
$K_2CO_3$ to $K_2CO_3 \cdot 2H_2O$

When anhydrous sodium carbonate, sodium carbonate monohydrate, or another alkali metal carbonate which is a hydrate forming compound is employed in the composition, the alkali metal carbonate serves the dual function of providing a source of carbon dioxide gas which causes the effervescence and also of serving as the hydrate forming compound. In such cases it is not necessary to employ other hydrate forming compounds in the tablet, although they may also be included in the composition as fillers and as additional means of holding any excess water used for binding the tablet.

The dichlorocyanurate in the tablet provides the source of available chlorine when the tablet is dissolved in water. The tablet also includes a solid acid which is capable of reacting with the carbonate component of the tablet to liberate carbon dioxide. When, as is preferably the case, the dichlorocyanurate is dichlorocyanuric acid, this one compound serves both as the source of available chlorine and as the source of the solid acid. However, the dichlorocyanurate may also take the form of an alkali metal salt of dichlorocyanuric acid, in which case another water-soluble solid acid or acid salt strong enough and in an amount sufficient to liberate carbon in the mixture. Examples of solid acid compounds including acids and acid salts which may be employed in the tablet composition include citric acid, alkali metal acid citrates, tartaric acid, alkali metal acid sulfates, lactic acid, malic acid, and maleic acid, alkali metal acid phosphates, alkali metal acid phthalates and p-toluene-sulfonic acid.

The molding pressures required to produce tablets of good tensile strength vary with the particular composition employed. Thus, using a mixture of 5 g. of dichlorocyanuric acid 2.1 g. of sodium bicarbonate and 0.35 g. of sodium sulfate decahydrate, and a die having a diameter of 1.25 inches, it was found that a pressure of at least 1,500 psi was required to form tablets having adequate mechanical strength. Pressures above about 3,500 psi are preferably avoided in tableting this composition inasmuch as the tablets so formed dissolve at an appreciably slower rate than those formed in the preferred 1,500 to 3,500 psi range. On the other hand, when tableting a mixture made up of 5 g. of dichlorocyanuric acid and 1.55 g. of sodium carbonate monohydrate, a pressure of at least 3,000 psi is required to provide adequate tablet strength, and preferably pressures of from 3,000 to 4,000 psi are employed.

Example 1: In this operation, tablets were formed from a powdered mixture of dichlorocyanuric acid (DCA), sodium bicarbonate and sodium sulfate decahydrate, each tablet having the following composition:

| | Grams |
|---|---|
| Dichlorocyanuric acid | 5 |
| $NaHCO_3$ | 2.1 |
| $Na_2SO_4 \cdot 10H_2O$ | 0.35 |

The tablets were formed at various pressures in a pelleting machine having a die diameter of 1.25 inches. All of the tablets so formed were then dried for 5 minutes at 175°C. Tests were then made to determine the length of time required to fully dissolve the tablet in 4 liters of water at 50°C. It was found that this time became increasingly long as higher molding pressures were employed, the data so obtained being given in the following table.

| Degree of Compression (psi) | Time for Solution (seconds) |
|---|---|
| 1,000* | 12 |
| 1,500 | 13 |
| 2,000 | 14 |
| 2,500 | 15 |
| 3,000 | 16 |
| 5,000 | 24 |
| 8,000 | 32 |

*This tablet had inadequate mechanical strength

Example 2: It is found that tablets having excellent bleaching characteristics, together with rapid solubility and good mechanical strength, can be formed at pressures of 2,500 psi using the following formulation with the amount in each tablet being specified.

| | |
|---|---|
| Potassium dichlorocyanurate | 4.8 g. |
| $Na_2CO_3 \cdot H_2O$ | 2.5 g. |
| $NaHSO_4$ | 1.7 g. |

## Fluidized Bed Coated Polychlorocyanurate Particles

J.H. Morgenthaler and T.D. Parks; U.S. Patent 3,112,274; November 26, 1963; assigned to The Procter & Gamble Company describe a process to manufacture solid, homogeneous and stable bleach compositions consisting of alkali salt coated polychlorocyanurate particles. The coating salts are alkali phosphates, carbonates or borates. To obtain uniform coating over the polychlorocyanurate particles, a fluidized bed process is used. The cyanurates include trichlorocyanuric acid, dichlorocyanuric acid and sodium or potassium dichlorocyanurate. The coating salts are used in an aqueous slurry.

The process can be made continuous by continuously supplying the fluidized bed with uncoated polychlorocyanurate powder, continuously spraying on the slurry of inorganic coating material and continuously drawing off partially dried coated particles which are then dried continuously, preferably in fluidized bed in a second fluidizing vessel, where finished dried product is continuously drawn off.

As used in connection with this process the expression "free moisture" means moisture other than that of the water of hydration existing in inorganic salt hydrates which are stable (retain their water of hydration) at least up to 120°F. For example, the water in sodium sulfate decahydrate is classed herein as free moisture because sodium sulfate decahydrate is stable only up to about 90°F. where its water of hydration is liberated. However, the water in sodium tripolyphosphate hexahydrate is not classed as free moisture because that hydrated salt is stable up to about 220°F. Figure 6.2 is the flow diagram of the process.

FIGURE 6.2: FLOW DIAGRAM OF COATING PROCESS

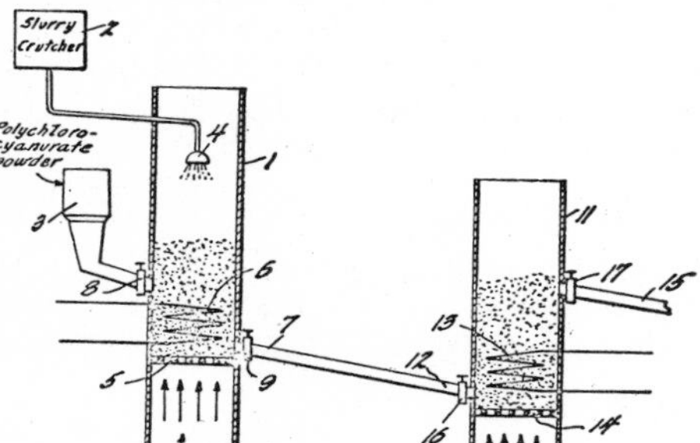


Source: J.H. Morgenthaler and T.D. Parks; U.S. Patent 3,112,274; November 26, 1963

The polychlorocyanurate powder is introduced from its storage (3) into the fluidizing vessel (1) in a quantity predetermined to form a fluidized bed of the desired size. Valve (8) is then closed. The powder is then formed into a fluidized bed by forcing a gas which is substantially inert to the powder such as air or nitrogen, through the grid (5). The grid (5) is preferably a porous plate but any grid can be used which provides substantially even distribution of the upward flowing gas over the cross-section of the fluidizing vessel (1).

If a porous plate is used, it should be porous enough to permit upward passage of air without an unduly large pressure drop but should not be so porous that substantially even distribution of the fluidizing gas is not obtained. The porous plate can be made of sintered stainless steel or of a porous ceramic material. Instead of porous plates, finely perforated plates, fine screens or bubble cap plates of the type used in distilling and deodorizing columns can be used as grids. A grid can be dispensed with in smaller fluidizing vessels which have a conical bottom which evenly distributes the fluidizing gas.

The velocity of the fluidizing gas should be sufficient to buoy the polychlorocyanurate particles into a fluidized bed but should not be so great that the upper free surface of the bed is lost either by extreme bubble formation or slugging within the bed (severe vertical surging of the bed) or by sweeping the particles out of the vessel. Polychlorocyanurate powders having a particle size in the range of about 5μ to about 5 mm., preferably 20μ to 850μ, can be formed into the fluidized beds which are useful in the process of this method. Particles smaller in size than about 5μ are too fine to form properly into a bed. It has been found that gas velocities in the range of about 1 ft./sec. to about 10 ft./sec., preferably 2 ft./sec. to 6 ft./sec., should be used with polychlorocyanurate powders in the above particle size range to form satisfactory fluidized beds. These velocities and the velocities hereinafter mentioned are measured after the fluidizing air has risen above the upper free surface of the bed and are more useful figures than velocities measured below or within the bed.

The height of the fluidized bed does not appear to be a critical factor; however, the best fluidization and processing results occur when the bed height is between 1 and 2 feet. Practical fluidizing vessel diameters range from about 1 to about 5 feet. Bed heights higher than 2 feet can be used in vessels having a diameter of about 3 to 5 feet. After the bed has been formed and it is at the desired temperature, the slurry of coating material from crutcher (2) is sprayed on to the bed through spray nozzle (4) from a point above the upper free surface of the bed. The spray nozzle (4) is a conventional atomizing nozzle using pressure atomization or preferably a two-fluid nozzle using air or other inert gas as the atomizing fluid. Liquid spray pressures can be in the range of about 10 to 100 psi with air or gas pressures in the range of about 10 to 100 psi.

The slurry of coating material should have a water content in the range of about 30 to 85%. Less than about 30% water results in a viscous slurry which is hard to spray, increases the tendency of the slurry to spray dry and makes difficult the coating of the polychlorocyanurate powder. More than about 85% water in the slurry results in poor coating characteristics and a high evaporative load in the process. The preferred range is 55 to 75%. The slurry temperature should be high enough to obtain a pumpable viscosity but not be so high that there is undesirable spray drying on atomization. Satisfactory slurry temperatures vary with slurry composition but the preferred temperatures for slurries containing sodium tripolyphosphate and/or sodium sulfate are in the range of about 70° to about 140°F. The temperature maintained within the bed during the coating step is in the range of about 100° to 170°F. A temperature within this range coupled with the evaporative effect of the fluidizing gas results in the formation of the inorganic coating around the polychlorocyanurate particles and in the partial drying of this coating. The temperature should not be less than about 100°F. in order to prevent the buildup of so much water in the coated particles that proper fluidization is prevented. If the temperature exceeds about 170°F. the atomized slurry of coating material has a tendency to be spray dried, which prevents effective coating of the polychlorocyanurate particles. The preferred temperature range is 120° to 150°F.

Sufficient coating material should be sprayed on the polychlorocyanurate particles to obtain a weight ratio of coating material to polychlorocyanurate in the finished product in the range of about 1:3 to 5:1, preferably 1:1.5 to 2:1, on an anhydrous basis. The tendency of the coated particles to agglomerate with each other in the form of clumps of coated particles (if it is desired to have a larger particle size in the finished product) can be increased by adding a small amount of an anionic organic detergent to the slurry of coating material.

Example 1: A slurry consisting of 35% sodium tripolyphosphate and 65% water was added to a crutcher and heated to 92°F. 10 pounds of sodium dichlorocyanurate powder was added to a circular fluidizing vessel which had a diameter of 12 inches. The powder had a particle size corresponding to the following sieve analysis (Tyler standard sieve):

| | |
|---|---|
| On 65 mesh | 14% |
| Through 65 mesh, on 150 mesh | 11% |
| Through 150 mesh, on 325 mesh | 24% |
| Through 325 mesh | 51% |

The vessel had a grid which consisted of a plate of sintered powdered stainless steel. The plate was 1/16 inch thick and had a porosity grade of "D" as described by J.E. Campbell in Materials and Methods, pages 98 to 101, April 1955. The powder was formed into a fluidized bed by passing air, evenly distributed by the grid over the cross-section of the vessel, at 168°F. upward through the grid. The velocity of the air as measured above the bed was 2.27 feet per second. The air had sufficient velocity to buoy all of the particles of powder in a fluidized bed having an upper free surface. Heating coils located within the bed area of the vessel were heated to 368°F. with steam.

The slurry was sprayed downward onto the bed through a two-fluid spray nozzle located 24 inches above the upper free surface of the bed. Air at 50 psi was the atomizing fluid and the slurry was pumped through the nozzle at 27 psi. The bed temperature was 125°F. The slurry was sprayed onto the bed until a coating of sodium tripolyphosphate had formed around the sodium dichlorocyanurate particles in the weight ratio, on an anhydrous basis, of coating to sodium dichlorocyanurate of 1.2:1. The coated particles had a free moisture content of 1.4%. The coated particles were then dried by increasing the bed temperature to 205°F. The heating coil temperature was 368°F. and the fluidizing air temperature was 171°F. The drying was stopped when the free moisture of the coated particles was completely eliminated. The final product had the composition shown on the following page.

| | |
|---|---|
| Sodium tripolyphosphate | 47.1% |
| Sodium dichlorocyanurate | 39.1% |
| Water (as hydrate) | 13.8% |

This product was a stable granular dry bleach composition which when dissolved in water provided active and efficient bleaching, sterilizing and disinfecting properties. The sodium tripolyphosphate provided the solution with the desired buffered alkalinity and sequestered water hardness ions. The coating of sodium tripolyphosphate around the sodium dichlorocyanurate salt particles provided greater storage stability and resistance to attack from air or moisture than either the untreated salt or the salt spray dried with an equivalent amount of sodium tripolyphosphate. The granules had only a very slight chlorine odor. The granules were substantially uniform in composition throughout the product and there was no noticeable tendency for the ingredients of the composition to segregate. A mixture of equal parts by weight of sodium tripolyphosphate and sodium sulfate can be substituted for the sodium tripolyphosphate in an equal amount by weight in Example 1 with substantially the same results.

## Boric Acid Stabilizer for Chlorinated Isocyanurate Tablets

F.N. Stepanek, Jr.; U.S. Patent 3,325,411; June 13, 1967; describes a process to stabilize chlorinated isocyanurates by the addition of boric acid which also serves as a lubricant to permit high speed tabletting. To render the chlorinated isocyanurates physically subject to rapid tabletting in high speed machinery, the initially synthesized finely divided crystals are dried until they contain approximately 1% water or less. The dried crystals are then compacted into thin sheets by rollers, for instance a Fitzpatrick compactor, under pressures ranging from 20,000 to 120,000 pounds per square inch. The most desirable range is 60,000 to 90,000 pounds per square inch. It is also desirable to dissipate heat from the rollers during the compacting process in order to maintain the temperature of the crystals at about 50°C.

The compacted sheets of crystals are then run through a mill or grinder for reduction to a usably sized particle. The mill or grinder discharge is classified to select granules between -20 to +200 mesh with oversized and undersized particles recycled through the compacting process. While the aforementioned mesh tolerances for classification of materials are acceptable, material classified within the range of -60 to +140 mesh are more desirable. Boric acid crystals are classified to -200 mesh or less. The boric acid crystals and the chlorinated isocyanurates are then formulated and dry mixed for a period of 15 to 30 minutes. The final tablet product contains approximately 97% or less of the chlorinated isocyanurate and 3 to 15% boric acid by weight. Other ingredients may also be added to enhance the breakup of tablets, control pH or function as fillers.

For instance, sodium sulfate is a common filler whereas certain phosphates can be added to increase the pH and enhance the solubility of the chlorinated isocyanurate. Alkali metal salts could also be substituted for this purpose. Corn starch can be added to assist the breakup of the tablet. In addition, olfactory agents, optical brighteners and dyes can be added. The blended material is fed into a high speed tablet press. The mixture flows freely and permits an even and constant flow through feed devices into the die cavity. It can be pressed at a rate of 1 to 60 tablets per minute based upon a single punch die machine without binding in the die cavity, sticking to the punch face, laminating or breaking on the ejection stroke. The pressure of the die must be at least 3 tons per square inch and it may be as great as 40 tons per square inch. The best results are obtained at approximately 15 tons per square inch. The tablet thus produced is ready for packaging. The boric acid functions to stabilize the chlorinated isocyanurate whether it is in the granular form or tablet form. In addition, the boric acid functions as a lubricant to prevent the chlorinated isocyanurate from sticking to the punch face or binding in the die cavity. When tabletting, the chlorinated isocyanurate should comprise 97% or less by weight of the final product and the boric acid comprise 3 to 15% by weight. If the chlorinated isocyanurate is not in the specified physical form, excessively large percentages of boric acid are required to obtain the necessary lubricating characteristics. This produces a formula having relatively large percentages of fine particles which cannot be tabletted without lamination.

The relative loss of available chlorine over varying periods of time was compared for two characteristic formulas. An examination discloses that Formula Number 1, composed of trichloroisocyanurate acid, 90.5%; powdered boric acid, 3.2%; and corn starch, 6.3% loses no available chlorine in the tablet form during the first 10 days. After 10 to 12 weeks, tablets of Formula Number 1 stored in the open lose but 1% of available chlorine whereas tablets in closed storage lose by 0.8% available chlorine and ground tablets or granular formulas of the Number 1 variety lose by 0.6%. Similar results are obtained by Formula Number 2, composed of trichloroisocyanurate acid, 93.7%; powdered boric acid, 1.9%; and corn starch, 4.4% in which even greater quantities of trichloroisocyanuric acid are employed with lesser quantities of boric acid. These results compare favorably with liquid sodium hypochlorite at 5.25% which loses 8% or 8 times as much available chlorine as the greatest loss in Formulas 1 and 2.

## Addition of Dendritic Sodium Chloride to Bleach Tablets

E.H. Krusius and R.R. Keast; U.S. Patent 3,338,836; August 29, 1967; assigned to FMC Corporation found that firm, strong and abrasion resistant tablets containing polyphosphate, detergents and chlorocyanuric compounds can be prepared by adding dendritic sodium chloride to the formulation. The sodium chloride employed in the tablets is a particulate material having a particle size of 10 to 325 mesh. The preferred sodium chloride is the dendritic material, having a particle size of 30 to 200 mesh, with about 85% being plus 100 mesh. The dendritic material is a form of sodium chloride having branched or starlike crystals as opposed to the cubical structure of regular vacuum salt or the crystal aggregates of flake type salt. It has the fine granulation of vacuum pan salt, typical dendritic salt having a density on the order of 56 pounds per cubic foot.

It is produced by pretreating brine with 12 to 20 parts per million of yellow prussiate of soda (sodium ferrocyanide), and producing the salt crystals by a vacuum pan evaporation process. The additive causes growth to take place on the corners, rather than at the faces, of the crystals providing the characteristic dendritic crystal. Use of sodium chloride having particle sizes substantially outside these ranges results in tablets not having the advantages of rapid dissolution and disintegration and strength. A variety of kinds of cleansing tablets are prepared from compositions containing the herein sodium chloride additive and water conditioning agent or agents. The following table presents a number of such tablet compositions, in terms of both useful and preferred ranges of the various additives. In some cases the ingredients are given in ranges which include 0; this means that the particular ingredient is not essential to the composition.

## Compositions by Tablet Type

| Major Components | Dry Bleach | | Laundry Detergent | | | |
|---|---|---|---|---|---|---|
| | | | Anionic | | Nonionic | |
| | Useful | Preferred | Useful | Preferred | Useful | Preferred |
| Sodium Tripolyphosphate | 0-95 | 20-50 | 25-75 | 50-65 | 25-75 | 50-65 |
| Sodium Chloride | 2-95 | 5-30 | 2-69 | 5-30 | 2-69 | 5-30 |
| Chlorocyanuric Compound | 0.5-98 | 10-50 | | | | |
| Anionic Synthetic Detergent (beads pref.) | 0-10 | 1-8 | 3-20 | 5-15 | | |
| Nonionic Synthetic Detergent | 0-1 | 0-0.5 | | | 3-20 | 5-15 |
| Sodium Silicates: | | | | | | |
| Dry $Na_2O:SiO_2$: | | | | | | |
| 1:2-1:3.2 | | | 2-12 | 4-7 | | |
| 2:1-1:1 | | | | | | |
| Liquid $Na_2O:SiO_2$, 1:2-1:3.2 | | | | | 1-10 | 2-5 |
| Sodium Sulfate | 0-50 | 0-25 | | | | |
| Sodium Bisulfate | | | | | | |
| Sodium Carbonate | 0-50 | 0-25 | | | | |

| Major Components | Detergent-Bleach | | Cleaners | | | |
|---|---|---|---|---|---|---|
| | | | Hard Surface | | Chlorinated Industrial | |
| | Useful | Preferred | Useful | Preferred | Useful | Preferred |
| Sodium Tripolyphosphate | 25-75 | 50-65 | 5-35 | 10-20 | 5-60 | 20-40 |
| Sodium Chloride | 2-74 | 5-30 | 2-94 | 30-75 | 2-93 | 30-50 |
| Chlorocyanuric Compound | 5-10 | 6-8 | | | 1-10 | 3-5 |
| Anionic Synthetic Detergent (beads pref.) | 3-20 | 5-15 | 1-20 | 3-10 | 1-20 | 3-10 |
| Nonionic Synthetic Detergent | 0-1 | 0-0.5 | 0-5 | 1-3 | | |
| Sodium Silicates: | | | | | | |
| Dry $Na_2O:SiO_2$: | | | | | | |
| 1:2-1:3.2 | 2-12 | 4-7 | | | | |
| 2:1-1:1 | | | | | | |
| Liquid $Na_2O:SiO_2$, 1:2-1:3.2 | | | | | | |
| Sodium Sulfate | | | | | | |
| Sodium Bisulfate | | | | | | |
| Sodium Carbonate | | | | | | |

| Major Components | Automatic Dishwashing | | Sanitizer | | Sanitizer-Cleaner | | Toilet Bowl Cleaner | |
|---|---|---|---|---|---|---|---|---|
| | Useful | Preferred | Useful | Preferred | Useful | Preferred | Useful | Preferred |
| Sodium Tripolyphosphate | 10-60 | 25-50 | 0-50 | 10-20 | 20-60 | 30-50 | | |
| Sodium Chloride | 2-80 | 5-20 | 2-99 | 5-88 | 2-72 | 5-30 | 2-39 | 5-15 |
| Chlorocyanuric Compound | 0-8 | 2-6 | 1-100 | 2-25 | 1-30 | 2-15 | 0-5 | 1-3 |
| Anionic Synthetic Detergent (beads pref.) | | | 0-2 | 0-1 | 2-20 | 5-10 | 1-6 | 2-4 |
| Nonionic Synthetic Detergent | 0-5 | 1-3 | | | | | | |
| Sodium Silicates: | | | | | | | | |
| Dry $Na_2O:SiO_2$: | | | | | | | | |
| 1:2-1:3.2 | | | | | | | | |
| 2:1-1:1 | 10-70 | 15-30 | | | 5-30 | 10-20 | | |
| Liquid $Na_2O:SiO_2$, 1:2-1:3.2 | | | | | | | | |
| Sodium Sulfate | | | | | | | | |
| Sodium Bisulfate | | | | | | | 50-90 | 65-75 |
| Sodium Carbonate | | | | | | | 10-25 | 15-20 |

The above tabletting compositions may contain also, as needed, other agents normally employed in providing the various tablets. These agents include bactericidal additives, fillers such as sodium carbonate and sodium sulfate, excipients such as sodium stearate and vegetable fats, optical brighteners, antiredeposition agents such as carboxymethylcellulose, dyes, pigments, dedusters such as mineral oil, foam stabilizers, tarnish inhibitors, ammonium chloride and phosphates such as trisodium phosphate, monosodium phosphate and the like. Polyphosphates found useful in the compositions include the sodium tripolyphosphate, tetrasodium pyrophosphate and the so-called phosphate glasses (including sodium hexametaphosphate) with those having chain lengths of about 10 to 16 being preferred.

Dry, particulate chlorocyanuric compounds useful in the tablets together with sodium chloride include trichlorocyanuric acid, dichlorocyanuric acid and the alkali metal and alkaline earth metal salts of dichlorocyanuric acid. The preferred chlorocyanuric compound is sodium dichlorocyanurate, a fast-dissolving compound which is highly effective as a bleach and a sanitizer, yet which does not damage fabrics and the like. The chlorocyanuric materials are normally highly stable in storage when compared with other active chlorine chemicals, however, they are decomposed with loss of active chlorine by contact with water in alkaline systems at elevated temperatures, and accordingly should be handled in a fashion to avoid exposure to this combination of conditions in formulation of tablets. The potassium magnesium and calcium dichlorocyanurates are also highly useful in the compositions, as are the other alkali and alkaline earth metal dichlorocyanurates. Synthetic detergents useful in the above compositions are the anionic and nonionic nonsoap synthetic detergents. The useful anionic nonsoap synthetic detergents may be designated as water-soluble salts of organic sulfuric reaction products having in their molecular structure an alkyl or acyl radical of carbon atom content within the range of about 8 to about 18 and a sulfonic acid or sulfuric acid ester radical. Important examples of these anionic detergents are: sodium or potassium alkyl benzene sulfonate in which the alkyl group contains from about 9 to about 15 carbon atoms in either a straight chain or a branched chain which is derived from polymers of propylene; sodium and potassium alkyl glyceryl ether sulfonates, especially those ethers of higher fatty alcohols derived from the reduction of coconut oil; the reaction product of higher fatty acids with sodium or potassium isethionate.

Nonionic nonsoap synthetic detergents useful in the tablets may be broadly classed as being constituted of a water-solubilizing polyoxyethylene group in chemical combination with an organic hydrophobic compound such as polyoxypropylene, alkyl phenol, the reaction product of an excess of propylene oxide and ethylene diamine, and aliphatic alcohols. The nonionic synthetic detergents have a molecular weight in the range of from about 800 to about 11,000. Effervescing agents may be used in the tablets, and include such materials as particulate sodium bicarbonate or carbonate which are used together with an acid which acts in solution on the bicarbonate or carbonate to liberate carbon dioxide gas which aids in tablet disintegration. Dry bleach tablets containing dendritic sodium chloride showed improved strength and reduced disintegration time when compared wtih similar tablets in which the sodium chloride is replaced with the common filler sodium sulfate.

## Dichlorocyanurate Base Bleach Packaged In Envelopes

H.E. Wixon; U.S. Patent 3,346,502; October 10, 1967; assigned to Colgate-Palmolive Company found that dry water-soluble bleaching compositions in powder form can be produced from organic active chlorine producing agents, ultramarine blue and optical brighteners. The composition may be packaged in envelopes of polyvinyl alcohol which contribute considerable soil antiredeposition properties to the system and thus the packets are specially suited for use in conjunction with soap and detergent products in connection with the washing and bleaching of soiled clothes. The use of water-soluble polyvinyl alcohol film is also advantageous in that it is thermoplastic and may readily be heat sealed, it offers high resistance to permeability by gases, it has excellent oil and grease resistance, long shelf life, and it can easily be printed with alcohol-type inks.

Hypochlorite-generating agents suitable for use in the compositions are those water-soluble dry solid materials which generate hypochlorite ion on contact with, or dissolution in water. Examples thereof are the dry, particulate heterocyclic N-chlorimides such as trichlorocyanuric acid, dichlorocyanuric acid and salts thereof such as sodium dichlorocyanurate and potassium dichlorocyanurate. Other imides may also be used such as N-chlorosuccinimide, N-chloromalonimide, N-chlorophthalimide and N-chloronaphthalimide. Additional suitable imides are the hydantoins such as 1,3-dichloro-5,5-dimethylhydantoin; N-monochloro-C,C-dimethylhydantoin; methylene-bis(N-chloro-C,C-dimethylhydantoin); 1,3-dichloro-5-methyl-5-isobutylhydantoin; 1,3-dichloro-5-methyl-5-ethylhydantoin; 1,3-dichloro-5-methyl-5-n-amylhydantoin, and the like. Other useful hypochlorite-liberating agents are trichloromelamine and dry, particulate, water-soluble anhydrous inorganic salts such as lithium hypochlorite and calcium hypochlorite. The ultramarine blue pigment is stable in the presence of the hypochlorite-liberating agent during both storage and use of the compositions. The pigment is in the form of particles substantially all of which exhibit a diameter of less than about 0.05 mm., and is characterized by the ability to impart a faint blue visible shade to fabrics treated therewith without staining such fabrics when used at recommended concentration and fashion, being generally considered to be nonsubstantive, or at least nonaccumulative, on fabrics.

The fluorescent brighteners or optical dyes used in the compositions are substantive to textiles, e.g., cotton, and are resistant to attack by the hypochlorite-liberating agent. Such brighteners are of particular assistance in connection with the bleaching of textiles or fabrics viewed under illumination wherein the visible bluing action supplied by the ultramarine is subject to being supplemented by a blue-fluorescent optical bleach. Examples of suitable such textile-substantive optical bleaches are triazole compounds such as

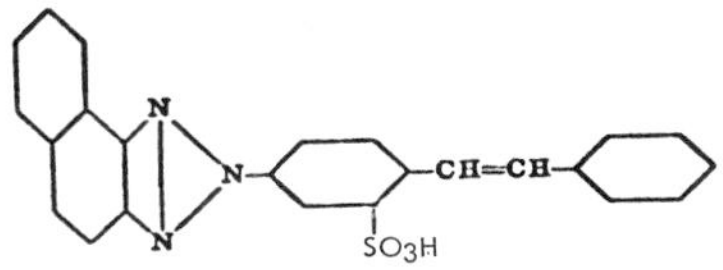


sulfonated 3,7-diaminodibenzothiophene dioxide such as

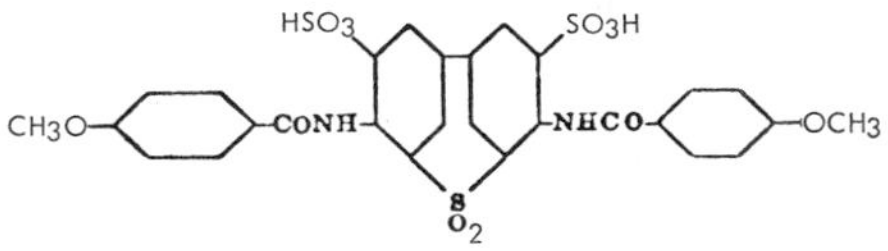


and bisbenzimidazoles such as

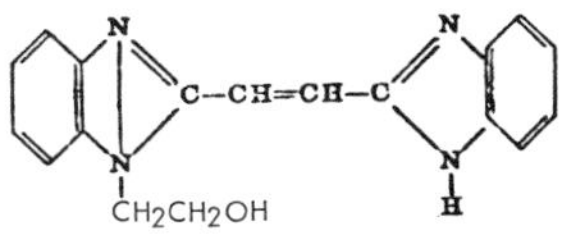


Example 1: The following product is highly effective in improving the appearance of both white- and bleach-fast colored cottons:

| | Parts by Weight |
|---|---|
| Pentasodium tripolyphosphate | 30.00 |
| Alkylaryl sulfonate (containing 76% sodium tridecyl benzene sulfonate, 9.9% sodium silicate, 10.8% sodium sulfate, and 3.3% moisture) | 2.62 |
| Fluorescent dye | 0.16 |

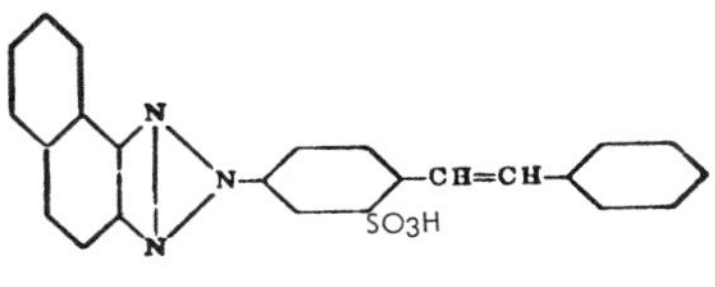


(continued)

| | Parts by Weight |
|---|---|
| Perfume | 0.25 |
| Limonene | 0.25 |
| Potassium dichlorocyanurate | 20.60 |
| Mineral oil | 0.25 |
| Ultramarine blue pigment | 0.25 |
| Sodium sulfate (anhydrous) | Balance to 100 |

The composition is a mechanically dry-mixed finely divided powder having a maximum particle size of less than 0.5 mm. The ultramarine blue pigment has an extremely small particle size, at least 99.5% by weight thereof having a particle diameter of less than 0.05 mm. Forty grams of this formulation are heat sealed within an envelope of commercial polyvinyl alcohol film which has been prepared by a hydrolysis of polyvinyl acetate. This polyvinyl alcohol contains about 15% of unhydrolyzed vinyl acetate, and is devoid of ethoxylation.

In use, the entire packet is dropped into a household washing machine (containing white household linens including bed sheets, pillow cases, and towels) at the start of a ten-minute washing period. The water used has a hardness of 150 ppm and a temperature of about 120°F. A commercial heavy duty detergent composition is present in a concentration of 0.15%, and the concentration of the composition is 0.05%. The packet promptly dissolves, forming a hypochlorite solution containing the bleach-stable fluorescent dye and the ultramarine blue, which cooperate to produce cotton textiles of substantially improved appearance, i.e., a highly whitened or brightened appearance to the eye.

Example 2: The powder composition of Example 1 may be replaced by the following, which exhibits the same particle size characteristics:

| | |
|---|---|
| Pentasodium tripolyphosphate | 30.00% |
| Sodium carbonate | 5.00% |
| Sodium sulfate, anhydrous | 40.615% |
| Fluorescent brightener of Example 1 | 0.16% |
| Ultramarine blue pigment | 0.375% |
| Alkylarylsulfonate of Example 1 | 2.62% |
| Perfume | 0.25% |
| Limonene | 0.25% |
| Potassium dichlorocyanurate | 20.60% |
| Mineral oil | 0.13% |

## Detergent Bleach Packaged in Water-Soluble Envelopes

F.W. Gray; U.S. Patent 3,528,921; September 15, 1970; assigned to Colgate-Palmolive Company describes a process to package bleaching and detergent compositions in a water-soluble, polyvinyl alcohol envelope. In addition to the hypochlorite generating bleaching agent, the envelopes contain surface active, amphoteric, cationic and nonionic detergents, soap builders and optical brighteners. The envelopes are suitable for direct additives to home laundry machines. The polyvinyl alcohol film employed in the preparation of the packets is water-soluble, that is, with mild agitation it dissolves substantially completely in water at temperatures of from about 80° to 140°F. within about 30 seconds. Commercial polyvinyl alcohol which is prepared by the hydrolysis of polyvinyl acetate may be employed; however, it has been found highly desirable to avoid the use of water-soluble polyvinyl alcohol which has been ethoxylated, i.e., reacted with ethylene oxide, inasmuch as such ethoxylated polyvinyl alcohol appears to be less stable on aging and storage in contact with the hypochlorite generating agents than does material which is free from ethoxylation. Polyvinyl alcohol which is devoid of ethoxylation and which contains on the order of from about 12 to 40% by weight of unhydrolized vinyl acetate has been found to be desirably stable towards the hypochlorite liberating agent during storage and use and to possess highly valuable flexibility characteristics and solubility characteristics in both hot (140°F.) and cold (80°F.) water. Hypochlorite generating agents suitable for use in the packets are those water-soluble dry solid materials which generate hypochlorite ion on contact with, or dissolution in, water.

Examples thereof are the dry, particulate heterocyclic N-chloroimides such as trichloroisocyanuric acid, and dichloroisocyanuric acid and salts thereof such as sodium dichloroisocyanurate and potassium dichloroisocyanurate. Other imides may also be used such as N-chlorosuccinimide, N-chloromalonimide, N-chlorophthalimide and N-chloronaphthalimide. Additional suitable imides are the hydantoins such as 1,3-dichloro-5,5-dimethyl hydantoin; N-monochloro-5,5-dimethylhydantoin; methylene-bis(N-chloro-5,5-dimethylhydantoin); 1,3-dichloro-5-methyl-5-isobutylhydantoin; 1,3-dichloro-5-methyl-5-ethylhydantoin; 1,3-dichloro-5,5-diisobutylhydantoin; 1,3-dichloro-5-methyl-5-n-amylhydantoin, and the like. Other useful hypochlorite-liberating agents are trichloromelamine N,N-dichlorobenzoylene urea, N,N-dichloro-p-toluenesulfonamide and dry, particulate, water-soluble anhydrous inorganic salts such as lithium hypochlorite and calcium hypochlorite.

Normally, chlorine-liberating agents are employed in a proportion, within the range, to yield a product which contains from about 1 to about 20% available chlorine on a total weight basis, although other proportions may be employed if desired. The remainder of the bleaching composition is a water-soluble diluent therefore, such as a water-soluble inorganic salt, the term "water-soluble" being used in the sense of water-soluble or dispersible. The diluent itself normally will constitute a material such as an alkaline inorganic buffer salt, e.g., a detergent builder salt, which does not react deleteriously with a hypochlorite-generating agent either during storage or use.

Example: An aqueous solution of commercially available polyvinyl alcohol resin of the type employed for the preparation of water-soluble polyvinyl alcohol films is prepared. The resin is of a grade which contains approximately 88% hydrolyzed polyvinyl acetate, and the temperature of the solution is about 30°C. Sodium nitrite is dissolved in the solution in a proportion equal to 7% by weight of the resin therein, the total solids content of the final solution being about 25% by weight. The uniform solution is then cast on the surface of a polished stainless steel band to form a film of uniform thickness. The band is heated to about 275°C. to drive off the water in the film. After drying, the cast film is peeled away from the band. The dried film, which is 1.5 mils uniform thickness, is then used to prepare packets having measurements of approximately 2.75 x 1.75 inches and weighing about 0.4 gram.

In the preparation of packets, one rectangle of film is placed over a shallow depression in a hollow block. The depression is evacuated, thus drawing the central portion of the film into the depression and forming a pouch or pocket therein, 40 grams o of the following compositions are placed in the pouch formed in the film, and the edge portions of the film which protrude from the depression in the block are moistened with water. A second piece of film is then aligned with the first piece and pressed against it to form a closed packet completely enveloping the composition contained therein. Hot air is applied to the surface of the packet to dry the moistened portion thereof, and the completed bleaching packet, after breaking the vacuum, is removed from the block on which it is formed. The composition encased within the package is as follows:

| | Parts by Weight |
|---|---|
| Potassium dichloroisocyanurate | 21.5 |
| Pentasodium tripolyphosphate | 30 |
| Sodium carbonate | 5 |
| Sodium sulfate | 40.5 |
| Sodium dodecylbenzenesulfonate | 2 |
| Sodium silicate | 0.5 |
| Perfume and coloring, balance to 100 | |

The packets so produced show excellent stability with respect to retention of original chlorine content, water solubility, color, odor, flexibility, and other properties on long storage and on accelerated aging at elevated temperatures, e.g., 130°F.

Bleach Packaged in Polyvinyl Alcohol Envelopes

J.H. Pickin; U.S. Patent 3,634,260; January 11, 1972; assigned to Colgate-Palmolive Company describes the use of water-soluble polyvinyl alcohol envelopes to package bleaching and cleaning compositions based on potassium dichlorocyanurate. The polyvinyl alcohol film employed in the preparation of the packets is water-soluble, that is, with mild agitation it dissolves substantially completely in water at temperatures of from about 80° to 140°F. within about 30 seconds. Commercial polyvinyl alcohol which is prepared by the hydrolysis of polyvinyl acetate may be employed, however, it has been found highly desirable to avoid the use of water-soluble polyvinyl alcohol which has been ethoxylated, i.e., reacted with ethylene oxide, inasmuch as such ethoxylated polyvinyl alcohol appears to be less stable on aging and storage in contact with the hypochlorite generating agents than does material which is free from ethoxylation.

Polyvinyl alcohol which is devoid of ethoxylation and contains on the order of from about 12 to 40% by weight of unhydrolyzed vinyl acetate constituents has been found to be exceptionally stable towards the hypochlorite liberating agent during storage and use and to possess highly desirable flexibility characteristics and solubility characteristics in both hot (140°F.) and cold (80°F.) water. Hypochlorite generating agents suitable for use in the packets are those water-soluble dry solid materials which generate hypochlorite ion on contact with, or dissolution in, water. Examples thereof are the dry, particulate heterocyclic N-chlorimides such as trichlorocyanuric acid, dichlorocyanuric acid, and salts thereof such as sodium dichlorocyanurate and potassium dichlorocyanurate.

It has been found that the formation of insoluble material during extended storage and aging of polyvinyl alcohol bleach packets which contain the hypochlorite generating agents may be alleviated by the presence, in such packets, of an olefin having a double bond containing a tertiary carbon atom. The use of such an olefin, in a packet prepared from the preferred acetate-containing polyvinyl alcohol film, facilitates the preparation of packets having long shelf life without undergoing formation of insoluble material. The preferred olefins constitute odoriferous terpenes such as those isoprenoid hydrocarbons containing two or more, usually from two to six, isoprene units in a cyclic or acyclic structure, i.e., terpenes, sesquiterpenes, diterpenes, triterpenes, and the like which broadly may be referred to as "terpenes." It is preferred to employ the normal liquid polyunsaturated terpenes containing at least two isoprenoid units, such as limonene and myrcene, as these materials have been found to be particularly suitable and effective in the packets. Other olefins having a double bond containing a tertiary carbon atom which may be employed in the packets include polymerized isobutylene, e.g., diisobutylene, polymerized propylene, e.g., propylene tetramer, and 5-butyl-4-nonene such as may be conveniently prepared by dehydration of tributylcarbinol. The proportion of the olefin employed in the packets typically constitutes about 0.1 to 40% by weight of the hypochlorite-generating agent which is present.

Example 1: The following composition is dry mixed in a rotating drum, the perfume and limonene being sprayed into the drum as the last two constitutents to be added:

| | |
|---|---|
| Pentasodium tripolyphosphate | 30.00% |
| Sodium dodecylbenzenesulfonate | 3.72% |
| Sodium carbonate, granulated | 5.00% |
| Sodium sulfate | 48.06% |
| Potassium dichlorocyanurate | 12.81% |
| Perfume | 0.25% |
| d-Limonene | 0.16% |

Approximately 42 grams of the foregoing composition are sealed in a packet or envelope made of a cast 1 1/2 mil thick film of polyvinyl alcohol prepared by hydrolysis of polyvinyl acetate and containing about 15% by weight of polyvinyl acetate. This polyvinyl is free from ethoxylation. The film is heat sealed to form a packet which consists of a centrally folded sheet of cast film heat sealed along three edges (two of which are heat sealed prior to filling and one subsequently thereto). The packet is then packed within an aluminum foil covered paperboard container which helps to protect the packet from exposure to moisture vapor during storage prior to use.

Example 2: The following constituents are dry blended in the proportions indicated, the limonene, perfume, and white mineral oil being sprayed on the other constituents of the formula while they are tumbled:

| | |
|---|---|
| Spray dried pentasodium tripolyphosphate | 30.00% |
| Sodium dodecylbenzenesulfonate (containing approximately 3% moisture) | 3.64% |
| Sodium carbonate | 5.00% |
| Sodium sulfate | 45.29% |
| Chlorine-bleach stable fluorescent brightener* | 0.16% |
| d-Limonene | 0.16% |
| Perfume | 0.25% |
| Potassium dichlorocyanurate | 15.25% |
| White mineral oil | 0.25% |

*

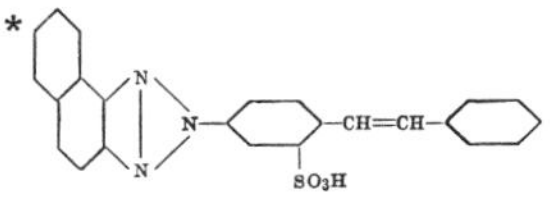

About 40 grams of the above dry blend are heat sealed within an envelope of the water-soluble polyvinyl alcohol film of Example 1. The envelope, when empty, measures 6 by 9 centimeters (the weight ratio of dry mix to envelope being 98.16 to 1.84). Ten of these packets are placed in an airtight pouch of polyethylene film 0.0045 inch thick which pouch has one open end. The open end of the pouch is multiply folded to provide an air and vapor tight reclosable seal, and the closed pouch is packaged in an aluminum-foil covered cardboard box which provides additional protection from atmospheric moisture vapor.

## TEXTILE MILL PROCESSING

### Use of Pyridinium Salt of Dichloromethyl Ether

F. Kocher; U.S. Patent 3,076,688; February 5, 1963; assigned to Traitements Chimiques des Textiles, S.A., France describes a process of wet and dry creaseproofing and bleaching cellulose with a pyridinium salt of dichloromethyl ether. The creaseproofing is obtained while keeping the decrease in tensile strength of the fabric to below 40% even in the filling. These properties are obtained by treating cellulosic textile fabrics with a pyridinium salt of a di- or other polymethyl ether (polymethoxy) halide having two or more ($-OCH_2Hal$) groups and being bifunctional or polyfunctional in nature, which makes it suitable for chemical modification of the cellulose upon impregnation of the fabric therewith and heating to a suitably high temperature.

A specific bifunctional form of this type of compound that has been used with good success for obtaining the above properties, is the pyridinium salt of ethanediol-1,2-dichloromethyl ether. Instead of pyridine, other water-soluble quaternary salts of a nitrogenous base might be used, but from the standpoints of availability and established function pyridine is preferred. The preferred compound, namely, the pyridinium salt of ethanediol-1,2-dichloromethyl ether, may be prepared by first reacting two mols of formaldehyde with one mol of ethylene glycol and saturating the resultant mixture with hydrogen chloride. This produces the ethanediol-1,2-dichloromethyl ether and the pyridinium salt is obtained by reacting with pyridine in stoichiometric quantity.

For imparting wash and wear and crease resistant properties to cellulosic fabrics they may be treated with an aqueous solution of the above mentioned pyridinium salt of the dichloromethoxy compound, followed by drying of the treated fabric and then heating sufficiently to effect chemical reaction between the cellulose of the fabric and the compound. For improving the dry crease resistance, and "hand" of the fabric and for minimizing the strength loss, the fabric may be treated with a latex in addition to the methoxy compound. The latex may be applied to the fabric alone, followed by application of the methoxy compound, and the latex may be composed of natural or synthetic rubber.

Example: In this example the process is carried out in two steps, first with a synthetic rubber latex and then with the methoxy compound. The rubber latex is used in the form of a thin aqueous dispersion of the following composition: 50 grams per liter synthetic rubber latex (Hycar 1562); 2 grams per liter of polyvinyl alcohol (Elvanol); and 10 grams per liter of polyethylene oxide condensate. The latex in the above formula is the principal ingredient and the other two materials are in the nature of additives. The polyvinyl alcohol is an aid for obtaining the desired "hand" in the fabric, and the polyethylene oxide condensate is a softening agent. The first step in this process is to pass the cotton cloth in open width through a bath containing the above latex composition so as to impregnate the cloth with the composition, and then pass the cloth between press rolls which remove the excess solution and leave in the cloth approximately 70% wet pickup. The impregnated cloth is then dried on a conventional tenter frame which is operated at approximately 120 yards (110 meters) per minute and on which the cloth is heated to approximately 110°C. for about 10 to 15 seconds. After drying on the tenter frame the cloth is passed over a cool roll maintained at about 50°C. and the cloth is wound up on a roll and at this point contains approximately 5% moisture.

The next step in the process is to treat the cloth, which is now impregnated with the latex, with a solution of the above mentioned ethanediol-1,2-dichloromethyl ether pyridinium compound. In the dry state this product is in the form of a white powder and a solution thereof is produced according to the following formula: Methoxy compound, 68 grams per liter; sodium acetate, 18 grams per liter; and water, 300 liters. The sodium acetate, above, functions as a buffer to avoid an excessive decrease in the pH during the course of the reaction which would otherwise deteriorate the cloth. The methoxy dry white powder, and the sodium acetate dissolve in the water very easily with stirring at room temperature and the aqueous composition is in the form of a water-like thin liquid.

This composition is padded onto the above mentioned impregnated dry cloth and the further impregnated cloth pressed between squeeze rolls so as to effect a wet pickup of approximately 70 to 75%. The cloth thus treated is next passed through a loop drier operated at about 80°C. and the cloth maintained in the drier for a period of about 7 minutes to dry it to a moisture content of approximately 5%. Drying at a relatively low temperature of about 50° to 80°C. is preferred to avoid premature decomposition of the methoxy compound. The next step in the process is mechanical in nature and is designed to improve the "hand" or feel of the cloth. This is accomplished by running the dry cloth between a moistened rubber belt and a heated metal cylinder and then around a drying reel from which the cloth may be either wound up or folded into a box. This treatment improves the softness of the cloth and removes some of the wrinkles that have occurred during the previous processing of the cloth.

In the next step the cloth is run through a heated oven in which it is subjected to a temperature of approximately 150°C. for about 8 minutes. This high temperature heat treatment is for the purpose of causing the methoxy compound to react with the cellulose of the fabric. It is also possible that this heat treatment causes some reaction between the methoxy compound and the rubber latex, which would improve the wash and wear properties. After heating of the cloth as just described, it is rolled up and permitted to stand for approximately two hours to provide further opportunity for the reaction or reactions above mentioned to take place, and to obtain greater uniformity in the product between the treated cloth which is first rolled up and the cloth which is finished last.

In the next step in the process the cloth after standing in roll form for about two hours is run through an open width washer to remove excess or unreacted chemicals. In this washer the cloth passes successively through several cold water baths and then through a bath containing sodium carbonate in a concentration of 10 grams per liter and a moistening agent in a concentration of 1/2 gram per liter maintained at 70°C.

for neutralizing the cloth. At the exit end of this washer the wet cloth is run through squeeze rolls and then rolled up in wet form. The wet roll is permitted to stand for about one hour so as to enable the remaining traces of pyridine to leave the cloth and thereby avoid any residual pyridine odor. After standing for one hour in the wet roll, the cloth is passed again through the open width washer in which the cloth is washed in several stages with hot water and is then run through a cold water bath containing sodium bicarbonate and finally the cloth is run through a bath containing two grams per liter available chlorine and is batched up in wet form. The cloth in wet roll form is permitted to stand for two hours so as to effect bleaching of the cloth with the chlorine which has been added in the last stage above mentioned. Following the chlorination during this two hour period, the cloth is washed again first in cold water then in sodium hydrogen sulfite in a concentration of 1 g./l. and finally in cold water.

Following this cold water washing stage the cloth is dried on a tenter frame and is then run over a conventional calendar to provide a smooth ironed effect, free of wrinkles, following which it is ready for shipment. In the above described process, the latex and methoxy compound impregnating solutions will contain usually about 5 to 15% by weight of active compounds. Also, the impregnating procedures may be carried out at room temperature or with heated solutions as desired but there is no particular advantage in using elevated temperaturees since the solutions are easily formed at room temperature and may be easily applied to the fabric at room temperature.

Drying of the impregnated fabric and reacting of the chemical impregnants with the cellulosic fabric are carried out at elevated temperatures as described above. The heat treatment for effecting chemical reaction will usually be carried out at temperatures between 130° and 160°C. and for a period of time of about 5 to 15 minutes. In most cases a temperature of approximately 145° to 150°C. and a treating time of 8 to 10 minutes is advantageous.

Use of Lower Alkyl Quaternary Ammonium Perhalides

J.F. Mills; U.S. Patent 3,156,521; November 10, 1964; assigned to The Dow Chemical Company describes a process to bleach textile fibers with lower alkyl quaternary ammonium perhalides such as tetraalkyl ammonium dichlorobromides. The process is particularly suitable for fibers having nitrogen atoms associated with their structure; for example, silk, wool, synthetic polyamides, or fibers coated or impregnated with nitrogen-containing materials such as melamine resin-finished cotton. Most effective are the perhalides having the structure $R_4NBr_nCl_{(3-n)}$, wherein each R is a lower alkyl radical containing 1 to 6 carbon atoms and n is an integer from one to three inclusive. The compounds included are therefore the tetraalkyl ammonium dichlorobromides, dibromochlorides, and tribromides.

Most preferred, particularly on economic grounds, are the dichlorobromides. Compounds wherein at least three R's are methyl are advantageous because of their relatively higher water-solubility. The perhalide compounds described above have been found to be effective and advantageous bleaching agents when applied in water dispersion to various substances having undesirable and bleachable color or stain. Straw, wood pulp, paper stock, leather, various natural fibers such as cotton, silk, wool, and linen, and synthetic polyamide, polyester, and other like fibers may be treated by this process to remove natural or foreign discoloration. These perhalides are crystalline solids which are stable while dry and which may be incorporated as bleaching components of various dry formulations for household and industrial use. The tetraalkyl ammonium perhalides are relatively soluble in water and, when so dissolved, readily liberate two atoms of halogen per molecule. Dichlorobromides liberate one atom each of chlorine and bromine while dibromochlorides and tribromides liberate two atoms of bromide. These compounds have been found to have particular advantage in the bleaching of fabrics which have been treated with nitrogen-containing resins for crease resistance. Specifically, melamine resin-treated cotton may be bleached by these compounds with less yellowing and weakening of the fabric than is possible with the commonly used chlorine bleaches. These results are obtained along with bleaching efficiencies at least as high as those shown by chlorine-liberating bleaches when used in equivalent halogen concentrations.

Example: One liter portions of water in stainless steel beakers were heated to 130°F., the bleaching compounds to be tested were dissolved in the respective beakers of water, and the pH of each test solution was adjusted to 9 to 10 where necessary by the addition of tetrapotassium pyrophosphate. Bleaching efficiencies were measured at concentrations of 300 ppm available halogen calculated as chlorine. Available halogen was determined by titration of iodine released from an acidified KI solution by a known quantity of perhalide. Three 5-inch squares of unbleached cotton sheeting were bleached for 10 minutes with agitation in each test solution, the cloths were rinsed 5 minutes in water, and they were pressed dry with a hand iron. The average reflectance of the three swatches of cotton sheeting as measured by a reflectometer was taken as a measure of bleach performance. The figures given are percentages based on a magnesium oxide block standard which is assigned a value of 100% reflectance.

| Bleaching Compound: | Average Reflectance |
|---|---|
| Sodium hypochlorite | 84.5 |
| $(CH_3)_3C_2H_5NBrCl_2$ | 84.3 |
| $(CH_3)_3C_2H_5NClBr_2$ | 85.0 |
| $(CH_3)_3C_4H_9NBrCl_2$ | 84.2 |
| $(CH_3)_4NBrCl_2$ | 84.3 |

## Chlorinated Cyanuric Acid and Hypochlorite for Polyacrylic Fibers

P. Lhoste and J. Bouvet; U.S. Patent 3,413,078; November 26, 1968; assigned to Office National Industriel de L'Azote, France describe a bleaching process for polyacrylic fibers using chlorinated cyanuric acid, cyanuric acid, sodium hypochlorite and their mixtures. The process consists of a room temperature presoak followed by the bleaching proper at 20° to 100°C. and a dechlorinating bath with sodium bisulfite or hydrogen peroxide.

Example: A bleaching bath is prepared as follows: A mixture of 9.3 grams per liter of trichlorocyanuric acid and 2.6 grams per liter of cyanuric acid is prepared by first homogeneously mixing these two ingredients in the dry state in the molar proportion of 2 to 1. The mixture is then dissolved in cold water in an amount of 11.9 grams per liter of water, in the presence in the water of 5.05 grams per liter of sodium carbonate. The pH of the bath is 5.85. A polyacrylic fiber fabric (made on the basis of fibers commercially available as Crylor) is first soaked in the bath at a temperature of about 20°C. The thus-impregnated fabric is then squeezed to an uptake of 100% of solution (100% wet) and entered into a steaming apparatus, advantageously a so-called "J" box conventionally used in bleaching with chlorine, wherein it is held for a period of 5 minutes at 100°C. whereupon the bleaching is complete. The thus-bleached fabric is then entered into a dechlorinating bath, heated to 60°C. and containing 8 g./l. of sodium bisulfite and 1 ml. per liter of 2P-21°Bé. hydrochloric acid. After squeezing the thus-soaked fabric, it is wrapped around a roller and the ensuing dechlorination reactions allowed to proceed for a period of one hour, whereupon the dechlorinated fabric is subjected to washing with water in a conventional washer to remove bleaching and dechlorinating materials as well as impurities still present therein. The fabrics thus bleached present a satisfactorily improved whiteness. If desired, the fabrics can be aftertreated with optical bluing agents.

Chlorocyanurates and Hydrogen Peroxide for Linen and Jute

J. Bouvet and P.L. Lhoste; U.S. Patent 3,552,907; January 5, 1971; assigned to Azote de Produits Chimiques S.A., France describe a rapid process for bleaching Liberian fibers consisting of a strongly alkaline bath under steam atmosphere at 100°C., followed by an aqueous solution of 4 to 5 pH containing trichlorocyanuric acid and cyanuric acid, at 20° to 50°C., and a final alkaline bath of hydrogen peroxide. The process may be continuous or semicontinuous.

Example: An untreated linen textile fabric is subjected to continuous bleaching at a speed of 70 meters per minute. This fabric has the following characteristics: Natural impurities soluble in water, 10%; hemi-cellulose, 13%; mean degree of polymerization, 3,000; the denier, 200 g. per square meter.

(1) Continuous Alkaline Treatment — The fabric treated in boiling water, then impregnated in a solution of caustic soda at 50 g./l., after having been wrung out at 100 to 110%, is subjected to a vaporized steam atmosphere at 100°C., then to ripening in the form of folds in a J-box for 15 minutes at the same temperature. It is then well washed and rinsed on a washing machine with 6 baths, at the outlet from the J-box.

(2) Continuous Treatment in Chlorocyanuric Solution — The wrung-out fabric is admitted into a saturator containing an aqueous bath heated to 50°C. having the following ingredients maintained constant: Trichlorocyanuric acid (ATCC), 11.1 g./l. of bath; cyanuric acid (AC), 3.0 g./l. of bath; sodium bicarbonate (BS), 6.1 g./l. of bath; acetic acid necessary to maintain the pH at 4.5. This solution in which the molecular ratio ATCC/AC/BS is substantially 2/1/3 has a concentration of available chlorine of 10 g./l. On leaving the saturator, the fabric is wrung out to about 100%, and is then subjected to ripening in a second J-box at 50°C. for 15 minutes before being washed in boiling water, rinsed and forcibly wrung out.

(3) Continuous Treatment in Alkaline Solution of Hydrogen Peroxide — At the close of the second treatment, the fabric is admitted into a second saturator containing an oxidizing aqueous solution the following composition of which is maintained constant: Hydrogen peroxide at 130 volumes, 30 $cm.^3$/l. of bath; sodium silicate at 37°Bé., 24 g./l. of bath; sodium hydroxide, 6 g./l. of bath. The fabric impregnated by exchange is wrung out at 100 to 110% at the outlet from the saturator. After passing into a vaporizer at 100°C. it is passed in the form of folds into a third J-box where it is subjected to ripening for 15 minutes at 100°C. The final washing is effected on a washing machine with 5 baths, the first two of which are fed with boiling water. After drying, the fabric displays a degree of whiteness of 90.3%, the degree of

percent of transmission being effected by means of a photocolorimeter, related to a standard of whiteness of $MgCO_3$. The degree of polymerization of the cellular fiber exceeds 1,950.

## Chlorocyanuric Acid with Ethylidene Diphosphonic Acid

X. Kowalski; U.S. Patent 3,579,287; May 18, 1971; assigned to Monsanto Company describes a process to bleach textile fabrics with chlorocyanuric acids, salts or complexes followed by contact with an aqueous acidic solution containing hydroxy ethylidene diphosphonic acid, sodium hexametaphosphate, ethylene diamine tetraacetic acid, nitrilotriacetic acid, alkali metal salts of the foregoing acids or mixtures of these compounds as well as the sulfite salts conventionally utilized in souring solutions. The process provides fibers characterized by improved appearance and reduced heat sensitivity.

Commercial bleaching operations involve removal of natural fiber color and/or stains and discolorations developed in fiber processing. Therefore, more severe conditions in terms of available chlorine concentration, are required than in household or commercial laundry operations designed to remove less tenacious applied stains. These more severe conditions require the use of a "souring" step to neutralize or remove excess bleaching agent. In the souring step, the fibers, after bleaching, are contacted with an aqueous acidic souring bath containing a sulfite salt such as sodium sulfite or sodium bisulfite.

In accordance with this process, at least 0.05% and preferably at least 0.1 to 2% by weight of hydroxy ethylidene diphosphonic acid, sodium hexametaphosphate, ethylenediamine tetraacetic acid, nitrilotriacetic acid, alkali metal salts of the foregoing acids, or mixtures of the foregoing is incorporated in the souring bath as a supplemental ingredient in addition to the conventional souring agents. With the exception of the addition of such supplemental agent to the souring bath, the souring operation is carried out in a conventional manner. The use of such supplementary agent has a twofold beneficial effect. First, fiber appearance as compared to fibers soured in a bath containing no supplement agent, is superior. Further, heat sensitivity (a tendency of the fibers to yellow when exposed to heat, for example in drying, ironing or storage) is reduced.

Example 1: Identical samples of fabric woven from acrylic fibers are bleached by contact for 30 minutes at 195°F. with an aqueous solution containing sufficient trichlorocyanuric acid to provide about 785 parts/million by weight available chlorine. The pH of the bleaching baths is adjusted from 3 to 6 by use of phosphoric acid or sodium hydroxide. The samples are then rinsed with water acidified to a pH of 3.0 with $H_3PO_4$. The samples are then soured using a souring bath containing 2.5 g./l. $NaHSO_3$ and 3 g./l. ethylene diamine tetraacetic acid. The sample shows superior appearance and lower heat sensitivity than an identical sample soured in a bath containing only 2.5 g./l. $NaHSO_3$.

Example 2: The procedure of Example 1 is repeated using a souring bath containing 3 grams per liter $Na_2SO_3$ and 3 grams per liter nitrilotriacetic acid. The bleached sample shows superior appearance and lower heat sensitivity than an identical sample soured in a bath containing only 3 grams/per liter $Na_2SO_3$.

Example 3: The procedure of Example 1 is repeated with samples of polyester/cotton fabric and cotton broadcloth being substituted for the acrylic fabric. Similar improvements in the appearance and heat sensitivity of these fabrics are obtained.

## Chlorocyanurates plus Fatty Acid Taurate

X. Kowalski; U.S. Patent 3,586,474; June 22, 1971; assigned to Monsanto Company found that the bleaching action of chlorocyanurates is improved and their corrositivity reduced by the addition of a fatty acid taurate, such as a coconut oil acid taurate. The process is applicable to any chlorine bleachable textile fiber which may be in the form of slivers, tops, yarns or fabric. The fibers are contacted with an aqueous solution of dichlorocyanuric acid, trichlorocyanuric acid, alkali metal salts thereof, complex compounds such as monotrichloro-tetra(monopotassiumdichloro)-pentaisocyanurate or trichloro-(monopotassiumdichloro)-diisocyanurate or mixtures of these. In accordance with conventional practice, the bleaching solution may contain various stabilizers for the bleaching agent, sequestering agents and the like.

The bath must also contain from 0.01 to 0.5% by weight of a taurate of a saturated fatty acid. Taurates of unsaturated fatty acids containing 10 to 18 carbon atoms, for example, N-methyl-N-coconut oil acid taurate (predominantly N-methyl-N-lauric acid taurate) are preferred. After bleaching, the fibers are generally "soured" in a sodium bisulfite or other conventional souring bath to remove chlorine, after which they are rinsed and dried.

Example 1: A swatch of acrylic jersry fabric is bleached for 30 minutes at 210°F. in an aqueous bleach bath having a pH of 3 and containing 0.58 g./l. trichlorocyanuric acid, 1.0 g./l. sodium-N-methyl-N-coconut oil acid taurate, and 0.5 g./l. sodium hexametaphosphate. After bleaching, the fabric is soured for 10 minutes at 150°F. in a bath containing 2 g./l. sodium bisulfite and then rinsed in water. The "whiteness number" of the fabric as determined with a Gardner Color Difference Meter is 87. (High "whiteness numbers" represent superior fabric appearance.)

For purposes of comparison, identical swatches of fabric are bleached according to the above procedure using bleach baths wherein the sodium-N-methyl-N-coconut oil acid taurate is replaced with (a) 1 g./l. linear alkylbenzene sulfonate (alkyl chain lengths $C_{11}$ to $C_{13}$); (b) 1 g./l. ethoxylated nonylphenol (containing 9 to 10 molecular proportions of ethylene oxide); (c) 1 g./l. ethoxylated dodecyl phenol (containing 9 to 10 molecular proportions of ethylene oxide); (d) 1 g./l. sodium-N-methyl-N-oleoyl taurate (a taurate of an unsaturated fatty acid); and (e) no surfactant. The whiteness numbers obtained by bleaching in these baths are respectively, 85; 80; 81; 83; and 85.

It is seen from the above experiment, that the use of the saturated fatty acid taurate in the bleach bath provides superior results as compared to baths containing no surfactant, (e); baths containing commonly employed surfactants (a), (b), (c); and baths containing unsaturated fatty acid taurate surfactants, (d). The appearance improvement in the fabric is significant in that a whiteness number difference of 1 is visually observable.

Example 2: The procedure of Example 1 is repeated with the exception that dichlorocyanuric acid is substituted for trichlorocyanuric acid in the bleach bath. Similar results are obtained.

Example 3: Bleach baths containing about 0.88 g. trichlorocyanuric acid and quantities of various surfactants are prepared and heated to about 212°F. Small coupons of stainless steel 316 are maintained in each bath and in the vapor above each bath for 5 hours. It was found that corrosivity of chlorocyanuric acid solutions (due to evolution of $NCl_3$) is effectively reduced by the addition to the bath of saturated fatty acid taurates. It is further seen that the use of conventional surfactants such as alkylbenzene sulfonates surfactants is significantly less effective.

# PERMANGANATE

## PERMANGANATE AND PHOSPHORIC ACID FOR JUTE BLEACH

H.V. Simpson and L. Bogan; U.S. Patent 3,384,444; May 21, 1968; assigned to Reeves Brothers, Inc. describe a process to produce lightfast jute by bleaching the jute with potassium permanganate and phosphoric acid, followed by sodium bisulfite treatment. Jute, treated by this process, does not have the tendency to darken on aging.

The process consists of: (a) bleaching a jute fabric by treating the fabric with an aqueous bleaching solution containing potassium permanganate and phosphoric acid at a temperature in the range from about 60° to 110°F., and at a pH below 3.0, the ratio of potassium permanganate to phosphoric acid in the aqueous bleaching solution being in the range between about 1:0.7 to 1:1.1; and (b) scavenging the bleached jute fabric by treating the bleached fabric with an aqueous solution of an inorganic reducing agent at a pH below 4.0, thereby forming a lightfast jute fabric.

Lightfast jute fabrics produced in accordance with the process possess a colorfastness equal to not less than 25 Standard Fastness Hours (and in most instances from 30 to 40 Standard Fastness Hours), as determined by AATCC Standard Test Method 16A-1963, with no progressive shade change thereafter. By way of comparison, jute fabrics which had been bleached by conventional techniques possess a colorfastness of only 5 Standard Fading Hours before the fabric undergoes a shade change after exposure to a standard light source.

The concentration of the aqueous bleaching solution used to bleach the jute fabric preferably should contain from about 3 to 15% by weight of potassium permanganate and from about 2 to 17% by weight of phosphoric acid (calculated on the basis of 100% $H_3PO_4$, or from about 3 to 22.5% by weight of 75% $H_3PO_4$), all percentages except those designating the phosphoric acid strength being based on the weight of the fabric being treated. The aqueous bleaching solution is effective for the purposes of the process only when the ratio of potassium permanganate to phosphoric acid (basis 100% $H_3PO_4$) is in the range between about 1:0.7 and 1:1.1, or on the basis of 75% strength $H_3PO_4$, in the range between about 1:1 to 1:1.5, and the pH of the solution is below 3.0, for any variation of these reaction conditions adversely affect the colorfastness of the fabric.

The jute fabric is immersed in the aqueous bleaching solution until the jute is completely bleached or the potassium permanganate completely reduced. Depending on the concentration of potassium permanganate and phosphoric acid employed in the aqueous bleaching solution, the jute fabric will be bleached for a period from about 20 minutes to 1 hour, although these time periods are not as critical to the process as the pH of the aqueous bleaching solution and the ratio of potassium permanganate to phosphoric acid in the bleaching bath.

Upon removal of the bleached jute fabric from the aqueous bleaching solution, the fabric is then scavenged by immersion in an aqueous solution containing an inorganic reducing agent, preferably sodium bisulfite, at a temperature in the range from about 80° to 160°F. and at a pH below 4.0. Although any acid may be used to adjust the pH of the scavenger solution, outstanding results have been obtained using phosphoric acid in this solution. Preferably, the scavenger solution should contain from about 4 to 22% by weight of sodium bisulfite, based on the weight of the fabric being treated, and sufficient phosphoric acid (generally from 40 to 60% by weight of the sodium bisulfite of 75% $H_3PO_4$) to lower the pH to below 4.0.

Example: A coarsely woven jute fabric, identical to the unbleached greige fabric sold commercially, was bleached in an aqueous bleaching solution containing 10% by weight of potassium permanganate and 12% by weight of 75% phosphoric acid, both percentages being based upon the weight of the fabric being treated. The fabric was immersed in the bleaching solution at a temperature of about 65°F. for approximately 45 minutes, after which the fabric was washed in cold water until the rinse water was clear and then washed in hot water (140°F.) until the rinse water was clear.

The bleached jute fabric was then immersed in an aqueous solution containing 10% by weight of sodium bisulfite and 5% by weight of 75% phosphoric acid, both percentages being based on the weight of the fabric. The temperature of this bisulfite scavenger solution was kept at about 120°F., and the fabric treated in this manner for approximately 45 minutes.

The fabric was then removed from the bisulfite scavenger solution, washed for 10 minutes with hot water (140° to 160°F.), and then scoured for an additional 10 minutes with hot water (180° to 190°F.) containing 0.5% by weight of a nonionic surfactant (Triton X-100, which is an alkyl aryl polyether alcohol). The fabric was then rinsed with water until no residual bisulfite odor remained and then scoured with 2% by weight of acetic acid which, after rinsing, left the fabric with a pH of about 6.0. After drying, the fabric had a very attractive white finish. Exposure of the finished jute fabric to the Fade-Ometer apparatus showed that a Class 4 (Gray Scale) shade change occurred between 30 and 40 hours of continued exposure, with no progressive shade changes even after 50 hours of continued exposure, following which the test was discontinued.

## INHIBITING YELLOWING OF BLEACHED JUTE

J.S. Panto and F.K. Burr; U.S. Patent 3,472,609; October 14, 1969; assigned to Nujute Incorporated describe a process of bleaching jute and inhibiting subsequent yellowing of the bleached jute under sunlight comprising treating the jute with an aqueous solution containing 8 to 12% by weight of the jute of potassium permanganate and 8 to 12% by weight of the jute of a strong mineral acid and thereafter rinsing the jute in an aqueous solution of a reducing agent, and the resulting product.

Example: A 900 pound roll of burlap fabric (300 lbs. dry weight of fabric, 600 lbs. water) was processed in open width on a standard open jig. The jig was prepared by adding to 50 gallons of water at 65°F. in the jig, 30 lbs. of potassium permanganate (dissolved in the minimum amount of water) and 30 lbs. of sulfuric acid, 66°Bé. The fabric was then run, back and forth, for 20 minutes, when the active permanganate was exhausted. The bath was dropped and replaced with water at 190°F. containing 60 lbs. of sodium bisulfite (dissolved in the minimum amount of water) and the fabric run for 15 minutes.

Upon drying, the fabric was found to be well bleached and when tested in a standard Fade-Ometer, according to AATC(2) Standard Test Method 16A-1963, showed no appreciable shade change after 10 hours, only a slight shade change after 20 hours, and no substantial further change after 80 hours. In contrast, the same original fabric when bleached only by a conventional hydrogen peroxide bleach showed a substantial and objectionable shade change after 10 hours exposure. When dyed by conventional procedure, as used in the dyeing of cotton, with Vat Red 13, color index No. 70,320, the fabric of this example when tested in the Fade-Ometer showed no appreciable color change after 10 hours, only a slight shade change after 20 hours and no substantial further change after 80 hours.

In contrast, the same original fabric bleached only by a conventional hydrogen peroxide bleach and dyed in the same way with the same dyestuff showed a substantial and objectionable shade change after 10 hours.

# SULFITES AND HYDRIDES

## WOOL BLEACH WITH BISULFITE AND BOROHYDRIDE

G.T. Gallagher and H.N. Tobler; U.S. Patent 3,250,587; May 10, 1966; assigned to FMC Corporation describe a process to bleach wool rapidly and without degradation to an excellent brightness by a two-bath process. The scoured wool, fibers, tow, yarn or fabric, is first saturated with an aqueous solution containing about 3 to 5% of an alkali bisulfite to provide 50 to 150% by weight of this first solution on the wool, immersing the saturated wool for about 5 seconds up to about 10 minutes in a second aqueous solution containing about 0.01 to 1.0% by weight of an alkali borohydride. This solution is used in an amount to provide about 0.2 to 2% by weight of the alkali borohydride based on the weight of the wool being treated and providing a pH of 5 to 7.

The process continues by withdrawing the treated wool from the second solution and within 10 minutes from commencement of the borohydride treatment, removing the borohydride therefrom either by washing it with water or by heating it to a temperature of about 130° to 160°F., for about 10 to 30 minutes to decompose the borohydride therein. Preferably the wool is treated mechanically to reduce the amount of second solution carried on it prior to the heating, for the obvious reason that this reduces the amount of heat required. Removal of borohydride is essential to avoid discoloration of the treated wool, which occurs if the borohydride is exposed to oxygen in the atmosphere or in water after the wool has been permitted to remain in contact with the borohydride and the bisulfite compound for more than 10 minutes before such exposure to oxygen. The bleaching steps of this process can be carried out in as little as 10 seconds, and provide excellently bleached wools having brightnesses fully equivalent to those produced by the best processes.

The alkali borohydride employed is an alkali metal or alkaline earth metal borohydride, preferably the sodium, potassium, lithium or magnesium borohydride. The alkali borohydride is a solid and is stable in this form. When it is introduced into the aqueous solution employed in the second step of the process, its pH is normally maintained above neutral to avoid decomposition of the borohydride. When the bisulfite-saturated wool is immersed in the borohydride solution, only that borohydride in contact with the wool, and therefore with the bisulfite carried thereon, is available for reaction. The amount of borohydride contacting the wool may be varied by agitation of the solution, for example, by moving the wool through the solution, but in the absence of such agitation the amount of borohydride contained in a given weight of solution about equal to the weight of the wool can be considered to contact the wool and bisulfite. With this in mind, one part by weight of borohydride should be made available for contact with wool containing about 2.5 to 9 parts by weight of bisulfite.

The wool is permitted to remain in contact with the borohydride solution for no more than 10 minutes, whereupon it either is heated to about 110° to 180°F. and preferably 130° to 160°F. for about 10 to 30 minutes, or it is washed with water, preferably with the aid of commonly used wool-washing detergents such as Triton X-100. This washing or heating step is carried out to destroy residual borohydride on the wool; if this agent is not destroyed in this time, when the wool is contacted with air, even that in water, the wool darkens. Where the heating step is used it is preferred to wash residual chemicals from the wool with water. Following washing, the wool can be dried in any normal way, for example in an oven at about 110° to 180°F.

The process is carried out at ambient temperatures, that is, in the range of temperatures of normal processing water, about 60° to 80°F. However, it is possible to carry the process out effectively at temperatures as low as about 40° to 50°F. or as high as 230°F., temperature not being critical in the process because the time-temperature relationship usually encountered with bleaching agents such as hydrogen peroxide is not found to be a serious factor in this process. Accordingly, it is desirable to operate the process without either cooling or heating the solutions; this has the added advantage of not resulting in accelerated decomposition of borohydride as sometimes occurs upon the application of heat.

The process is particularly suited to continuous operation because of the short times required for the treatments in the baths involved. Furthermore, no complicated equipment or temperature controls are necessary for operation of this process and the bleaching agents employed are quite stable and the baths may be employed continuously with simple replenishment of the reactants as necessary. It is possible, however, to carry the process out batchwise, and this is done where only a few pieces are to be bleached, or equipment is not available which is suitable for continuous processing.

All percentages referred to in the examples are percentages by weight. Reflectance values were determined on a Hunterlab Reflectometer Model D-40, employing a blue filter. The wool samples employed were scoured before treatment for 30 minutes at 120°F. in a 0.2% aqueous solution of Triton X-100, followed by a water wash and drying at 150°F.

Example 1: A scoured lightweight apparel-grade wool fabric having a reflectance of 37.3% was impregnated in a 5% aqueous solution of sodium bisulfite having a pH of 4.8. It was then squeezed to leave in the wool an amount of solution equal to the weight of the dry wool, and the saturated wool was placed immediately into a solution containing 0.1% of sodium carbonate and 0.01% of sodium borohydride maintained at 80°F. and having a pH of 9.2. The solution was permitted to remain in this bath for 15 seconds during which time the pH of the solution in the immediate environment of the wool was between 5 and 7.0. The fabric was then removed, washed with water and dried at 160°F. for 15 minutes. The reflectance following bleaching was 48.7%. By way of comparison, a 3-hour bleach with a 0.3% solution of hydrogen peroxide containing also 0.17% of ammonium hydroxide, and 0.17% of tetrasodium pyrophosphate carried out at 135°F. provided a reflectance in the wool of 46.2%.

Example 2: A sample of wool fabric taken from the same batch as that used in Example 1 was immersed in an aqueous solution containing 5% of sodium bisulfite and having a pH of 4.8, and maintained at 120°F. for 15 minutes. The fabric was then removed from this bath and squeezed to provide a retained weight of solution on the fabric equal to the weight of the fabric. The saturated fabric was then immersed directly in an aqueous solution containing 0.1% of sodium carbonate and 0.1% of sodium borohydride and having a pH of 9.3, for 15 seconds. The pH of the borohydride solution adjacent the wool was about 5 to 7.0, and the temperature of the solution was 70°F. The fabric was then removed from the bath and cut into two equal portions, one portion was washed with water and dried and the other was squeezed to 100% saturation and then dried at 150° to 160°F. for 15 minutes without washing. The portion which was heated and dried without washing had a reflectance of 55.7% and the washed portion, 54.2%.

## BOROHYDRIDE PLUS OZONE FOR CELLULOSE

R.C. Wade; U.S. Patent 3,318,657; May 9, 1967; assigned to Metal Hydrides Inc. found that the oxidation of cotton, jute, linen, rayon and wood fibers by ozone in a bleaching process can be prevented if the fibers are given a prior treatment in an alkaline aqueous borohydride solution. The amount of borohydride used should be between about 0.01 and 2% based upon the dry weight of the fiber. While greater amounts of borohydride can be used, it is unnecessary and becomes uneconomical. The concentration of borohydride in the water is not critical but is entirely dependent upon the method of application to the fiber. Thus, when treating wood fibers very large volumes of water relative to the wood are used and the borohydride concentration can be very low.

In treating woven textile fibers, the solution can be padded on the fabric and since the normal weight pick-up of solution is about 100%, the concentration of borohydride might be 0.1 to 0.5%. When the fibers are sprayed with relatively small amounts of solution based upon the weight of the fiber, more concentrated solutions of borohydride can be used.

The pH of the borohydride solution should be above 9 and in many cases it should be applied with at least 1% of sodium hydroxide present in the solution. Wood fibers in particular should be treated with borohydride solutions stabilized with excess caustic soda, caustic potash, or lime. These bases prevent excessive hydrolytic decomposition of the borohydride while it is in contact with the fiber. The time and temperature for fiber treatment with borohydride can vary widely. In general this may be from 4 hours at 20°C., 1 to 2 hours at 30° to 35°C., to 2 to 10 minutes at 80° to 100°C. Higher temperatures can be employed by operating this treatment under pressure, thereby reducing the contact time necessary.

After treatment with the alkali metal borohydride the fibers are prepared for treatment with ozone by lowering their pH to at least 7 and, preferably, to a pH of about 4. This may be accomplished by treating the fibers with a solution of an acid, such as hydrochloric, sulfurous, sulfuric, acetic acid, etc., or by washing the fibers to remove the alkaline materials prior to treatment with the acid solution.

The ozone treatment can be carried out in an aqueous solution having a pH between 2 and 7, and preferably between 2 and 4, or by exposing the wet fibers to an atmosphere containing ozone in a concentration of up to 6% by weight, the preferred concentration being about 2%. The time of exposure to give the desired bleaching must be determined for each fiber and is related to ozone concentration and temperature. These should be adjusted so that about 0.1 to 3% of ozone based on the weight of the dry fiber is consumed during the process. Exposure time may require 2 hours at room temperature to 2 minutes or less at 100°C. Again the exact conditions for each fiber must be tested individually. After ozone treatment the fiber is again washed and if desired adjusted to a pH required for further processing or finishing.

In the above-described technique bleached cellulosic fibers are obtained which are generally 5 to 10% stronger or less degraded than fibers which have been treated with ozone but have omitted the borohydride pretreatment.

Example: A sample of cotton cloth 1 1/2" wide and 13" long, weighing 2.2 grams, was placed in an aqueous solution containing 0.2% of sodium borohydride by weight, the sample remaining in the solution for 1/2 hour. The sample was then wrung out to approximately 100% wet pick-up and placed in a resin kettle on a wire hanger and ozone was passed through at a rate of 1.4 liters per minute for 1/2 hour, the ozone concentration being 14 mg. per liter. The amount of sodium borohydride picked up by the cloth was 4.4 mg. The treated cloth had a brightness of 70.5 points and a tensile strength of 55.6 pounds per square inch, the original brightness being 64 points and the original tensile strength being 54.6 pounds.

In the same experiment described above, another piece of cotton which was not pretreated with sodium borohydride but only wetted with water to 100% weight pick-up was placed in the resin kettle and exposed to ozone as described above. The average brightness of the sample was only 65.5 points and the tensile strength was only 53.2 pounds.

## STRIPPING DYES WITH HYDROSULFITE AND QUATERNARY AMMONIUM COMPOUNDS

I. Sapers; U.S. Patent 3,591,325; July 6, 1971; assigned to Arkansas Co., Inc. describes a process to remove organic dyes and pigments, such as phthalocyanine, vat and azo pigments, from textiles by quaternary ammonium compounds. The stripping agents of this process are obtained from primary aliphatic amines or araliphatic amines or diamines, which have been ethoxylated with 4 to 12 mols of ethylene oxide and then reacted with an aralkyl halide. They may be represented by the following formulae:

(1)
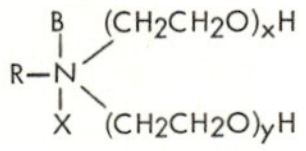


(2)
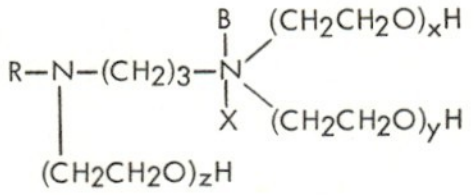

In these formulas R is a long aliphatic chain, preferably alkyl, containing 12 to 20 carbon atoms, which may or may not be substituted with a mononuclear aryl group such as phenyl, and where x+y or x+y+z is 4 to 12; B is benzyl, halo benzyl, xylyl or xylylhalide; and X is a halogen such as chlorine, bromine or iodine. The preferred range of ethylene oxide (ETO) is 5 to 10 mols, and the halogen substituents of B may be chlorine, bromine or iodine.

Example 1: A mixture of 125 parts ethoxylated tallow propylene diamine, 10 mols ETO (Ethoduomeen T/20) and 37 parts benzyl chloride was heated to 125°C. in a 3-neck reaction flask equipped with stirrer, condenser and thermometer and the reaction was continued at 120° to 140°C. for 3.5 hours. Heating was then discontinued and the viscous liquid reaction product was discharged.

Example 2: 200 parts ethoxylated stearyl amine, ETO-10 mols and 38 parts benzyl chloride were heated as in Example 1 except that the reaction mixture was heated 30 minutes at 125° to 130°C. and 120 minutes at 110° to 115°C. The total reaction time was 5 hours.

Example 3: A bath was prepared as follows: 2.0 parts of product of Example 2, 20.0 parts sodium hydroxide, 10% solution, 2.0 parts sodium hydrosulfite and 176.0 parts water. A cotton upholstery fabric printed over its entire surface with patterns of greenish-blue, green, red, and violet, and in which the greenish-blue predominated, was immersed in this bath, heated to the boil and boiled for 30 minutes. The bath was dropped and the fabric rinsed well and dried. The red color was stripped (removed) completely, the greenish-blue and green were stripped to light beige, and the deep violet was stripped to a light lavender.

## BLEACHING MECHANICAL WOOD PULP WITH HYDROSULFITE

V.N. Gupta; U.S. Patent 3,709,779; January 9, 1973; assigned to Canadian International Paper Company describes a process in which mechanical pulps are bleached with sodium or zinc hydrosulfite in the presence of sodium silicate. The addition of silicate increases the brightness gain obtained by hydrosulfite alone and reduces the grinding power consumption. Silicate can thus replace sodium tripolyphosphate for this use and thereby eliminate phosphate pollution due to newsprint or other paper mill effluents.

The hydrosulfite bleaching of mechanical pulps has to be regulated because of two characteristics of hydrosulfite. This material is a strong reducing agent and quite stable in an alkaline solution, but in weakly acidic or acidic solutions, it decomposes at a very fast rate. The bleaching of mechanical pulps has to be carried out in an acidic medium because of the darkening of mechanical pulps at pH above 6.5. Considering the above two factors, laboratory experiments have shown that the optimum pH range for hydrosulfite bleaching of mechanical pulps is 5.5 to 6.0. In the mills, therefore, where the pH of mechanical pulps is in the 4.0 to 4.5 range, the addition of a buffering agent is necessary to raise the pH although it normally does not reach the optimum range.

Another problem with hydrosulfite bleaching is the catalytic effect of heavy metal ions on the rate of decomposition of the bleaching agent. To counteract this effect, addition of sequestering agents to the mechanical pulp during bleaching is desirable.

The most common material added for this purpose is sodium tripolyphosphate (STPP). Although STPP is quite satisfactory for use, it is lost in the effluent waters of the mill. On hydrolysis it yields the phosphate ion which, in combination with carbon and nitrogen, is believed to be responsible for algal growth in natural water systems, and the waters became fouled by a green scum. Hence, for ecological considerations it is essential to replace STPP by a suitable substitute. The cost of other satisfactory materials such as ethylenediaminetetracetic acid (EDTA), sodium citrate, etc. is a hindrance to their use.

It has been found that the ecological problems attendant the hydrosulfite bleaching of mechanical pulps in the presence of sodium tripolyphosphate can be avoided completely by eliminating the use of sodium tripolyphosphate and using in its stead an alkali metal silicate, such as for example, sodium, potassium or ammonium silicate.

Mechanical pulp of standard production, either stone or refiner groundwood, has a fiber concentration within the range of from about 1 to 15%. Based upon equipment limitations and to expedite handling, it is preferred, however, to employ an aqueous pulp suspension having a consistency of from about 1 to 6%, based on the weight of the moisture-free pulp, and even more preferably a pulp having a consistency of from about 3 to 4%. The consistency of 3 to 4% mentioned as preferable is usually the consistency in the deckers where the hydrosulfite bleaching is normally carried out in the mill.

Either zinc or sodium hydrosulfite may be employed in accordance with this process, in a concentration ranging from about 0.05 to 1.0%, based on the weight of moisture-free pulp. In most cases it is preferable to use from about 0.3 to 0.5%, by weight, of sodium or zinc hydrosulfite. In the preferred embodiment of this process, sodium silicate and the hydrosulfite are mixed with the aqueous suspension of mechanical pulp and heated for a period of from about 10 minutes to about 6 hours at from about room temperature to about 100°C. Preferably, however, the reaction is carried out at a temperature of from about 50° to 70°C. for a period of about 1 to 3 hours.

To assess the usefulness of the process in a mill operation, a mill trial was run using 15 to 20 pounds of 41°Bé. sodium silicate per ton of stone groundwood. An unrelated and unexpected result was noticed. It was found that the energy required for grinding the wood with stone grinders was reduced by 5 horsepower day/air dry ton of groundwood, which corresponds to 6 to 7% of the total energy used. This is a substantial saving in the energy consumption of the mill. This advantage of power requirement reduction is opposed to what is observed with STPP. Use of 3 to 4 pounds STPP per ton of groundwood usually raises the power requirements by about 3 to 4 horsepower days/air dry ton of mechanical pulp.

Example 1: A 40 g. (25% solids content) sample of commercial stone groundwood prepared from a mixture of black spruce and balsam fir was suspended in 303 ml. of water by stirring to give a suspension of groundwood in water of 3% consistency. The slurry was heated to 60°C. in a constant temperature water bath during 5 minutes. The heated slurry was treated with 0.1 g. sodium hydrosulfite and the closed jar shaken for 30 seconds to mix the pulp slurry with bleaching agent. An immediate brightening effect was visible. The complete the bleaching the jar was then replaced in the 60°C. water bath and held there at 60°C. for one hour. After this interval, the pulp was filtered off and washed with water. Two handsheets weighing 5 g. each and 12.5 cm. in diameter were prepared from the original unbleached pulp and two similar sheets were prepared from the bleached pulp. The sheets were pressed between cellulose blotters and stainless steel plates for 2 minutes at a pressure of 240 lbs./square inch. After drying between rings for 16 hours, their brightness (reflectance) was measured using an Elrepho reflectance meter. Relative to the magnesium oxide standard with a reflectance of 100%, the unbleached pulp had a brightness of 59.6% and the bleached pulp 64.2%. Therefore treatment of this groundwood with 1% sodium hydrosulfite increased the brightness by 4.6 percentage points.

Example 2: A 40 g. sample of (25% solids content) groundwood (same as used in Example 1) was suspended in 303 ml. of water containing 20 mg. of sodium tripolyphosphate (0.2% by weight on moisture-free pulp) to give a 3% consistency. This suspension was heated during 5 minutes to 60°C., treated with 100 mg. sodium hydrosulfite (1% by weight on moisture-free pulp), mixed and heated at 60°C. for one hour. After this interval the pulp was filtered, washed with water and two handsheets made as described in Example 1.

The brightness of the bleached pulp was 64.8% showing a brightness gain of 5.2 percentage points due to sodium hydrosulfite and sodium tripolyphosphate. Since the same amount of sodium hydrosulfite alone produces a brightness gain of 4.6 percentage points, the brightness gain due to 0.2% sodium tripolyphosphate is 0.6 percentage points.

# CONTINUOUS TEXTILE BLEACHING

## WITH HYPOCHLORITE FOLLOWED BY HYDROGEN PEROXIDE

J. Lindsay; U.S. Patent 3,030,171; April 17, 1962; assigned to Pittsburgh Plate Glass Company describes a process for the bleaching of sized cotton fabrics in a continuous process including a preliminary hypochlorite treatment followed by a hydrogen peroxide bleach. In accordance with this process, a sized woven or knitted vegetable fabric such as cotton is introduced, without desizing, into an aqueous solution of an alkali metal hydroxide and an alkali metal hypochlorite and/or an alkaline earth metal hypochlorite. Sodium, potassium and calcium hypochlorite are typical of the hypochlorites employed. The fabric is permitted to remain in the solution for a period of time sufficient to substantially impregnate the fabric with the solution.

After nipping to give 50 to 150% liquor pickup basis the weight of the dry goods, the goods are heated in steam for a period of time sufficient to disperse the motes and render the nonfibrous content of the fabric water-extractable. Thus, starches are solubilized and oils and fats emulsified during the steaming operation. Fibers in the fabric treated swell and softening and dispersion of the motes occurs. In addition, the hypochlorite content of the solution impregnated fabric is substantially removed. The material after the heating operation is water washed and introduced into an aqueous solution of hydrogen peroxide.

The fabric is permitted to remain in the hydrogen peroxide solution for a period of time sufficient to impregnate substantially the woven fabric with the hydrogen peroxide solution. After the fabric has been impregnated to 50 to 150% of liquor on weight of dry goods, it is then heated a period of time sufficient to bleach it to the required degree. In the treatment of extremely heavy fabric such as cotton poplins and the like, a desizing step may conveniently be employed to insure adequate bleaching of the material.

Reference is made to Figure 9.1, which is a diagrammatic illustration of the method and apparatus utilized in bleaching woven vegetable fabrics according to this process. In the drawing is shown the fabric (1), flameburners (2) and (3), a washing

FIGURE 9.1: CONTINUOUS BLEACHING EQUIPMENT

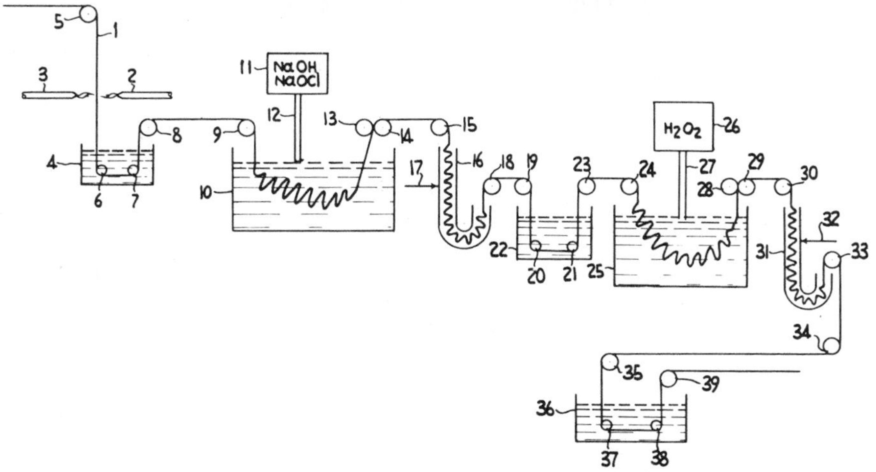


Source: J. Lindsay; U.S. Patent 3,030,171; April 17, 1962

tank (4), the caustic-hypochlorite saturator (10), J-box (17), washer (22), hydrogen peroxide saturator (25), J-box (31), washer (36). In the operation of the process in conjunction with the equipment shown in the drawing, a woven fabric is drawn over rollers (5), (6) and (7) so that the fabric is essentially intermediate between the positioning of burners (2) and (3) and the flames produced by these burners. Passage of the cloth intermediate the burners effectively singes lint, fuzz and other like material from the cloth surface. The fabric is then drawn through washer (4) and, after washing or quenching, is delivered by way of rollers (8) and (9) to caustic saturator (10). In saturator (10) the fabric is contacted with an aqueous alkali metal hydroxide solution containing an alkali metal hypochlorite or an alkaline earth metal hypochlorite.

A holdup of cloth in saturator (10) is permitted to accomplish a substantial saturation of the material with the solution contained therein. Upon leaving saturator (10), the fabric is passed through rollers (13) and (14) and excess solution expressed or squeezed therefrom. The fabric is then passed over roller (15) and introduced into the J-box (16). Located at a point on the J-box is a steam inlet line (17). Steam is introduced into the J-box at a temperature of approximately 212°F. at atmospheric pressure, and the cloth is permitted to remain therein for a substantial period of time. After the steam treatment, the cloth is drawn over rollers (18), (19), (20), (21) and (23) through a washer (22). The cloth is then introduced into saturator (25) where it is contacted with an aqueous solution of hydrogen peroxide.

A holdup of the cloth in saturator (25) is permitted for a period of time sufficient to accomplish a substantial saturation of the cloth with the hydrogen peroxide solution. After the cloth has been thoroughly saturated with the hydrogen peroxide solution contained in saturator (25), it is drawn through rollers (28) and (29) over roller (30) into the J-box (31). Rollers (28) and (29) function to express excess solution from the cloth leaving the saturator (25). Steam is introduced into the J-box (31) through a steam inlet (30) at a temperature of approximately 212°F., and the cloth is permitted to remain therein for a period of time sufficient to accomplish bleaching of the impregnated cloth. Upon completion of the bleaching operation in the J-box (31), the cloth is drawn over rollers (33), (34), (35), (37), (38) and (39) through a washer (36) where it is thoroughly washed with water and removed from the tank for further processing, for example, dyeing operations or merely to be dried and utilized as such.

Example 1: Three samples of cloth, one a print cloth, one a sateen and one a broadcloth, were bleached following the conventional bleaching process and compared with results obtained by following the bleaching process as hereinabove described. In the first series of runs, one sample of print cloth, one sample of sateen cloth and a sample of broadcloth were passed successively through two caustic saturators containing an aqueous sodium hydroxide solution of 3% by weight sodium hydroxide concentration at 140°F. A pickup of about one pound of solution per pound of cloth was obtained in the saturators and the weight concentration of the solution in the cloth controlled by nipping to 100% by weight of solution basis the weight of the dry cloth as it was removed from the saturator and placed in the J-box.

The samples were permitted to remain in the J-box for a period of one hour and contacted with steam at 210°F. (during this time period). The cloth upon removal from the J-box was rinsed in water at 210°F. in a two-compartment washer. The samples upon removal from the washing tank were passed into a bleaching bath at 122°F. containing 1% hydrogen peroxide by weight and 1% sodium silicate [$Na_2O(SiO_2)_{2.5}$] by weight. A pickup of about one pound of solution per pound of cloth was obtained and the solution expressed from the cloth as it was removed from the box to provide a solution content of 100% by weight basis the weight of the dry cloth.

Each of the samples were then placed in another J-box operated at 210°F. for one hour and steamed therein. After the one hour period, the samples were removed from the compartments, and washed in water at 210°F. A series of tests were then conducted on the samples to determine reflectance, tensile strength and absorbency. The results of these tests are shown in the following table.

Example 2: Samples of the same print cloth, broadcloth and sateen as treated in Example 1 were treated by passing the samples through successive caustic saturators operated at 140°F. The sodium hydroxide concentration of the caustic saturators were 3% by weight. In addition to the caustic concentration, each saturator contained 0.5% sodium silicate [$Na_2O(SiO_2)_{2.5}$) by weight and 0.1% sodium hypochlorite by weight. The samples were treated in a J-box for one hour at 210°F. under the same conditions obtained in the treatment of the first samples.

The hydrogen peroxide bleaching operation was conducted in a hydrogen peroxide bleaching bath at 122°F. containing 0.67% by weight hydrogen peroxide and 0.67% sodium silicate. The same pickup and expression of solution procedures were followed in Example 1, and this set of samples was also treated in a J-box for one hour at 210°F. After completion of the treatment, the same series of tests were run on these samples as were run on the samples of Example 1. The results show that by use of the hypochlorite as shown in Example 2, 2/3 the amount of hydrogen peroxide is required.

| | | Reflectance, % Relative to MgO | Tensile Strength, Units Warp, lb./in.$^2$ | Tensile Strength, Units Filling, lb./in.$^2$ | Absorbency, (sec.) |
|---|---|---|---|---|---|
| Example 1 | 20 | 84.4 | 22.7 | 18.0 | 2 |
| | 21 | 86.8 | 55.3 | 23.6 | 1 |
| | 22 | 87.3 | 28.9 | 16.6 | 2.5 |
| Example 2 | 20X | 85.0 | 23.7 | 20.6 | 2 |
| | 21X | 85.7 | 58.1 | 23.0 | 1 |
| | 22X | 86.0 | 33.9 | 17.6 | 2 |
| Example 3a | 20D | 86.0 | 22.6 | 16.8 | 2 |
| | 21D | 87.4 | 52.1 | 24.3 | 1 |
| | 22D | 86.6 | 27.3 | 18.9 | 1 |
| Example 3b | 20DX | 85.9 | 22.6 | 18.4 | 2 |
| | 21DX | 86.0 | 53.6 | 25.0 | 1 |
| | 22DX | 85.5 | 33.2 | 13.7 | 2.5 |

All D samples were desized before treatment in the caustic saturator. All X samples were treated with 0.5% silicate in an 0.1% solution of hypochlorite before the peroxide bleaching step. All #20's are print cloth. All #21's are sateen cloth. All #22's are broadcloth. Reflectance was measured on a Hunter multipurpose reflectometer. Absorbency was measured by dropping water from a pipet held 2 inches from the surface of the unstretched cloth. The time required for the disappearance of the specular reflectance from a drop as visually observed is the absorbency.

Example 3: Two more sets of runs were conducted on desized samples of cloth of the same types as described in Examples 1 and 2. One set of runs (a) was conducted with a simple caustic wash at 140°F. while a second set of runs (b) was conducted in caustic saturators at 140°F. containing sodium hypochlorite in 0.1% by weight concentration basis the weight of the solution. In addition those runs employing the hypochlorite treatment in the caustic saturator employed 0.67% hydrogen peroxide in the bleaching bath at 122°F. while those not employing the sodium hypochlorite in the caustic solution were treated with a 1% by weight aqueous solution of hydrogen peroxide at 122°F. The same series of tests as applied to the samples of Examples 1 and 2 were conducted and the results are listed in the above table.

## FOR SIZED, WOVEN OR KNITTED FABRICS

J.A. Anderson and R.R. Currier; U.S. Patent 3,056,645; October 2, 1962; assigned to Pittsburgh Plate Glass Company describe a continuous bleaching process for sized, woven or knitted cotton fibers or mixtures of cotton with wool, rayon or nylon using a preliminary hypochlorite treatment, followed by a second hypochlorite bleaching step and a final hydrogen peroxide bleach. The process used is the same as in U.S. Patent 3,030,171 assigned to the same company and reviewed above.

The only difference is the inclusion of an additional hypochlorite bleaching step between the preliminary hypochlorite bleach and the hydrogen peroxide bleach. The equipment is the same as shown in Figure 9.1 above except that an additional hypochlorite bath and a J-box are included in the equipment.

## IN SITU HYPOCHLORITE FORMATION

H.G. Smolens, O.S. Sprout, Jr. and V.C. Lane; U.S. Patent 3,077,372; Feb. 12, 1963; assigned to Pennsalt Chemicals Corporation describe a process to utilize the residual caustic alkali remaining in cotton goods after the washing step and prior to the hydrogen peroxide bleach, to form with chlorine hypochlorite as a preliminary bleaching step. Cotton cloth goods usually are processed in a continuous manner. Generally, the cloth speed varies from about 100 yards per minute for open cloth bleaching to 300 yards per minute for rope bleaching.

A normal finishing operation for cotton goods comprises, first, a water wash at a temperature of from 60° to 120°F. The hot water washing will remove some impurities from the cloth and assist in the removal of foreign particles adhering to the goods. After the water washing operation, the cloth is directed into a desizing bath wherein the cloth is exposed to the action of enzymes. After the desizing treatment, the cloth is moved into hot water washers to remove excess enzymes and to effect removal of the impurities produced by the enzyme treatment. The goods then move into a caustic alkali saturator. The caustic solution is generally within the range of 2 to 5% and is held at a temperature of about 120° to 180°F.

Following saturation with the sodium hydroxide sufficient time is allowed for the caustic to act upon the goods, usually in a J-box, after which the cloth moves into several water washing stages to remove the caustic alkali and to effect solution and removal of the impurities which have been rendered by the sodium hydroxide. Squeeze rollers following the caustic saturator and following each washer assist in the removal of the excess carry-over of liquid from one stage to the next. Even with repeated washings with hot water at temperatures up to 200°F., it has been impossible to remove the alkali from the cloth below a value of about 0.5 to 0.05% by weight. The extensive washing will not remove all of the caustic soda from the cloth, and the pH of the last washing tank will vary from 9 to 11 depending upon whether copious amounts of water are used. This residual caustic interferes with bleaching and promotes decomposition of hydrogen peroxide.

The apparatus in which the neutralization of the residual caustic alkali is effected may conveniently be similar to that in which the caustic saturation is accomplished with some modification to accommodate the introduction of gaseous chlorine. Normally, the gaseous chlorine is introduced at the botton of the saturator through a diffuser such as a ceramic diffusing plate. No agitation is necessary in the saturator since the rapid movement of cloth through it accomplishes agitation of the liquid. Occasionally some water may be added to keep the chlorinator full, but normally, the water leaving the chlorinator is made up by that coming into the chlorinator with the cloth. The most effective means of controlling the chlorine addition is to provide a pH controller in the chlorinator and have it regulate the introduction of the chlorine.

It is necessary that the aqueous chlorinator be maintained within a pH of 6.6 to 7.4 for neutral bleaching or within the range of 7.4 to 10.0 for alkaline bleaching. The neutral pH range is preferred for rapid bleaching and elimination of holdover of the cloth in a J-box for 20 to 40 minutes. If the cloth is saturated with the hypochlorite at a pH within the range of about 6.6 to 7.4, then the bleaching will take place within a matter of a few seconds. However, if the pH is allowed to rise to within the range of about 7.4 to 10.0, then the bleaching action is considerably slowed down and as long as 40 minutes may be required to obtain bleaching from the hypochlorite solution. Bleaching at a pH below 6.6 must be avoided because of the release of noxious chlorine containing vapors from the solution.

Another advantage of the process is the introduction of chlorine gas into water in the presence of the cloth accomplishes a type of bleaching which cannot be obtained by sodium hypochlorite or peroxide. Thus, in this process of peroxide bleaching there is the bleaching action by the chlorine gas, the bleaching action by the hypochlorite produced from the retained alkalinity in the cloth and the bleaching action of the peroxide itself. Consequently, the cloth whiteness is far superior to the conventional processes. Alternatively, this increased bleaching can be used to effect savings in peroxide by bleaching to the whiteness normally produced by the peroxide alone. The process is set forth in flowsheet form in Figure 9.2.

FIGURE 9.2: FLOWSHEET OF CONTINUOUS BLEACHING

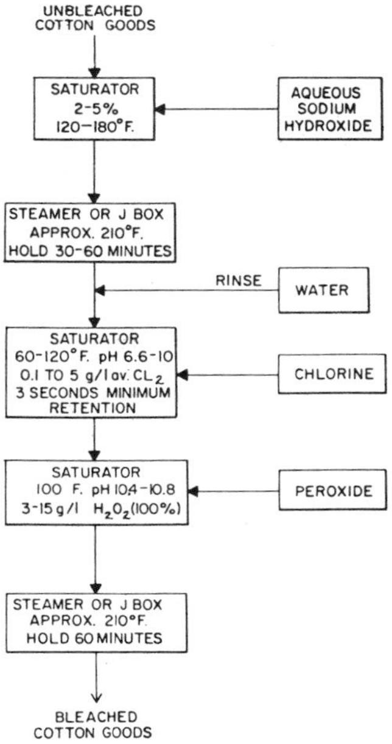


Source: H.G. Smolens, O.S. Sprout, Jr. and V.C. Lane; U.S. Patent 3,077,372; February 12, 1963

Example: In a cloth processing train comprising a caustic soda steamer, a four compartment water washer, a two compartment water washer, a hydrogen peroxide saturator, a hydrogen peroxide steamer and a two compartment water washer, 1.72 yards per pound cotton twill was introduced to the process at a rate of 90 yards per minute or 5,400 yards per hour. Following the caustic soda steamer, the water washing tanks were maintained with hot water at a temperature between 180°F. and approaching 212°F. The wet cloth leaving the fourth water washing tank had a pH of approximately 11 indicating large amount of residual caustic soda. Chlorine gas was introduced into the bottom of the first compartment of the two compartment washer which followed the four compartment washer. This compartment was filled with 400 gallons of water and maintained at a temperature of about 60° to 80°F. The warm cloth from the previous water washing stage maintained at a temperature of about 200°F. warmed the water to about 110°F. A chlorine diffuser was connected by piping through a rotameter and control valve to a chlorine cylinder.

As the cotton cloth progressed through the chlorinator, the chlorine flow was adjusted to produce a substantially neutral solution in the water in the compartment. Analysis of the solution indicated an appreciable available chlorine content but no titratable alkalinity. When the chlorine addition was interrupted, the available chlorine content of the solution quickly vanished and titratable alkalinity appeared, the pH of the solution rising to 10.7. During the addition of chlorine no fumes or odors of chlorine were detected from the solution being chlorinated indicating good stability for the chlorine-containing solution at the neutral pH point.

About 5,400 yards or 3,000 pounds of cotton goods were processed in 1 hour during which time 6 pounds of chlorine was introduced. The cloth leaving the chlorinator was lighter in color and cleaner in appearance than the cloth entering the chlorinator. The following table shows the temperature conditions, the chlorine feed rate and the presence of available chlorine or titratable alkalinity in the chlorinator together with pH for various time intervals. Six pounds of chlorine was added during the 1 hour period.

| Time | $Cl_2$ Rotameter Reading (Uncorrected) | Temp. of Aqueous Solution, °F. | g./l. Av. $Cl_2$ | g./l. NaOH | pH | Cloth Whiteness [1] | Tensile Strength Scott Tester (average of ten breaks) | |
|---|---|---|---|---|---|---|---|---|
| | | | | | | | Warp | Filling |
| 2:10 start | 7 lb/hr. Increased | | | | | | | |
| 2:20 | 13 lb./hr. Decreased | | .13 | .24 | | | | |
| 2:30 | 9 lb./hr | 110 | .3 | 0 | 6.4 | 1.8 | 133 | 55 |
| 2:45 | 8 lb./hr | 110 | .42 | 0 | 6.8 | 2.1 | 144 | 56 |
| 2:55 | 6 lb./hr | 110 | .3 | 0 | 6.8 | | | |
| 3:10 | 6 lb./hr | 110 | .3 | 0 | 8.5 | | | |
| 3:12 | $Cl_2$ off | | | | | | | |
| 3:25 | $Cl_2$ off | 110 | .04 | .4 | 10.7 | 2.8 | 136 | 50 |

[1] Tristimulous measurements on Hunter reflectometer using green, blue and amber filters. Lowest positive numeral value is most white while a highest positive numerical value is most yellow.

## CHLORITE PLUS SULFITE FOR WOOL

J. Meybeck and C. Schirle; U.S. Patent 3,097,049; July 9, 1963 describe a process to bleach and shrinkproof wool without staining or discoloration. The process consists of sodium chlorite bleach, followed by a reducing treatment in a hydrosulfite solution. Wool is characterized by polypeptidic chains formed by the union of a relatively large number of elementary α-aminoacids. These aminoacids are at least four in number and include cystine, methionine, tryptophane and tyrosine. It was found that, by the action of a solution of acid chlorite, some of these aminoacid components of the wool are oxidized and become transformed into new compounds having very characteristic properties. In a cold acid medium, the solutions of chlorite cause the wool to turn pink. Under these conditions, cystine is transformed into a colorless sulphonic derivative. Methionine leads to a colorless oxidation product (very likely it is a sulfoxide). Tryptophane gives pink oxidation products, which may be brought to a colorless state and tyrosine appears to be unaltered. Under the same conditions of pH and concentration, but at a temperature higher than 50°C., a different alteration of the wool is observed, the wool assuming in this case a yellow-brownish coloration. Cystine, methionine and tryptophane are oxidized in the same manner as in the cold but, in addition, tyrosine is also attacked, leading to brown products, of the nature of indole, in proportions which are greater the higher the temperature the longer the duration of the reaction or the lower the pH. The brown products are not reductible into colorless products. As a result, a wool so treated can no longer be rendered colorless by a reducing treatment or by an oxidizing treatment without risk of deterioration.

In this process, wool is introduced into and submerged in a body of water and there is then gradually added to the water a dilute aqueous solution of an acid with continuous agitation of the body of water until a pH of 2 to 4.5, preferably 3.5, is reached. With continued agitation, there is then gradually introduced into the acidified treating bath an aqueous chlorite solution of pH 8 to 9 and free from $ClO_2$ until the wool turns pink. Under these conditions the pH of the chlorite solution is lowered without evolution of $ClO_2$ and the treating bath remains substantially colorless throughout the treatment. During this first operation, the wool turns pink as a result of the oxidation of the tryptophane. The pink wool is then rinsed with water and treated with a reducing solution which advantageously comprises either a bisulfite or a hydrosulfite, or a mixture of both.

The body of water into which the wool is immersed and the resulting treating bath are substantially at room temperature, i.e., a temperature of about 13° to 25°C., preferably 18° to 20°C. The treatment with the reducing solution, however, is carried out at a higher temperature and the reducing solution is advantageously at a temperature of 45° to 60°C., preferably about 50°C. The acid employed to acidify the initial treating bath is a mineral acid, such as phosphoric acid or sulfuric acid, or an organic carboxylic acid, such as acetic acid or formic acid. In the case of acetic acid, it is advantageously used in combination with its alkali metal salt, e.g., sodium acetate, to serve as a buffer at the indicated pH.

The acid is employed in the form of a dilute aqueous solution and is suitably in a concentration of about 5 to 20%, preferably about 7 to 10%, by weight. When a buffer is used, generally 5 to 40% based on the weight of the undiluted acid is sufficient. The chlorite introduced into the acidified bath is suitably an alkali metal chlorite, e.g., sodium chlorite, and is employed in aqueous solution of a concentration of 5 to 20%, preferably about 10%, by weight.

The reducing solution may be in the form of a bath into which the wool is immersed or it may be applied by impregnation. The reducing solution is an aqueous solution of a reducing agent such as an alkali metal bisulfite, e.g., sodium bisulfite, or an alkali metal hydrosulfite, such as sodium hydrosulfite, or a mixture of both bisulfite and hydrosulfite. The reducing agent is advantageously present in the solution in a concentration of 0.2 to 1.5%, preferably 0.4 to 0.6%, by weight. After treatment with the reducing solution, the wool is thoroughly rinsed with water and dried.

Conventional wetting agents, up to 0.1%, and emulsifying agents, may be incorporated in the treating bath and bleaching solution. In a preferred form of the process, hydroquinone is incorporated in the acidic chlorite treating bath. Hydroquinone, which is used in the amount of 0.005 to 0.05% based on the weight of the original body of water from which the bath is formed, has been found to have an activating action upon the chlorite solution. When hydroquinone is used, a fully white wool is obtained as in the case when hydroquinone is absent and at the same time even the slightest tendency to deterioration of the wool is substantially eliminated. This is of particular value when the wool is to be subjected to subsequent treatments, e.g., to dyeing treatments or to an additional treatment to render it substantially fully non-shrinkable. For this purpose, a proteolytic enzyme such as papain, trypsin, etc. may be employed.

It is particularly advantageous in this case to combine the reducing treatment and the activating treatment with the enzyme, which considerably simplifies the sequence of the operations, and makes it possible to obtain, with the least expenditure, a wool which is perfectly white and unshrinkable. When an enzyme is used, there is suitably added to the bath 0.2 to 1% by weight of sodium bicarbonate for the purpose of neutralizing the solution.

Example 1: 25 kg. of wool fabric are introduced into a stainless steel tank provided with a stirring device and containing 500 liters of water and 0.250 kg. of a wetting agent (sodium lauryl sulfonate). After 5 minutes of agitation, 1.5 liters of 50% phosphoric acid diluted in 15 liters of water are added to provide a pH of 3.5. Agitation is continued for 10 minutes, so as to uniformly distribute the acid in the wool. Then 1 kg. of sodium chlorite in 10 liters of water is added. Circulation in the bath is continued for 1 hour. Then the fabric is thoroughly rinsed with water and treated for 30 minutes at 50°C. with a reducing solution containing 10 cc per liter of a 30% solution of sodium bisulfite of soda, and 2 grams per liter of sodium hydrosulfite. The wool is then thoroughly rinsed with water and dried.

Example 2: 50 kg. of wool strands in small tanks are added to a stainless steel tank containing 1,500 liters of water, 350 g. of a wetting agent (alkylarylsulfonate) and the necessary quantity of acetic acid and sodium acetate to bring the pH to 3.5. This mixture is left at rest for a few minutes in order to distribute the acid in the wool, then 3.75 kg. of sodium chlorite in 35 liters of water are added and the resultant mixture left at 20°C. for one hour. The strands are then rinsed in water and treated for half an hour at 45°C., in a reducing bath containing 4 grams per liter of stabilized sodium hydrosulfite. Finally, the yarns are rinsed, passed through a diluted (0.05%) sulfuric acid bath, again rinsed, and dried.

Example 3: 50 kg. of wool yarns are introduced into a stainless steel tank containing 1,000 liters of water, 0.5 kg. of a wetting agent (lauryl polyethoxyether) and 3 liters of 50% phosphoric acid to provide a pH of 3.5. After waiting 10 minutes which delay is necessary to attain equilibrium in bath wool acid, 2.5 kg. of sodium chlorite are added and the reaction is left at rest for one hour. The strands are rinsed with water and treated for 30 minutes in a solution containing 10 cc per liter of 30% sodium bisulfite, and 2 grams per liter of sodium hydrosulfite. To this bath, maintained at 60°C., are added 4 grams per liter of bicarbonate of soda, and then 0.2 gram per liter of papain. The reaction is allowed to proceed for 45 minutes. The yarn is then rinsed, and dried.

Example 4: 100 kg. of worsted wool are introduced into a stainless steel apparatus provided with a bath circulation device. 1,500 liters of water are introduced, then 0.5 kg. of a wetting agent (sodium lauryl sulfonate) and 5 kg. of 50% phosphoric acid diluted in 35 liters of water. After 10 minutes' agitation, there are added 3.5 kilos sodium chlorite in solution in 35 liters of water. The reaction is continued for one hour at room temperature, after which the wool is rinsed and treated at 50°C. in 1,500 liters of water to which are added 15 liters of a 30% sodium bisulfite solution and 3 kg. of sodium hydrosulfite of soda. The reaction is allowed to proceed for 30 minutes, after which 6 kg. of bicarbonate of soda and 300 grams of papain are successively added, the treatment lasting one hour at 50° to 60°C.

Finally, the wool is rinsed and dried. The wool subjected to the treatments described in detail in the preceding examples is white and is of a much higher grade than that which is obtained by treatment with acid bisulfite, without having the drawback inherent to the latter, i.e., wools which become yellow when stored. The white obtained by the process is also comparable or even superior to that obtained by treatment with hydrogen peroxide, whereas it deals with the wool more gently. Thus, the loss of 9% with soda for raw wool never exceeds 15% after treatment according to the process, and even remains much lower than this latter figure when hydroquinone is used, whereas the loss with soda in the hydrogen peroxide treatment is never less than 19% for an equivalent white.

The dynamometric strength decreases very little and remains on the average greater than that of wool bleached with hydrogen peroxide. Furthermore, the decrease in shrinkage on washing is substantial and by combining the chlorite treatment with a standard chlorination or with an enzyme treatment as shown in the examples, a wool is obtained which is very white and totally unshrinkable.

## HIGH-SPEED, TWO-STAGE BLEACH FOR COTTON

S.M. Rogers; U.S. Patent 3,265,462; August 9, 1966; assigned to Allied Chemical Corporation describes a high-speed, two-stage bleaching of cotton cloth utilizing sodium hypochlorite and hydrogen peroxide. The process eliminates the J-box, an expensive component of previous bleaching processes. In accordance with the process, high-speed two-stage hypochlorite-hydrogen peroxide bleaching to produce cotton cloth of overall superior properties is accomplished by treating the cotton cloth in a sequential manner under carefully controlled conditions involving: (A) immersing the cotton cloth for a period of time from 4 to 6 seconds, in an aqueous solution containing the combination of 0.3 to 0.5 grams per liter sodium hypochlorite and 0.4 to 0.8 grams per liter of a nonionic surface-active wetting agent at a temperature between 170° to 185°F., the aqueous solution having a pH value of about 9.0 to pH 12.0 and being substantially free of carbonate ions.

(B) The cloth impregnated with the solution containing sodium hypochlorite and the nonionic surface-active wetting agent is squeezed to 80 to 120% saturation; (C) thereafter directly and without steeping or washing, the cloth is immersed in hydrogen peroxide solution containing 3.0 to 5.0 grams per liter of sodium silicate as an additive and having a pH value of 10 to 11, at a temperature within the range of about 50° to 100°F. (D) The cloth impregnated with the hydrogen peroxide solution is squeezed to 80 to 120% saturation and the thus squeezed cloth is maintained at a temperature within the range of about 190° to 240°F., for a period of 3/4 to 1·1/2 hours; and (E) the thus treated cloth is washed with water. The bleaching process is highly advantageous, particularly in the following respects: (1) elimination of the steeping operation and the requirement of large and expensive retention equipment such as the J-box between the hypochlorite stage and the peroxide stage; (2) a two-stage hypochlorite-hydrogen peroxide bleaching process in which the cloth from the hypochlorite stage need only be squeezed before the hydrogen peroxide stage permitting direct and immediate transfer from the first stage to the second stage and thus providing high-speed processing; (3) a high temperature hypochlorite first stage in which not only substantial bleaching is rapidly accomplished but also one in which cloth degradation is not a problem affecting practical operation.

(4) In addition to elimination of the J-box, substantial economic benefit and low cost bleaching of cotton cloth are realized by low reagent requirements, the presence of the minor amounts of the nonionic wetting agent in the cloth transferred to the second stage hydrogen peroxide treatment effecting even a further reduction in reagent requirements; and (5) there is produced a bleached cotton cloth of overall superior properties including high whiteness, high absorbency and low fluidity, with fluidity and absorbency values improved over those obtained by the above referred to prior two-stage process. Figure 9.3 diagrammatically illustrates the process.

FIGURE 9.3: FLOWSHEET OF HIGH-SPEED TWO-STAGE BLEACHING OF COTTON CLOTH

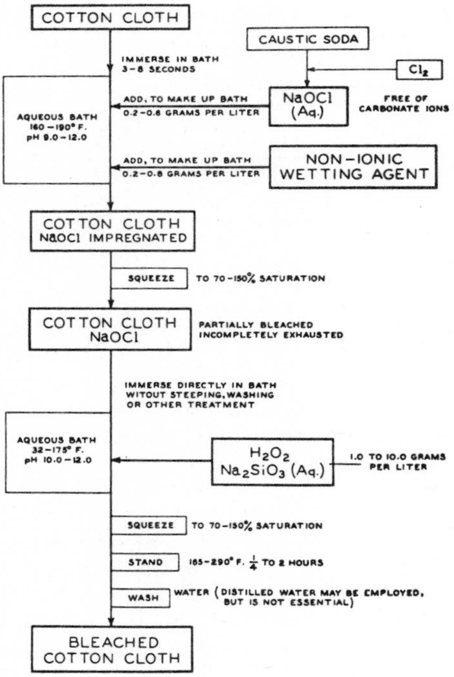


Source: S.M. Rogers; U.S. Patent 3,265,462; August 9, 1966

The whiteness of bleached cloth can be determined on a General Electric reflectometer with a blue filter. The results obtained are expressed as percentage reflectance as compared with reflectance of magnesium carbonate. A whiteness of 86% or higher is considered good for medium weight fabric.

Example: Cotton cloth of 80 x 80 thread count, 4.00 yards per pound weight, was prepared for bleaching by singeing, desizing, caustic treatment and washing in the conventional manner. A 5" x 7" sample of the prepared cloth was wet-out by immersing with mild agitation in an aqueous solution maintained at a temperature of 180°F. and containing 0.5 grams per liter sodium hypochlorite and 0.75 grams per liter Triton N-100 (isooctyl phenyl polyethoxy ethanol). Total immersion time in the hypochlorite solution was about 5 seconds. The hypochlorite solution had a pH of about 10.5. The cloth removed from the hypochlorite liquor was squeezed to about 120% saturation and without washing or other treatment was passed immediately into an aqueous hydrogen peroxide solution containing 1.68 grams per liter hydrogen peroxide and 8.5 grams per liter sodium silicate.

The peroxide solution had a pH of about 10.8. Retention time in the peroxide liquor which was at room temperature was about one minute. The saturated cloth removed from the peroxide liquor was squeezed to about 120% liquor saturation and allowed to stand for one hour at a temperature of about 212°F. which was maintained by indirect heating. The bleached cloth was then washed ten times in lukewarm water, pressed dry between blotters, pressed again between fresh blotters at 600 psi for about 30 seconds, and dried on a paper sheet drier for about 10 minutes. The cloth was tested for the properties of absorbency, fluidity and brightness with the results as follows. The absorbency of the fabric was 44.0 points, with an absorbency of 25 to 30 being considered acceptable and above 30 being considered excellent. The fluidity was 2.29 rhes, with values above 6 indicating degradation of the cloth fibers, and values below 4 indicating superior fiber strength. The brightness was measured as 87.7 percent reflectance with 86 or higher being considered good.

From the above data, it will be evident that the cotton cloth bleached in accordance with this process is of particularly high quality based on all properties desired in the cloth. Specifically, the brightness of the cloth attained a high level of 87.7 or greater than that obtained with straight peroxide bleaching. Of particular importance, the absorbency of the cloth attained a high value of 44 or greater than that obtained by straight peroxide bleaching or previous two-stage hypochlorite-peroxide processes.

Fluidity of the cloth bleached in accordance with the process was reduced to a substantially lower level of 2.29 indicating an exceptionally high quality cloth and improvement over both the conventional peroxide bleaching and prior two-stage processes. These results were attained during a short 5 second hypochlorite treatment with as little as 1/3 the concentration of hypochlorite previously employed in two-stage hypochlorite-peroxide treatment and with about 1/2 the normal concentrations of peroxide and silicate conventionally employed in straight peroxide treatment.

## SCOURING AND BLEACHING CELLULOSE

W.E. Helmick and W.H. Cooper; U.S. Patent 3,281,202; October 25, 1966;

assigned to Pittsburgh Plate Glass Company describe a process of scouring and bleaching cellulose textiles in four steps, an alkaline scouring, a hypochlorite bleach, a bisulfite bath and a final peroxide bleach. Cotton, linen and mixed textiles, dyed and undyed, scoured and bleached by this process give a high degree of brightness, maintain the required tensile strength and show no bleeding of the dyes. In the sequential alkaline bath treatments, the goods are subjected in the first bath to an alkaline solution containing small quantities of ammonium hydroxide. The alkaline solution is an aqueous solution of sodium carbonate or sodium hydroxide or a combination of the two.

In this bath the goods are boiled for periods of time ranging from fifteen minutes upwards. After completion of the initial boiling step, the goods are passed to a hypochlorite bleaching bath maintained at a pH of 9 to 12. Treatment in this latter bath is usually conducted at room temperature and for a period of time ranging from one-half hour upwards. After completion of the hypochlorite bleaching step, the goods are passed into a bath containing sodium bisulfite and maintained in an acid pH range of between 1 and 5. From this sodium bisulfite solution, the goods are passed into a hydrogen peroxide bleaching tank where they are bleached to final whiteness. For a more complete understanding of the process, reference is made to Figure 9.4.

Example 1: A desized piece of linen sheeting having a colored ingrain pattern was placed in a scouring bath containing 3.74 liters of solution, having 1.5% $NH_4OH$

FIGURE 9.4: FLOWSHEET OF SCOURING AND BLEACHING PROCESS

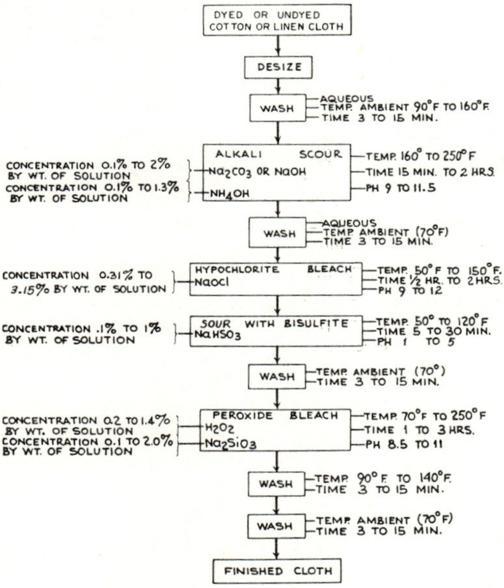


Source: W.E. Helmick and W.H. Cooper; U.S. Patent 3,281,202; October 25, 1966

by weight, 1% $Na_2CO_3$ by weight, and 0.1% Sandopan DTC (a sulfonated alcohol wetting agent). The souring bath was contained in a laboratory apparatus designed to simulate a commercial jig. The linen material was rotated continuously in the jig and was padded on the return side of the cycle and run at 200°F. for a period of thirty minutes. Upon completion of the thirty minute contact of the solution with the cloth, the solution was discharged from the machine and the material washed with cold water at ambient temperature (70°F.). A second aqueous solution of a 3.75 liter volume was prepared and contained 1.5% NaOCl by weight. The solution was introduced into the jig and the material rotated as before for a period of one hour at 95°F. Upon completion of this treatment, the solution was discharged and the material rinsed with cold water at 70°F.

A third solution of 3.74 liters volume was made up containing 0.1% sodium bisulfite by weight of solution. This solution was introduced into the apparatus and contacted with the cloth contained therein for a period of ten minutes at 70°F. Upon completion of this step, the solution was discharged from the apparatus and the material given a rinse in cold water at 70°F. A final aqueous solution of 3.74 liters by volume and containing 2% $H_2O_2$ (35%) and 0.5% $Na_2SiO_3$ was placed in the apparatus. The material was rotated with the solution at 180°F. for a period of two hours. This solution was then discharged and the material washed with hot water at 140°F. for a period of five minutes, and washed a second time with a cold water wash at 70°F. for a further period of five minutes. The material treated was designated as Sample Number 1.

Example 2: Utilizing the procedure of Example 1 and the solutions, times and temperatures remaining the same, a piece of cotton sheeting having a colored ingrain pattern was placed in the laboratory apparatus and subjected to the same treatment as the linen cloth of Example 1. The cloth so treated was designated as Sample Number 1A.

Example 3: Utilizing the solutions identical with those prepared in Example 1, a sample of linen sheeting having a colored ingrain pattern and a sample of cotton sheeting having a colored ingrain pattern were subjected individually to the identical bleaching sequence of Example 1, but all operations were conducted in a Gaston County Package dyeing machine. The linen sample obtained in this example was designated Number 2 and the cotton sample was designated Number 2A.

The samples obtained from Examples 1, 2 and 3 were subjected upon completion of the bleaching steps to tests for brightness, absorbency and tensile strength. Brightness was determined with a Hunter multipurpose reflectometer. Absorbency was measured by dropping a drop of water from a pipe held 2 inches from the surface of the unstretched cloth. The time required for the disappearance of the specular reflectance from a drop, as visually observed, is a measure of the absorbency.

| Sample No. | Reflectance Blue | Tensile Strength, lbs./in. | Absorbency, sec. |
|---|---|---|---|
| 68255-178- | | | |
| 1 | 70.2 | 88.5 | 1.0 |
| 1A | 84.3 | 62.7 | 1.0 |
| 2 | 69.7 | 82.5 | 1.0 |
| 2A | 85.4 | 58.2 | 1.0 |

## SODIUM CHLORITE FOLLOWED BY HYDROGEN PEROXIDE

Y. Sando and S. Mori; U.S. Patent 3,481,684; December 2, 1969; assigned to Sando Iron Works Co., Ltd., Japan describe a process and equipment to bleach singed and desized cotton fabric in a continuous process with sodium chlorite followed by hydrogen peroxide. In this process the fabric is first padded with a surface active agent containing acidified chlorite solution of a pH of about between 3 to 3.5. The thus padded fabric is then heated in a steam atmosphere of about 80° to 85°C. in a substantially closed system, whereupon the fabric is again passed through a heated chlorite solution of a pH of about 3 to 3.5 in folded, unstretched condition in the same closed system, thereby maintaining the pH value substantially constant throughout. After washing, the fabric is padded with an alkaline hydrogen peroxide solution, heated in a steam atmosphere in a substantially closed system and passed through a hot alkaline hydrogen peroxide solution, washed and dried.

In order successfully to carry out the process it is essential that the pH value of the two chlorite baths or solutions with which the fabric comes into contact is between about 3 to 3.5 and that this pH value is maintained throughout the treatment. Further, it is an important feature of this process that the steaming of the fabric and the second chlorite treating step are carried out in the same closed system. The processing plant consists of a number of vessels for the various steps, such as for the washing, the chlorite and hydrogen peroxide bleaching, the final washing and drying.

Example: A previously singed fabric is subjected to the following successive steps: (1) Simultaneous Desizing and Refining — In this step the fabric is maintained in an extended unfolded state. A length of cotton stuff is washed by passing through hot water in a tank at about 80° to 90°C., the time of stay in the bath amounting to 15 to 30 seconds. After squeezing to a pickup rate of less than 70%, the fabric is conducted into a saturator containing:

| | Percent |
|---|---|
| Sodium hydroxide | 1.5 to 2.5 |
| An alkali salt or an ester of Caro's acid | 1 to 2 |
| An anion type of surface active agent as detergent | 0 to 3.5 |

The percentage was based upon the starting weight of the fabric. The fabric is then squeezed to a pickup rate of about 100%, conveyed into a closed type of reaction tower and folded therein and subjected to steam heating for a period of 20 to 45 minutes at about 100°C. Thereafter, the fabric is conducted into a second tank arranged in the lower part of said reaction tower and communicating with the lower space. The tank holds a solution of the aforementioned chemicals at a concentration of about 1/3 to 1/5 of that in the first tank; the addition of chemicals to the tank content is performed only once at the start. The fabric is subjected in this solution to a temperature of about 90° to 100°C. for a period of 5 to 10 minutes. Then, the fabric is drawn out from the tank, washed, brought to neutral reaction, again washed and squeezed.

(2) Bleaching by Chlorite — After finishing the desizing and refining step of 1 above, the fabric is fed into a saturator containing a solution which is prepared so as to contain the following ingredients.

| | Percent |
|---|---|
| Sodium chlorite (as 85% solution) | 1.0 to 1.5 |
| An anionic or nonionic surface active agent or a mixture of both | 0.1 to 0.3 |

The percentage was based on the starting weight of the fabric, and adjusted to a pH of approximately 3 to 3.5 by the addition of formic acid or another organic acid. The fabric is then squeezed to a pickup rate of about 100% and steam heated at 75° to 90°C. for 15 to 25 minutes in a reaction tower of the same construction as employed in step 1, the heating time being variable according to the kind of fibers constituting the fabric. After transferring into a tank in the bottom of the tower space and containing a solution of the abovementioned chemicals at a concentration of 1/3 to 1/5 of that in the saturator and steeping in this solution at about 90° to 100°C. for 5 to 10 minutes, the fabric is drawn out from the tank, washed and squeezed (in some cases, it is possible to finish both the steps of washing and squeezing in a total period as short as 5 to 6 minutes).

(3) Treatment with Hydrogen Peroxide — The object of this step is to obtain various effects such as dechlorination, further increases of bleaching effect, securing of permanent whiteness and higher degree of grease removal (degree of refining). Thus, the fabric emanating from the process of step 2 above is fed into a saturator containing a solution of:

| | |
|---|---|
| Hydrogen peroxide (as 35% solution), percent | 0.3 to 0.5 |
| Sodium hydroxide, g./l. | 0.06 to 0.1 |
| Sodium metasilicate (water glass), g./l. | 0.3 to 0.5 |

The percentage was based on the starting weight of the fabric, and had a pH adjusted to about 11. After squeezing to a pickup of about 100%, the fabric is steam heated at about 90°C. for 15 to 20 minutes in the aforementioned reaction tower, cooked therein at a temperature of about 100°C. for 5 to 10 minutes in a hot solution containing the abovementioned chemicals at a concentration of 1/3 to 1/5 of that in the saturator and then washed and dried. (Also with regard to this process step, it is possible in certain circumstances to finish both steps of washing and drying in a total period as short as 5 to 6 minutes.)

## DESIZING AND BLEACHING CELLULOSE

H. Grunow; U.S. Patent 3,572,987; March 30, 1971; assigned to Entreprise Miniere et Chimique, France describes a continuous process to desize and bleach cotton textiles by successive exposure to amylase, trichloroisocyanuric acid and finally alkaline sodium peroxide solution. The process can be operated at a feed rate of the textile of 50 to 100 meters per minutes.

Example: A starch-sized, cotton poplin fabric 140 centimeters in width and weighing 200 grams per meter was subjected to this process without previous desizing. The fabric was fed to a castor-tank having a capacity of 20 meters of the fabric at a rate of 80 meters per minute. The bath contained 5 grams of amylase, 10 grams of sodium chloride, and 1 gram of a wetting agent based on the octylester of sulfosuccinic acid, per liter. The temperature of the bath was 70°C. and the immersion time 15 seconds. The fabric was squeezed to a 70% wet takeup after leaving the tank and then fed to a washing tank containing water at 100°C. and having a capacity of 15 to 20 meters of the fabric. Amylase starch size were removed by this washing. The fabric was then squeezed to a 55% wet takeup and fed to a second castor-tank containing a solution at 30°C. of 8.25 grams per liter of trichloroisocyanuric acid, 2.29 grams per liter of cyanuric acid, and 4.46 grams per liter of sodium bicarbonate. The fabric remained in this tank for 15 seconds and was then squeezed to 70% wet takeup and passed to a washing tank containing water at 100°C., where it was washed for 10 seconds and the excess isocyanuric acid derivative was removed. The fabric was then squeezed to 55% wet takeup and passed to a third tank; it was soaked in a solution at 30°C. containing 15 grams per liter of sodium hydroxide, 25 grams per liter of 36° Bé. sodium silicate solution, 7 grams per liter of a wetting agent based on a condensate of a fatty acid with a depolymerized protein, and 30 milliliters of 35% aqueous hydrogen peroxide.

The concentration of this bath was kept constant by adding to it as necessary a solution containing the same ingredients in triple concentration. The fabric after leaving this bath was squeezed to a 75% wet takeup and then passed vertically through a steamer at 100°C. It was then fed to a maturing chamber having a capacity of 1,000 meters and kept therein for 10 minutes at 100°C. The fabric was then passed to a series of five washing tanks, the first of which contained sodium hydroxide solution in a concentration of 10 grams per liter at 100°C. and the other four hot water. Very white, completely desized fabric of excellent hydrophilicity was obtained. The polymerization degree of the cellulose was reduced only from 2,700 for the unbleached cotton to 2,100 for the bleached fabric.

## CHLORINE DIOXIDE PLUS HYDROGEN PEROXIDE

R.J. Borezee; U.S. Patent 3,619,110; November 9, 1971; assigned to Ugine Kuhlmann, France describes a bleaching process to improve water absorbency and whiteness of cotton fabrics by a pretreatment with chlorine dioxide followed by a hydrogen peroxide bleaching step. The treatment with chlorine dioxide involves contacting the cellulosic material with a small amount of chlorine dioxide in the gaseous state or in solution for a very short period of time. If the chlorine dioxide is used in the gaseous state, it is advantageously diluted with air or inert gas, such as nitrogen, for example, and can be applied by passing the cellulosic material which had been previously moistened with water through a chamber containing the gaseous medium in proportions so that the cellulosic material picks up between about 0.05 to 0.5% by weight chlorine dioxide under the particular operating conditions employed.

The temperature and residence time of the cellulosic material in the chamber containing the diluted chlorine dioxide gas can be varied, depending upon the particular concentration of the chlorine dioxide gas contained therein, the form of the cellulosic material, as well as the type of cellulosic material, as will be apparent to those skilled in the art. Generally, the temperature in the chamber when the diluted chlorine dioxide gas is used can be between about 10° and 130°C. and the residence times can vary from between a few seconds to about 30 minutes. The cellulosic materials can also be treated with an aqueous solution of chlorine dioxide by soaking the cellulosic material in the chlorine dioxide aqueous solution. Ordinary bleaching and dyeing equipment, such as vats, jiggers, washing machines, and so forth, can be used for treating the cellulosic materials with chlorine dioxide in aqueous solutions. The chlorine dioxide aqueous solution can also be applied to the cellulosic materials by impregnation with foulardage and then let the cellulosic material stay in equipment of the J-box type, a pad-roll type maturation chamber, or any other suitable accumulation system. The application of the chlorine dioxide to the cellulosic material can also be carried out in an autoclave to permit heating to higher temperatures and at pressures of from about 1 to 4 atmospheres if this ever becomes desirable. When utilizing an aqueous solution of chlorine dioxide, the amount of chlorine dioxide used should result in a pickup on the cellulosic material of from about 0.05 to 0.5 parts by weight of chlorine dioxide per 100 parts of the textile material during its passage through a residence in the chlorine dioxide containing aqueous solution under the particular operating conditions employed.

When using an aqueous solution of chlorine dioxide, it has been found to be advantageous to use an aqueous solution of chlorine dioxide having a pH of between about 2 and 7 and to carry out the process at a temperature of between about 10° and 130°C. The cellulosic material can be passed through a hot solution of chlorine dioxide or, if desired, it can be immersed in a cold solution of chlorine dioxide, and the temperature of the solution raised to the desired temperature. For example, in the pad processes the cellulosic material can be impregnated with a cold solution of chlorine dioxide and after drying or removal of excess solution, the cellulosic material can be heated with steam or some other heating system and placed in a J-box, for a period of time which can vary from a few seconds to 30 minutes, depending upon the particular cellulosic material being treated and the process conditions.

After pretreatment with chlorine dioxide the textile is treated, with or without intermediate washing, with the peroxide bleaching solution. Apart from peroxide, the latter contains the usual alkaline agents: caustic soda and sodium carbonate, the usual stabilizers: sodium silicate and magnesium silicate, and/or organic sequestrants, such as, ethylenediamine tetracetic or diethylenetriamine pentacetic acid and their alkaline or alkaline-earth salts. Bleaching can be effected in ordinary equipment kiers with or without pressure, vats, jiggers, impregnation tanks and any thermal accumulation system of the J-box and pad-roll types or in equipment for pressure bleaching in a steam medium. The duration of bleaching with hydrogen peroxide can vary but generally takes between a few minutes and 30 minutes. The temperature at which bleaching is carried out can also vary and can generally lie between 80° and 140°C.

Example: An Egyptian cotton cloth, previously desized with malt diastase and washed in boiling water, was impregnated at the ambient temperature with a chlorine dioxide solution prepared by the action of sulfuric acid on a sodium chlorite solution and containing 2 g. per liter of chlorine dioxide. The cloth was then dried by passing between the rolls of a pad so that it retained only its weight of liquid, after which its temperature was rapidly brought to 100°C. by passing it through a steam-heated chamber. It was then stored for 5 minutes in a thermally insulated J-box. The cloth was then washed in cold water, then squeezed out and impregnated with a cold solution of hydrogen peroxide containing, in addition, caustic soda and sodium silicate.

The concentration of this solution and the degree of squeezing (drying) of the cloth were so calculated that, after passing between the rolls of the foulard, there remained on the cloth 1% hydrogen peroxide, 0.5% caustic soda, and 2% sodium silicate at 36°Bé. relative to the weight of the cloth. The latter was heated by passing through a steam-filled chamber and then kept for 10 minutes in the J-box. Finally, it was washed first in hot, then in cold water. The hydrophilic property of the cloth was checked by the ASA method No. L 14115-1961 which consists in letting a drop of water fall from a buret fixed 1 cm. above the stretched surface of the cloth and in measuring the time taken by the textile to absorb the drop. Cloth treated as described above absorbs the water in 2 seconds (mean of 20 tests) whereas the same cloth which had only been desized before being treated with hydrogen peroxide only absorbed the water in 114 seconds (mean of 20 tests). By way of comparison a sample of the same cloth desized, boiled in soda, and then bleached with peroxide absorbs the water in 6 seconds. The degree of whiteness, measured with the Zeiss Elrepho electrophotometer using a 457μ filter, was 82.2 (relative to a magnesium oxide standard) for the cloth pretreated with chlorine dioxide and 80.1 for the nonpretreated cloth. The degree of polymerization of the cloth pretreated with chlorine dioxide was 1,830, that of the nonpretreated cloth 1,850.

## LAUNDRY PROCESS USING PEROXIDE AND HYPOCHLORITE

L.F. Luechauer; U.S. Patent 3,650,667; March 21, 1972; assigned to Steiner American Corporation describes two variations of a laundry process to give bacteria and spore free wash even on heavily soiled cloth. The following are the two methods: Method (A) for laundering fabrics comprises washing fabrics in hot water at a medium water level in the presence of soap and a sequestering agent; flushing the fabric load; washing the fabric at a low water level for about 10 to about 15 minutes at a temperature of about 175° to about 195°F. in the presence of a strong alkali and soap or a detergent; flushing the fabric load; bleaching the fabric load in hot water at a medium water level with an oxygen-containing bleach which is hydrogen peroxide or a material that forms hydrogen peroxide or nascent oxygen in the presence of water at a concentration of about 0.1 to about 0.2% by weight of the dry fabric weight, the bleaching being carried out at a temperature of at least about 175°F. for about 6 to 10 minutes.

This is followed by rinsing the fabric load; souring the fabric load in the presence of a souring agent and a water-soluble germicide at a concentration level sufficient to inhibit the growth of mildew in the laundered fabric, and extracting the water from the soured fabric load. Method (B) for laundering fabrics comprises washing fabrics in hot water at a medium water level in the presence of soap and a sequestering agent; flushing the fabric load; and treating the fabric load at a low water level at a water temperature of about 140° to about 160°F. in the presence of an oxygen-containing bleach which is hydrogen peroxide or a material which forms hydrogen peroxide or nascent oxygen in the presence of water for a time sufficient to obtain substantially uniform penetration of the oxygen-containing bleach into the fabric with the concentration of oxygen-containing bleach ranging from about 0.1 to about 2% by weight of the dry fabric.

This is followed by increasing the temperature of the fabric load to at least about 175°F. in the presence of a strong alkali and soap or a detergent for about 10 to about 15 minutes; rinsing the fabric load; souring the fabric load in the presence of a water-soluble germicide at a concentration level sufficient to inhibit the growth of mildew in the laundered fabric, and extracting the water from the soured fabric load. In either of the above methods (A) and (B) a hypochlorite bleach step may be used subsequent to the treatment with the oxygen-containing bleach and prior to the souring treatment to remove stains from the fabric.

## OXYGEN-PERACETIC ACID-CHLORINE DIOXIDE FOR WOOD PULP

S.K. Roymoulik; U.S. Patent 3,695,995; October 3, 1972; assigned to International Paper Company describes a three-stage bleaching process for chemical wood pulps which comprises a first stage of bleaching with oxygen in the presence of $Na_2S_2O_4$ or $Na_2S_x$, wherein x is an integer from 1 to 4, and optionally NaOH, followed by a peracetic acid second stage, resulting in a pulp having a brightness of about 60% or above which is suitable for newsprint furnish. A further bleaching stage employing chlorine dioxide results in a pulp having a brightness of about 80% or above which is suitable for dissolving pulp and writing and bond papers. It has been found that when a chemical pulp is bleached in the first stage with oxygen in the presence of a compound selected from the group consisting of $Na_2S_2O_4$ and $Na_2S_x$ and, optionally, sodium hydroxide, followed by a second stage employing peracetic acid, a pulp having a brightness of about 60% or above is obtained.

If the second stage, i.e., peracetic acid bleach stage, is followed by a third stage employing chlorine dioxide, either gaseous or in solution, the resulting pulp has a brightness in excess of 80%. In the first stage an unrefined, digested chemical pulp, prepared by either the sulfate or sulfite process, and having a consistency of from about 3 to 35% and preferably above 20%, based on the weight of the oven-dried pulp, is contacted or reacted with oxygen, at a temperature of from about 70°C. to about 140°C., preferably about 100°C., and a pressure of from about 60 psi to about 130 psi, preferably about 100 psi, for approximately one-half hour to about eight hours in the presence of at least 0.2%, based on the weight of the pulp, of a "protector" compound selected from the group consisting of polysulfides having the formula $Na_2S_x$, wherein x is an integer from 1 to 4, and $Na_2S_2O_4$.

In the description of this process the use of the term "protector" compound shall mean either an oxidizing or reducing agent which is capable of preventing the degradation and depolymerization of the cellulosic portion of the pulp by reacting with the carbonyl end groups, to convert them into hydroxyl groups when $Na_2S_2O_4$, a reducing agent is used, or carboxyl groups, when $Na_2S_x$, an oxidizing agent is used. $Na_2S_4$ is the preferred polysulfide. The concentration of $Na_2S_2O_4$ or $Na_2S_x$ should be at least about 0.2% and preferably from about 2% to about 10%, based on the weight of the oven-dry pulp. If concentrations in excess of 10% are employed, costs are increased and no additional benefits are obtained.

It should be noted that since the protectors contain only sodium and sulfur ions, both of which are present in sulfate or sulfite pulping systems, the effluent can be fed back into the chemical recovery system thereby preventing stream pollution which would otherwise occur if the effluent were emptied into the stream, as is conventionally done. If pulps having a lignin content of 5% or above are employed, the addition of from about 1 to about 5% of sodium hydroxide to the reaction mixture can speed up the reaction rate. Also, if the concentration of $Na_2S_2O_4$ or $Na_2S_x$ is 2% or less, sodium hydroxide can be employed to good advantage.

After water washing followed by centrifugation or filtration, the partially bleached pulp of the first stage is subjected in the second stage to a bleach with peracetic acid. The concentration of the peracetic acid can be from about 0.5 to about 2.0%, preferably 1.0%, based on the weight of the oven-dried pulp, and the pulp consistency can be from about 10 to about 20%, preferably between about 10 and about 15%. The reaction can be carried out at a temperature of from about 50° to about 90°C., with 70°C. being preferred. During the peracetic acid second stage, which can take from about 30 minutes to 120 minutes, the initial or starting pH must be within the range of from about 8.0 to about 12.0, with a range of from about 8.0 to about 10.0 being preferred. The pH can be adjusted by adding $Na_2CO_3$ and/or NaOH. The initial pH must be adjusted to 8.0 or above to optimize the bleaching results. The pulp after the first and second stages, namely, oxygen (O) and peracetic acid (Pa), has a brightness of 60% or above and can be utilized as newsprint furnish. In order to improve pulp brightness to 80% or above, so that it can be used as a dissolving pulp or in the manufacture of bond papers, the pulp is treated in a third stage with chlorine dioxide. Either gaseous chloride dioxide or chlorine dioxide in solution can be employed.

When a solution of chlorine dioxide in water is used, the consistency of the pulp can be from about 10 to about 20%, preferably between 10 and 15%, based on the weight of the oven-dried pulp. The concentration of the $ClO_2$ can be from about 0.5 to about 2.0%, based on the weight of the oven-dried pulp. The reaction can be carried out at a temperature of from about 50°C. to about 90°C., preferably 70°C., and at a starting pH between 5.0 and 6.0 for a period of from about 30 minutes to about 120 minutes. When gaseous chlorine dioxide is used, the pulp consistency can be from about 10 to about 40%, with a consistency above 30% being preferred. The reaction time can be reduced to about 5 to 10 minutes and can take place at a temperature of from about 20° to about 90°C., with 70°C. being preferred.

Example: A 100 gram sample of a hardwood pulp, which had been prepared in accordance with the kraft process and which had then been washed with deionized water, was bleached as follows. (The first stage was carried out in an electrically heated pressure reactor.)

1st Stage

| | |
|---|---|
| Reaction temperature | 100°C. |
| Reaction pressure ($O_2$) | 100 psi |
| Reaction time | 3 hours |
| Pulp consistency | 20 to 22% |
| $Na_2S_2O_4$ (hydrosulfite) | 2% on O.D. pulp |
| NaOH | 2% on O.D. pulp |

The pulp was then thoroughly washed with water, centrifuged and further bleached in accordance with stage 2 below.

2nd Stage

| | |
|---|---|
| Peracetic acid | 1% on O.D. pulp |
| Pulp consistency | 10% |
| Starting pH (adjusted with NaOH) | 8.0 |
| Reaction temperature | 70°C. |
| Reaction time | 2 hours |

The semi-bleached pulp was thoroughly washed with water and centrifuged. The brightness of the pulp was found to be 66% as measured in accordance with Tappi Standard Method T217.

3rd Stage

| | |
|---|---|
| Chlorine dioxide | 0.5% on O.D. pulp |
| Pulp consistency | 10% |
| Starting pH | 5.5 to 6 |
| Reaction temperature | 70°C. |
| Reaction time | 2 hours |

The fully bleached pulp was then washed with water, centrifuged and evaluated. This process results in a fully bleached pulp of a brightness comparable to that obtained by a five-stage bleaching sequence and having a higher opacity. The physical properties are equivalent to those obtained by a conventional bleaching sequence.

# MISCELLANEOUS BLEACHING AND STAIN REMOVAL

## SPECIAL BLEACHING PROCESSES

### Steam and Nitrogen for Textiles

C.A. Meier-Windhorst; U.S. Patent 3,411,862; November 19, 1968; assigned to Artos Dr. Ing. Meier-Windhorst KG, Germany describes a process for bleaching and dyeing of textiles in heated chambers filled with steam and nitrogen. The process provides for a continuous refining treatment of lengths of textile materials of all types, where the length of textile material which is continuously impregnated with a treating medium, is introduced into a heat treating chamber which is filled uniformly with a mixture of steam and an inert gas, such as nitrogen, the ratio of the mixture being such that the actual saturation temperature of the partial pressure of steam corresponds to the real treating temperature of the refining process.

Preferably, a mixture of steam and inert gas is continuously introduced into the heat treating chamber, whereby the desired partial steam pressure is preferably maintained by suitably varying the amount of the inert gas by measuring the temperature indicated by a wet bulb thermometer close to the inlet of the length of material. It was also found advisable to conduct a certain excessive amount of the mixture of steam and inert gas in countercurrent over a heating zone for the material which is located close to the inlet of the material.

The excessive amount of the mixture of steam and inert gas should leave the treating chamber or the heating zone through a very narrow inlet slit or, when the material continuously passes through the treating chamber, through an equally narrow outlet slit, whereby the actual flow resistance in the treating chamber produces a slight excess pressure which prevents uncontrolled penetration of air, and consequently of oxygen. To diminish consumption of inert gas the length of material can be introduced into the treating chamber through a special seal serving slit and also, preferably, withdrawn through such a slit. It was found that a dyeing process following this process completely eliminates many difficulties heretofore encountered in bleaching, formation of sodium cellulose, and in dyeing with vat dyes, for example when dyeing close mercerized poplins.

## Hydrogen Peroxide for Rawhide

L.E. Nordstrom; U.S. Patent 3,450,483; June 17, 1969; assigned to Superior Pet Products, Inc. found that rawhide can be bleached to a color similar to animal bone by hydrogen peroxide at 9 to 11 pH. The bleached rawhide is used to make simulated animal bones. The bleaching process is as follows.

Bellies are cut from hides which have been treated in a beam house operation which would make the hides suitable for vegetable tanning. These operations include the steps of washing, soaking, liming, unhairing, fleshing, scudding, and washing and bating (partly deliming). These operations are carried out to the point that the hides have a pH of approximately 10 to 10.5. The stated pH is only approximate, because hides which have been only partly delimed do not have a uniform cross-sectional pH. If the hides at this stage fall within this pH range, then the strongest bleaching effect will be obtained in the steps that follow. Bating may be done, and if so, the key is to take the hides for the cutting and subsequent steps of the method at the point where the hide cross section pH is within the above range.

The bellies and their contained water are then weighed and placed in a rotatable drum having means to permit the flow of water from the drum, such as by using a conventional drum with a slotted door. Means are also provided for flowing water into the drum, either by the same slotted door or by other means. This drum with its bellies is now rotated at a speed of approximately 12 rpm, this speed not being critical. This speed can be varied from 3 to 25 rpm, but at the higher speeds there is a tendency for the bellies to tie together in knots. At lower speeds there will be less of a washing effect, and therefore washing must continue for a longer period of time. During this drum washing, water at a temperature of 60° to 70°F. is flowed into and out of the drum continuously, the temperature of the bellies and water being maintained within this temperature range.

Basically, the purpose of the washing is to get rid of excess enzyme residues, and while probably not all enzyme residues are removed, nevertheless sufficient residues will be removed so that the bellies will not decompose during the remainder of the operations, or in the finished product of this process. As an example of a practical operation, the stated speed of 12 rpm with water at a temperature of 60° to 70°F., washing being done for approximately 30 minutes, will provide sufficient enzyme residue removal where the drum size is 8' x 8', the hide load is approximately 4,000 pounds, and the water flow into the drum and out is approximately 50 gpm. The values given of load weight, water flow, temperature and drum revolution speed will give excellent production from the indicated drum size of 8' x 8'.

After the above washing is finished, excess water is drained from the drum and its contents. There is then added to the drum and the contained hides 12% (by weight of the bellies and their contained water) of hydrogen peroxide of 35% strength in water. If the drainage has been complete, then enough water should be added to obtain a final 17% concentration of hydrogen peroxide in the drum.

The above given figures of 12 and 17% are critical for optimum results. The final concentration of the hydrogen peroxide may fall, if desired, within the range of 9 to 15%, the final color and substance of the finished product being dependent on the actual percentage of the hydrogen peroxide concentration. Having obtained the 17% hydrogen peroxide concentration in the drum, the latter is then rotated (with a tight door in order to prevent loss of the bleaching solution) intermittently for 4 hours. The procedure is to rotate the drum for 5 turns during a period of approximately 30 seconds, stop the drum for a period of approximately 14 1/2 minutes, and then rotate 5 turns again, and so forth, this intermittent procedure being carried out for 4 hours. The speed of rotation of the drum is 12 rpm.

At the end of the 4 hours of drumming time, water is added at 105°F. temperature, the weight of water added being equal to the weight of the bellies with their contained water ascertained just before the above-described washing operation. The 100% figure is not critical, and can range from 50 to 105%. Thereafter, the drum is again rotated intermittently for another 4 hours, using the same schedule of 5 rapid turns in approximately 30 seconds, a rest period of approximately 14 1/2 minutes, and then successive periods of like kind during the 4 hours.

The bellies are then removed from the drum and transferred to a suitable area (for example, to a storage area), where they are stored for a period of 8 to 12 hours. The purpose of this is to allow drainage of the water and $H_2O_2$ from the bellies, and this drainage procedure is recommended as a practical matter for a production scale operation. However, if desired, the bellies could be wrung out, for example twice, and the same result as storage for the 8 to 12 hours could thus be obtained.

After drainage, the excess gas and water are pressed out and skiving off of any excess flesh is done in a wringer-type setting out machine. The purpose of wringing the bellies is to put them in proper condition for the folding and wrapping operations used in making simulated animal bones. It is not necessary to skive off any excess flesh, but the absence of excess flesh aids in the operations of folding, gives less trouble in the subsequent drying operation, and results in a final product (the simulated bone) which needs no cleaning. At this point, for optimum results in the finished bone product, the hide cross-sectional pH should be 7 to 8.5.

Bleaching Cotton Without Oxidizing Agents

U. Kirner and B. Hartmark; U.S. Patent 3,476,505; November 4, 1969; assigned to Badische Anilin- & Soda-Fabrik AG, Germany describe a process of bleaching fibrous material of natural cellulose in the absence of oxidizing agents by treating the fibrous material with a liquid medium consisting essentially of an aqueous solution of certain amounts of (1) alkali metal hydroxide, (2) alkali metal polyphosphates and/or alkali metal salts of aminopolycarboxylic acids, and (3) anion-active and/or non-ionic wetting agents.

The process is carried out by treating fibrous material of natural cellulose in the absence of oxidizing agents at a temperature of 90° to 150°C. for a period of from half a minute to 5 hours, preferably 1 minute to 5 hours, with an aqueous liquor which contains 4 to 13% by weight of alkali metal hydroxide together with 1 to 4% by weight of alkali metal polyphosphate and/or complex-forming alkali-metal aminopolycarboxylate, 0.5 to 3.0% by weight of anion-active and/or nonionic wetting agent which is active in alkaline solution, the treatment temperature t (measured in °C.) within the specified range being so correlated to the treatment period r (measured in minutes) that it lies within the limits $t_l$ and $t_u$ ($t_l$ denoting the lower limit, $t_u$ denoting the upper limit, both limits inclusive) defined by the equations listed on the following page.

(1) $t_l = 100 - 20(\log r - 1.5)$ (2) $t_u = 100 - 20(\log r - 4)$

The material thus treated is freed from alkali metal hydroxide, all percentages being with reference to the liquor. Instead of the alkali metal aminopolycarboxylate the free aminocarboxylic acid can be used inasmuch as it is converted into the alkali metal salt in the alkaline liquor. In principle the hydroxides of all the alkali metals may be used for the process. Potassium hydroxide and sodium hydroxide are preferred because of their easy accessibility and their very good efficacy.

Alkali metal polyphosphates are defines as those which contain two or more phosphorus atoms in the molecule, combined together linearly or reticularly. Polyphosphates having the general formula: $M_{n+2}P_nO_{3n+1}$ (in which M denotes an alkali metal ion, preferably a sodium or potassium ion, and n denotes one of the integers from 2 to 10), for example sodium or potassium pyrophosphate, tripolyphosphate, tetrapolyphosphate and hexapolyphosphate and commercial mixtures of these polyphosphates are particularly emphasized. The polyphosphates of this type may be used alone in the amounts stated above. These alkali metal polyphosphates are often referred to as alkali metal molecularly dehydrated polyphosphates. Among the complex-forming aminopolycarboxylic acids, those having the general formula:

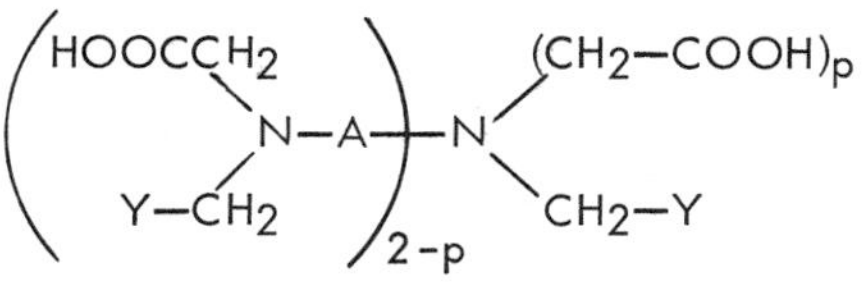


are preferred. In this formula p represents one of the integers 1 and 2, A denotes a radical having the formula:

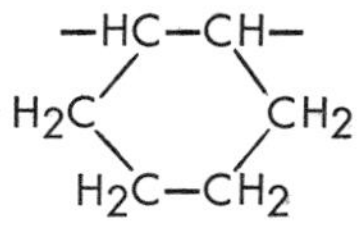


or preferably

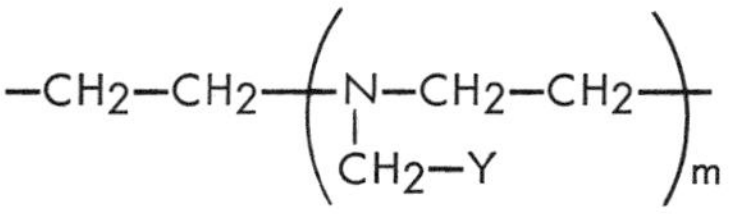


m denotes zero, 1 or 2 and Y denotes identical or different groups having the formula —COOH or $—CH_2OH$.

Examples of complex-forming aminopolycarboxylic acids are: N-hydroxyethylethylene diamine triacetic acid, o-cyclohexylene diamine tetracetic acid, diethylene triamine pentacetic acid, triethylene tetramine hexacetic acid, N-hydroxyethyldiethylene triamine tetracetic acid, nitrilotriacetic acid and particularly ethylene diamine tetracetic acid. The above-mentioned compounds may be used as alkali metal salts or as free acids, inasmuch as the acids are converted into the salts in the alkaline liquor.

The complex-forming aminopolycarboxylic acids or their alkali metal salts may be used alone instead of the polyphosphates in the amounts stated above. It has proved to be particularly suitable however to use the polyphosphates together with complex-forming aminopolycarboxylic acids or their alkali metal salts, the total concentration of the polyphosphates and aminopolycarboxylic acids or their salts in the treatment liquor being 1 to 4% by weight. Especially good bleaching results are achieved by using alkali metal polyphosphates and complex-forming aminopolycarboxylic acids or their salts in the weight ratio of 1:4 to 4:1. The wetting agents may be conventional substances of this type which are active in alkaline solution.

Examples of wetting agents which have proved to be very suitable are the following: sodium salt of the disulfonic acid of kogasin, diethanolamine salt of dodecylbenzenesulfonic acid, sodium salt of bis(decanesulfonimide), sodium salt of a sulfonated $\alpha,\beta$-olefin having 12 to 14 carbon atoms, sodium salt of the sulfuric acid half-ester of an adduct of 1 mol of nonyl phenol and 4 mols of ethylene oxide, adducts of 1 mol of colophony to 25 mols of ethylene oxide, of 1 mol of octyl phenol to 8 or 9 mols of ethylene oxide, of nonyl phenol to 10 mols of ethylene oxide, and of fatty alcohols of medium chain length, such as coconut fatty alcohol, to 7 to 9 mols of ethylene oxide, the condensation product of 2 mols of $\beta$-naphthalenesulfonic acid and 1 mol of formaldehyde, and particularly mixtures of several of these substances.

The term "kogasin" denotes a hydrocarbon fraction from the Fischer-Tropsch synthesis having a boiling range of about 200° to 300°C. These mixtures are therefore preferred. The wetting agents may furthermore contain antifoaming agents, such as triisobutyl phosphate, in the usual way.

It is essential for the process that the treatment temperature should be within the range of from 90° to 150°C. and should be correlated to the treatment period so that it does not fall short of or exceed the limits set by the above-mentioned equations (1) and (2). This means that for a given treatment period r, the higher of the two lower limits "90°C." and "$t_l$" and the lower of the two higher limits "150°C." and "$t_u$" determine the temperature range to be used. When falling short of or exceeding the working range thus defined, the effect of the treatment falls off so rapidly and so markedly that any appreciable whitening of the fibrous material no longer takes place. Following the treatment, the fibrous material is freed from alkali metal hydroxide by conventional methods, for example by thorough rinsing and if necessary by acidification.

The process may be carried out batchwise or preferably continuously. In continuous operation the procedure may preferably be to carry out the treatment for 30 seconds to 15 minutes, preferably from 1 to 15 minutes, at temperatures of from 120° to 150°C.

The process permits the achievement, without oxidizing bleaching agents, of a bleaching effect which is better than that with a medium oxidizing bleach without damage to the fiber which occurs in oxidizing bleaching. An even better degree of whiteness may be achieved by adding reducing bleaching agents, such as sodium dithionite, $\alpha$-hydroxyalkanesulfinic acids and their salts, reaction products of $\alpha$-hydroxyalkanesulfinic acids with ammonia or amines and salts of such reaction products, in amounts of at least 2.5% by weight, to the bleaching liquor but for reasons of economy it is preferred not to use reducing agents in amounts which produce an appreciable additional bleaching effect.

The addition of small amounts of reducing agent which are insufficient to effect bleaching, preferably of amounts up to 1.3% by weight, will however provide a very effective protection of the fibrous material from the destructive influence which atmospheric oxygen may exert under the working conditions. On the other hand, to achieve a strong additional bleaching effect it is very much more advantageous to carry out an additional reducing bleach following this process.

Example 1: Desized and dried unbleached cotton cloth is impregnated with a solution having the following composition: 7.7% of sodium hydroxide, 2.0% of the tetrasodium salt of ethylenediamine tetracetic acid, 2.0% of a mixture of 35 parts of the diethanolamine salt of dodecylbenzenesulfonic acid and 18 parts of an adduct of 25 mols of ethylene oxide to 1 mol of colophony and 1.0 part of pine oil, and 0.3% of a mixture of 59 parts of sodium kogasin disulfonate and 8 parts of triisobutyl phosphate.

The material treated with this solution is squeezed out on a padding machine to 100% liquor retention and treated with saturated steam at 135°C. for 10 minutes in a cottage steamer. It is then rinsed with desalted water twice at 100°C. and once at 60°C. It is then acidified with an aqueous solution which contains 3 to 5 ml. per liter of concentrated hydrochloric acid and again rinsed at room temperature. A material entirely free from husks is obtained having a degree of whiteness which varies from 83.0 to 86.0% (measured on the Elrepho with filter R 46 T) depending on the quality of the cotton.

Example 2: Desized and dried unbleached cotton cloth is impregnated with a solution having the following composition: 7.7% of sodium hydroxide, 2.0% of the trisodium salt of nitrilotriacetic acid, and as a wetting agent 2.0% of a mixture of 35 parts of the diethanolamine salt of dodecylbenzenesulfonic acid and 18 parts of an adduct of 25 mols of ethylene oxide to 1 mol of colophony and 1 part of pine oil, and 0.3% of a mixture of 59 parts of sodium kogasin sulfonate and 8 parts of triisobutyl phosphate.

The material treated with this solution is squeezed out on a padding machine to 100% liquor retention and treated with saturated steam at 130°C. for 15 minutes in a cottage steamer. It is then rinsed twice at 100°C. and once at 60°C. with desalted water, then acidified with an aqueous solution containing 3 to 5 ml. per liter of concentrated hydrochloric acid and rinsed again at room temperature. Material completely free from husks is obtained having a degree of whiteness of 81.0 to 85.0% (measured as in Example 1) depending on the quality of the cotton.

The same result is obtained using an impregnating liquor which contains 8.0% of sodium hydroxide, 2.0% of the tetrasodium salt of ethylenediamine tetracetic acid and one of the following wetting agents:

(1) 1.8% of the sodium salt of the sulfuric acid hemiester of an adduct of 4 mols of ethylene oxide to 1 mol of nonyl phenol.

(2) 2.5% of a mixture of 5 parts of the adduct of 10 mols of ethylene oxide to 1 mol of nonyl phenol, 57 parts of the sodium salt of the sulfuric acid hemiester of an adduct of 4 mols of ethylene oxide to 1 mol of nonyl phenyl and 10 parts of pine oil.

(3) 2.0% of a mixture of 10 parts of an adduct of 7 to 9 mols of ethylene oxide to 1 mol of fatty alcohol, 54 parts of the sodium salt of the sulfuric acid hemiester of an adduct of 4 mols of

ethylene oxide to 1 mol of nonyl phenol and 5 parts of pine oil.

(4) 1.5% of a mixture of 10 parts of an adduct of 7 to 9 mols of ethylene oxide to 1 mol of fatty alcohol and 54 parts of the sodium salt of the sulfuric acid hemiester of an adduct of 4 mols of ethylene oxide to 1 mol of nonyl phenol and 0.3% of a mixture of 59 parts of sodium kogasin disulfonate and 8 parts of triisobutyl phosphate.

(5) 1.5% of a mixture of 5 parts of an adduct of 10 mols of ethylene oxide to 1 mol of nonyl phenol and 57 parts of the sodium salt of the sulfuric acid hemiester of an adduct of 4 mols of ethylene oxide to 1 mol of nonyl phenol and 0.3% of a mixture of 59 parts of sodium kogasin disulfonate and 8 parts of triisobutyl phosphate.

Example 3: Desized and dried unbleached cotton cloth is impregnated with an aqueous solution having the following composition: 7.7% of sodium hydroxide, 2.0% of a mixture of 12 parts of an adduct of 8 to 9 mols of ethylene oxide to 1 mol of octyl phenol, 16.5 parts of the diethanolamine salt of dodecylbenzenesulfonic acid, 32 parts of sodium kogasin disulfonate, 5 parts of triisobutyl phosphate and 1.6 parts of a condensation product of 1 mol of formaldehyde and 2 mols of $\beta$-naphthalenesulfonic acid, and 2.0% of one of the following substances or mixtures of substances:

(1) Sodium pyrophosphate.
(2) Sodium tripolyphosphate.
(3) Sodium hexametaphosphate.
(4) A mixture of 50 parts of sodium pyrophosphate and 50 parts of the tetrasodium salt of ethylenediamine tetracetic acid.
(5) A mixture of 50 parts of the trisodium salt of nitrilotriacetic acid and 50 parts of sodium hexametaphosphate.

The impregnated material is squeezed out to 100% liquor retention on a padding machine and treated in a pad-roll plant with saturated steam for three hours at 100° to 103°C. The plant should be equipped to operate free from air as far as possible. The material is then rinsed twice at 100°C. and once at 60°C. with desalted water. It is then acidified with an aqueous solution containing 3 to 5 ml. per liter of concentrated hydrochloric acid and rinsed again at room temperature. A completely husk-free material is obtained having a degree of whiteness varying from 79.0 to 83.0% (measured as in Example 1), depending on the quality of the cotton.

## Dithionite Stabilized with Zinc Compound for Wood Pulp

A. Janson, F. Poschmann and G. Wittmann; U.S. Patent 3,653,804; April 4, 1972; assigned to Badische Anilin- & Soda-Fabrik AG, Germany found that the stability of dithionite bleaching solutions for wood pulp is greatly extended by the addition of sparingly soluble zinc compounds. Examples of suitable compounds are zinc oxide, zinc hydroxide, zinc carbonate, basic zinc carbonate, basic zinc silicate and basic zinc phosphate. Mixtures of these compounds are similarly effective.

It is advantageous to add the zinc compounds to the bleaching liquor in an amount of at least 5% by weight with reference to the dithionite added. The zinc compound may be added to the dithionite prior to addition of the latter to the bleaching mixture.

Addition of the sparingly soluble zinc compounds (which are capable of adsorbing hydrogen sulfide and sulfur dioxide) to the solid dithionite improves the odor when the solid dithionites are stored. For the two known types of pulp (namely groundwood pulp which is formed by simple grinding of wood, and chemical pulp which is made by first clipping the wood and then softening the chips with chemical reagents, as for example sodium sulfite, at elevated temperature and pressure) the best bleaching effect is achieved in the pH range of from about 4.0 to 6.0, because bleaching at pH values of 7 or more results in yellowing of the pulp.

Bleaching of both types of pulp is carried out at about 60°C. with addition of dithionites. On a commercial scale it lasts about 15 minutes. The most favorable results are obtained by carrying out bleaching for about 1 hour provided the reducing agent remains active during the whole period. A consistency (i.e., content of dry fiber) of 0.5 to about 12% by weight, preferably from 2 to 4% by weight, is used in the bleaching of groundwood pulp, and usually about 0.5 to 1.0% by weight (with reference to the amount of dry fiber) of dithionite is added as bleaching agent. This means that in 1 liter 20 to 40 g. of wood fiber is suspended and that at the beginning of the bleaching the amount of dithionite added, either as solid or as solution, is such that 200 to 400 mg. of dithionite is present per liter at the commencement of bleaching.

The durability of sodium dithionite for this concentration of bleaching agent is indicated below without additions and with additions of sparingly water-soluble zinc compounds in buffer solutions of the Michaelis buffer acetic acid, sodium acetate. The pH value of 3.5 to 5.6 can be kept constant, depending on the mixing ratio, by mixtures of 0.2 N acetic acid solution and 0.2 N sodium acetate solution. The buffer is not attacked by dithionites. Durability is investigated under nitrogen at 60.0° ± 0.5°C. and the content of sodium dithionite is determined in dependence on pH value and on time by titration according to the indigo carmine method [A. Binz, H. Bertram, Angew. Chem., 18, 168 to 170 (1905)]. The tests show that increasing amounts of zinc oxide or basic zinc carbonate have a progressively increasing effect on the durability of sodium dithionite.

Degradation of sodium dithionite with an addition of 20 mg. of zinc oxide (equal to 5% with reference to 400 mg. of sodium dithionite) requires 21 minutes and with an equivalent addition of basic zinc carbonate 16 minutes, whereas sodium dithionite without additive is destroyed after only 6 minutes. Surprisingly this effect is not observed in the case of zinc acetate or other water-soluble zinc salts. It evidently requires the solid phase of a sparingly soluble zinc compound to bind traces of hydrogen sulfide formed, whereas in the case of zinc acetate the solubility product first has to be exceeded before sulfide ions can be bound from the solution. Sulfide ions have a strongly destructive effect on dithionites in the weakly acid pH range.

The favorable effect of the additives on the result of the bleaching is clear from a comparison of bleachings of wood pulp with sodium dithionite which have been carried out with an addition of a sparingly soluble zinc compound and without such an addition. The series of experiments given in the following table apply to the bleaching of wood pulp having a consistency of 2% at 60°C., a pH value of 4.5 and a period of 1 hour with a concentration of bleaching agent of 0.5 to 1.0% with reference to the dry solids. The whiteness of the sheets obtained is determined with an Elrepho spectrophotometer, filter R 46T (Zeiss, Oberkochen). The unbleached pulp has a whiteness of 59.4.

| Percent of Bleaching Agent | | Whiteness Achieved with Bleaching Agent 0.5% | 1.0% |
|---|---|---|---|
| Zinc oxide: | 0 | 64.0 | 66.4 |
| | 5 | 64.8 | 68.6 |
| | 10 | 65.2 | 69.4 |
| Basic zinc carbonate: | 0 | 64.0 | 66.4 |
| | 5 | 64.6 | 68.4 |
| | 10 | 65.4 | 69.2 |

The favorable affect of additions of sparingly soluble zinc compounds on the bleaching process in the weakly acid range is maintained even in the case of sodium dithionite which has been stabilized by small additions of urea against spontaneous thermal decomposition upon the action of water or by small additions of macromolecular substances, such as polyacrylamide or sodium polyacrylate, against thermal decomposition. The use of additions of sparingly soluble zinc compounds also has the advantage that discoloration and corrosion by hydrogen sulfide are avoided and, for example, in the bleaching of wood pulp, action of hydrogen sulfide on the bronze screen used after bleaching in the separation of the pulp is also avoided.

## STAIN REMOVING AND DYE STRIPPING

### Poly-N-Vinyl-5-Methyl-2-Oxazolidinone

W.E. Walles, W.F. Tousignant and R.J. Axelson; U.S. Patent 3,097,048; July 9, 1963; assigned to The Dow Chemical Company describe a method for stripping dyes from cellulosic fabrics by aqueous solutions of poly-N-vinyl-5-methyl-2-oxazolidinone, designated as PVO-M.

PVO-M is a water-soluble polymer described in U.S. Patent 2,919,279. PVO-M may be prepared as high polymers having molecular weights, for example, in the range from 10 to 50 thousand higher (as determinable from Fikentscher K-values of about 10 or more to as high as 75 to 100, or so) in order to provide a wide variety of polymer material than can be advantageously employed in these compositions. Thus, PVO-M in broad molecular weight ranges up to 100,000 to 200,000, is quite soluble in water and can be beneficially used in formulating the dye-stripping compositions of this process.

Highly efficient stripping can be effected when PVO-M is incorporated in the stripping bath, for instance, stripping a vat dye from cotton such as to leave the cotton essentially completely white in a single stripping operation without any fiber damage is possible when PVO-M is employed in a typical stripping bath for vat dyestuffs described before.

The amount of PVO-M necessary to use in the stripping bath will, of course, depend on the intended results. Usually the stripping efficiency will increase with the amount of PVO-M employed and the upper limits of PVO-M that can be beneficially utilized would only be controlled by the solubility of the polymer in the particular stripping bath. Accordingly, stripping baths containing up to at least 20% of the polymer, and frequently up to as much as 50 or more weight percent of the polymeric solute, based on the weight of the resulting solutions, are capable of being used in this process. The exact requirements for any one stripping will depend on the particular fiber or fabric, for example, whether it be cotton or viscose rayon; the particular dyestuff employed, some being more easily stripped than others; the amount of dyestuff on the fiber; the degree of strip required; and, to some extent, the polymer molecular weight. Generally, an amount of polymer about equal to the amount of dye originally used to dye the fiber provides highly beneficial and efficient stripping. Thus, for a light dyeing or shade about 2 to 3% of the dyestuff based on the weight of the fiber (OWF) may be used, whereas, for a heavy dyeing about 10 to 15% of the dyestuff (OWF) might be used. Correspondingly then, for an essentially complete strip of the dyestuffs from the fiber 2 to 3% PVO-M (OWF) would be used in the first instance, and 10 to 15% PVO-M (OWF) would be used in the second instance. Lesser or greater amounts may be used as the needs require. Indeed, excellent results can be obtained when only a small amount of the polymer is used, as little as a fraction of a percent (OWF) PVO-M can be significantly beneficial.

The PVO-M dye-stripping compositions can be used to strip vat dyes, including those more specifically referred to as anthraquinone vats, indigoid vats and indanthrene vats; sulfur dyes; and direct dyes from pure cellulosic fibers, i.e., cotton and viscose or cuprammonium rayon. Stripping may be accomplished whether the fibers were dyed according to normal procedures or slight modifications thereof, or printed. The physical characteristics of the fiber are of no importance in carrying out the process. Raw stock, yarn, or fabric are equally well stripped. Or, the cotton or rayon may be blended with other fibers in yarn or fabric form.

A sample of plain white cotton toweling was scoured with 1% on the weight of the fiber (OWF) Dupanol Wa, an anionic detergent, for 15 minutes at 71°C. in a 30:1 liquor to goods bath. After scouring, the cotton was rinsed in cold water and then dyed with 12% (OWF) Calcosol Navy Blue Paste, a vat dye (Color Index 59810). A 30:1 liquor to goods dye bath was used containing about 30% sodium hydroxide (OWF) and 15% sodium hydrosulfite (OWF). The cotton was dyed at 60°C. for about 30 minutes, treated with about 1% acetic acid (OWF) and 1.5% sodium dichromate (OWF) at 65°C. for 30 minutes, rinsed and dried. The dyed cotton toweling was then cut into smaller samples each of which was treated differently as described below.

(1) A sample of the initial white cotton was retained for comparison.

(2) A sample of the dyed cotton was retained without further treatment.

(3) A sample of the dyed cotton is given a standard stripping treatment, i.e., with an aqueous bath containing about 30% sodium hydroxide (OWF) and about 15% sodium hydrosulfite (OWF) and treated at about 70°C. for about 40 minutes, rinsed and dried.

(4) A sample was treated the same as (3) excepting to add about 50% based on the weight of the solution, of ethylene glycol methyl ether (2-methoxy ethanol).

(5) A sample was treated the same as (4) excepting to add 10% PVO-M (OWF) having a K-value of 30.

(6) A sample was given the treatment of (5) twice.

(7) A sample was given the treatment of (5) followed by bleaching with 30 ml. of a 5% aqueous solution of sodium hypochlorite in 1,000 ml. of water.

(8) A sample was treated the same as (4) excepting to substitute 20%,

*based on the weight of the solution, of ethylene glycol ethyl ether* (2-ethoxyethanol).

(9) A sample was treated the same as (8) excepting to add 10% PVO-M (OWF).

Visual observation readily made apparent the effectiveness of each treatment. (2) untreated was a deep blue; (3) and (4), which are about the same, exhibited some stripping but the cotton is still decidedly blue; (5) was considerably lighter than (3) and (4), being only a light grey; (6) appeared slightly lighter than (5); (7) was essentially completely stripped appearing only slightly off white; (8) was about the same shade as (4); and (9) approximated the shade of (5)

## Dye Removal by Polar Solvents

C. Davis; U.S. Patent 3,103,405; September 10, 1963 found that polar solvents, such as the plasticizers used for vinyl resins, readily remove dye stains from textiles. The same solvents are useful in reverse printing and in the preparation of reverse masters.

The dyes are generally known as hydrogen bridge complexes and are prepared by dissolving the crystalline phenol and the dye base to give a colorless or weakly colored solution in a polar organic solvent. Upon removal of the polar organic solvent from a surface coated with this solution, a highly colored hydrogen bridge complex remains on the surface. The dye bases employed in general have logarithmic dissociation constants below 6.5.

This process provides a method for eradicating the above highly colored hydrogen bridge complexes from any surface by contacting them with a polar organic solvent containing nitrogen, oxygen, or sulfur in one or more of the following groupings: alcohol, ketone, ester, aldehyde, nitro, amide, substituted amide, amine, sulfoxide, etc. Exemplary of the polar organic solvents which are useful for eradicating highly colored hydrogen bridge complexes formed from a weak dye base and a crystalline phenol are the following: methanol, ethanol, isopropanol, butanol, acetone, methylethylketone, methylisobutylketone, dioxane, pyridine, dimethylformamide, dimethylsulfoxide, ethyl acetate, methyl salicylate, methyl butyrate, tributyl phosphate, trioctyl phosphate, dibutyl phthalate, dioctyl phthalate, and the like. A class of solvent which are particularly useful for the purposes of this process are the group known as the plasticizers or plasticizing fluids, in particular those which are useful as plasticizers for vinyl resins.

Examplary of this preferred group of solvents are tributyl phosphate, tri(2-ethylhexyl) phosphate, dibutyl phthalate, dioctyl phthalate, tributyl citrate, dioctyl adipate, dioctyl azelate, tricresyl phosphate, and the like. The action of the above polar organic solvents in dissolving and dissociating highly colored hydrogen bridge complexes makes possible the eradication of these hydrogen bridge complexes from a variety of surfaces. These surfaces include those prepared from fibers of both animal and vegetable origin as well as the newer synthetic fibers and plastics. Thus the process is useful for removing the highly colored hydrogen bridge complexes from human skin, wool, cotton, rayon, clothing or rugs, tile surfaces, glass, paper, plastics, and similar surfaces where they have spilled accidentally.

Example: Removal of Ink Stains from Cotton Shirting — A patterned fabric cotton broadcloth shirting was sprayed with an ethanolic solution of Michler's Hydrol Diethylbenzensulfinate and phenolphthalein. Bright blue spots of the colored hydrogen bridge complex of Michler's Hydrol Diethylbenzenesulfinate and phenolphthalein appeared on the shirt upon the evaporation of the ethanol. The stained areas were swabbed with trioctyl phosphate; the stains disappeared immediately. After remaining overnight the shirt was washed in a standard detergent solution to remove the excess trioctyl phosphate. The clean shirt showed no residual ink color, no fading of the patterned fabric, and no deleterious effect upon the fiber.

A white cotton shirting was sprayed with a dioxane solution of N-phenyl rhodamine B lactam and dihydroxydiphenylsulfone. Upon evaporation of the solvent, magenta-colored spots of the hydrogen bridge complex of N-phenyl rhodamine B lactam and dihydroxydiphenylsulfone appeared on the white shirt. The stained areas were swabbed with dioctyl phthalate, and the stain disappeared. The shirt was washed in a standard detergent solution to remove the excess dioctyl phthalate. The clean shirt showed no residual ink color and no yellowing of the white fabric.

## Ball Point Ink Eradicator

G.J. Burstein; U.S. Patent 3,160,468; December 8, 1964 describes a process using an acid, a solvent and a bleach to remove ball point ink. The acids used include orthophosphoric, citric and acetic acids. The solvents include kerosene, ethylene glycol, turpentine, a liquid detergent, a volatile petroleum distillate, naphtha and liquid solvent for plastics. The bleach is sodium hypochlorite used in aqueous solution in concentration of 3 to 12%. The acids are used in 10% concentration. The solvent is needed because the ball point ink contains greasy or resinous matter.

The acid, solvent, and bleach may be employed in the order named, with intermediate blotting, as liquids and as a three-step process. However, the solvent is compatible with the acid and is also compatible with the bleach, so it may be combined with one and/or the other to allow for eradication as a two-step process; that is, applying a mixture of an acid and a solvent, then, after blotting a mixture of a solvent and a bleach, or if a solvent was used with the acid, only a bleach need be used, and then blotting. If the bleach does not entirely remove the color, then it may be necessary to reapply and blot again.

Example 1: As a three-step process, the acid is applied first, blotted; then the solvent is applied and blotted; finally the bleach is applied and blotted. If necessary, the bleaching step is repeated.

Example 2: As a specific example of the materials which may be used in a three-step process, dissolve orthophosphoric acid, about 10% by weight, in water and apply the solution to the ink to be eradicated, blotting thereafter; kerosene is applied and blotted. As a final step, about 5.25% by weight solution of sodium hypochlorite in water is applied and finally blotted.

Example 3: Apply a solution of about 10% by weight of orthophosphoric acid and about 0.7% by weight of ortho-benzyl-parachlorophenol in water, blot, then apply Lestoil, blot, then apply about 5.25% by weight solution of sodium hypochlorite in water, and finally blot.

Dye Removal from Foam Carpet Backing

S.A. Fisher; U.S. Patent 3,432,251; March 11, 1969; assigned three-fourths to Robinette Research Laboratories, Inc. and one-fourth to Techniservice Corporation found that carbonyl-containing solvents remove stains from the foam backing of carpeting.

When solid foam-backed textile material, such as carpeting, is dyed in a bath the foam backing is likely to sorb dyestuff from the bath, and normal rinsing procedure is ineffective to remove the sorbed dyestuff from the backing. On some floor surfaces the retained dyestuff may transfer from the backing to the floor surface to produce an objectionable staining. This is a particular problem with disperse-dyed, polyurethane foam-backed, nylon carpets on vinyl tile floors.

A primary object of this process is the prevention of carpet staining of vinyl floor tile or other flooring material. Another object is selective desorption of dyestuff from the backing component of dyed solid foam-backed textile material without disturbance of the dyestuff on the textile component thereof. Suitable solvents are the ketones such as acetone, diacetone alcohol, methyl isoamyl ketone, and methyl isobutyl ketone.

The solvent may be applied by immersing the textile material or by spraying the solvent on the material, in which case the operating temperature may be fixed at or above (usually not below) room temperature. A modified form of hot application may be accomplished by volatilizing the solvent, condensing it onto the textile material, which is kept at room temperature or otherwise relatively cool, and draining off the condensed solvent with dyestuff removed from the foam backing dissolved therein.

Example: A section of nylon carpet backed with polyurethane foam and disperse-dyed to a deep shade of blue was cut up into small squares. These squares were dipped into a variety of anhydrous organic solvents for a sufficient length of time to wet out both the foam and the face fibers. The solvent-wet squares of carpet were then squeezed between white paper towels, and the success of the extraction was evaluated in terms of the removal of color from the foam back and the absence of evidence of color removal from the face fibers as shown by the absence of color on the paper squeezed in contact with the face of the test sample.

Typical results from the variety of solvents employed in the Example are tabulated in the table below. Solvents rated as very good in extraction are those where the foam color was changed from a deep blue to white in a single extraction. Good solvents are those in which the removal left the foam light or faintly blue, a fair extraction showed some lightening of the foam color, a poor rating is indicated by only traces of color on the paper during the extraction process, and a very poor rating indicated no evidence of extraction.

| Solvent | Dye Removal from Backing |
|---|---|
| Acetic acid | Very good |
| Acetone | Very good |
| Ethyl acetate | Very good |
| Methyl isobutyl ketone | Very good |

(continued)

| Solvent | Dye Removal from Backing |
|---|---|
| Salicylaldehyde | Good |
| Diacetone alcohol | Good |
| Methyl isoamyl ketone | Good |
| 2-Ethylhexoic acid | Good |
| 1,3-Butyl glycol | Fair |
| Diethylene glycol | Poor |
| Ethylene glycol | Poor |
| Hexyl alcohol | Poor |
| Isopropyl alcohol | Very poor |

## Pet Stain Removal from Carpets

E.M. McIntyre; U.S. Patent 3,607,760; September 21, 1971 describes a composition of solvents, hydrogen peroxide and a chelating agent to remove pet stains from carpets.

Example: A cleaning composition was made from the following ingredients in the following parts by weight.

| Ingredients | Parts by Weight |
|---|---|
| Butyl cellosolve (2-butoxy ethanol) | 13 |
| Isopropyl alcohol | 10 |
| Hydrogen peroxide (3.5% solution) | 2 |
| EDTA (sodium salt) | 1/2 |
| Water | 103 |

Preferably, the above ingredients are mixed by first mixing the EDTA in the water, thereafter adding the alcohol and cellosolve and finally the hydrogen peroxide solution. After thorough mixing, the composition is ready for use. In the event that any solid particles appear in the mixture, the latter can be simply strained to remove these particles.

When it is desired to remove a pet stain from a carpet or the like, a small quantity of the above composition is sprayed over the stain. The composition is worked into the carpet by agitating or scratching until the stained area is thoroughly wetted. For the average small stain, this should take not more than 1 minute. Thereafter, the wetted area is rubbed dry with a towel. If all of the stain is not removed on the first application, the above steps are repeated. Occasionally, one will run into a pet stain which has been in the carpet for a considerable period of time. For these more difficult stains it is desirable to spray the composition over the stained area, to rub it into the stain, and then to cover the stain with diatomaceous earth or similar powder capable of absorbing the liquid composition. After the powder has dried, the latter can be simply lifted away with a vacuum cleaner.

The composition of this process can be used for removing all types of pet stains including vomit, urine or feces. Butyl cellosolve may be replaced by ethyl or methyl diethylene glycol; ethyl alcohol may be used instead of isopropyl alcohol; and for the sodium salt of EDTA, Versene of Dow Chemical may also be used.

## Silver Nitrate Stain Removal

F. Colombo and E.V. Babcock; U.S. Patent 3,434,796; March 25, 1969; assigned to United States Borax & Chemical Corporation describe a process to remove silver nitrate stains from fabrics by an aqueous solution of equal parts of sodium or ammonium persulfate and urea.

Example 1:

| | Parts by Weight |
|---|---|
| Thiourea | 1 |
| Sodium persulfate | 1 |

The thiourea is in a particulate condition as is the sodium persulfate. The particle size of each is approximately the same so that stratification is avoided. The ingredients are blended in conventional mixing apparatus. To utilize the resulting mixture, five parts by weight thereof is added to one thousand parts of water at approximately room temperature. The water is stirred for a sufficient period to dissolve the persulfate-thiourea mixture. Thereafter, a garment such as a hospital gown which has become stained with silver nitrate is dipped into the aqueous bath. The garment is stirred to insure complete wetting. After a relatively short period of time of only 30 seconds the garment is removed from the bath with the concomitant result that the stain due to silver nitrate is completely removed.

It has been found that once the stains have been removed, the garment should be immediately removed from the bath so that the system is not needlessly exhausted. The bath may be used for a considerable number of garments before exhaustion is noticed. In such an event it is possible to replenish the strength of the solution by adding an additional quantity of the persulfate-thiourea mixture. On the other hand, it is relatively simple to prepare an entire fresh batch of the solution to insure uniform treatment.

Example 2:

| | Parts by Weight |
|---|---|
| Thiourea | 1.5 |
| Ammonium persulfate | 5 |

The ingredients are compounded and utilized in the same manner as in connection with Example 1.

Example 3:

| | Parts by Weight |
|---|---|
| Thiourea | 2 |
| Potassium persulfate | 3 |

Again the ingredients are compounded and utilized in the same manner as in connection with Example 1. At the end of the treatment for stain removal the garment is laundered in the conventional manner usually by automatic washing machines employing standard commercially available detergents and soaps.

Often the apparatus in which the stain removing operation is accomplished contains parts fabricated from brass or copper. In order to avoid deposition of copper salts on the treated garments, they are treated with an aqueous solution of sodium nitrite or oxalic acid. This may be undertaken by either adding a quantity of the sodium nitrite or oxalic acid equal to the amount of persulfate compound utilized directly to the stain removal bath. On the other hand, the stain removal solution may be drained from the washing apparatus and substituted with an aqueous solution of the sodium nitrite or oxalic acid. The latter solution may be made by preparing the solution before supplying it to the washing apparatus or it may be prepared in situ. The garments are treated for a period of 5 to 10 minutes with suitable agitation. Thereafter, the garments are washed with detergents in the usual manner.

Again, turning to the examples, in connection with Example 1 the process is further demonstrated by adding one part by weight of sodium nitrite directly to the solution of thiourea and sodium persulfate after the garment has been freed of silver nitrate stains. After approximately 5 minutes the garment is removed for further washing.

In a modification of Example 2, 5 parts by weight oxalic acid is added to the aqueous bath containing thiourea and ammonium persulfate and the garment is handled as in the foregoing. In connection with modifying Example 3, 3 parts by weight of sodium nitrite is dissolved in a quantity of water equal to that used in dissolving the thiourea and potassium persulfate. The aqueous bath of thiourea and potassium persulfate is drained from the washing apparatus. The prepared sodium nitrite solution is then added to the washing apparatus. The garment is treated with the sodium nitrite solution for approximately 10 minutes. Thereafter the garment is laundered in a conventional manner.

## Dye Stripping from Acrylics

B. Kissling; U.S. Patent 3,582,255; June 1, 1971; assigned to Sandoz Ltd., Switzerland describes a process for stripping dyeings or prints produced with non-fiber-reactive disperse dyes on hydrophobic fibers, which process consists in treating the dyed fibers at temperatures above 90°C. with an amount of at least 0.5% (on the weight of the dyed fibers) of a nonionic compound of the polyglycol ether series which is soluble to at least 3% in water at 20°C., and which contains per polyglycol ether chain a hydrophobic radical having 12 to 26 carbon atoms.

All dyed hydrophobic, fully synthetic or semisynthetic organic fibers can be stripped of dye by this process. They include the aromatic polyester fibers, e.g., polyethylene terephthalate or condensation products of terephthalic acid and 1,4-bis-(hydroxymethyl)cyclohexane; the polyamide fibers, e.g., the nylon fibers, the polyurethane fibers, e.g., the condensation products of 1,4-butanediol and hexamethylene-1,6-diisocyanate, the polyacrylonitrile fibers, e.g., polyacrylonitrile itself and the copolymers of acrylonitrile and vinyl chloride, vinyl acetate or vinylpyridine; the polyvinyl chloride fibers; the cellulose ester fibers, e.g., cellulose triacetate and cellulose triacetopropionate; and the polyolefin fibers, e.g., polypropylene.

Dyeings or prints produced with any nonfiber-reactive disperse dyes can be stripped. The disperse dyes can belong, e.g., to the monazo, disazo, nitrostyryl, quinophthalone, anthraquinone or naphthazarine series.

The stripping agents are the nonionic compounds of polyglycol ethers. These are adducts of butylene oxide, propylene oxide or preferably ethylene oxide. The number of ethylene oxide groups must be at least great enough to render the polyglycol ether soluble in water at 20°C. to at least 3%. For a hydrophobic compound, such as lauryl alcohol, the number of ethylene oxide groups is about 5 to 6; for nonylphenol, about 8. The maximum number is about 100 to 120 per hydrophobic radical. The terminal hydroxyl group of the polyglycol ethers may be replaced by an alkoxy, arylalkoxy, or carboxyalkoxy group. Examples of some suitable polyglycol ethers are: tridecyl-$(O{-}C_2H_4)_8{-}OH$ and cetyl-$(OC_2H_4)_{12}{-}OH$.

The preferred polyglycol ethers are the adducts of about 8 to 100 mols of ethylene oxide on unsaturated alcohols, such as oleyl alcohol or on unsaturated acids, such as oleic acid, e.g., the oleyl polyglycol ethers having 20 to 60 $C_2H_4O$ groups, such as oleyl-$(OC_2H_4)_{25}{-}OH$, oleyl-$(OC_2H_4)_{40}{-}OH$, and oleyl-$(OC_2H_4)_{60}{-}OH$, and the oleoyl polyglycol ethers having 8 to 50 $C_2H_4O$ groups, such as oleoyl-$(OC_2H_4)_{13}{-}OH$; further the adducts of 25 to 200 mols of ethylene oxide on 1 mol of castor oil, e.g., castor oil + $32(C_2H_4O)$, castor oil + $40(C_2H_4O)$, castor oil + $46(C_2H_4O)$ and castor oil + $110(C_2H_4O)$. These castor oil derivatives are triglycerides and each contains 3 hydrophobic unsaturated radicals, each of which has 18 carbon atoms, plus 3 polyglycol ether chains, each of which has 8 to about 70 ethylene oxide groups.

Example 1: A poplin fabric of 100% polyester fiber (polyethylene terephthalate) dyed with 0.8% of brominated 1,5-diamino-4,8-dihydroxyanthraquinone is padded with an aqueous solution of 20 g./l. of oleyl-$(O{-}C_2H_4)_{12}{-}OH$, expressed to 60% pickup, dried and dry heat treated for 2 1/2 minutes at 215°C., after which it is washed off with hot water and dried on a hot flue. The stripping effect is about 70% and the stripped light blue dyeing can be redyed without difficulty.

Example 2: 100 parts of a fabric of polyacrylonitrile dyed with 1% of the blue dye 1-methylamino-4-(2'-hydroxyethylamino)-anthraquinone are treated for 1 hour at 98° to 100°C. in 4,000 parts of a liquor containing 20 parts of di(isoamyl)-phenyl-$(OC_2H_4)_{12}{-}OH$, with subsequent rinsing and drying. The dyeing is stripped to a very much lighter depth and can be redyed and/or corrected to the desired depth with the required amount of dye. In contrast to normal carriers, such as chlorinated benzenes, treatment of dyeings on polyacrylonitrile fibers with di(isoamyl)-phenyl-$(OC_2H_4)_{12}{-}OH$, or one of the polyglycol ethers listed above does not produce an unpleasant odor and does not adversely affect the fastness properties of the dyeings in any way.

Example 3: A cellulose triacetate fabric dyed with 1% of the red dye 4-nitro-4'-N-ethyl-N-β-hydroxyethylamino-1,1-azobenzene is padded with a solution of 200 grams per liter of castor oil+$32(C_2H_4O)$ at 18° to 20°C., rinsed with hot and cold water and dried. The dyeing is stripped to 80%.

Example 4: A polyamide 66 fabric dyed with 1% of the dye of Example 3 is treated as described in Example 3. The dyeing is stripped to practically 100%. When the dyeing on polyamide 66 fabric is replaced by a dyeing on polyamide 6 fabric and the treatment is carried out for 1 minute at 180°C., the dyeing is stripped to about 90%, i.e., only about 10% of the dye is left on the fabric.

## BLEACHING FLOOR AND WALL COVERINGS

### Hydrogen Peroxide-Alkali Metal Bicarbonate

N.J. Stalter; U.S. Patent 3,017,236; January 16, 1962; assigned to E.I. du Pont de Nemours and Company found that if an alkali metal bicarbonate, preferably sodium bicarbonate, is used in slightly acidic peroxide bleaching solutions, a more effective bleaching action is obtained when these solutions are applied to linoleum, wood, paperboard and other solid surfaces susceptible to the bleaching action of peroxide. Using the preferred bicarbonate, $NaHCO_3$, in an amount of at least 5% by weight together with 10 to 50% hydrogen peroxide, a solution can be prepared which can be applied to practically any linoleum surface without damage to the surface. The compositions are readily prepared simply by dissolving the required amount of the sodium bicarbonate compound in a commercial aqueous hydrogen peroxide, e.g., of about 27 to 50% strength. No cooling is required during their preparation to prevent peroxide decomposition.

Solutions were prepared as follows: (A) 10 parts of sodium bicarbonate ($NaHCO_3$) were added to 90 parts of 35% aqueous hydrogen peroxide plus 1 part of a wetting agent, alkyl aryl sodium sulfonate. This solution had a pH of 6.2. (B) 20 parts of ammonium bicarbonate ($NH_4HCO_3$) were added to 80 parts of 35% aqueous hydrogen peroxide plus 1 part of wetting agent, alkyl aryl sodium sulfonate. This solution had a pH of 6.4. (C) 128 parts of water were added to 20 parts of sodium hydroxide ($NaOH$), one-fifth of 1 part of calcium hydroxide [$Ca(OH)_2$], 1 part of aqueous sodium silicate 42° Bé. and 1 part of the same wetting agent. This solution was then added to 450 parts of 35% aqueous hydrogen peroxide. Each of solutions (A), (B) and (C) were sponged onto separate sections of a piece of linoleum. An application of 0.12 lb./sq. yd. was used. The samples were then dried for 2 minutes at 205° to 220°F., followed by a rinse in cold tap water.

The samples bleached with solution (A) had a superior bleach to samples bleached with solutions (B) and (C). Solution (A) spread more easily and evenly on the linoleum than solution (B), a decided advantage in the application of bleach solution on a production basis. Certain samples of linoleum showed damage in the form of pitting when bleached with solution (C). There was no damage or pitting of samples bleached with solution (A).

### Hydrogen Peroxide and Ammonium Bicarbonate

R.B. Dustman, Jr.; U.S. Patent 3,034,851; May 15, 1962; assigned to E.I. du Pont de Nemours and Company describes a bleaching composition for solid surfaces such as floor coverings of rubber tiles or linoleum consisting of hydrogen peroxide with ammonium carbonate additions. Any of the ammonium carbonates can be used in this process. These include ammonium bicarbonate, ammonium sesquicarbonate and ammonium carbonate. The bicarbonate is preferred. The concentration of the ammonium carbonate or bicarbonate in the bleaching composition should be from 10 to 23% by weight based upon the weight of the composition in order to obtain the rapid and effective bleaching generally desired.

The concentration of hydrogen peroxide should be from 20 to 40% of the weight of the composition. The compositions are readily prepared simply by dissolving the required amount of the ammonium carbonate compound in a commercial aqueous hydrogen peroxide, e.g., of about 27 to 50% strength. No cooling is required during their preparation to prevent peroxide decomposition. This is particularly so when making the preferred peroxide-bicarbonate compositions, since ammonium bicarbonate dissolves in aqueous hydrogen peroxide with a negative heat of solution.

Example: Two solutions were prepared as follows: (A) 1 part of solid ammonium carbonate, $(NH_4)_2CO_3 \cdot H_2O$ (30% $NH_3$ equivalent), was dissolved in 9 parts of 35% aqueous hydrogen peroxide. The resulting solution had a pH of 7.75. (B) 1 part of 28% aqueous ammonia was added to 9 parts of 35% aqueous hydrogen peroxide. The resulting solution had a pH of 8.75. The surfaces of samples of an unbleached kraft paperboard were sprayed uniformly with one or the other of the above solutions at an application rate of 3 lbs. of solution per 1,000 sq. ft. of board surface. Immediately after spraying, the sprayed surface was contacted for about 5 seconds with a smooth aluminum surface previously heated to a temperature of 325° to 350°F., whereby the surface was rapidly dried and bleached.

Although uniform bleaching resulted in each instance, the degree of bleaching was much greater when solution (A) was used. The brightness (percent reflectance of blue light) of the samples as measured by a Hunter Multipurpose Reflectometer was 15 for the unbleached board, 29 for the board bleached with solution (A) and 24 for the board bleached with solution (B). In similar trials it was found that other ammonium salts such as the acetate, the citrate and nono- and diammonium phosphates, when used in place of ammonium carbonate in solution (A), gave distinctly poorer bleaching results in that bleaching was much less and uneven and the dried bleached surfaces were gritty due to residual salt.

# COMPANY INDEX

The company names listed below are given exactly as they appear in the patents, despite name changes, mergers and acquisitions which have, at times, resulted in the revision of a company name.

## INVENTOR INDEX

## U.S. PATENT NUMBER INDEX



### NOTICE

# Small Boat Navigation

# *Small Boat Navigation*

Lt. Cdr. Pat Hepherd

STANLEY PAUL, LONDON

STANLEY PAUL & CO LTD
*3 Fitzroy Square, London W1*

AN IMPRINT OF THE HUTCHINSON GROUP

London Melbourne Sydney Auckland
Wellington Johannesburg Cape Town
and agencies throughout the world

*First published 1973*



*This book has been set in Baskerville type, printed in Great Britain on antique wove paper by Anchor Press, and bound by Wm. Brendon, both of Tiptree, Essex*

ISBN 0 09 114410 8

# Contents

# Illustrations



*Photographs are reproduced by kind permission of Kelvin Hughes*

# Introduction

This little book is not intended as an exhaustive treatise on the subject of Navigation, but as a help to a beginner who wishes to go out in his own boat and venture away from his normal sailing grounds. It attempts to explain, as simply as possible, some of the well-tried techniques and practices that have been used by marine navigators over the years. It is only concerned with the use of the simplest equipment, in other words that likely to be available to the small boat sailor.

Perhaps it will help to apply that 'Left-hand-down-a-bit' at the right moment. I hope so!

# 1

# The Magnetic Compass

Since the earliest sailors set off in their tiny craft on long distance voyages, some method of establishing direction on the Earth's surface has been a necessity. It seems fairly certain that the early pioneers, such as the Phoenicians in Europe and the Polynesians in the Pacific, used their knowledge of the sun and the stars, the prevailing winds and currents to guide them. They had no instruments, but generations of accumulated experience enabled them to cover immense distances and eventually return from whence they came. By modern standards, however, such methods as they used were bound to be crude and chancy, particularly in bad weather.

The discovery of a natural ferrous rock, lodestone, led to the first instrument that would give a visual indication of direction. Lodestone is magnetic and could therefore be seen to attract pieces of iron. Soon, no doubt, it was discovered that stroking a piece of iron with the rock would induce the iron in its turn to become magnetic. Then someone found out that a piece of iron or lodestone, suitably shaped and suspended, would point in a fixed direction approximately North or South. No-one knows, for certain, in which country a primitive magnetic compass was first used, but it is definitely recorded that compasses were used in both Europe and China between A.D. 1000–1100.

The earliest compasses probably consisted of a needle or double-bow shaped piece of iron or lodestone floated on a piece of wood in a water-filled vessel; alternatively a thin magnetised needle was pricked through a piece of straw which kept it afloat on the water surface. Whichever method was used, it must have been appreciated that the North-seeking force on the needle was weak and therefore an almost friction-free method of suspension was needed.

*The Earth's Magnetism*

The cause of the phenomenon which showed that a piece of freely suspended iron would point North remained a mystery to the early astronomers and explorers. At one time it was thought that the North end might be attracted by the Pole Star (Polaris) which lay close to the Earth's axis of rotation and was therefore known to give a good general indication of North. There were other theories, but matters became even more difficult to explain when it was discovered that a properly balanced needle not only indicated a different North in different areas of the world but also tended to dip its North end down from the horizontal as it travelled into higher Northerly latitudes.

In about the seventeenth century, it was suggested, correctly as we now know, that the Earth itself was a vast magnet whose poles attracted one end of a magnetised needle (in accordance with the normal laws of magnetism, a magnet always has two poles; the North-seeking/Red end of one will always attract the South-seeking/Blue end of the other; or unlike poles attract: like poles repel).

At that time, however, there was no explanation for the apparent difference between Magnetic North (as defined by compass) and True North, the direction of the geographical North Pole. Also, in different places the angle between the two was known to vary. In fact, in the middle of the sixteenth century, some European compass makers deliberately misaligned the needle when it was fixed to a graduated card so that when the needle pointed at Magnetic North the card indicated True North. This was all very satisfactory whilst the ship remained in home waters, but as soon as it ventured away considerable inaccuracies became apparent. It slowly became obvious that the magnetic poles could not be in the same place as the geographical poles, and furthermore that the magnetic poles were not stationary but changing their positions slowly over long periods of time. Not until 1831 was it definitely established that the Earth's Magnetic Pole was over 1000 miles from the Geographical North Pole in the approximate position 70° North 97° West. The South Magnetic Pole has since been accurately located in one of the most inaccessible places in the world on the bleak snow plateau of Antarctica (about 72° 30′ South 155° East).

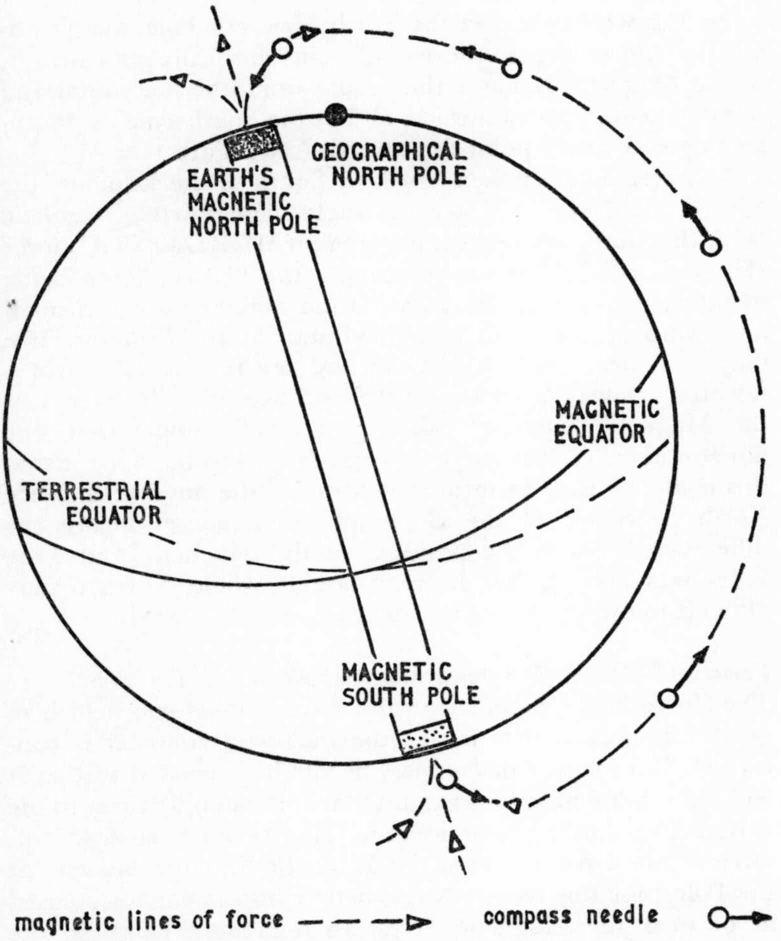


Figure 1. The Earth's Magnetic Lines of Force

This concept of the Earth having its own Magnetic Poles explained the problems that had been worrying scientists over the years. As with an ordinary bar magnet, magnetic 'lines of force' emanate more or less vertically out from the North Magnetic Pole, bend over the Earth's surface, and then descend vertically into the South Magnetic Pole. At any point on the Earth's surface, therefore, a simple magnet (or compass needle) suitably suspended will align itself to these lines of force. If it were held over the North Magnetic Pole, the North-seeking end of the needle would point vertically downwards, at the Magnetic Equator the needle would be horizontal and parallel to the Earth's surface, and at the South Pole its South-seeking end would point downwards (see Figure 1).

At intermediate points North or South of the Equator, the needle tilts down at a certain angle. The Earth's magnetic field, therefore, has two components; a HORIZONTAL force (H) and a VERTICAL force (Z), the vertical force being maximum at the Poles and zero at the Equator; the horizontal force zero at the Poles and maximum at the Equator. If a magnet is suspended so that one end can tilt up and down—towards or away from the Earth's surface—it will (except at the Magnetic Equator) take up a certain angle from the horizontal called 'the Angle of Dip'. It used to be thought that this could be used to measure Latitude (the angular distance North or South of the Geographical Equator) before the difference between the positions of the Magnetic and True Poles was known, but in any case it would be extremely difficult to observe this angle at sea in a rolling boat.

*Variation—The Angle between Magnetic North and True North*

The Horizontal component of the Earth's magnetic field is of greater importance as far as the magnetic compass is concerned. If a magnet or compass needle is suspended so that it will only rotate in the horizontal plane, it can only react to the earth's Horizontal magnetic force (H). As has been seen, this force will be a maximum at the Magnetic Equator and zero at the Poles; for this reason, a magnetic compass cannot be used at or near the Magnetic Poles. In a modern compass, the needles are carefully balanced and pivoted to gain the maximum effect from the horizontal force and to nullify the effects of the vertical component of the Earth's field. In normal latitudes, with one exception which will be mentioned later, the Earth's vertical force (Z) can be ignored.

Due partly to the physical difference in position between the Magnetic and Geographical Poles, at most places on the Earth's surface there is an angular difference in direction between the two as seen from the observer's position. This angle, between TRUE NORTH and MAGNETIC NORTH, is known as the

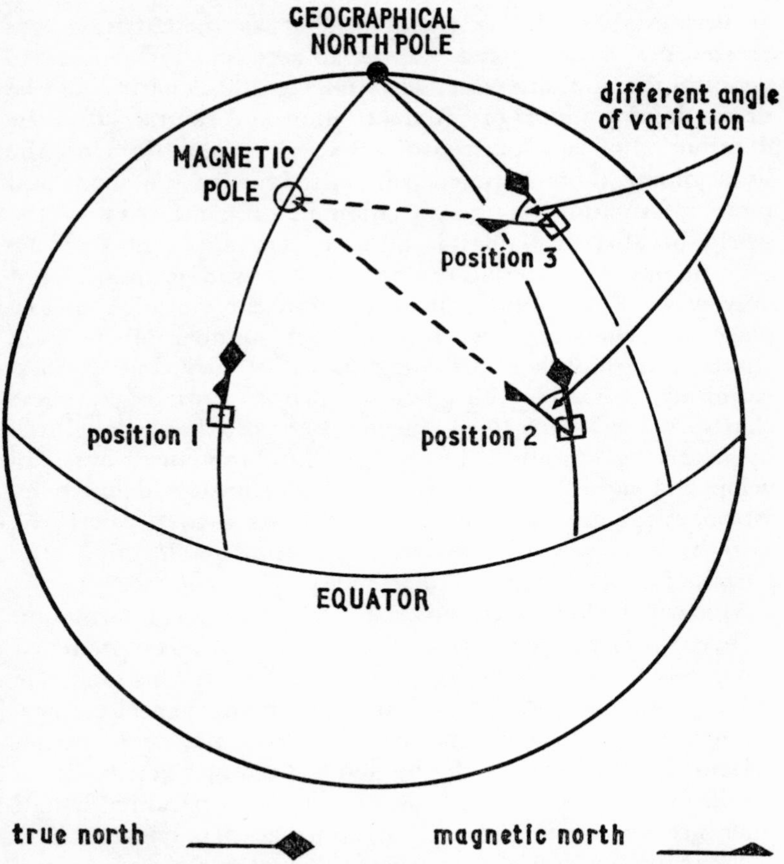


Figure 2. The Angle of Variation

ANGLE OF VARIATION. In Figure 2, Position 1 the angle of Variation is zero, as the direction of the Geographical and Magnetic Poles is coincident. However, in Positions 2 & 3, the two Poles lie in different directions from the observer and the angle between these directions in each case is the Angle of Variation.

The early navigators had tried to associate Dip with Latitude; they also tried to associate Variation with Longitude, i.e. the angular distance East or West of an arbitrary meridian.*

* A Meridian is a Great Circle passing through both geographical poles. The Greenwich Meridian is that which passes through Greenwich, and is historically the one from which Longitude is measured.

It was established, for example, that as the Atlantic was crossed the variation first reduced to zero and then increased again in the opposite sense. They inferred that charts could be drawn joining places of equal variation and assumed that the distribution of these places would be regular, and therefore the lines joining them symmetrical. As time went on, more and more information on the variation in different parts of the world became available; it became very apparent that the construction of a Variation chart was no simple task, as not only was the distribution irregular, but the variation in any place was changing over the years. A number of variation charts were published in the eighteenth century, but the first Admiralty Variation chart was not printed until 1858. These charts enabled ships to obtain a very approximate longitude by observing variation, but when the chronometer was developed longitude could be more accurately defined by astronomical means. The variation charts then began to be used as a means of correcting the compass heading—the purpose for which they are used today.

Modern techniques, and the vast store of information garnered over the years, now enable Variation to be predicted accurately for any given place and time. The earth's magnetic field, in other words, has been precisely plotted and its yearly change established. At the present time, the variation in Britain is changing annually by about 8 minutes of arc. There are also small seasonal and daily changes in variation although these are not of significance to the practical navigator. The peculiar distribution of places having equal variation has been charted and places where local magnetic effects are strong (due to a large amount of ferrous rock near the surface) discovered.

To the modern seaman, variation is no mystery but allowance must still be made for it if an accurate TRUE COURSE is to be steered at sea. Remember it is the angle between TRUE NORTH and MAGNETIC NORTH and is defined as WESTERLY VARIATION if Magnetic North is to the West of True, and EASTERLY if Magnetic is East of True North. (See Figures 3A and 3B).

The value of the variation in any particular place is shown on the chart of the area; on ocean charts 'isogonic' lines, that is to say lines joining places of equal variation, are drawn at

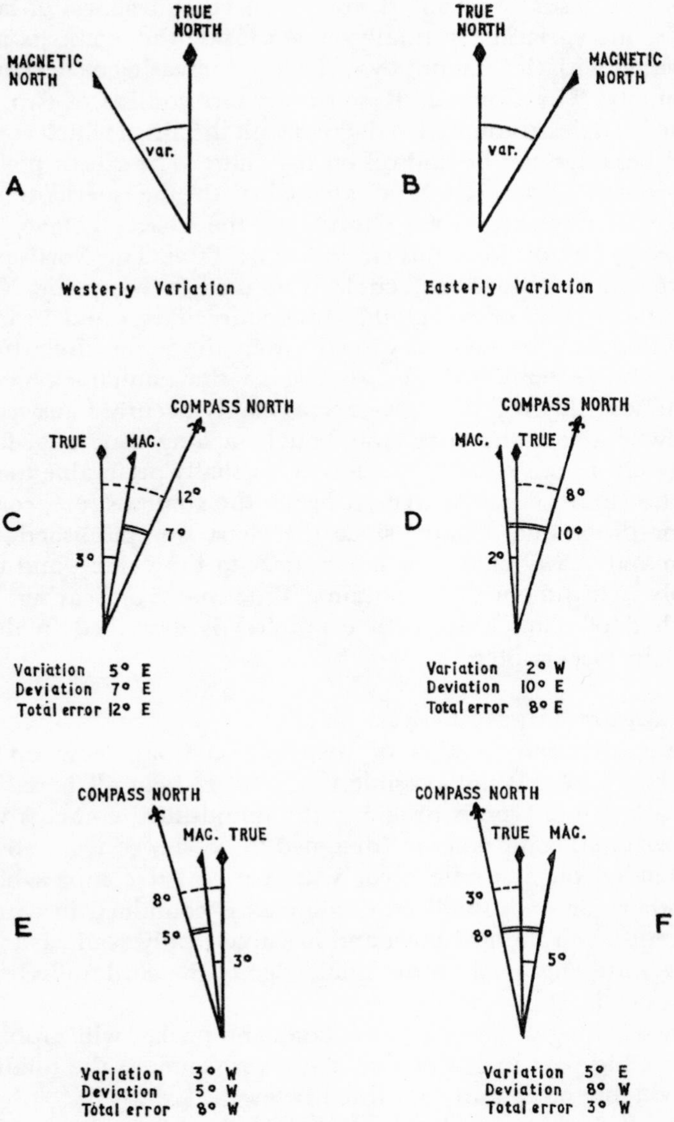
TRUE NORTH
MAGNETIC NORTH
var.
A
Westerly Variation
TRUE NORTH
MAGNETIC NORTH
var.
B
Easterly Variation
COMPASS NORTH
TRUE
MAG.
12°
7°
5°
C
Variation 5° E
Deviation 7° E
Total error 12° E
COMPASS NORTH
MAG.
TRUE
8°
10°
2°
D
Variation 2° W
Deviation 10° E
Total error 8° E
COMPASS NORTH
MAG.
TRUE
8°
5°
3°
E
Variation 3° W
Deviation 5° W
Total error 8° W
COMPASS NORTH
TRUE
MAG.
3°
8°
5°
F
Variation 5° E
Deviation 8° W
Total error 3° W

Figure 3

intervals (usually of one degree). On coastal charts of larger scale, the variation is usually shown inside the 'compass rose', together with the amount by which it is increasing or decreasing annually. The Compass Rose on a chart consists of two concentric circles graduated in degrees with the aid of which courses and bearings can be laid off on the chart. The North point of the outer (True) circle is aligned with the meridian and therefore indicates True North on the chart; courses and bearings laid off from this circle will be True. The North point of the inner (Magnetic) circle is aligned to Magnetic North *at the date the chart was printed.* Magnetic courses and bearings can therefore be laid off directly from the inner circle but it must be remembered that, owing to the annual change in variation, this will only give a reasonably accurate answer for a few years; if the chart was bought a long time ago, fairly large errors can result. Therefore it is usually preferable to look at the variation value written inside the compass rose, correct it for the annual change since the chart was published—the date and yearly change is shown next to the value—and then apply it arithmetically to obtain a True course or bearing. The method of doing this (with examples) is described in detail later in this chapter.

*The Modern Magnetic Compass*

There are many makes of magnetic compass now on the market, and it is not possible to describe them all here. The most common type is probably the liquid-filled compass with a horizontal compass card (designed to be viewed from above) suspended on a needle pivot within a sealed compass bowl. Another common small craft compass is contained in a transparent hemispherical bowl and has a vertically read card; the figures are engraved on the thick edge of the card and viewed from eye level.

In selecting a compass for a boat, the pocket will probably play a big part in the choice. However, some of the qualities to look for (or against) are listed below:

(a) *It should preferably be gimballed,* i.e. suspended in two rings which enable the bowl to keep level and comparatively still however much the craft pitches or rolls. If a compass is not gimballed, the card may foul the bowl as the boat moves and thus give an inaccurate and erratic heading. In a gimballed compass, the clearance between the card and the bowl allows a tilt of about 10 degrees which is quite adequate to cope with a sudden, sharp movement. Some compasses used in boats are of the un-gimballed type, similar to (or indeed may be) aircraft compasses. Most simple aircraft compasses are un-gimballed, but sprung in some fashion underneath by felt pads, springs, rubber cushions and the like. They also have a greater clearance between the card and the cover glass; normally about 20°. This is sufficient in an aircraft for as the aircraft banks steeply (say 90°), centrifugal force causes the compass card to take up a false horizontal, dipping one side down in the direction of bank. However in a boat, the motion is different and unless there is adequate clearance, the card may foul. An analogy can be made with a bucket of water; if this is swung round and round by the handle (aircraft turning) the water will remain in the bucket: put the bucket on a swing, however, where the direction of motion is reversed (boat rolling), and it will slop over.

(b) *Dry Card or Liquid Compass*

*Dry Card.* In this type, a graduated card made of paper or mica, with magnets attached beneath it, is suspended by a cap and pivot bearing. The lighter the card, the less friction and wear is imposed on the pivot. There is a limit to the amount the weight can be reduced, as the magnets must be strong enough to move the card physically. Secondly, more than one magnet must always be used, as a single needle, suspended at its centre, would tend to align itself in the direction of any motion imposed on it; for example, if the boat rolled, a single bar would try to swing in the direction of roll due to its moment of inertia.

Apart from the inevitable wear on the pivot, the main disadvantage of a dry card compass is its comparative instability, particularly in a small craft with a short period of roll. This wandering of the compass can be reduced to some extent by making the compass have a long period of oscillation—by attaching an aluminium or other alloy ring to the perimeter of the card, or by surrounding it with a copper bowl in which magnetic eddy currents set up by the magnets tend to oppose movements of the card.

The main advantages of a Dry Card compass are its cheapness and simplicity.

*Liquid Compass.* If the bowl containing the compass card is made watertight and filled with a liquid, some immediate benefits accrue. The liquid has a frictional damping effect on the card thus reducing any unwanted quick oscillations. The card and magnets can also be made just sufficiently buoyant to reduce the weight and friction on the pivot to a minimum. These are overriding advantages, and it was for this reason that many large ships went over to liquid compasses in about 1906.

There are, however, certain difficulties which must be overcome in the design of a good liquid compass. Obviously the liquid must not be allowed to evaporate or freeze. The liquid normally used is a mixture of water and alcohol (about 47% alcohol). To prevent evaporation, the bowl must be completely sealed, but must allow the liquid room to expand as the temperature rises and then contract when the temperature falls without any bubbles forming. In good compasses, these factors are allowed for by the fitting of expansion chambers or bellows, usually made of copper, which expand and contract as the pressure changes inside the bowl.

Another problem which the designers had to face was that of liquid 'swirl'. As the boat alters course, some of the liquid is drawn round the bowl by friction. The effect is mainly felt by the liquid near the edge of the bowl. To prevent this swirl affecting the card, a good liquid compass has a card of considerably smaller diameter than the bowl. This also reduces the effect of another undesirable feature which could be encountered in bad weather. With a heavy roll or pitch, the card becomes tilted relative to the bowl (particularly in a compass with no gimbals) and liquid tries to pass from below the card to above it. If the gap between the bowl and card is adequate, this liquid movement will have no serious effect. Some compasses have domed, transparent top glasses and this shape of case tends to reduce the swirl effects, whilst at the same time providing some magnification of the card.

Occasionally, if a compass is old and its joints perished, or perhaps it has been damaged in some way, a bubble will appear in the liquid. It may be that the compass is still usable in this condition, but there is always a danger that it may read incorrectly, particularly if the bubble is large and has been allowed to remain in the compass for some time. There is no doubt that the best solution is to take the compass to a reputable compass maker for the removal of the bubble. However in emergency, the bellows cover or the top of some compasses can be removed by undoing the securing screws, lifting the top glass and rubber seal, and then refilling the bowl with alcohol and water. It is not easy to do this without trapping a certain amount of air. If alcohol cannot be had, distilled water is probably the best alternative additive, but dilution will, of course, raise the freezing point of the liquid. Neat alcohol, on the other hand, should not be used as too strong a solution will probably damage or raise the paints in the bowl or on the card.

*Compass Graduations*

Nowadays most compass cards are graduated from 0 degrees (North) right around clockwise to 359 degrees—the 360 degrees of a circle. Thus 0° is North—090° East—180° South—270° West—and back again to North or 360°. But, partly as a tradition from sailing ship days, and partly because it is difficult to steer a boat to within 1 degree, many cards are still marked with some of the old compass points.

The old system was to divide the angles between the Cardinal points (North—East—South—West) first into quadrants and then further into points. A point was of 11¼ degrees, and there were therefore 32 points in the full circle. Each point was named—'Boxing the Compass'—as shown in Figure 4.

Even on a modern compass, the Cardinal and Quadrantal points (NE—SE—SW—NW) are usually marked on the card inside the degree graduations. To set a particular course, the appropriate graduation on the card is brought opposite the Fore-and-Aft mark on the compass bowl—this is called the 'Lubber's Line'. For example, if North East is the course required, NE (045°) must be brought opposite the Lubber's Line by use of the helm.

*Alignment of the Compass*

Great care must be taken when the compass is first installed in a boat to ensure that the Lubber's Line is truly aligned fore and aft, i.e. parallel to the boat's keel. This is a comparatively simple thing to do if the compass is on the centre line and high enough for a sighting to be taken from it of the bow or forestay. Cockpit compasses and those offset from the centre of the boat are not so easy to align. One way of dealing with the alignment

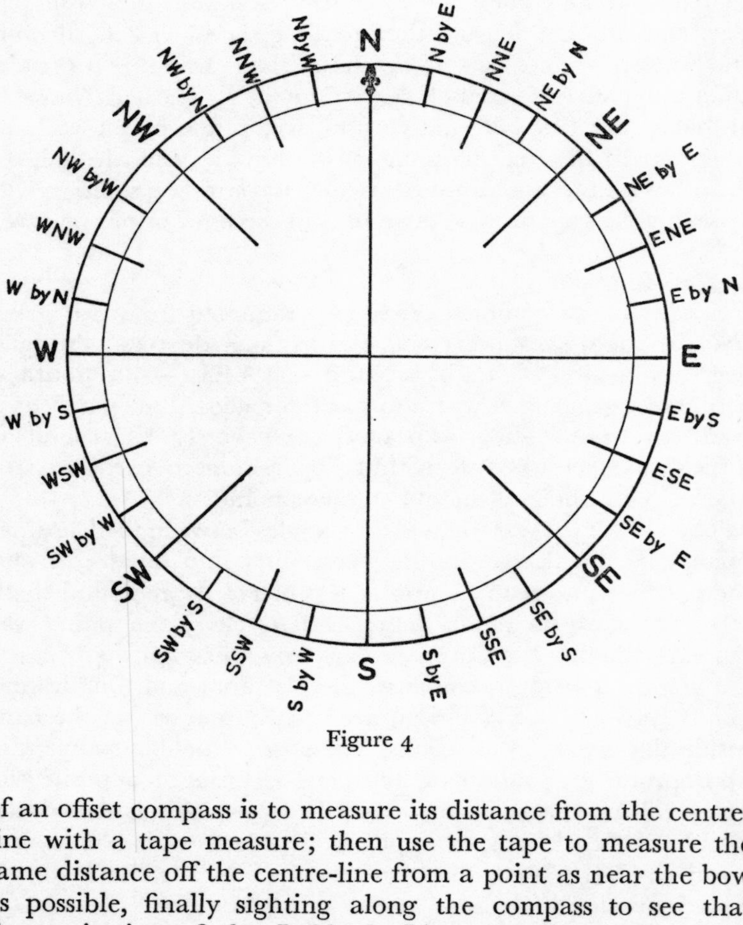


Figure 4

of an offset compass is to measure its distance from the centre-line with a tape measure; then use the tape to measure the same distance off the centre-line from a point as near the bow as possible, finally sighting along the compass to see that the projection of the Lubber's Line cuts the tape at this measurement. If it does not, move the compass bowl until it does!

Cockpit compasses are most easily checked by comparison with other compasses (Chapter 2—Coefficient A).

*The Compass in a Boat*

Up to now, only the effects of the Earth's magnetic field on a compass has been described. However, when the compass is installed in a boat, it is bound to feel the magnetic field from any magnetic material present in the craft, whether this be part of the hull or portable articles and fittings. The refrigerator in a luxury cabin cruiser might be, for instance, not too healthy an object to site too close to the compass position; similarly, in a dinghy, an iron bolt near the compass could cause errors.

The old wooden sailing ships had little magnetic material on board—even the three-decker at Trafalgar was comparatively free of disturbing influences on its compasses. In fact it was not until iron ships were built that mariners and scientists began to become concerned with the troubles caused by placing the compass near or within large masses of iron. (In the old days, it was not unknown for pistols and other weapons to be stowed in the binnacle (pedestal for the compass) as highly disturbing bed-mates.

Before going further, you might like to be reminded of the magnetic properties of ferrous metal. A magnet can be made by stroking or vibrating a piece of hard iron in a strong magnetic field. Similarly, when a vessel is built, the hard iron/steel in the boat is subjected to a lot of hammering or vibration whilst the craft is lying for a long period in a fixed direction on a slip or building berth. As a result, the hard iron is slowly magnetised by the Earth's magnetic field. The direction of magnetisation will depend on the heading of the vessel at the time of building. The magnetism in hard iron is called PERMANENT magnetism. This iron will not lose its magnetism unless it is subjected to similarly violent treatment to that which magnetised it in the first place.

Other iron in the boat is of a softer nature. This type of metal can be easily magnetised but does not retain its magnetism once the magnetic field affecting it is removed. It remains a hazard to the compass because if a piece of soft iron is placed close enough to affect the needle, its energisation by the Earth's field will cause a variable effect on the compass as the vessel alters course. Soft iron is therefore associated with INDUCED magnetism.

Even more difficult a problem is caused by the magnetisation of iron that is neither hard nor soft. This is sometimes called 'Intermediate Iron', and the magnetic effect associated with it SUB-PERMANENT magnetism. A piece of iron of this nature can be magnetised quite easily, but only retains this magnetism over a limited period.

*Deviation*

The modern boat owner is therefore faced with the prospect that a good deal of iron, of one sort or another, may be within spitting distance of the compass. In a well designed craft, this factor will have been taken into account and the best possible position chosen for the compass, non-ferrous metals used near it, etc. However, there is a practical limit on what can be done, particularly in a craft with a metal hull. Tables, called 'Safe Distance' tables, exist which list a wide number of objects—instruments, echo sounders, radio sets etc.—against their minimum safe distances from the compass bowl. These tables are principally of assistance to the boat builder, but may be of interest to the owner should he wish to add additional equipment or stow portable/potable objects on board.

Any magnetic material not outside these safe distances may cause a significant deflection or deviation of the compass needle. The amount of deviation caused will vary on different courses depending on the relative position of the compass and the object, and of the type of iron of which the latter is made.

The resulting 'Angle of Deviation' on a particular *heading* is defined as the angle between Compass North and Magnetic North. An angle of Deviation described as 2 degrees West means that Compass North is 2° W of Magnetic North. This element of error is therefore defined in a similar way to Variation, although its cause is different and (unlike Variation) its value will vary as the boat alters course.

The TOTAL COMPASS CORRECTION is a combined allowance for the errors caused by VARIATION and DEVIATION. Examples are shown in Figures 3C, 3D, 3E & 3F.

*Deviation and 'Swinging'*

The best magnetic position for a compass in a boat would probably be on top of the mast, but this would be a little inconvenient for the owner. The boat builder therefore chooses the best practical position for the compass consistent with ease of use at sea. Boats vary a great deal in their magnetic properties, but even in the best circles some magnetic influences are likely to be present. The Deviation of the compass on ALL headings must therefore be found.

This is normally done by 'Swinging Ship', i.e. swinging the boat slowly through North—East—South—West (but not necessarily in that order) through the whole 360° whilst checking the compass against a known True bearing.

Qualified Swinging Officers are employed for checking the magnetic compasses of big ships, but part of the job is well within the capabilities of the average boat owner. It is not difficult to find the deviation of a compass although a bit more knowledge of the subject is needed to remove the errors once they have been found. (The latter process, Compass Correction, will be described in Chapter 2). Once established, the Deviation values can be used to correct the compass heading so that a true course can be steered.

*WHEN*—Ideally, a boat should be swung to observe compass Deviation:

(a) On first delivery to the owner.
(b) At the beginning of a new sailing season.
(c) On a large change in Magnetic Latitude (say 10°).
(d) If any new fittings or equipment are put into the boat . . . remember that new outboard motor.
(e) If you have any cause to mistrust it!
(f) Finally, and most important—if any of the correctors have been changed or moved. (Note small son's movements).

*WHAT STATE*—Before a craft is swung, it is as well to have everything on board in its normal seagoing position—the anchor stowed, engine (if outboard) in its normal place, Mum's knitting needles firmly entrenched in her knitting and not near the compass—in fact all portable articles such as knives, keys, cans of beer kept well away from the scene of operations. Similarly, keep the boat away from other iron hulled craft, jetty piles and large iron mooring buoys during the swing.

*WHERE*—It is normally easier to swing underway out in open water, providing the weather is reasonable, and a distant charted object (church—lighthouse—edge of a cliff) can be seen at a great enough range. If the boat is kept moving within a circle of about 200 yards diameter over the ground, then the distant object used need not be more than about six miles away—a maximum bearing error of 1° is possible at this range. If closer objects are used, then the possible errors are obviously magnified. If visibility is bad, close range transits can be used instead.

Another method is to circle tightly round a buoy or other navigational mark, or even to secure the boat by the bows to the mark. The stern can then usually be swung by use of the engine or with the aid of a 'volunteer' in a dinghy; in these circumstances the object being used for bearings need not be more than about 2 miles away. The danger of using this method lies in the fact that the mark, and particularly a heavy steel navigational buoy, may be a considerable magnet in its own right.

Finally, a swing can be carried out at anchor—the stern of the boat being pulled around by a dinghy or by 'warping'. The latter is done by passing a rope outboard clear of all guardrails, stays and stanchions. One end of the rope is attached to the anchor cable under the bow; the other end is secured to a cleat or bollard in the stern. If the anchor chain is veered (let out), the stern will be pulled towards the side around which the rope has been passed. Obviously it is not possible to turn the boat through the whole circle by this method—the rope must be changed to the other side at some stage; neither must there be much of a current or wind. Either way, dinghy or warping, the anchor not being in its sea-going position is a possible cause of error here.

*WHAT MARKS*—The use of distant objects has already been mentioned. To obtain the true bearing, the bearing of the object from your swinging position is merely read off the chart—providing, of course, you know precisely where you are! (this is one reason why the circle round a navigational mark has its attractions).

Another way of obtaining an accurate bearing is to use a TRANSIT. A transit is merely the bringing of two objects exactly into line. If the objects are individually charted, it follows that the craft must lie on an extension of the line joining the two objects and that the bearing of the objects from the boat must also be known. Transits are used a good deal in pilotage, and it is not usually difficult in a well charted area to select suitable marks. The distance from these objects is not vital. Ideally, the

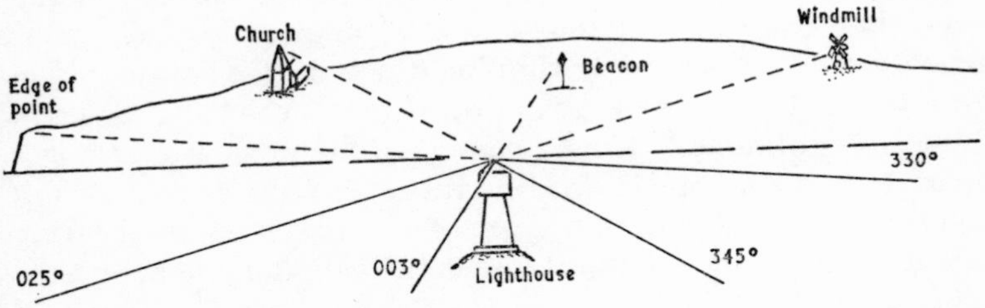


Figure 5

Note. The large angles shown here between each transit are too great for accurate interpolation between them. In practice, choose transits separated by only a few degrees.

craft should be twice as far from the nearest object as the distance between the two objects, but this is not important. All that need be remembered is that a transit becomes more sensitive the closer you get to the marks and conversely, the further away, the less sensitive it will be.

When swinging underway, it is clearly tricky to keep altering course whilst staying 'on' the transit. It can be done by backing and filling, but this takes time, and a better dodge is to use a single charted front mark and a number of charted back marks. The bearing of successive transits can then be read off the chart, and if the boat at any given time is not exactly on any of them, it is comparatively simple to interpolate the true bearing. (See Figure 5.)

If plenty of transits are available, and they are well chosen, the boat can be swung easily from heading to heading over a large expanse of water without the craft ever getting outside the area covered by these transits.

*How to take the Bearing*

Whatever method is used, an accurate compass bearing of the charted distant object, transit etc. must be taken. The larger craft are often fitted with a compass from which a bearing can be taken directly; some form of sighting attachment is fitted on top of the compass for this purpose. The earliest designs took the form of a notched sight, not unlike the back-sight on a rifle. Nowadays, these bearing devices, normally called azimuth circles, have a prism (and sometimes a telescope) mounted on a ring fitted over the compass card. The object is viewed through the 'V' notch on top of the prism, and at the same time the reflected image of the compass card is visible beneath the 'V'. A hairline running down the face of the prism from the bottom of the notch appears to cut the card image at the bearing obtained. The advantage of the prismatic arrangement over the old open sight is that the observer's eye need not be exactly aligned with the centre of the card when the bearing is taken. Providing the object is seen through the notch, and the reflected image of the card in the prism, the bearing will be accurate.

If the compass in your boat is fitted with an azimuth circle, well and good. The process of taking bearings for ordinary navigation as well as swinging for compass adjustment is straightforward. However, an azimuth circle is not much use if the compass itself is not placed in the boat so that a good all round view of the horizon can be obtained from it. Many compasses, such as those used for steering alone, are sited well down below the gunwale. The point here is that if the boat has one compass with good horizon visibility, bearings can be taken with its azimuth circle and any other compasses compared with it. If the boat has no compass with a view, or azimuth circle, then life is more difficult. In this event, a bearing plate should be used for swinging. Briefly, this consists of a graduated non-magnetic card which can be turned manually by means of a knob on its face-plate, over which is fitted an azimuth circle. The instrument is placed high up on its tripod on the centre-line of the boat, and its Lubber's Line accurately aligned with ship's head (by taking a sight on the bow). The next boat's compass heading to be checked—e.g. North East (045°—is pre-set on the bearing plate opposite the

Lubber's Line. As the boat reaches this heading (ship's head 045° by the compass being checked) the shore bearing is taken from the bearing plate. The bearing obtained at that instant is directly equivalent to that which would have been obtained, were it possible, from the compass direct.

Perhaps hand bearing compasses ought to be mentioned here, as they provide an alternative method of taking bearings; albeit a fairly unreliable method in this context. They are simple hand held magnetic compasses with some form of sighting device for taking bearings. The marine variety are normally fitted on top of a wooden handle. The snag with them is that, being portable, they suffer a varying and unknown amount from any local magnetic influences in the boat and are therefore not really reliable as checking instruments. The hand bearing compass could suffer a worse deviation than the fixed compass, and certainly its deviation will alter as it is moved from place to place in the boat; it will always be an unknown quantity. However, if these limitations are fully appreciated, the hand bearing compass can be an invaluable aid to yachtsment for day to day use, and could be used, at a pinch, for checking another compass. When using it for this purpose, select a place in the boat where it is least likely to feel any magnetic influences. In a new boat, this will be largely a matter of guesswork, but in a known craft it will soon become apparent—from compass checks on passage and practical fixing—in which position the hand bearing compass can best be used and how good its results have been. Obviously, if it is known to perform accurately on board, there is nothing against using it for checking another compass.

*The Swing*

The business of swinging the boat round from one heading to the next should be done as slowly as possible. Ideally, one complete circle should take about an hour, but patience, or the call of less mundane things, may not permit this. Do not, however, be tempted to swing too fast as this may have an adverse effect due to certain induction influences on the compass. Forty minutes should be the minimum. This process has only to be done once, if it is done well, and you will sleep the sounder for it later on.

With a well placed compass, the taking of bearings should also be fairly simple. All you need is one bearing on every two points, or better still, every ten degrees; thus sixteen or thirty-six headings and their related bearings.

You may find that on a particular heading the shore object is 'wooded', i.e. obscured from the compass by some part of the boat's structure. Therefore, if possible, have up your sleeve the true bearing of an alternative object. You will also find, with experience, that using the modern azimuth circle it is quite easy to take a good bearing 'through' some obstruction, providing this does not have too large a spread across the field of view.

*Procedure.* In calm weather, and with the use of an engine, it is possible to carry out a swing alone and unaided. However, it is much simpler if there is another person on board to handle the boat whilst the 'swinger' takes the bearings and notes down results. Before starting the swing, prepare a notebook listing in one column the sixteen (or thirty-six) ship's heads and in the other leave spaces for the compass bearing of the object against each ship's head. If using more than one distant object or interpolating between transits, another column will be needed for true bearings. It does not matter on which heading you start, nor does it matter in which direction, clockwise or counter-clockwise, the boat is swung. Steady the boat therefore on the nearest convenient heading and take an accurate bearing of your chosen object. Write this bearing down opposite the ship's head concerned. Then swing slowly on to the next heading and repeat the process, noting each time the compass bearing (and the true bearing if using more than one object). Continue the swing until all headings have been checked.

A few tips—you must take great care in getting accurate bearings but it is not vital that the ship's head be exactly on the chosen heading when the bearing is taken. A couple of degrees either way will not matter. Secondly, if you decide to use transits, and interpolate between them whenever necessary, it is as well to draw a small diagram in your notebook showing the true bearings of the transits so that you can interpolate between them on the spot and not have to keep referring to a chart.

*Tabulating the Results.* Having completed the swing, the results must be tabulated for future use. An example is given below, followed by explanatory notes:

| A | | B | C | D | E |
|---|---|---|---|---|---|
| *Ship's Head by Compass* | | *True Bearing of Distant Object/ Transit* | *Magnetic Bearing of Object* | *Compass Bearing of Object* | *Deviation* |
| Deg. (°) | | Deg. (°) | Deg. (°) | Deg. (°) | Deg. (°) |
| N | 000° | 223° | 228° | 232° | 4° W |
| NNE | 022½ | 223 | 228 | 233½ | 5½ W |
| NE | 045 | 223 | 228 | 234 | 6 W |
| ENE | 067½ | 223 | 228 | 233 | 5 W |
| E | 090 | *279 | 284 | 287 | 3 W |
| ESE | 112½ | 223 | 228 | 230 | 2 W |
| SE | 135 | 223 | 228 | 229 | 1 W |
| SSE | 157½ | *279 | 284 | 282 | 2 E |
| S | 180 | 223 | 228 | 223 | 5 E |
| SSW | 202½ | 223 | 228 | 221 | 7 E |
| SW | 225 | 223 | 228 | 223 | 5 E |
| WSW | 247½ | 223 | 228 | 224 | 4 E |
| W | 270 | 223 | 228 | 225 | 3 E |
| WNW | 292½ | 223 | 228 | 226 | 2 E |
| NW | 315 | 223 | 228 | 228 | Nil |
| NNW | 337½ | 223 | 228 | 230 | 2 W |

(1) * Denotes second distant object used, as the first was 'wooded' on these bearings.
(2) Variation—at place of swing—5° West.

*Explanation of this table*

COLUMN 'A'—SHIP'S HEAD BY COMPASS

The successive Ship's heads—by the compass being checked—are written in this column.

COLUMN 'B'—TRUE BEARING OF OBJECT

Two distant objects have been used in the course of this swing. The True bearings of the objects from the swinging position are taken off the chart and entered in this column.

COLUMN 'C'—MAGNETIC BEARING OF OBJECT

This is the bearing of the object after the *Variation* has been applied to the True Bearing. The Variation in this locality was 5° West (as read off the chart). This means that Magnetic North is 5° West of True North, and a Magnetic bearing of 228° is therefore the equivalent of a True bearing of 223°.

To expand on this point a little—if going from Magnetic bearings or courses *to* True bearings or courses you *subtract Westerly* and *add Easterly* Variation. The converse applies if going from True to Magnetic. A True bearing of 223°—Variation 5° West—is the equivalent of a Magnetic bearing of 228°. A Magnetic bearing of 284° = True bearing of 279° with the same Variation applied.

### COLUMN 'D'—COMPASS BEARING OF OBJECT

As the boat swings through each successive ship's head by compass (Column A) a compass bearing of the shore object is taken. These are noted in Column D.

### COLUMN 'E'—THE DEVIATION TABLE

Variation has already been applied to give Magnetic bearing (Column C) from the True. The difference between Columns C and D must therefore be the Deviation of the compass on each particular heading. Taking two examples from the table:

(a) Ship's head by compass 000°
Compass bearing of object 232°
Magnetic bearing of object 228°
As the compass is over-reading, its North must be to the West of Magnetic North, and therefore the Deviation is Westerly—4° West in fact.

(b) Ship's head by compass 202½°
Compass bearing of object 221°
Magnetic bearing of object 228°
The compass is now under-reading and therefore the Deviation must be 7° East.

*Practical application of Variation and Deviation*

Once the Deviation has been obtained by swinging, it can and should subsequently be used for correcting compass courses and bearings. Needless to say, these corrections must be properly applied—otherwise shipwreck becomes handsomely probable! The rules for applying variation and deviation have already been mentioned, but to help remember them a mnemonic is frequently used. This is the word CADET. 'Compass Add East to get True'. Thus whenever a Compass course has to be converted to a True course, and the variation/deviation is Easterly, then the value must be added.

Some examples are given below:

(a) GOING FROM COMPASS HEADING TO TRUE HEADING

| | | | |
|---|---|---|---|
| Variation | 6° W | Compass Course | 355° |
| Deviation | 4° W | | |
| | 10° W = | Total Correction = | −10° |
| | | True Course | 345° |

| | | | |
|---|---|---|---|
| Variation | 14° W | Compass Course | 012° |
| Deviation | 6° E | | |
| | 8° W = | Total Correction = | −8° |
| | | True Course | 004° |

(b) GOING FROM COMPASS BEARING TO TRUE BEARING

Corrections are applied to Compass bearings in exactly the same way, but it is important to remember that the value of the *Deviation* used must be that for the *Ship's Head by Compass* and not that for the direction of the bearing.

Compass course 000°—Deviation on this course (from table) 4° W

| | | | |
|---|---|---|---|
| Variation | 10° E | Compass bearing of object | $224\frac{1}{2}$° |
| Deviation | 4° W | | |
| | 6° E = | Total Correction = | +6° |
| | | True bearing of object | $230\frac{1}{2}$° |

Compass course 180°—Deviation on this course (from table) 5° E

| | | | |
|---|---|---|---|
| Variation | 10° E | Compass bearing of object | $224\frac{1}{2}°$ |
| Deviation | 5° E | | |
| | 15° E = Total Correction = | | +15° |
| | | True bearing of object | $239\frac{1}{2}°$ |

In the above examples, the compass bearing of the object is the same, but due to the difference in boat's heading when the two bearings were taken, the *True* bearing is different.

Compass course 180°—Deviation on this course 5° E

| | | | |
|---|---|---|---|
| Variation | 12° W | Compass bearing of object | 003° |
| Deviation | 5° E | | |
| | 7° W = Total Correction = | | −7° |
| | | True bearing of object | 356° |
| | | (Course of 003° = 363° | |
| | | 363 − 7 = 356°) | |

(c) GOING FROM TRUE COURSES TO COMPASS COURSES

This is a converse case when, for instance, you wish to know the Compass course to steer to make good a True course read off the chart. When going from True to Compass you must apply the corrections in the opposite sense.

| | | |
|---|---|---|
| True Course needed | 000° | Variation 20° W |
| Variation (20° W) | +20° (sign reversed) | |
| Magnetic Course | 020° | |

You will note that the Variation has been applied first and separately here. This is because the variation is large and therefore you will arrive at a compass course that will not be near North. In this case, the value of *Deviation* to be read from the table should be that for a heading of NNE ($022\frac{1}{2}°$).

Magnetic Heading 020°
Deviation ($5\frac{1}{2}$° W) $+5\frac{1}{2}$° (sign reversed)

Compass Heading $025\frac{1}{2}$° To make good 000° True.

The second example combines these two steps in one:

True Course needed 185° Variation 10° E

Variation 10° E—Therefore Magnetic Heading 175°.
Deviation 4° E—for heading of 175°

14° E = Total Correction.
True Course needed 185°
Total Correction 14°

Compass Heading 171°

If the deviation was large, it might be necessary to make a second approximation as the calculated compass heading could require the application of a different deviation value.

(d) GOING FROM TRUE BEARING TO COMPASS BEARING

It is sometimes necessary to convert True bearings into Compass bearings, e.g. when running in on a bearing. Again the Deviation value for the Heading must be used:

True bearing of object 293° Ship's Head by compass 270°

Variation 12° W
Deviation 3° E (for heading of 270°)

Total Correction 9° W
True bearing of object 293°
Total Correction +9°

Compass bearing 302°

# 2

# Correcting the Compass

As has been described, it is fairly simple to obtain a deviation table for your boat's compass by carrying out a swing. If the deviations obtained are fairly small (say less than 5° on any heading) it may be as well to let sleeping dogs lie and merely use the known deviations to correct your courses and bearings at sea. However, it may be that your compass is badly sited in your boat through being placed near some magnetic material which is part of the boat's structure. Deviations of more than 40° have been known!

These deviations can be removed by correcting the compass.

Compass correcting is a bit of an art which needs a good depth of knowledge and practice if a perfect result is to be achieved; but there are some things that the amateur can do with a good chance of improving matters and without much risk of a debacle. This chapter is therefore addressed to those whose compass points South when it should be pointing North, and who consequently would like to do something about it. To those in happier situations, whose compasses never tell a lie, a quick shift to the next chapter is recommended.

First study the deviation table produced as a result of the swing, as this will furnish some good clues to the cause of any trouble. The table previously given in Chapter 1 will be used as an example for this analysis—although the deviations given in that table are not too bad.

The various groupings of errors are normally broken down into co-efficients. Do not worry about this term—it is merely a convenient way of expressing the total error due to a particular cause or group of causes.

*Coefficient A*

This is found by taking the sum of all the deviations found during a swing and dividing the sum by the number of headings.

Easterly deviations are considered to be plus, and Westerly minus. So if you add up all the deviations (given in the table on page 31 being used as an example) you will find that:

Easterly (positive) total $+28°$
Westerly (negative) total $-28\frac{1}{2}°$

Sum $= -\frac{1}{2}°$

Therefore Coefficient A $= \dfrac{-\frac{1}{2}}{16} = \dfrac{-1°}{32}$ (Insignificant)

If, however, you do come out with a significant quantity for Coefficient A, this may be caused by:—

1. The Lubber's Line of the compass not being truly fore and aft. (A fairly common complaint with offset and steering compasses where there is no direct method of checking).

2. Friction in the pivot of the compass.
(Deflect the card a few degrees with a magnet. It should, when the magnet is removed, return to its original settling position—providing the boat's heading is kept steady during this procedure.)

3. Swing conducted too fast.
(Various induction effects can cause trouble if the boat is swung too fast.)

4. Other defects (e.g. a bubble) in the compass.

*Coefficient B*

Found by taking the sum of the deviation on East and West courses, changing the sign on West, and then dividing by two. Thus, from the table again:—

Deviation on East $3°\ W = -3°$
Deviation on West $3°\ E = -3°$ (sign changed)

Sum $= -6°$

Therefore Coefficient B $= \dfrac{-6}{2} = -3°$

Coefficient B is caused by the boat's FORE AND AFT PERMANENT magnetism. The diagrams in Figures 6A and 6B illustrate this.

In these examples, the boat is shown as having a 'Red' bow and 'Blue' stern. This is no reflection on the owner! Merely that the boat was magnetised in this sense on building, and the polarity could well be the opposite in another case. The North (Red) end of the compass needle is attracted by the Blue pole of the Boat's magnetism and repelled by the Red. Therefore, on an Easterly course the compass needle is pushed away from Magnetic North towards the West causing Westerly Deviation. On a Westerly course, the needle is pushed in the opposite direction, towards the East. On North and South courses, there is no deviating force as the boat's magnetism is overcome by the Earth's field (the compass needle senses either a gain or loss of directive force). On intermediate courses, only a proportional effect will be felt.

To correct for any Coefficient B, Permanent Magnets must be placed with their ends pointing fore and aft in opposition to the boat's field. The general rules for placing corrector magnets are given in a later paragraph.

*Coefficient C*

Found by taking the sum of the deviations on North and South, changing the sign on South, and dividing the sum by two. Thus:–

Deviation on North $4°W = -4$
Deviation on South $5°E = -5$

Sum $= -9$

Therefore Coefficient C $= \frac{-9}{2} = -4\frac{1}{2}°$

Coefficient C is caused by the boat's ATHWARTSHIPS PERMANENT magnetism. Its effect is very similar to that caused by Fore and Aft magnetism (Figures 6C and 6D).

Athwartship permanent magnets are placed in opposition to the boat's permanent magnetism to correct for the latter.

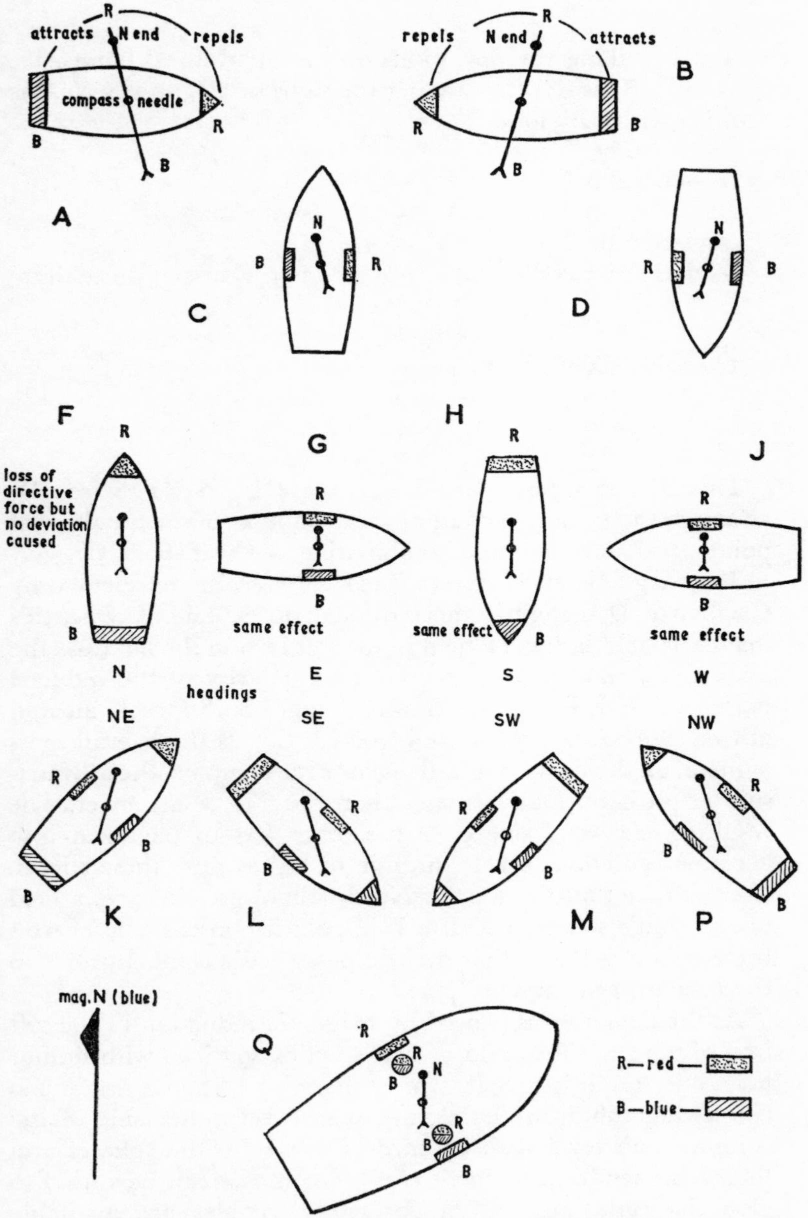


Figure 6. Deviation due to boat's permanent magnetism (A–D)
Deviation due to Induced Effects of Earth's field in boat's soft iron (F–Q)

*Coefficient D*

Found by taking the deviations on the quadrantal points, i.e. NE—SE—SW—NW, changing the signs on SE and NW. The sum is divided by four. Thus:

| | | | |
|---|---|---|---|
| Deviation on NE | 6° W | = −6° | |
| Deviation on SE | 1° W | = +1° | (sign changed) |
| Deviation on SW | 5° E | = +5° | |
| Deviation on NW | Nil | = 0° | (sign changed normally) |
| | Sum | = 0° | |

$$\text{Therefore Coefficient D} = \frac{0}{4} = 0°$$

This is a fortunate case where there is no Coefficient D, although there was a deviation value on most of the quadrantal points (probably due to a combination of Coeff. B. & C).

The magnetic effect responsible for the errors represented by Coefficient D is slightly more difficult to explain. The Earth's magnetic field induces a field in the soft iron in the boat. As the boat alters course, the direction and polarity of the induced magnetic fields in the metal also change. Both fore & aft and athwartship components are involved, but as the overall mass of metal on the beam is usually nearer the compass, the athwartship effect is normally greater than the fore & aft effect. The resulting excess of one over the other has to be taken into account and corrected. Figures 6F to P illustrates these effects. You will see that the Earth's North (Blue) pole induces a field in the boat's soft iron with a Red polarity on the side nearest the North pole; also, that the compass needle is much closer to the induced athwartship 'poles'.

As the deviation is caused by magnetic induction in the soft iron of the boat, it should be corrected by soft iron with similar magnetic qualities. Soft iron spheres, or sometimes boxes containing soft iron chains, are placed on either side of the compass and level with its card. The size of the spheres and their distance from the centre of the compass can be varied to give the right amount of correction. Tables are available (Admiralty Compass Department pamphlet CD 13B is one) which show the correct size of sphere to be used and their distances from the centre of the compass for various values of Coefficient D.

Figure 6Q illustrates the correction effect of such spheres on one particular heading. The North (Red) end of the Compass needle is repelled by the 'Red' Port side of the ship—from induced magnetism. This repulsion is balanced out by the side of the spheres nearest the compass being induced with the opposite polarity.

*Heeling Error*

This has no related coefficient, but it is due to the combined influence of the boat's Permanent Vertical magnetism and Induced Vertical magnetism. When the compass needles are horizontal, they are not affected by the boat's vertical magnetism. However, as the boat rolls, a component of its vertical magnetism has its effect on the needle. This is not easy to illustrate graphically, but the result is a pull on the compass needle one way when the boat rolls to port, and in the opposite direction on a starboard roll. The consequence is an oscillation of the compass in rough weather making it difficult to use for steering or for taking bearings.

Against the normal principles of correcting like with like, e.g. permanent magnetism with permanent magnets, heeling error is corrected solely with permanent magnets. The amount of correction applied is only, therefore, right for one magnetic latitude as the boat's induced magnetism will vary as the latitude changes (it has not been found possible to produce a satisfactory soft iron heeling error corrector due to induction effects from the other corrector magnets). Permanent vertical magnets are therefore employed, usually contained in a bucket on an adjustable chain hung underneath the compass.

There are two ways of finding the correct number of magnets to use and their optimum distance from the compass. The first (and most accurate method) is to use a Heeling Error Instrument—briefly, this consists of a magnetised needle balanced on a knife edge which allows it to see-saw freely in the vertical plane. The North seeking end is marked with an engraved circle near its tip. On the other end (in North latitudes) one or more small, adjustable collar weights are placed capable of being slid along the needle—which is graduated in divisions. The instrument is taken ashore, hung from some non-magnetic object at least three feet clear of the ground, and swivelled until the North seeking end is pointing approximately North. The weight is then adjusted at the other end until the needle is truly horizontal (a bubble at the bottom of the instrument indicates the level). A count is then taken of the number of divisions separating the weight from the centre of the needle. The instrument is then taken back on board, the compass bowl removed, and the H.E. Instrument installed in its place, again aligned with North. On board, the earth's vertical field will probably be weaker than ashore due to some screening by the vessel's structure. To allow for this, the weight must be moved in a small amount, an average factor being 0.9. For example, if the weight was balancing the needle at 16 divisions ashore, when on board it should be placed at $16 \times 0.9 = 13.4$ divisions. Vertical magnets are then placed in the bucket and this is moved up and down, more magnets added etc., until the needle is once again horizontal.

However, you may not have a Heeling Error Instrument in your pocket or be expecting one for Christmas. Therefore, the second less accurate but probably more attainable method is to correct the Heeling Error at sea when the boat is rolling or pitching. This is best done on East—West courses and the drill is quite simple; taking bearings of some object, change the direction (Red up or Blue up), number, or position of the vertical magnets until the compass oscillation disappears and the bearing remains steady. It is best to use a bearing as a check, for the boat's heading may be altering without it being noticed amongst the oscillations. On the other hand, if the compass concerned is a steering compass, with no azimuth circle, the only solution is to watch the ship's head carefully, and by experiment, steady it as far as possible with the magnets.

Whichever way it is done, the Heeling Error correction must be done *before* any other corrections are made. This is because the vertical magnets may induce magnetism in any soft iron, including the spheres, which will have an effect on the final deviations of the compass. So—if you have to adjust the Heeling Error magnets at sea because the compass is oscillating too badly to be of much use—you may well have affected the *deviations on all courses*, and a subsequent swing to find the new deviations will be necessary.

*The Corrector Magnets*

These can be obtained in many strengths, lengths, and diameters, and it is a question of choosing the size appropriate to the compass concerned. Some compasses are mounted on a binnacle, and the latter normally contains compartments designed to take the appropriate size of magnet. Other types of compasses (particularly those originally designed for aircraft) have corrector magnets built into the compass bowl which are moved, for adjustment, by a screw and cantilever arrangement.

However, it may be that the compass in your boat has no built-in arrangement for correction, and magnet holders must be improvised to do the job. There are certain essential points to watch in doing this:

1. Once placed, the magnets must be held securely in place. (Brass clamps or drilled wood blocks.)
2. They should be level with the compass needle.
3. It is not recommended to put a magnet closer than twice its own length from the compass. Bigger magnets further away are better than small ones close in.
4. It is better to use two magnets placed symmetrically either side of the compass rather than a single magnet on one side.

Blocks of wood drilled to take the appropriate sizes of magnets are probably the best 'home-made' corrector boxes. They must have a wooden or brass flap or plug to prevent the magnets moving or slipping out of their holes.

*Summary of Compass Correction*

1. Remove Heeling Error with vertical magnets, by either using a Heeling Error Instrument or by damping out the compass oscillation at sea. The boat should preferably be on an East—West heading when doing this.
2. Put the boat onto NE—SE—SW—NW headings in succession and observe the deviations to obtain a Coefficient D. If the sum—after changing the signs on SE and NW and dividing by four—is small (say less than 1–2 degrees) there is no

need for any soft iron spheres or chains. However, if the sum is large, it means that correction is needed and spheres should be placed on each side of the compass before going any further with the correction process.

3. Put the boat onto an East or West course, and correct the deviation obtained there by placing Fore & Aft magnets on either ahead or astern of the compass. Then turn the boat until it is on the opposite heading (West if you started on East) and remove half the deviation remaining—if any.

4. Put the boat onto a North or South heading, check the deviation against a shore bearing, and remove the deviation with athwartships magnets placed on either side of the compass. Then turn the boat onto the opposite heading and remove half the deviation remaining.

5. Make sure that all corrector magnets are securely in place, spheres screwed down etc.

6. Finally, having done all these things, *swing the boat and obtain a deviation table.* If the deviation is now zero on all headings, open a bottle of champagne.

*Notes*

In (3) and (4) above, you will see that on the final heading in each direction only half the deviation remaining is removed. This is because the deviations obtained on the Cardinal points may not be entirely due, in fact, to the boat's permanent magnetism; to simplify the issue, other possible causes have not been explained here. For instance, most ships and some large boats have a soft iron corrector, the Flinders Bar, placed in front of the compass to counteract the effects of induction in the funnel or other superstructure; although basically a vertical induction, it has a horizontal effect which can produce deviation on the Cardinal points, as well as on other headings. Detailed analysis is required to separate this effect from the boat's fore & aft and athwartships permanent magnetism. If you try to remove all the deviation on the final heading of each pair, you may find that you have produced a larger error on the opposite heading once again. And you can go on fiddling ad infinitum! Therefore be content and leave the small error remaining.

If as a result of your corrective efforts, the subsequent swing shows no deviations larger than 2 or 3 degrees, you may well be content. In big ships, it is possible with a well sited compass to get the deviation down to below $\frac{1}{2}$°, but the majority of yachts and small boats pose greater problems and absolute perfection may not be achieved.

# 3

# Distance, Course, Speed and Direction on the Earth's Surface

A good compass is an essential piece of equipment. However, it is not much use having a dependable compass but no idea of the proper course to steer to reach a particular destination or how long it will take to get there.

We are fortunate these days in that practically the whole of the Earth's surface has been either charted or mapped, and therefore the business of finding the course between two places is a simple matter. It was the British Admiralty that gave the major impetus to the science of cartography (map-making) in the 18th century by commissioning such men as Cook to carry out explorations and meticulous surveys and to record the observations they made. Nowadays, many nations produce charts from their own or other nation's hydrographic information, but the Hydrographic Department of the British Admiralty alone produces over 6000 different charts and diagrams in common use by seamen all over the world.

*Distance on the Earth's Surface*

The basic unit for the measurement of distance at sea is the nautical mile. This is a distance of 6080 feet. Why this awkward figure? It is in fact the distance subtended at the Equator by an angle of 1 minute of arc at the Earth's centre. The significance of this chosen distance is that 1 minute of Latitude = 1 Nautical Mile and therefore, as will be seen later, distance can simply be measured from the latitude scale on a chart without conversion or a separate scale being necessary.

For general purposes, the mile is considered to be 2000 yards long, divided into 10 cables of 200 yards apiece. The old fashioned term 'league' was a distance of just under three miles but this is seldom used at sea today; nor are the statute mile of 1760 yards, furlongs etc.

*Speed*

A knot is defined as a speed of one nautical mile per hour. Therefore, do not fall into the old landsman's trap of saying in airy fashion 'Oh! I was doing 20 knots per hour' as any bloke around might presume you were serving your time. Twenty knots means 20 nautical miles per hour.

If you have difficulty converting time and distance into speed and vice versa, remember this simple formula:

$$\text{Distance (miles)} = \text{Speed (knots)} \times \text{Time (hours)}$$
$$D = ST$$

Therefore, if you have covered 10 miles in 40 minutes your speed must be:

$$10 = S \times \frac{40}{60} \quad \text{or} \quad S = \frac{600}{40} = 15 \text{ knots.}$$

Another very useful dodge to remember is the 'six minute' rule. Six minutes is, of course, one tenth of an hour. Therefore, if you know or can estimate the distance you have covered in six minutes, a straight multiplication of this distance by ten will give your speed direct.

e.g.

You travel 0.2 miles ( 2 cables) in six minutes. Speed 2 knots
You travel 0.5 miles ( 5 cables) in six minutes. Speed 5 knots
You travel 1.2 miles (12 cables) in six minutes. Speed 12 knots
You travel 2.5 miles (optimist!) in six minutes. Speed 25 knots

Obviously you can adapt this rule in your head to cover different times, e.g. for a three minute run multiply by 20, and for twelve minutes' run multiply by 5, etc.

*Direction on the Earth's Surface*

This is not quite as simple as it may seem at first sight as the Earth's is not a plane surface; nor is it a sphere but a spheroid flattened at the poles. In comparison with the Earth's diameter this flattening is not great (a difference of about 20 miles between polar and equatorial diameters) and is of no great significance from the point of view of practical navigation.

The fact that the Earth is almost spherical does, though,

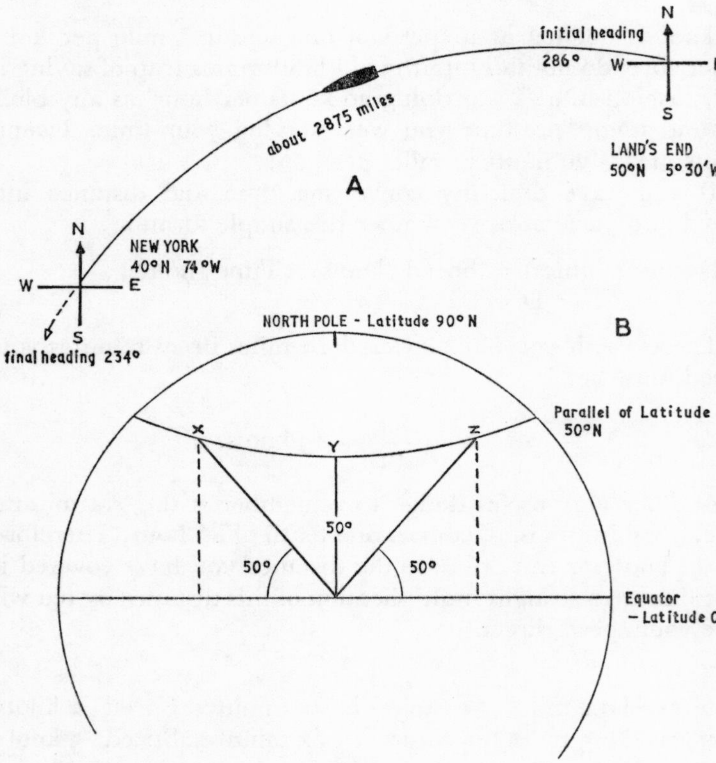


Figure 7

impinge directly on the navigator's problems and calculations. Firstly, the shortest distance between two places on the Earth's surface is not a straight line but a curve—called a Great Circle (whose plane passes through the Earth's centre). This is the path followed by radio waves. Secondly, the heading steered by a boat following a Great Circle track cannot be constant. The initial heading to follow the curve will differ from the final heading by an appreciable amount, as shown in Figure 7A. In this figure the curve is exaggerated for illustrative purposes, but the courses are correct for an uninterrupted Great circle track.

Thirdly, the fact that the Earth's spherical surface has to be translated onto a flat sheet of paper sets a problem for cartographers and some distortions are inevitable.

Before discussing the practical solution of these problems it is necessary to explain the geographical reference framework that is superimposed over the Earth's surface to define position —the Latitude and Longitude grid that has been in use for centuries.

*Latitude*

As the Earth is almost spherical, distances between a point on its surface and a reference axis must be measured in angular units. The reference used for Latitude is the Equator, the great circle midway between the poles. Latitude is measured North and South from 0° at the Equator to 90° N at the North Pole and 90° S at the South Pole. The Latitude of intervening places is defined by the angle that it subtends at the Earth's centre from the Equator. e.g. in Figure 7B places X, Y, and Z all subtend the same angle to the Equator and are therefore at equal Latitude. The small circle (not passing through the centre of the Earth) joining these places is called a Parallel of Latitude, in this case that of 50° N.

The Latitude of a particular point is given in degrees, minutes and seconds; for instance 50° 00′ 00″ North (60 minutes in a degree—60 seconds in a minute). Remember also that 1 minute of Latitude = 1 Nautical Mile. Therefore, the parallel of 50° N must be $50 \times 60 = 3000$ miles North of the Equator.

*Difference of Latitude—D'Lat.*

The difference of latitude between two places both North of the Equator or two places both South of the Equator is the straight arithmetic difference between them, e.g.

| | |
|---|---|
| First point | 56° 29′ 35″ N |
| Second point | 31° 53′ 27″ N |
| D'Lat. = | 24° 36′ 08″ |

Seconds are not often used in practice but are expressed in points of a minute as shown below:

| | |
|---|---|
| First place | 30° 20.2′ S |
| Second place | 52° 18.5′ S |
| D'Lat. = | 21° 58.3′ |

Where one place is North of the Equator and the other South, the difference of latitude is the sum of the two values, e.g.

| | |
|---|---|
| First place | 25° 10.2 N |
| Second place | 01° 25.8 S |
| D'Lat. = | 26° 36.0 |

From the difference in latitude alone, it is easily possible to work out the distance between two places quickly if they lie due North or South of each other; but this is of minor usefulness. D'Lat. must be used in various calculations, and this is its raison d'être.

*Longitude*

This is also measured in angular units. As the Earth is circular, it could have been defined from an arbitrary place on the surface from 0°—360° in one direction. However, the terms East and West had been in use for centuries, and as England was in the forefront of the maritime powers in the 18th century, the position of the Greenwich Observatory was chosen as the datum line at Longitude 0° or zero. From this place, Longitude is measured 180° East and 180° West. The 'join' at 180° Longitude occurs in a line that runs through the Pacific (the Date Line). Lines joining places of equal Longitude are called MERIDIANS—these circles (Great circles) start from one pole, cut the Equator at right angles and pass down to the other pole. Figures 8A and 8B illustrate the Longitude grid superimposed on the Earth's surface.

It can easily be seen that a person standing at the Pole can travel from the Greenwich Meridian through 180° East and

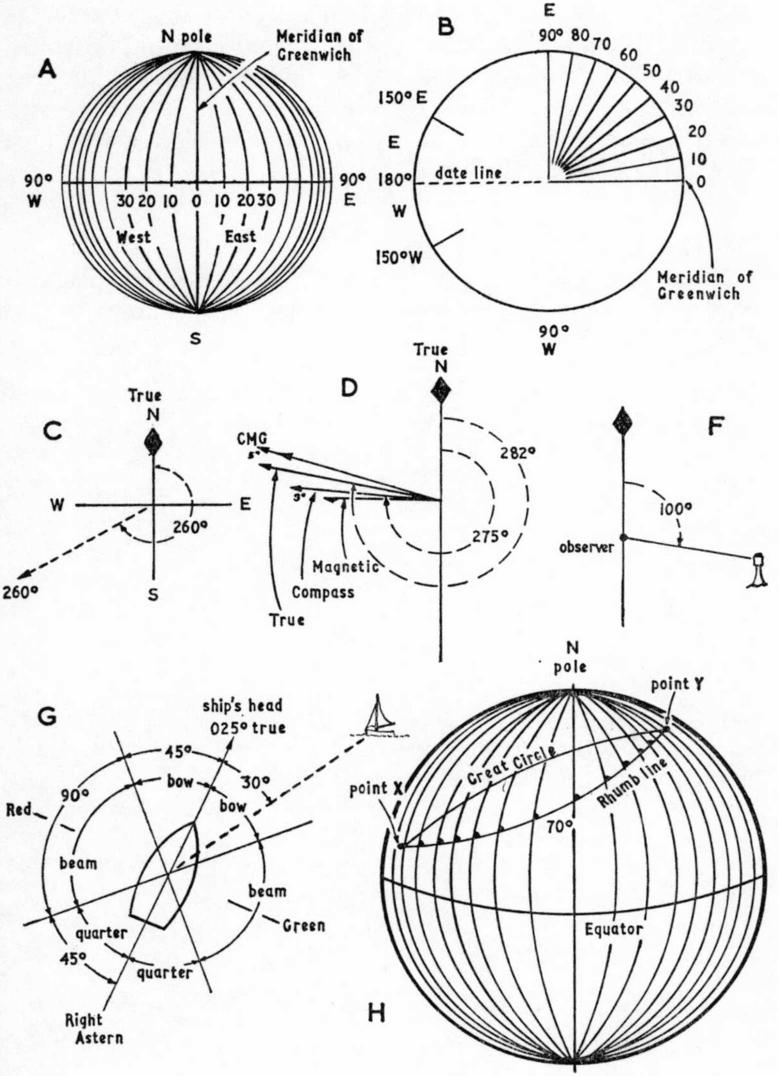

A
N pole
Meridian of Greenwich
90° W
90° E
30 20 10 0 10 20 30
West
East
S
B
E
90° 80 70 60 50 40 30 20 10 0
150° E
E
180°
date line
W
150°W
Meridian of Greenwich
90° W
C
True N
W
E
260°
S
D
True N
CMG
282°
275°
Magnetic
Compass
True
F
100°
observer
G
ship's head 025° true
45°
bow
30°
bow
90°
Red
beam
beam
Green
quarter
quarter
45°
Right Astern
N pole
point Y
point X
Great Circle
Rhumb line
70°
Equator
H


Figure 8

back through West merely by turning on his heel, but he would have to walk around the whole Earth's circumference to change longitude by the same amount at the Equator. In other words, the meridians converge at the poles, and the distance between one meridian and another on the Earth's surface varies with latitude. Also, if you travel up or down one meridian, you must be travelling North or South.

*Difference of Longitude*

In much the same way as D'Lat., difference of longitude is measured in degrees and minutes (and seconds if necessary) as an algebraic difference, e.g.:

| | |
|---|---|
| First place | 123° 00.2 E |
| Second place | 25° 43.6 E |
| D'Long. = | 97° 16.6 |

or

| | |
|---|---|
| First place | 22° 10.4 E |
| Second place | 02° 08.6 W |
| D'Long. = | 24° 19.0 |

A slight trap arises when the two places in question are nearly 180° apart in longitude, maybe across the Date Line. The shortest way round must be taken—thus the D'Long between

110° 05.0 E

and

70° 00.0 W

is NOT 180° 05.0 but 179° 55.0′ (180° − 110° 05.0 + 180° − 70°)

*Departure*

This is the distance *in miles* travelled in an East or West direction over the Earth's surface.

Departure = D'Long. × Cosine Mean Latitude.

*Direction and Bearing*

Direction is basically measured from the meridian, i.e. the direction of True North if referring to the geographical meridian. Both True Course—and the True Bearings of objects—are measured clockwise from the meridian and expressed as an angle between 0° and 360°.

*Course*

THE TRUE COURSE

The angle between the boat's fore & aft line and the meridian is her course. So if the boat is steering 260° True she will be heading through the water in a direction 10° South of West. (Figure 8c.)

If there is no tidal stream, current, wind, or compass error, she will also be making good this course over the ground.

COURSE MADE GOOD

Life, however, is seldom so simple that one or more of these disturbing effects does not exist in practice. Therefore though the boat may be heading 260° through the water, a current setting to the South may result in the boat making a 'course made good' of 250°. This 'course made good' can only be precisely established by fixing, i.e. finding the geographical position of the boat at a number of successive times, or it can be estimated from knowledge of the currents and the amount of leeway the boat makes in a wind of a given strength and direction.

The course steered and the 'course made good' are, therefore, more often than not different directions, the first related to the water and the second to the ground.

*Compass Course and Magnetic Course*

As explained in the last chapter, the Magnetic course is the angle between the Magnetic Meridian and the boat's fore & aft line. Similarly, the Magnetic Compass course is the angle between the direction of Compass North and the boat's fore & aft line. To arrive at a *True Course*, both variation and the compass deviation (on that particular heading) must be applied.

The following example may help to summarise this:—

| | | |
|---|---|---|
| Compass Course | 275° | |
| Deviation (from table) | 3° W | |
| Magnetic Course | 272° | |
| Variation (from chart) | 10° E | |
| TRUE COURSE | 282° | (see Figure 8D) |
| Tidal Stream/Leeway setting boat 5 degrees to the North | 5° | |
| COURSE MADE GOOD (CMG) | 287° | |

*Bearings*

True Bearings and Compass Bearings are defined in very much the same way. The True Bearing of an object is its angle measured clockwise from the meridian.

In Figure 8F, the lighthouse bears 100° True from the observer. The True Course to steer for the light is also 100°.

As a reminder—when using a magnetic compass for taking the bearing of an object, the compass bearing must be corrected for deviation and variation to obtain True Bearing (again remember to use the deviation for the boat's heading and NOT the direction of the object). Example:

Compass Course 300°—(Deviation on this course 3° E)

| | |
|---|---|
| Compass Bearing of object | 125° |
| Deviation (3° E) | + 3° |
| Variation (12° W) | −12° |
| TRUE BEARING = | 116° |

*Relative Bearing*

Relative bearings are commonly used for indicating the position of another object or craft relative to one's own vessel.

A relative bearing is measured in degrees from one's own ship's head, 180° in each direction from the bow. The Starboard (right) side is designated GREEN and the Port (left) side RED.

In Figure 8G, the relative bearing of the other boat is Green 30°, i.e. 30° on the Starboard bow. As the boat's course is 025°, the True Bearing of the other boat would be 055°

(025 +30). Were the other vessel 30° on the Port bow, the Relative Bearing of it would be Red 30°, and the True Bearing 355° (360 +25 −30).

The terms Bow, Beam, and Quarter are also commonly used. Strictly speaking, an object said to be 'on the Port Bow' without any number of degrees specified should be bearing Red 45°, i.e. 45° to port of right ahead—'on the Port beam' Red 90°—and 'on the Port Quarter' Red 135°. However, these terms are often more loosely used to indicate a general area, 'Port Quarter' meaning anywhere on the port side from right astern to Red 135°, 'Port Beam' meaning anywhere from Red 135° to Red 45°, and 'Port Bow' anywhere from Red 45° to right ahead.

*The Great Circle and the Rhumb Line*

Earlier in this chapter it was stated that the Great Circle route is the shortest distance between two places on the Earth's surface, and that steering a Great Circle course means a constantly changing heading. In practice, on a leg of less than about 200 miles, the distance lost in steering a steady course and not a Great Circle is not significant. Even when following a Great Circle route, it is not a practical proposition to alter the boat's heading continuously and gradually to follow the curve, as the rate of change of heading is so small even at high speed.

*The Rhumb Line*

In order to steer a steady course over the Earth's curved surface, the boat must follow a track that cuts successive meridians at the same angle (otherwise it would be altering course). Such a track is called a RHUMB LINE—by definition a line that cuts all the meridians at the same angle. On the Earth's surface it is, in fact, a spiral (Figure 8H).

In this example, a Rhumb Line course of 070° is shown. In practice, the navigator always steers a succession of Rhumb Line courses, even when following a great circle track. This is explained in Appendix II.

# 4

# Charts

The chart is to the seaman what a map is to the landsman—a pictorial representation of the Earth's surface. The chart, however, shows features both above and below the water line.

Different projections, or methods of transferring the Earth's curved surface onto a flat piece of paper can be and are used in chart construction. But by far the most common projection used for charts is the Mercator's projection. This is used for practically all charts of a natural scale smaller than 1½ inches = 1 mile.

*Mercator's Projection*

The simplest way to explain the Mercator's projection (which came into use at sea in about 1630) is to imagine a paper cylinder wrapped around the world. Imagine also that the Earth is fairly transparent and that a lamp is burning at the Earth's centre. The various land and sea features would then be shown up as shaded images on the surrounding paper.

Figure 9A is a representation of a standard Mercator projection where the cylinder touches the Earth at the Equator.

A Transverse Mercator projection is similarly constructed but in this case the paper cylinder is turned through ninety degrees and touches the Earth at the poles.

In fact, the simple explanation so far given is not entirely accurate, as the projection is mathematically adjusted to give the chart certain properties. The principal ones are:

1. Rhumb lines are straight lines on the chart. Therefore courses can be laid off on the chart or read off without difficulty.

2. The Equator, which is both a Rhumb line and a Great circle on the Earth, is a straight line on the chart.

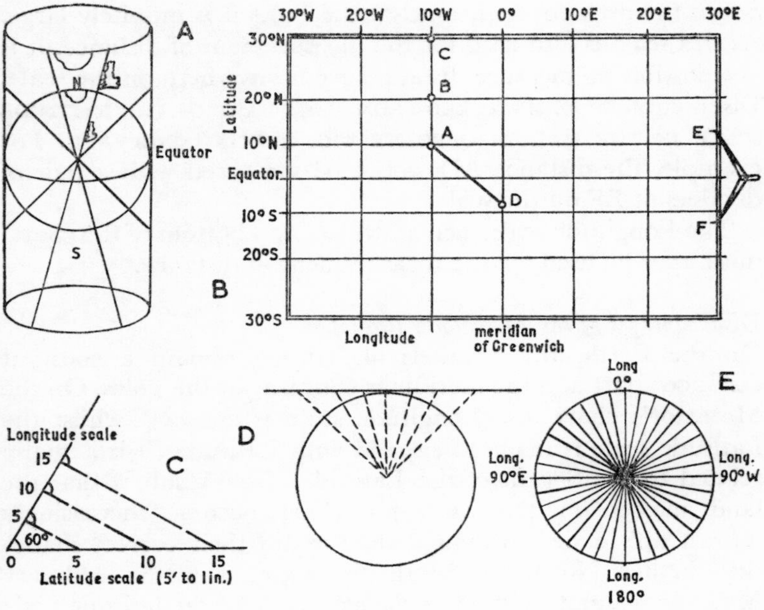


Figure 9

3. The Meridians (great circles converging at the poles on the Earth) are straight, equidistant parallel lines on the chart, cutting the Equator at right angles.

4. The Parallels of Latitude (small circles on the Earth) are straight, but not equidistant, lines parallel to the Equator on the chart.

5. Great circles, on the chart, are curves whose apexes lie towards the poles.

Thus the Latitude and Longitude grid (or graticule as it is sometimes called) appears on the chart as a lattice of straight lines, with the meridians equally spaced, but with the parallels of latitude increasingly far apart the further they are from the Equator.

You will see in Figure 9B that the length of the line BC is greater than AB on the chart although both correspond to a difference of 10° in latitude. The latitude scale on a Mercator chart is thus variable, steadily expanding the further North or

South the area covered, until at the Poles it is infinitely large. As this scale is also used for the measurement of distance, it is not possible to measure distances from anywhere on the scale. Distance MUST BE MEASURED FROM THAT PART OF THE LATITUDE SCALE IN THE MARGIN OPPOSITE THE POINTS CONCERNED. For example, the distance AD would be measured with a pair of dividers at EF on the scale.

The Longitude scale (across the top and bottom of the chart) must *never* be used for the measurement of distance.

*Disadvantages of the Mercator's Projection*

On the Earth, the parallels of latitude remain a constant distance apart and the meridians converge at the Poles. On the Mercator's chart, the Longitude scale is constant whilst the Latitude scale is made to expand with Latitude (it is actually related to the Secant of the Latitude). The result is that the land masses on the Mercator chart become increasingly 'blown up' in size (although they retain their correct shape) the further North or South they are, and the ordinary Mercator cannot be used at the Poles where the latitude scale would be infinite. Except in the Polar regions this distortion does not matter practically. For instance, although Iceland and an Equatorial island of the same actual size on the same Mercator chart would appear totally out of proportion in size (Iceland seeming to be three times bigger than the other), the Latitude scale opposite to Iceland would also be three times bigger. Thus shape, distance, and bearing are preserved, but areas are distorted.

*Constructing a Latitude and Longitude grid on a plain sheet of paper*

In ocean areas well away from land, it quite often happens that the scale of the only chart available is very small—a pencil point may be the equivalent of one mile. In such situations it is impossible to plot an astronomical fix accurately. It is therefore necessary to draw a Latitude/Longitude grid on a plain plotting sheet using a much larger scale, for instance 5 miles equals one inch.

This is easily done (as illustrated in Figure 9C) by following this procedure:—

(i) Draw your Latitude scale horizontally across the paper.

(ii) From the zero of the Latitude scale, draw a line at an angle equal to the Latitude of your position—in this case 60°.

(iii) Drop perpendiculars to this line from the Latitude graduations. Your Longitude scale is then given along the inclined line.

It is fairly easy to see from this that at Latitude 90°, the Longitude scale would be zero, whilst at the Equator (Latitude 0°) the two scales would be the same This is the 'mirror picture' of the Mercator chart graduations (Longitude scale constant and Latitude scale infinite at the Poles). The main thing to remember is that, except at the Equator, the latitude graduations will always be further apart than the Longitude ones. (Note: This construction is only reasonably accurate over a comparatively small area—about 200 miles each way.)

*The Gnomonic Projection*

The main property of this projection is that Great Circles appear on a gnomonic chart as straight lines.

The chart is drawn by projecting the Earth's surface onto a plane touching the Earth's surface at a chosen tangent point. If the tangent point selected is a Pole, the Latitude and Longitude grid produced is quite simple, as shown in Figure 9D and E.

The Parallels of Latitude are concentric circles round the Pole, and the Meridians straight lines radiating from it.

Unfortunately, Rhumb lines are not straight lines on a Great Circle chart, and therefore course and direction cannot be determined directly from a small scale gnomonic projection.

On a much larger scale they are, however, used for harbour plans, where the small area covered can safely be assumed to be 'flat'. Courses can be laid off on these charts, and a separate distance scale is normally given on gnomonic charts of this type, together with the Latitude and Longitude of a given 'spot' on the chart.

The other uses of Great Circle charts and diagrams are for Polar charts (where the Mercator projection becomes useless) and for Great Circle sailing.

## CHART CHARACTERISTICS

*British Admiralty Charts*

The ordinary navigational charts published by the British Admiralty cover all the areas of the world. They are grouped in Folios, each folio covering a particular geographical area and containing charts of both large scale (harbour plans) and small scale (passage and oceanic charts). Each chart is given an individual distinguishing number (e.g. Chart No. 1824a shows the East Coast of Ireland together with the Irish Sea and St. George's Channel).

The numbers of all the charts published by the Admiralty, their folio groupings, and their prices are given in the Catalogue of Admiralty Charts, which is itself re-published annually.

These charts are fairly expensive, averaging about 83 pence for an ordinary chart to £1.20 for one that is overlaid with a radio aid grid—such as a Decca lattice. The average yachtsman will not want to be too lavish when buying charts from a chart agent. The best way, therefore, to see what charts are available and to obtain the necessary area coverage without excessive cost, is to consult the diagrams in the chart catalogues showing the individual limits of each chart and then to choose accordingly. A reputable chart agent will hold copies of the catalogues and help with the selection, so a personal purchase of a catalogue is not essential. References to charts (and their limits) are also contained in the Admiralty Sailing Directions (or Pilots). However, the catalogue is the better reference, being republished every year whereas the Sailing Directions are not.

*Yachtsman's Charts*

Until recently, the Admiralty also published a series of yachtsman's charts and these are now produced commercially by firms such as Stanford and Imray. They have the big advantage for small boat use of being much smaller in size, which allows them to be used on small chart or saloon tables—or even on a portable board. Their current price is about 90 pence per chart.

*Sizes of Charts*

The ordinary Admiralty charts are designed to fold into a uniform size for convenient stowage in a folio, but they are not all identical in overall size; for instance some charts (mainly harbour plans) are printed on a single unfolded sheet; others (passage charts for example) are folded double and must be opened for use. If the space is available, an important point in a boat is to have a chart table of large enough size to allow a comfortable working surface—the minimum area needed to take an ordinary Admiralty chart unfolded is about 38 inches × 25 inches. Some Yachtsman's charts are 13 inches long × 8 inches deep, other about 21 inches × 15 inches.

*Selection of Charts*

To sum up—there are these two basic types of chart: the Admiralty charts in use by seamen, in ships large and small, all over the world and the commercial adaptations of these charts specifically designed for use in small craft in limited areas. Charts are not cheap and you will not want to spend more than is necessary. However, when planning a voyage, it is essential to have sufficient charts to cover all eventualities—for example, if going on passage down the Spanish coast it would be unwise to rely merely on a small scale passage chart; in the event of a blow, you might have to run for cover and it might prove difficult to trip your way through the rocks on the back of a small postage stamp. Therefore, you would need the coastal charts and harbour plans on that coast. Combine prudence with thrift!

*Chart Titles and Numbers*

Each chart has its own individual title and number. These chart numbers are geographically somewhat haphazardly distributed, so in ships carrying complete folios it is customary to give the charts another number—the consecutive number. The top chart of the folio would be given consecutive number 1, and so on. Most boats will not be carrying complete folios, but it is not a bad idea to adopt a similar numbering system for the charts carried. A folio list on the front of the chart cover showing these consecutive numbers against each title much simplifies the task of finding or stowing away a particular chart.

*Chart Scale and Type of Projection*

The natural scale of the chart and the type of projection on which it is drawn is shown just underneath the title. The natural scale is a straightforward comparison of distance on the chart against distance on the Earth's surface. With a scale of 1/150,000, an inch on the chart therefore represents 150,000 inches on the Earth's surface, or about 2.08 miles.

This is of general interest, but of more practical importance is the scale of distance. Remember that on Mercator projections *one minute* of LATITUDE equals *one mile*, and that when using dividers to measure distance on the chart, that part of the scale (at the *side* of the chart) opposite the distance being measured must be used. NEVER use the Longitude scale, running across the top and bottom of the chart, for distance measurement.

On harbour plans—quite often drawn on the Gnomonic projection—a special distance scale is provided.

## INFORMATION ON CHARTS

It is impossible to list here all the conventional symbols that are used on charts and getting to know them is largely a matter of experience. However, a special chart is published by the Admiralty (No. 5011) which gives all the symbols used on British charts and this is an easy means of reference.

*Land Features*

The amount of information given on land features is limited to that which is of practical use to the navigator at sea. Thus a considerable amount of detail is shown of the coastal areas, whilst inland areas are largely left blank—except where, for instance, a hill or mountain can be seen from seaward. In recent years, with the increased use of radar for navigation, contour lines and spot heights (e.g. 9̇90—a hill 990 feet in elevation) have become of more direct interest and greater detail is shown of these.

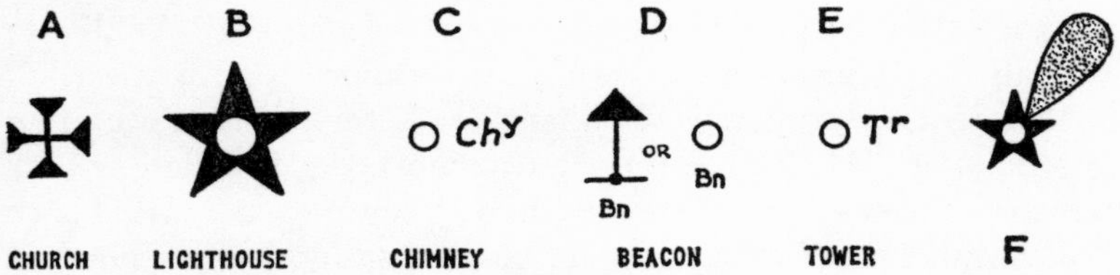


Figure 10

Just a few of the more common land symbols are shown in Figure 10 (A–E).

An object that stands out boldly when viewed from seaward may have the notation (conspic.) written on the chart against its symbol—a conspicuous object.

*Lighthouses*

These are a minor subject in themselves. They are shown by the conventional symbol for a lighthouse surmounted in the case of a major light with a magenta coloured flash (Figure 10F) so that they readily catch the eye. Obviously on small scale ocean charts it would be impossible to include all the navigational lights on a seaboard, so only the biggest and best are shown.

The British Admiralty publishes a book called the Admiralty List of Lights (in a number of different volumes covering different parts of the world) which gives complete details on all the navigational lights. Their characteristics, heights, power etc. are all listed and also whether they are attended (lighthouse keeper) or unattended (nobody around!). In some places, a light may only be switched on at particular times or for a particular event. Do not therefore expect a light that is marked (Occas.)—occasionally—on the chart or in the Light Lists to be always switched on when you desperately need it—it may be that the local luminary has gone home, and one is not normally handily placed to ring him up and ask that it be lit.

The lighthouses along a particular stretch of coastline all have different characteristics, so that one may not easily be confused with another. The main varieties of lights are Fixed, Flashing or Occulting. A fixed light, self-evidently, is one that is on all the time; a flashing light is one whose characteristics are such that the period of darkness exceeds the period of light; and an Occulting light is one that is ON more than it is off.

If no colour is specifically mentioned, then it is white. The letters (W) (R) (G) are used on charts to indicate colour whenever necessary; white, red and green respectively. There is a complication in the existance of Alternating Lights (marked Alt. on the chart). These successively flash the colours indicated.

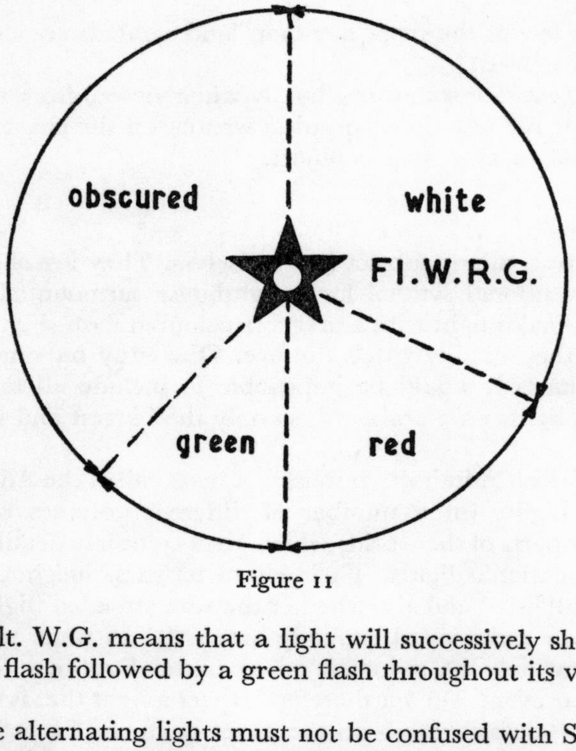


Figure 11

Thus Alt. W.G. means that a light will successively show first a white flash followed by a green flash throughout its visibility range.

These alternating lights must not be confused with Sectored lights, which show different colours continuously in different sectors. These sectors are normally marked on the chart by dotted lines, and the Red sector of a light is usually chosen to cover some particular hazard in that area. One other thing to remember about sectors—in the Light Lists, the sectors covered by a particular colour are always given in BEARINGS FROM SEAWARD. For example, a light showing White from 180° to 290°, Red from 290° to 000°, Green from 000° to 030°, and

Obscured from 030° to 180° would be shown on the chart as shown in Figure 11.

Some other points on lights that need to be mentioned. The 'Period' of a light is the time it takes to complete one whole cycle of its operation, measured from the start of its first flash. Some examples again—a light that is shown on the chart as:—

| | |
|---|---|
| *Gp.Fl.* (2) 10 *sec.* 29*M* | is a light that gives two white flashes in quick succession—followed by an interval of darkness—every 10 seconds. The 29M means that it is visible on a clear night 29 miles away from an observer whose height of eye is 15 feet above sea level. |
| *Occ.W.R.* 10 *sec.* | This is a sectored light (white shown in one sector, red in another) which goes out for a brief period once every 10 seconds. |
| *Gp.Fl.* (3) 10 *sec.* 125 *ft.* 17*M F.R.* 112 *ft.* 10*M* | This lighthouse shows two lights. The major light flashes three times in quick succession every 10 seconds. It is at an elevation of 125 feet above sea level, and visible 17 miles.<br>The second is a Fixed Red light at a slightly lower elevation and only visible 10 miles. |
| *Gp.Occ.* (4) *W.R.* 16 *sec.* 17,14*M* | This light goes out four times in quick succession every 16 seconds. The white light is visible for 17 miles and the red for 14. Again it is a sectored light. |
| *Alt. W.R.* 20 *sec.* 15*M* | This gives a white flash followed by a red flash every 20 seconds all round the horizon and is visible from 15 miles. |

These are but a few of the characteristics of lights that may be met with. The majority of the markings against lights on the charts are fairly self evident. If further information on a particular light is required, a glance at the Light List will give the full details, for example how long each successive flash is in a Group Flashing light. In practice it is not often necessary to consult the Light List unless you do not possess a good chart of the area concerned.

Providing the visibility is good, the range at which a light is sighted can give a good indication of your distance from it. The loom of very powerful lights can normally be seen long before the light itself is sighted, but it should be remembered that lights at very high elevation may be obscured by cloud; and weak lights may be obscured by haze.

*Depths on a Chart*

At the present time, the depths on existing British Admiralty charts are normally shown in fathoms and feet (a fathom = 6 ft.) On some large scale charts such as harbour plans, the depths may be shown in feet alone. It is normally self evident which depth scale is being used, as with the fathom scale, the odd number of feet between whole fathoms are shown as a small figure beside the fathom digit. A depth of $5_2$ on a chart indicates that the depth in that particular spot is 32 feet i.e. $5 \times 6 + 2 = 32$). However, if there is any doubt, check the notation under the chart Title—it is always stated whether the depths are given in fathoms, feet or soon (perish the thought) metres.

For at the present time British charts are being converted to the metric system already in use on many foreign charts. The soundings will then be shown in metres and points of a metre. A conversion table for putting fathoms and feet into metric quantities is given below.

CONVERSION TABLE—FATHOMS AND FEET INTO METRES

| Feet | | 6 | 12 | 18 | 24 | 30 | 36 | 42 | 48 | 54 | 60 |
|---|---|---|---|---|---|---|---|---|---|---|---|
| Fathoms | | 1 | 2 | 3 | 4 | 5 | 6 | 7 | 8 | 9 | 10 |
| Feet | — | 1.8 | 3.6 | 5.5 | 7.3 | 9.1 | 10.9 | 12.8 | 14.6 | 16.4 | 18.3 |
| 1 | 0.3 | 2.1 | 3.9 | 5.8 | 7.6 | 9.4 | 11.3 | 13.1 | 14.9 | 16.7 | 18.6 |
| 2 | 0.6 | 2.4 | 4.2 | 6.1 | 7.9 | 9.7 | 11.6 | 13.4 | 15.2 | 17.0 | 18.9 |
| 3 | 0.9 | 2.7 | 4.5 | 6.4 | 8.2 | 10.0 | 11.9 | 13.7 | 15.5 | 17.3 | 19.2 |
| 4 | 1.2 | 3.0 | 4.9 | 6.7 | 8.5 | 10.3 | 12.2 | 14.0 | 15.8 | 17.7 | 19.5 |
| 5 | 1.5 | 3.3 | 5.2 | 7.0 | 8.8 | 10.6 | 12.5 | 14.3 | 16.1 | 18.0 | 19.8 |

*Chart Datum*

The depths shown on a chart cannot be taken as absolute depths, i.e. ones that never change, because of the rise and fall of the tide; indeed, meteorological conditions in some parts of the world can alter the level of a whole sea area at different seasons of the year.

When a survey is carried out, therefore, the depths found at all the different points on the sea bottom have to be corrected for the tidal conditions appertaining at the time and referred to a common datum before they are plotted and the chart drawn. This datum is known as Chart Datum, and for safety reasons is so chosen that the sea level seldom falls below it (it approximates to the lowest predictable tide level under average meteorological conditions).

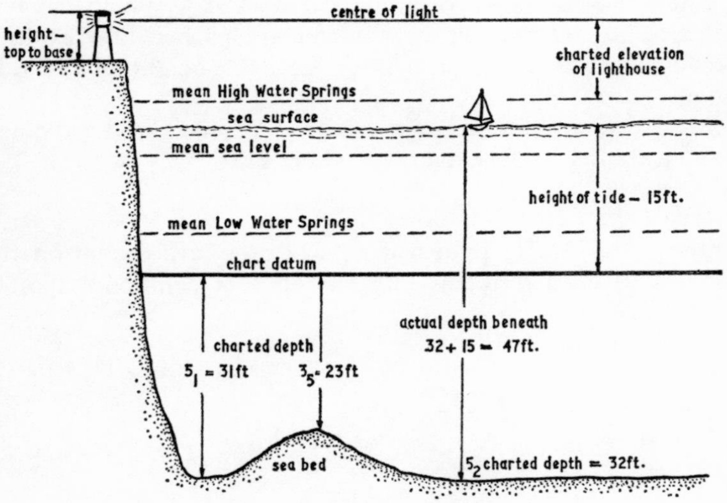


Figure 12

From a practical point of view, it is not necessary to know the actual datum level chosen—i.e. 52.9 feet below an ordnance survey bench mark set into the Vicarage wall at Little Belchington; it is, though necessary to appreciate that the depth of water over a patch marked $2_1$ (two fathoms one foot) will seldom be exactly 13 feet. It is likely for the majority of the time to be more than that, but there may be a few occasions when it is less. To find the actual depth over that patch at a particular time, the height of the tide must be calculated and added to it. Again, to err on the side of caution, the elevation of land features on a chart are referred to a different plane—Mean High Water Springs. (An explanation of Spring and

Neap Tides is given in the next chapter.) This reference is chosen so that their charted elevation will rarely be less than the figure shown. This point is perhaps of less importance these days in large ships when ships' navigators use radar instead of vertical sextant angles for taking the range of an object—but a vertical sextant angle can still be useful to the yachtsman not possessing any other ranging device.

Figure 12 may help summarise the difference between the various levels and datums:

The yacht is shown over a spot marked $5_2$ on the chart; however the actual depth of water under her keel at that particular moment in time is 47 feet, the height of the tide being 15 feet.

Mean sea level is also shown; this can be taken as the average level of the sea surface over the whole year.

*Fathom Lines*

These lines, joining points of equal depth, are drawn on the chart with each different line having a conventional symbol

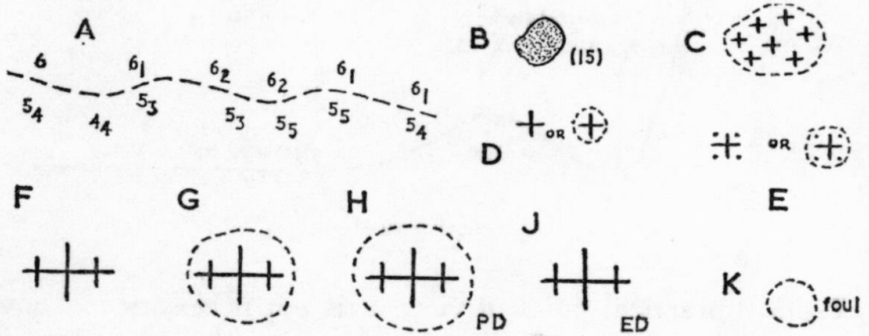


Figure 13

*C* Rocky patch which does not uncover—dotted line encloses danger area

*D* Individual dangerous rock with less than 6 feet of water over it

*E* Rock awash at the level of Chart Datum

*F* Position of a wreck that is NOT dangerous to surface navigation

*G* Wreck that *is* considered dangerous to surface navigation

*H* Dangerous wreck—Position Doubtful

*J* Wreck—Existence Doubtful

*K* Foul

composed of dots or dashes (sometimes combined with colour shading) to indicate its value. However, it is not necessary to remember the individual symbol for each fathom lines as the soundings on either side of the line indicate its numerical value. For instance, the 6 fathom line might appear as shown in Figure 13A.

*Drying Heights*

Certain features are covered and uncovered by the sea, depending on the state of the tide; reefs or rocks near the shore, for example, or sandbanks out to sea. A patch that dries 2 feet above Chart Datum is shown by the figure 2 with a line drawn underneath it, e.g. $\underline{2}$. Sometimes features are marked with the words 'Dries . . . feet' written alongside. The latter method is normally used with individual features—a rock for example, and the former to indicate a small area or patch.

An above water rock that does not cover at all would be marked as in Figure 13B. The numeral alongside gives its elevation above M.H.W.S.

*Dangers*

Some of the most important chart symbols for the dangers to navigation that lie beneath or on the surface are given in Figures 13C–J.

*Foul Ground*

The remains of a wreck or other debris on the bottom which is not dangerous to surface navigation, but which should be avoided by vessels anchoring. (Figure 13K.)

*Type of bottom*

The nature of the bottom is indicated by various letters written alongside the soundings. These are primarily of interest from the point of view of anchoring, as the nature of the sea bed will affect the holding power of the anchor. The best holding ground is thick mud or sand and the worst, rock and weed. The types of holding ground are (with the chart symbols) in approximately the correct descending order of preference:

| | | |
|---|---|---|
| S | ......Sand | Good Holding Ground |
| M | ......Mud | |
| Cy | ......Clay | |
| G | ......Gravel | Moderate |
| Sh | ......Shingle | |
| P | ......Pebbles | |
| St | ......Stones | |
| Sh | ......Shells | |
| R | ......Rock | Poor |
| Wd | ......Weed | |

*Buoyage*

Navigational buoys are used where it is not possible or economic to build permanent marks such as lighthouses or beacons. Some dangers, such as sandbanks, shift their position quickly and therefore the marks showing the limit of these hazards have to be moved quite frequently. The major hazards in busy shipping lanes out to sea are sometimes marked by Light Vessels instead of Buoys.

The charted position of a buoy is the position of the small circle in its base. However, it must always be remembered that buoys, which are moored to the bottom by a length of chain, may be shifted by heavy weather, and therefore total reliance should not be placed in their position.

Two systems of buoyage are used—called the LATERAL and CARDINAL systems. The former is used in United Kingdom waters (and in the tidal waters of many other countries) and is the one described here. The other, CARDINAL system, is more commonly used in areas where there exist a large number of offshore islands, rocks or shoals, and the shape of a particular buoy is governed by its true bearing from the danger.

*The Lateral System of Buoyage*

With the Lateral system, the character of the buoys used is governed by the direction of approach, i.e. a port hand buoy (of a particular shape) must be left to Port WHEN ENTERING A CHANNEL OR GOING *with* THE FLOOD STREAM.

The Flood Stream, as its name implies, is the Tidal Stream associated with a rising tide (the EBB stream with a falling tide). Around the British Isles, the Flood Stream flows up the Channel as far as Dover. The Flood also flows up the Irish Sea around the North of Scotland and down the East coast of England. There is thus a meeting point near Dover where the water coming from two directions—up Channel and down the North Sea—meet.

The offshore buoyage round the British Isles is governed by this pattern of water movement. Coming up Channel from the West, a buoy marking the limit of a shoal extending from the Sussex coast would therefore have Port Hand Characteristics. A buoy in a similar position off Newcastle would have a Starboard Hand characteristic, and be left to Starboard.

The most important buoys used in the Lateral system are shown in Figure 14.

### MIDDLE GROUND BUOYS

(Used to mark a shoal in the middle of a channel)
The topmarks on the buoys at the inner end of the channel are different from the outer marks. The lights shown by these buoys are distinctive whenever possible but follow the normal rules, e.g. red flashing or an even number of white flashes for the 'Main channel to the Right or Channels of Equal Importance buoys' and white flashing with an odd number in the group for 'Main channel to the Left' buoys.

### *Mid Channel Marks*

These buoys are sometimes used to mark the *centre* of a channel—particularly at the entrance to some harbours where the position of the channel cannot otherwise be marked effectively. They can be of any shape, but must be easily distinguishable from the Port, Starboard, and Middle Ground buoys. The characteristics of their lights must also be distinctly different from neighbouring lights. See Figure 14.

### *Spoil Ground and Quarantine Anchorage Marks*

Spoil grounds may either be the result of the dumping of refuse in an area or may be made unhealthy by the outflow of sewers, etc. Quarantine anchorages are used by ships with cases of infectious disease on board. In either event, the area is not

*Colour*: Red or Red & White/Yellow
*Light*: Red flashing/occulting or white flashing/occulting in even groups (2-4-6)

*Colour*: Black or Black & White/Yellow
*Light*: White flashing/occulting in odd groups (1-3-5). Occasionally green (but see Wreck buoys)

*Colour*: Red and White Horizontal stripes

*Colour*: Black and White Horizontal stripes

*Colour*: Red and White Horizontal stripes

*Colour*: Black & White or Red & White Vertical stripes

*Colour*: Yellow above—black below

*Colour*: Yellow

*Colour*: The buoy itself—wide red and black horizontal bands sometimes divided by a narrow white band.
*Lights*: May flash white or red.

*Colour*: Can buoy—green with white letter W. Spar buoy—green upper, red lower.
*Flashes*: Green, even number flashes.

*Colour*: Conical—green with white W. Spar—green upper, black lower.
*Flashes*: Green, odd number.

*Colour*: Spherical buoy—green with white letter W. Spar—green.
*Light*: usually single occulting.

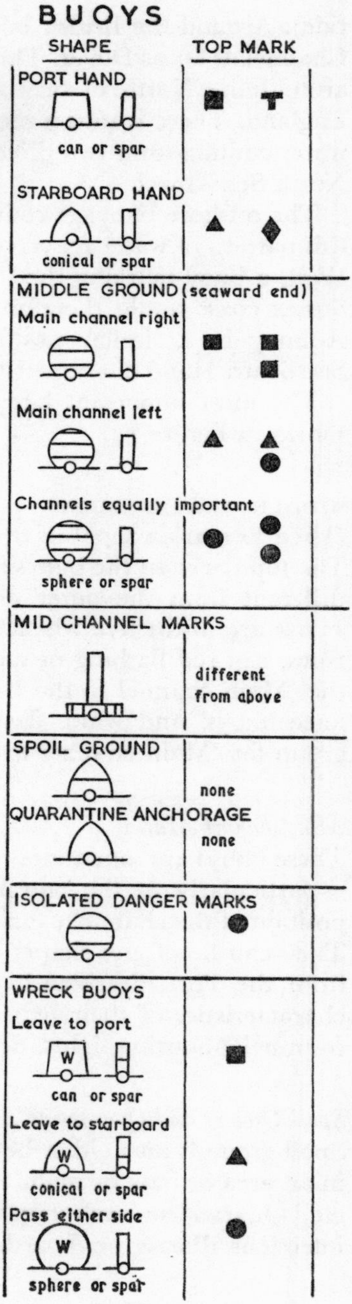


Figure 14

usually a salubrious one in which to anchor and pass the time of day.

The shape of the buoys is optional and chosen to be different from others in the vicinity. These buoys are not always lit at night.

*Isolated Danger Marks*
This type of buoy is used to show the position of an isolated danger which may be passed on either side.

*Wreck Buoys*
Green in colour—see Figure 14.

## TIDAL STREAM AND CURRENT INFORMATION

A limited amount of information on currents is shown on charts; this is in the form of a current 'arrow' showing the general direction in which the current normally sets and sometimes also gives its rate (in knots). For instance Figure 15A.

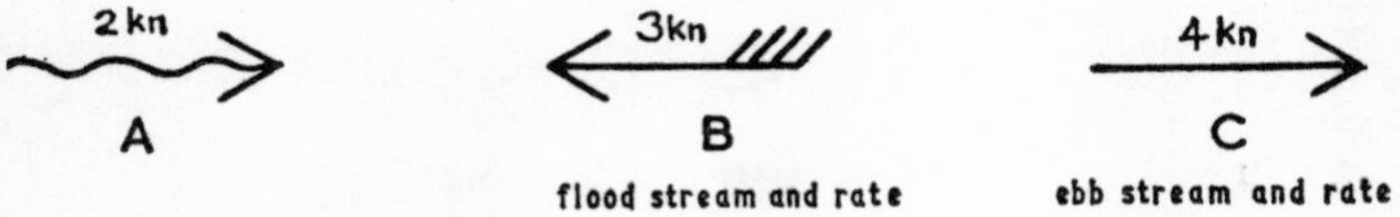


Figure 15

More detail on the currents to be expected in particular places, estuaries, channels, rivers and the like can be gleaned from the Pilots (Admiralty Sailing Directions). For currents in the open oceans, the best reference is probably 'Ocean Passages of the World' (also published by the Admiralty).

On some charts, and particularly those of British waters, a good deal of data on Tidal Streams is given directly on the chart. This information is given in two forms; on the older charts (and on, incidentally, most charts of foreign waters) arrows are again used to indicate the direction and rate of the Flood and Ebb streams, e.g. Figures 15B and 15C.

This method of presenting tidal stream information is not

altogether satisfactory as no indication is given of the varying rates at which the tidal stream increases in strength or slackens. This depends on the interval in time from the local High or Low water and whether it is a Spring or Neap tide.

The modern way of showing tidal information on a chart is by the use of Tidal Diamonds. Inscribed on various places on a chart are capital letters enclosed in a diamond, for example ◇A. Elsewhere on the chart, normally near the edge, will be found a block of tables, each with its alphabetical indicator above it. A typical example is:

◇A 49° 52.2 N
5° 10.9 W

◇B

| | Hours | Dirn. | Rate Sp. | Rate Np. |
|---|---|---|---|---|
| Before H.W. Devonport | 6 | 256 | 1.8 | 0.9 |
| | 5 | 254 | 1.2 | 0.6 |
| | 4 | 234 | 0.4 | 0.2 |
| | 3 | 045 | 0.4 | 0.2 |
| | 2 | 054 | 1.0 | 0.5 |
| | 1 | 059 | 1.8 | 0.4 |
| | H.W. | 067 | 2.3 | 1.1 |
| After H.W. Devonport | 1 | 075 | 1.8 | 0.9 |
| | 2 | 082 | 0.8 | 0.4 |
| | 3 | 203 | 0.4 | 0.2 |
| | 4 | 233 | 1.3 | 0.7 |
| | 5 | 247 | 2.3 | 1.1 |
| | 6 | 257 | 1.9 | 0.9 |

All that is necessary for the use of this table is to look up in Tide Tables the time of High Water on that particular day.

Let us assume that H.W. Devonport was at 1315 (GMT) and it was Spring Tide period. The tidal stream at 0715 would be setting in the direction 256° at 1.8 knts and at 1510–082° at 0.8 knts.

*Chart Corrections*

When originally bought, the charts should normally have been corrected up-to-date for any details that have changed since the chart was published, e.g. a buoy moved, or newly discovered shoal inserted etc.

The date of publication of a chart is shown outside the bottom margin. However, if a lot of detail has changed in the course of time, a new edition of the chart may be printed—again the date of this new edition is shown near the date of publication. Then, if important information comes to light which is too complex to insert by hand on the chart, a Large Correction may be made. This, once again, involves the re-printing of the chart by the Admiralty—the date of the Large Correction being shown under the date of the new edition.

Outside these reasons for reprinting the chart in toto and discarding the old copies, a small correction system is used for keeping charts fully up to date. Every week, Admiralty Notices to Mariners are issued which give all the important corrections that need to be made to all charts. The corrections are suitably indexed so that each individual chart affected by a correction can be seen at a glance. These permanent corrections are normally made in pen and waterproof violet ink on the chart copy held by the individual owner. Sometimes, if the detail is sufficiently complicated to merit it, 'blocks' are enclosed in each copy of the Notice to Mariners; these can be cut out and pasted on the chart. It is a good tip to align these blocks into the correct position on the chart before applying any glue, and tick the corners with a pencil mark—then apply the glue and stick them on.

Some corrections are classed as Temporary or Preliminary corrections when a (T) or (P) is shown against the particular notice affected. These corrections should only be applied in pencil for, as their names imply, they concern impermanent changes—a light vessel being removed temporarily from its station for maintenance would warrant a temporary notice to

Mariners. Preliminary notices may later need to be inked in. The numbers of these small corrections are listed at the bottom of the chart concerned outside the bottom margin on the left, against the year concerned, e.g. 1968–22. 156. 273. means that in 1968, Notices to Mariners 22, 156 and 273 affected that particular chart and the necessary corrections to it have been made.

Chart correcting can be a tedious business, particularly in a small boat with limited space, but for general navigational safety in strange waters, it is a very necessary chore.

Admiralty Notices to Mariners can be obtained via any chart agent.

# 5

# Tides and Tidal Streams

For practical purposes, it is not necessary to have a detailed knowledge of the complex astronomical factors that give rise to the tide generating forces. The Moon and Sun, in that order, have a differential attractive or repellent effect on the water covering the Earth. It is largely the relative position of these bodies to points on the Earth's surface which determines the amount of attraction they impose on the water surface at a particular place and time.

As you know, the Earth rotates anti-clockwise about its Polar axis once every 24 hours. The Moon orbits around the Earth—in the same direction as the Earth's rotation—once every 29½ days or so. Thus the lunar day—from an earthworm's point of view—is longer than the ordinary day by a period of about 50 minutes. If a spot on the Earth was directly underneath the moon at 1400 on the 17th July, it would not be directly underneath the Moon again at 1400 on the 18th July, because the Moon would have moved on in its orbit and the transit will occur at a later time.

The Earth is also in orbit round the sun, completing a full revolution once every 365.3 days (hence the Leap Year every four years). To an observer on the Earth, the Sun appears to be rotating around the Earth once every 24 hours; this is due to the Earth's rotation. But it is obvious that the Sun does not rise or set at the same times or on the same bearings every day of the year. This is principally due to the fact that the plane of the Earth's orbit round the Sun is not aligned to our Equator, but makes an angle of about 23½ degrees with it. The apparent orbital path of the Sun around the Earth (the reverse way of looking at it) is called the ECLIPTIC. (See Figure 16A).

Twice during the year, on about March 21st and September 21st, the sun is directly over the Equator and the period of light equals the period of darkness—12 hours each. These are called the EQUINOXES (equal day and night). At other times of the year, the Sun's Declination—its angle above or below the Equator—varies until it reaches a maximum of 23½ South on December 21st (the Winter Solstice) and a maximum declination North on about June 21st (the Summer Solstice). This is speaking as a Northern Hemispheric man! Down under, no doubt, they would want to arrange things differently.

Similarly, the Moon's Declination, or its position above or below the plane of the Equator, is not constant but varies between 28½ North to 28½ South approximately every 13 days.

More important still, as far as the tides are concerned, is the phase of the Moon, its position relative to the Sun as seen from the Earth. A new moon—only a thin crescent visible—occurs when the Moon approaches a position between the Earth and the Sun. (When the Moon and Sun have almost the same declination, an Eclipse of the Sun may occur as the Moon gets between Sun and Earth.)

About 7 days later, the Moon will be at 90° to the Sun—as seen from the Earth—and half of it will be illuminated. The Moon is then said to be in its First Quarter and it will be waxing —the illuminated portion increasing each night with the horns of its crescent pointing to the left.

Another 7 days and it will be a Full Moon on the opposite side of the Earth from the Sun. A full Moon will always, therefore, start rising over the eastern horizon near sunset.

For the next seven days, the Moon wanes until it reaches its Last Quarter—the other half illuminated—and in a further seven days it will be a New Moon once more.

When the Moon is in its New or Full phases, the Sun's and Moon's gravitational forces combine to exert the maximum pull on the Earth's water surfaces.

Supposing Figure 16B shows the situation at a particular instant in time. The Moon and Sun—drawn wildly out of scale—are in 'conjunction' and on the Earth's Equator, i.e. both have zero declination. In accordance with the normal gravitational laws, both the Moon and the Sun will exert an attractive force on the Earth. This force is considered as being directed from the centre of one body to the centre of the other and is inversely proportional to the square of the distance

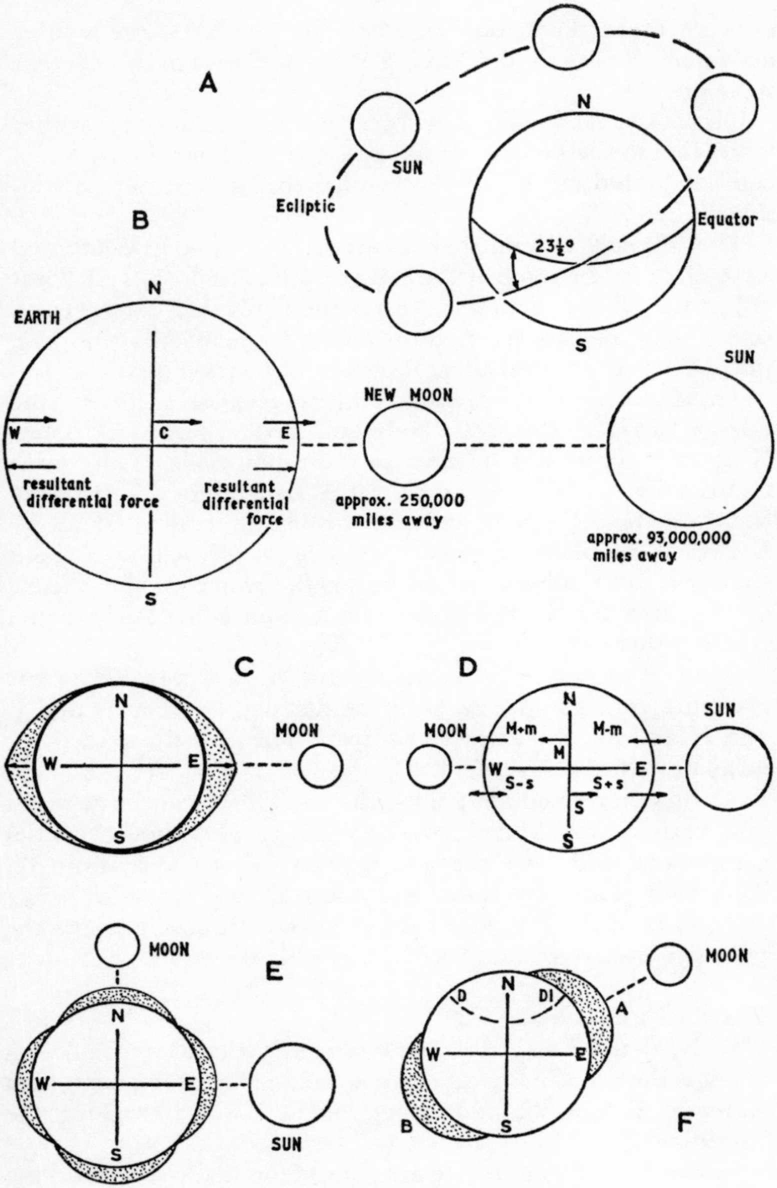
A
SUN
N
Ecliptic
Equator
23½°
S
B
N
EARTH
W
C
E
resultant differential force
resultant differential force
NEW MOON
approx. 250,000 miles away
SUN
approx. 93,000,000 miles away
S
C
N
W
E
S
MOON
D
MOON
N
M+m
M
M-m
SUN
W
E
S-s
S+s
S
S
MOON
N
W
E
S
SUN
E
MOON
N
D
DI
A
W
E
B
S
F

Figure 16

between them. Therefore, although the Sun has a far greater mass than the Moon, the Moon's attractive force in this context is the greater.

Obviously, these attractive forces must be balanced by other forces due to the bodies' orbital speed around one another or a collision would result, but these other forces need not be considered here.

The Moon and Sun, then, exert an attractive gravitational force of a certain value at the centre of the Earth (C). At Point (E) in the Diagram they will exert the equivalent force PLUS a certain quantity, as that point is nearer to them. At Point (W), they will exert the equivalent force MINUS a certain quantity.

These differential forces have their greatest value on the Earth's surfaces nearest the body and farthest away—at points N and S they would be zero as those points lie at the same relative distance as the centre of the Earth. The effect of the forces on the solid land surface is negligible, and as far as the Earth's atmosphere is concerned they merely cause a slight change in atmospheric pressure. But the water on the Earth's surface, like the air, is free to respond, and being fairly dense, reacts to them.

As a result, water is drawn towards E and away from the heavenly bodies at W. At other points, e.g. between N and E and N and W, the attractive force is not directly towards or away from the local vertical.

At these intermediate points, the total force can be resolved into vertical and horizontal components, and the horizontal components will move the water towards E and away from W. The result of all these forces is to cause a 'bulging' of the water surfaces at E and W which progressively reduces towards the points 90° away—N and S as shown in Figure 16c.

*The Earth's Rotational Effect*

Clearly, if the Earth did not rotate, nor the Moon and Sun change their declination, there would be a permanent high water at E and W, and a permanent low water along the Meridian of N to S.

But the Earth does rotate and the Moon likewise around the Earth. In one Lunar Day (of about 24 hours 50 minutes) any particular point on the Earth's surface will therefore feel (Moon and Sun having zero declination) two High Waters and two Low Waters. This is called a SEMI-DIURNAL Tidal Pattern, i.e. in ordinary time, a High Water followed by a Low Water about 6 hours 12 minutes later, another High Water at 12 hours 25 minutes, a Low Water at 18 hours 37 minutes, and the cycle completed by the following day's High Water at 24 hours 50 minutes.

*Spring Tides*

The earlier diagram illustrated the situation with the Sun and Moon working in conjunction on the same side of the Earth—at the period of a New Moon. The tide raising forces are pulling together and are at a maximum. These are occasions when SPRING TIDES, those that rise highest and fall lowest, will occur. The example also illustrated the particular time of the year when the Sun had zero Declination, at the Equinox, when the highest and lowest tides of a semi-diurnal nature are likely to occur in the whole year—hence the expression EQUINOCTIAL SPRING TIDES.

Spring tides will also occur when the Sun and Moon are in 'opposition' at the opposite sides of the Earth at the time of a full moon, as shown in Figure 16D.

In this context, the term 'opposition' is misleading as, in fact, the tide raising forces of the two individual bodies are still additive. The Moon's gravitational pull exerts Force M at the Earth's centre, M+m at W and M+m at E. The Sun exerts Force S at the centre, S+S at E and S—s at W. They both exert therefore a differential force towards E and W as before. SPRING tides result, and when a FULL Moon occurs at the Equinox, EQUINOCTIAL SPRING TIDES will occur.

*Neap Tides*

These are tides which have the smallest rise and fall in height. They occur when the Moon is in its Quarters, i.e. Half-Moon. The differential forces of the Sun and Moon are now applied at 90° to each other.

As a result, the Moon is trying to raise High Water at N and S, whilst the Sun is trying to raise High Waters at E and W as shown (diagrammatically only) in Figure 16E. The Moon, being closer, will win, but smaller amplitude tides will follow.

*The Effect of Declination*

So far, the Moon and Sun have been shown at zero Declination. However, in the case of the Sun this occurs but twice a year, and for the Moon every 6½ days. To consider the Moon alone for a moment, and the diagram in Figure 16F.

Having a certain Northerly Declination, the Moon will raise the maximum 'humps' in the water at points A and B. A point 'D' as it rotates with the Earth's rotation will experience no Tide Raising force at D, but a maximum at $D_1$. Thus only one High Water and one Low Water per day will occur. These are called DIURNAL TIDES.

The Sun's Declination will have a similar, but smaller effect. The Semi-Diurnal Tide Raising effect is therefore greatest when the Sun/Moon's Declination is zero, and the Diurnal Tide Raising effect greatest when both have maximum declination.

Each ocean area has a natural period of oscillation—the 'slop' in a basin. The Atlantic responds readily to semi-diurnal tide generating forces, whereas the Pacific is most affected by those of a diurnal nature. The tides round the British Isles, therefore, are mainly semi-diurnal in character.

*Summary*

To summarise these effects: At New and Full Moon, the attractive forces of the two bodies combine to produce SPRING tides, i.e. tides with the greatest difference in height between High and Low Water. When the Moon is in its Quarters, NEAP tides result, tides that have the smallest difference in height between High and Low water. (The Spring tides actually occur usually about 1 day after the Moon is New or Full as there is a time lag in the force taking effect.)

At certain times of the year, all the factors attracting the tide generating power of the bodies—such as Sun's declination, Moon's declination, phases of the Moon, combine to produce the highest tides of the year. This occurs about the time of the Equinoxes—hence the expression Equinoctial Spring Tides.

In European waters, the tides are almost invariably semi-diurnal in character; two high and two low waters a day, with the time of the first high water becoming about fifty minutes later each successive day. However, although the astronomical forces are the basic cause of tides, their effect in particular areas is much affected by the shape of the sea or ocean, the gradient of the sea bed, etc. If a particular ocean has a natural oscillatory frequency near to the period of the tidal forces, then high tides will result. The North Atlantic is subject to some of the highest tides in the world; on the other hand, there is very little tidal movement in the smaller enclosed basin of the Mediterranean.

In some other areas, notably in the Pacific, the tide is diurnal in character, i.e. only one high water and low water a day.

In other places, local anomalies may cause a stand of the tide; in Southampton water, for example, there may be as many as four high waters a day, the shape of the Solent and Spithead causing double high tides.

Over the years, a great deal of information has been gathered to enable the tides, particularly in European waters, to be predicted with considerable accuracy. However, these predictions only apply when the normal meteorological conditions for the time of year pertain. Large falls or rises in barometric pressure can alter the height of the sea to quite a marked extent, and strong, persistent winds can build up the sea level against a coast line (remember the floods on the East Coast). Any tidal prediction is subject to unforeseen factors such as this, and cannot therefore be infallibly correct.

*Tide Tables*

The British Admiralty publishes annually three volumes of Tide Tables, and based on much the same data, a good number of other tide tables are produced, either for local use, or for inclusion in Almanacs, such as Reed's.

Part I of the Admiralty Tide Tables is published in three volumes:

Vol. 1—covers European Waters (including the Mediterranean)
Vol. 2—covers the Atlantic and Indian Oceans
Vol. 3—covers the Pacific Ocean

These tables give the times and heights of High and Low waters for a large number of the major ports—called Standard Ports. They also provide methods of working out the heights and times of tides at an even greater number of smaller places (Secondary Ports) based on a Standard Port in their vicinity.

Tables are also provided for working out the height of the tide at times between high and low water. The method of doing these calculations is not identical in each volume of the Tide Tables. It is not intended to go into these calculations here, as a full explanation of the calculations and examples of their practical solution are given in each volume.

Apart from the Admiralty Tide Tables, there exist a large number of other tables produced by publishers and local authorities. These are normally perfectly satisfactory for finding out the heights and times of high and low water, but do not always give a facility for calculating the height of a tide at intervening times. There is not much doubt that the Admiralty Tide Tables will give the widest range of ports and the most accurate predictions, but there are others, notably those contained in Reed's Nautical Almanack, which may suffice for the average yachtsman's needs.

Whichever Tide Tables you use, always be careful to check the *Zone Time* of the predictions. For example, a lot of the U.K. tables are published in G.M.T. and therefore in England, when Summer Time (Zone —1) is being kept, one hour needs to be added to the predicted times, e.g.:

10.00 G.M.T. = 11.00 B.S.T. (Zone -I or A) (1100A)

## DEFINITIONS

A few simple definitions may be of help in de-cyphering Tide Tables.

*Height of a Tide*—is its height above or below Chart Datum. If the tide falls below chart datum, the height is shown in the tables as a minus quantity, e.g.—1.2 (feet).

*Range of a Tide*—is the difference in feet between the heights of successive High and Low waters, e.g.:

| | | | |
|---|---|---|---|
| High Water | Height | 16.5 feet | 19.5 feet |
| Low Water | Height | 0.2 feet | —0.6 feet |
| Range of Tide | = | 16.3 feet | 20.1 feet |

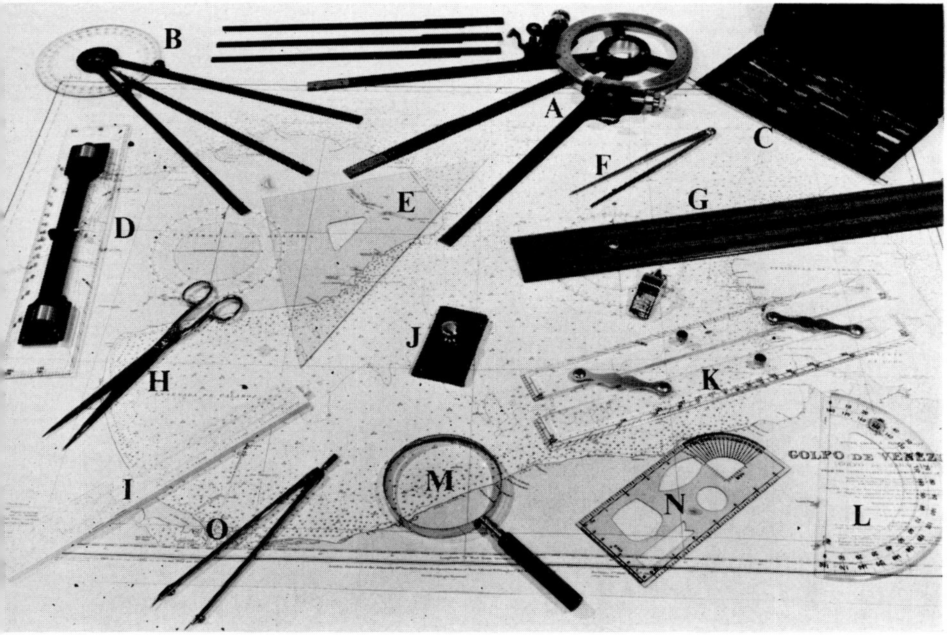


*Above*) CHARTROOM INSTRUMENTS:
& B—Station pointers; C—Drawing sets; D—Rolling rules;
—Set squares; F—Brass dividers with stainless steel points;
—Straightedge; H—9-in scissors; I—Slide rule; J—Chart weight;
—Parallel rules; L—Protractor; M—Chart magnifier; N—Mariner's rule;
—Pencil compass.

*Below*) Chart, binoculars, telescope and sextant.

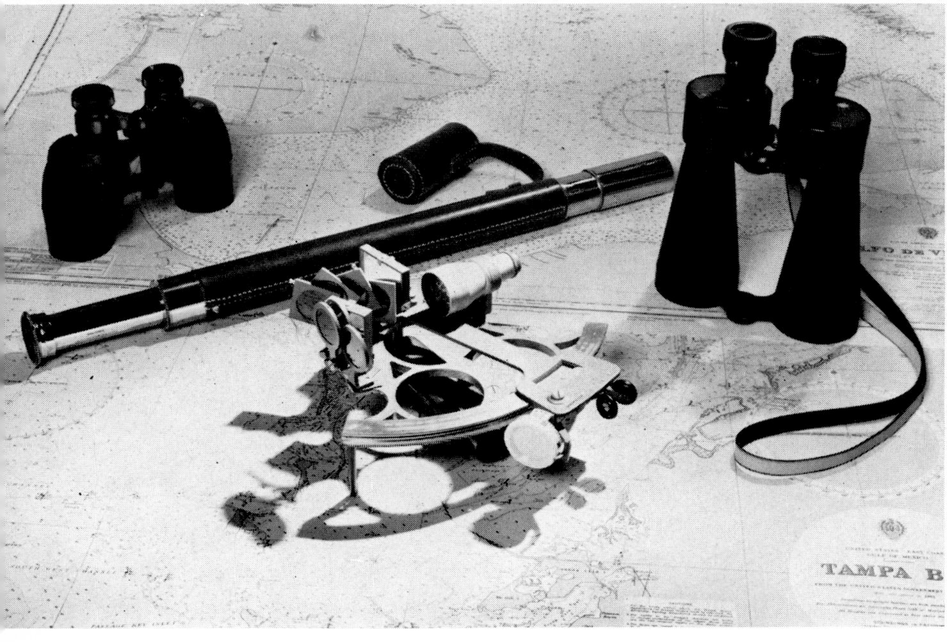

(*Above*) Yachtsman's sextant (normally cheaper but slightly less accurate than Marine sextant).

(*Below*) Marine sextant.

(*Left*) Hand-bearing magnetic compass.

(*Below*) Magnetic compass in gimbal ring.

(*Right*) Fixed D/F Loop and Goniometer; with a fixed D/F Loop, incoming signals are picked up by each section of the double Loop. The Goniometer, in conjunction with an ordinary radio set, compares the amount of signal received by each section and thus obtains a bearing of the transmitter.

(*Below*) Cockpit of motor yacht showing wheel, engine instruments, echo sounder and radar displays, with radar aerial above.

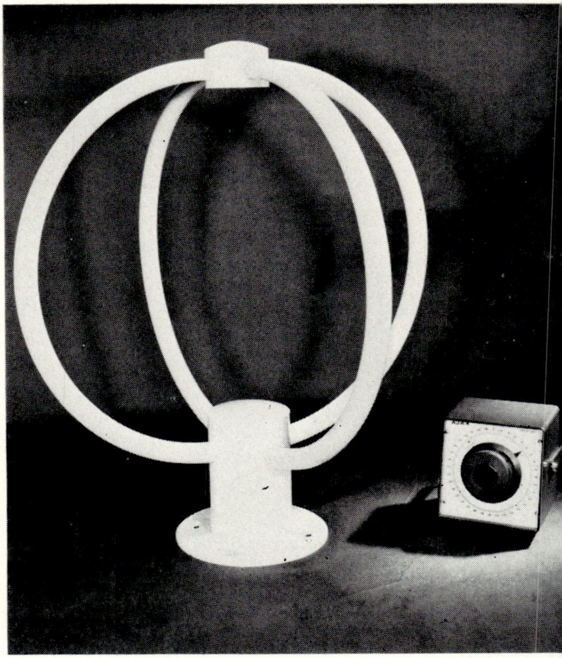

*Duration of a Tide*—is the time interval between successive High and Low waters.
*The Interval*—is the time difference between High (or Low) water and the time at which it is desired to predict the height of tide.
*Spring Tides*—Occur roughly once a fortnight—normally in the day succeeding a New or Full Moon. They are tides with the greatest range, and consequently produce the strongest Tidal Streams.
*Neap Tides*—occur at the intervening weekly intervals. Smallest range, and therefore weakest Tidal Streams.
*Mean High Water Springs and Mean High Water Neaps*—are the average heights of high water at springs and neaps, taken over a long period. Mean Low Water Springs and Mean Low Water Neaps are calculated on the same basis.

*Other Methods of Tidal Prediction*

The Tide Tables are produced by the careful analysis of a large number of harmonic constituents related principally to the movements of the Moon and Sun, but this sort of work needs the services of a tidal machine or computer, and only the results are of interest to the seaman. Next down the ladder is a method of prediction using a few of these harmonic constituents to produce a graph of a particular tide at a particular place. This gives a very accurate result in working out the height of a tide at a secondary port, but it is fairly complex and slow and would not normally be used by small boat sailors. (Those interested will find it in Part III (not Volume III, Part I) of the Admiralty Tide Tables.) Next in accuracy come the predictions for the Secondary ports given in the Tide Tables.

There are, however, a couple of rough and ready, and not very accurate, methods of predicting tides where these are semi-diurnal in character. These are by using the non-harmonic constants shown on some charts. The constants are:

*Mean High Water Lunitidal Interval* (terrible mouthful—so M.H.W.I.)—This is the interval between the Moon's transit at Greenwich and the next following High Water. To find the time of the Moon's Meridian passage you will need to look in a Nautical Almanac; to this time you must add M.H.W.I. to find the time of High Water. To find the time of Low Water, add or subtract 6 hours 12 minutes from the High Water time.

*High Water Full and Change* (*H.W.F. & C.*)—This term is sometimes given on the older charts. It means that when the moon is full or changing, the next high tide in the area concerned will take place at the stated interval after the Moon's meridian passage. It is therefore similar to M.H.W.I. above.

Both these constants can be made use of, without reference to the time of the Moon's meridian passage, if the chart you are using states the constant and you know the time of High Water at another place along the coast—and the constant there. The difference between the two constants, added or subtracted from the time of High Water at the other place, will give a reasonable approximation for the time of High Water in your own area.

There is one other useful approximation concerning the likely height of a tide at intermediate times between High and Low Water, assuming that there is a regular semi-diurnal tidal pattern. The level is likely to change in the following way:

| | |
|---|---|
| During the First hour (after H.W. or L.W.) | By 7% |
| During the Second hour (after H.W. or L.W.) | By 18% |
| During the Third hour (after H.W. or L.W.) | By 25% |
| During the Fourth hour (after H.W. or L.W.) | By 25% |
| During the Fifth hour (after H.W. or L.W.) | By 18% |
| During the Sixth hour (after H.W. or L.W.) | By 7% |

So, assuming that the Range of a particular tide is 20 feet, and the height at High Water was 24 feet, the expected heights of tide would be:

| | |
|---|---|
| At the end of the First hour after H.W. | 22.6 feet |
| At the end of the Second hour after H.W. | 19.0 feet |
| At the end of the Third hour after H.W. | 14.0 feet |
| At the end of the Fourth hour after H.W. | 9.0 feet |
| At the end of the Fifth hour after H.W. | 5.4 feet |
| At the end of the Sixth hour after H.W. | 4.0 feet |

*Note*: This method is only *very* approximate. The decimal points are entirely unrealistic in practice.

### *Tidal Stream Tables*

Apart from the Tidal Stream and Current information shown on charts by means of tidal diamonds and arrows, other books exist which give additional information. For Tidal Streams, some of the most useful are the Pocket Tidal Atlases covering the waters around the British Isles. There is one, for example, for the Channel, another for the Solent, and another for the Irish Sea. They display, by means of arrows, the expected set and rate of the tidal stream for each hour before or after High Water at the port on which they are based (in most cases—Dover). These are particularly useful books for yachtsmen as they are small, comparatively cheap, easy to use and contain a lot of information in condensed form.

Unfortunately their equivalents do not exist, to my knowledge, outside U.K. waters. However, The Admiralty Tide Tables for foreign waters (Vols. II and III) include some Tidal Stream tables amongst the other tidal predictions.

# 6

# Chartwork, Fixing and Pilotage Hints

This is the nub of the business of navigation and pilotage; messy and unmethodical chartwork can quite quickly lead to that 'indefinable feeling of impending doom' and the hire of a prayer mat. So decide that you are not going to wear out the latter from the start by being accurate and neat in your workings on a chart. Speed is sometimes also necessary, but should not normally be sacrificed for accuracy.

The simple instruments required for use on a chart are:

1. *A Soft pencil*—2B or softer. Hard pencil marks are difficult to erase and the chart will not remain usable for long if it is deeply scored with hard pencil marks. A clutch pencil, with a box of spare leads, is ideal as it can be sharpened easily without spraying the chart table, boat, and the soup with chippings.

2. *A good soft rubber*.

3. *A pair of dividers*—these can be bought in various shapes (bow or straight legged) and sizes; a pair with 6″ legs is convenient for general use, although you may occasionally need a longer pair. Personally, I prefer the straight type as they seem more easily handled with one hand. They should be workable with one hand, but not so slack that they will not stay fixed at a set spread.

4. *A pair of compasses*—any pair will do providing they work satisfactorily and will hold pencil or lead.

5. *A Station Pointer or Douglas Protractor*. The former may be quite expensive unless you can pick up a second-hand one cheaply but is extremely useful for plotting horizontal sextant angle fixes. It consists of a central circular scale, engraved in degrees, to which is attached three long legs; the central leg is

fixed to the scale but the outer legs can be set and clamped at any angle. Good instruments of this type are normally made of brass or gunmetal and have a vernier arrangement to allow the angles to be set accurately to within a few minutes of arc.

You will probably be familiar with the Douglas Protractor—a large square plastic protractor engraved with parallel lines across its face and in degrees around the edge. This is cheap, and also useful for plotting fixes; its limitation lies in its size for when plotting fixes on a large scale chart you may 'run out of protractor' whereas the Station Pointer will still cope. Also, of course, angles cannot be marked very accurately on a Douglas Protractor.

6. *A Parallel Ruler*. This is essential. It is used for transferring courses or bearings from one part of the chart to the compass rose, so that they can be read off, and for the plotting of position lines. Parallel rulers are available in two types, roller or hinged. The roller type is much the quickest and easiest to use if the chart table is sufficiently large to allow the chart to be laid out flat upon it. A sketch of this type of ruler is shown in Figure 17A.

It simply consists of a rectangular strip of brass, wood, or plastic. At each end is a roller projecting slightly beneath the under surface, the two rollers being joined to a common axle (in some newer types with nylon rollers there is no axle but a double roller arrangement which serves the same purpose). These rollers only allow the ruler to move in one direction unless they are lifted off the chart surface, and therefore parallel lines can be transferred from one area of a chart to another. This type of ruler does not, however, work satisfactorily on uneven surfaces.

The second type is the hinged variety (Figure 17B) wherein two rectangular strips are moved by two pivoted metal arms. The ruler is moved across the chart by successively keeping one strip held to the paper whilst pivoting the other. It is much slower and more fiddly to use than the roller on a flat surface, but it is lighter and easier to operate when the chart is folded or laid on a bumpy surface.

For laying of a course or bearing with a parallel ruler, see Figure 17C.

With either type in hand, suppose that you want to lay off a course of 030° from Point X on the diagram above. The parallel

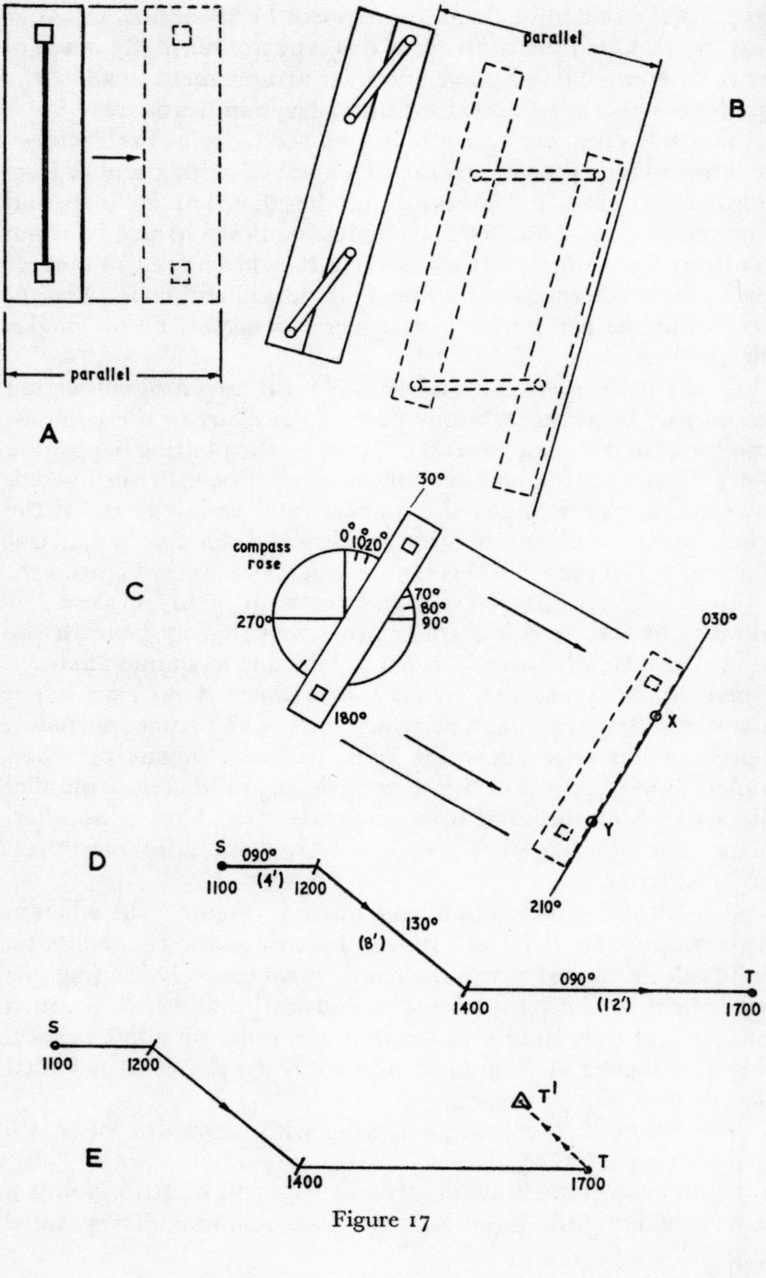
parallel
A
parallel
B
30°
compass rose
0°
10°
20°
70°
80°
90°
270°
180°
C
030°
X
Y
210°
D
S
090°
(4')
1100
1200
130°
(8')
090°
(12')
1400
T
1700
S
1100
1200
T'
E
1400
T
1700

Figure 17

ruler is laid on the nearest 'compass rose' on the chart and one edge aligned through the centre of the rose to the 030° graduation. The ruler is then slid across until an edge passes through Point X. A pencil line is then drawn from this edge and can be extended as necessary.

Additionally, it will be necessary to find the bearing between two points on a chart, or the course between them. The ruler is then first aligned with one edge through both points and subsequently moved across to the nearest compass rose. The edge joining the centre of the rose to its circumference will give the course or bearing required. (Point Y bears 210° from Point X in the diagram above.)

7. *A Notebook for 'keeping the reckoning'*. A lay-out for this notebook is discussed at the beginning of Chapter 9.

The easiest way of getting oneself lost at sea is to keep no record of the boat's position on the chart, and no written details of course alterations, fixes, etc., in a notebook. The starting position is always known (presumably!) but a plot must be kept on the chart of the boat's subsequent movements.

*The Dead Reckoning Position*

The Dead Reckoning or D.R. position is arrived at by allowing only for the boat's true course and speed *through the water* since leaving a known point. Looking at Figure 17D, let us say a boat left Point 'S' at 1100. It steered 090° for 1 hour, altered course to 130° at 1200, and altered course again to 090° at 1400; then its D.R. position at 1700 allowing for a boat's speed of four knots is Point 'T':

There are certain conventions in plotting the D.R. The courses steered are sometimes written against the track as shown, and the D.R. position is always shown as a small 'tick' across the line with a time against it. The distances (bracketed here) would not normally be written on the chart, but merely measured off with a pair of dividers.

It is also conventional to use a *single* arrow to indicate the course being steered through the water.

*The Estimated Position (E.P.)*

The Estimated position allows for the best estimate of tidal stream and leeway, etc., being applied to the D.R. position

In other words, it is the best estimate of the boat's position *over the ground* allowing for course, speed, current or tidal stream and leeway, since the last known position of the boat.

The symbol for Estimated position is a triangle with a dot in it, viz. △ 1300. (The time is also written alongside—as with a D.R. position.)

Figure 17E shows much the same example as before; the D.R. position at 1700 is Point 'T'—but the E.P. at 1700 is Point 'T¹' allowing for an estimated set to the North West of T–T¹ over the six hours since 1100. Note that the direction of a set is normally indicated by three arrows along the line.

*A Fix*

A fix is the positive establishment of the boat's position at a given time (although some fixes may be more positive and honest than others). The conventional sign for a fix is a circle with a dot in it—and a time against it, viz. ⊙ 1400.

A fix may be obtained by a variety of means and its accuracy will depend on many factors. The methods of obtaining a fix given below are those mostly commonly used in a small boat without radio aids or radar.

## I. THE TERRESTRIAL FIX

*(Visual bearing of charted objects)*

This is the simplest form of fix and one of the most accurate under normal circumstances with a good compass.

If a bearing is taken from the boat of a single charted object on shore and this bearing (corrected for compass errors) is plotted on the chart, it has been established that the boat must lie somewhere on a line reciprocal to that bearing. For example, in Figure 18A, the compass bearing of a lighthouse at 1000 was 085°. The Variation was 10° W and the Deviation (for the boat's heading) was 2°E.

| | |
|---|---|
| Compass Bearing | 085° |
| Variation | —10 |
| Deviation | +2 |
| True Bearing | 077° |

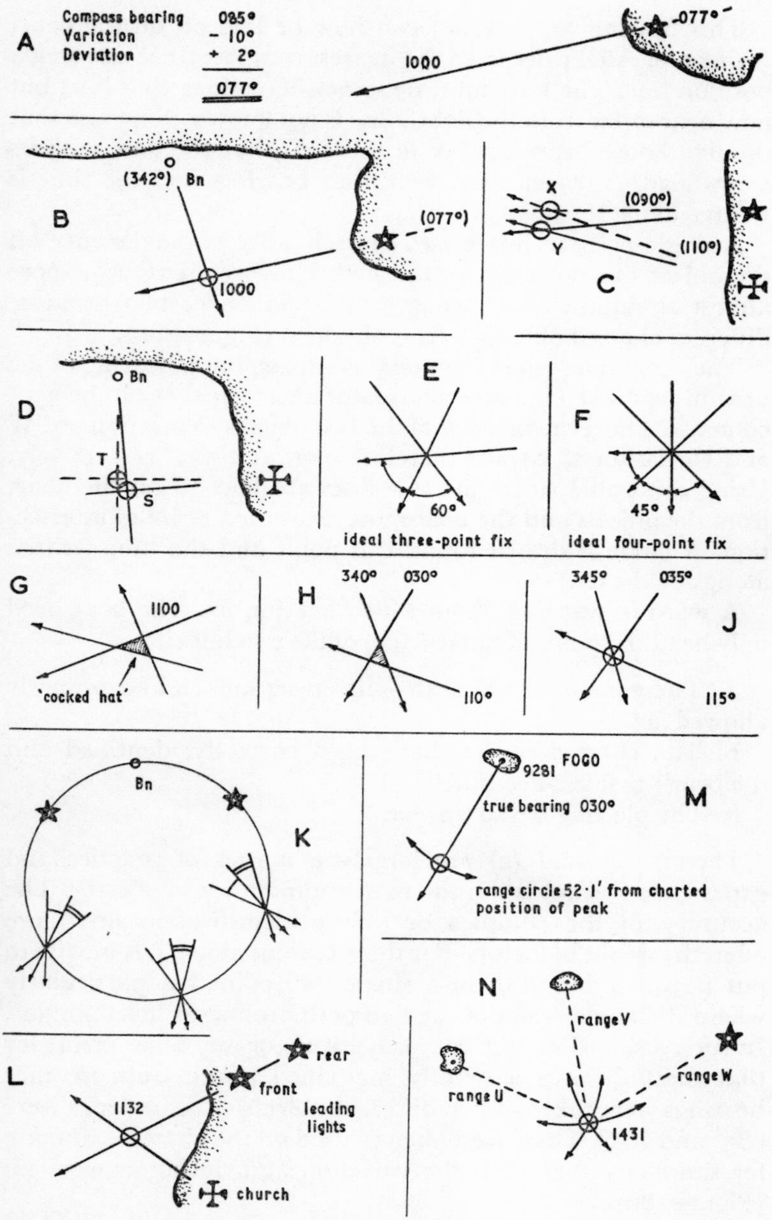
A
Compass bearing 085°
Variation − 10°
Deviation + 2°
077°
1000
077°
B
(342°) Bn
(077°)
1000
C
X
(090°)
Y
(110°)
D
Bn
T
S
E
60°
ideal three-point fix
F
45°
ideal four-point fix
G
1100
'cocked hat'
H
340°
030°
110°
J
345°
035°
115°
K
Bn
M
9281 FOGO
true bearing 030°
range circle 52·1′ from charted position of peak
L
rear
front
leading lights
1132
church
N
range V
range U
range W
1431

Figure 18

This bearing (077° True) can now be laid off on the chart (using a parallel ruler from the nearest compass rose) as a single position line. The boat must lie somewhere along that line, but how near or far from the lighthouse is not known. Note also that the lighthouse bears 077° from the boat, but the line, with its arrowhead, is drawn *away* from that bearing, and the time is written along the line.

A single position line is not, in itself, a fix as the distance off the object has not been established. However if TWO or more almost simultaneous bearings can be taken of two or more different charted objects, a fix is obtained (Figure 18B).

The lighthouse again bore 085° compass, but a bearing of the beacon was taken a few seconds later and turned out to be 350° compass. The True bearing of the two objects (Variation 10° W and Deviation 2° E) was therefore 077° and 342° respectively. Using a parallel ruler, the two lines are laid off on the chart from the objects and the boat must lie at their point of intersection. A circle is drawn round this point and the time written alongside the fix.

A word of warning about a two bearing fix. The boat need only be at the point of intersection of the two lines if:

a. The compass is accurate—its errors known and properly allowed for.

b. The charted objects have been correctly identified and the bearings taken accurately.

c. The plotting is also correct.

Factors (b) and (c) are largely a matter of practice and experience and (c) is not the most common cause of error. The accuracy of the compass or object identification are more often the doubtful factors. For these reasons alone, it is unwise to put unwarranted faith in a single two point fix, particularly where the angle of cut of the two position lines is less than 30°. In such a situation, any compass errors (or any other errors for that matter) have a greatly magnified effect. Suppose that bearings were taken of two different objects—the bearings were 085° and 105°. These were then plotted on the chart—assuming for simplicity that both the variation and deviation were nil (Figure 18C).

By all appearances the boat should be at Point Y. But there is only 20° difference between the two bearings, and suppose that the compass was a few degrees in error . . . say 5° when the bearings were taken. The plotted bearings should have been 090° and 110°. The boat is, in reality, at Point X, a considerable distance away from its supposed position.

On the other hand, a 90° cut between the two bearings taken is ideal; whatever the compass error, its effect is minimised. Assuming the same error . . . 5° and bearings of 355° and 085° (Figure 18D).

Point S is the supposed position and Point T the true position. T is much closer to S than in the previous example.

Therefore, when taking a two-point fix:

1. Try and obtain bearings of objects about 90° apart.
2. Do not use objects less than 30° apart if this can possibly be avoided—if you must, treat the resulting fix with some suspicion.

It is usually far better, if sufficient objects are visible, to use a 3 or 4 point fix. Again, with a three-point fix the optimum choice is three objects each 60° apart, and for a four-point 45°, viz. Figures 18E and F.

However, these separations are not critical; the aim must be to keep a difference of at least 30° between position lines.

*The 'Cocked Hat'*

The great advantage of using three or more position lines in a fix is that the quality of the fix is immediately apparent when it is plotted on the chart. The three lines should pass through a point, and if they do, you can be pretty sure that the boat was in the position shown at the time of the fix. (There is one exceptional circumstance when this does not hold true and this is mentioned below.)

Perhaps though, the three lines do not pass through a single point when they are plotted. In this event, the area of uncertainty produced is called a 'Cocked Hat' (see Figure 18G).

All that can be assumed initially in this case is that the boat is somewhere within the area enclosed by the three lines. In open water it is normal to assume that you are in the centre of the area of doubt. But close inshore or in the proximity of danger, it is safer to assume that your position is in the corner nearest the danger.

A 'Cocked Hat' may be caused by:

a. Compass error, or the wrong variation/deviation applied.
b. Charted objects incorrectly identified.
c. Inaccurate taking of bearings.
d. Bad plotting.
e. Unlikely, but possible—distortion of the chart.

If the compass is suspect, it is sometimes possible to 'close' the 'cocked hat' by looking at the plotted fix and seeing whether the lines can be made to pass through a point by adding or subtracting the *same amount* to/from each bearing.

In Figure 18H, a cocked hat exists. If 5° is added to each bearing as in Figure 18J, it may be found that the position lines pass through one point. In practice, of course, it would be necessary to rub out the original fix and re-plot it using the adjusted bearings.

It should be emphasised that this practice of 'closing a cocked hat' should be used with caution, and principally as a method of evaluating an unsuspected compass error. It is most dangerous to continually try to 'fudge' fixes to make them pass through a point. If a fix shows up as a cocked hat, the safest thing to do is to check that the right objects have been taken and then fix again. If another cocked hat appears, then the compass may well be suspect . . . the immediate response should be to steer a safe course on the assumption that the boat is in the most dangerous corner of the fix—and then check the compass on a transit, or by some other means, as soon as possible.

Earlier, it was said that if three (or more) position lines pass through a point, the boat must be at their intersection and the fix an absolute indication of the boat's position. The one exception to this assumption occurs when the boat and the three objects taken all lie on the circumference of the same circle. In this particular case, even if the compass has an unsuspected error, the position lines will always pass through a point as shown in Figure 18K. Note that the angles between the objects always remain the same. Thus although the fix appears to be brilliant on the chart, perhaps 'fings ain't what they ought to be!'

## II. FIXING BY TRANSIT AND BEARING

The great advantage of a transit is its certainty. If the charted marks are visibly in alignment, then the boat must lie on the extension of the line joining them. Secondly, if a compass bearing of the aligned marks is taken, the total compass error, on that heading, is established. If too, a bearing of a different charted object is taken almost simultaneously, this bearing can be corrected with the transit-found error. Thus a reliable fix is obtained and the compass checked at the same time. (Figure 18L.)

The two leading lights in line bore 032° by compass and the church 115. The charted bearing between the lights, found with a parallel ruler on the chart, was 045°. Therefore, the compass error at that moment (a combination of variation and deviation) must have been 13° (045—032). It must also be Easterly, or compass reading low, as the True bearing was the greater. The transit line can be plotted straight away, but before the position line from the church is plotted, this correction must be applied to the church's bearing; 115+13=128°. This is the true bearing plotted and the actual position of the boat at 1132 is shown by the fix on the chart. Plotting the 'neat' compass bearings would have given a very different answer.

Transits are not, of course, always available when you want them, either for fixing or checking the compass. However, an intelligent glance at the chart will often show that if a suitable back mark can be found that is visible over a wide range of bearings, an assortment of front marks are quite often available. The transit marks need not be man-made charted objects such as leading lights, beacons, towers, etc., but can be natural features such as the left-hand edge of a point being in transit with the right-hand edge of an island. Beware, though, of using gently shelving points of land or low spits. Their charted position approximates to the high water mark and if the tide level is different from this, a bearing of a low point may be considerably in error vis-à-vis its charted position. Also, never use buoys for transits; they move position slightly with the tidal stream or current and may drag well out of position as the result of heavy weather.

## III. FIXING BY RANGE AND BEARING

This is a common method of fixing in use at sea today; a compass bearing and simultaneous range is taken of a charted object. In ships, the range is normally obtained by radar—outside the capabilities of the normal small boat. Some yachtsmen may possess a small optical range finder but these are comparatively expensive instruments. The majority of the larger cruising boats will, however, have a sextant on board. The principal use of this instrument is in finding the altitude of the sun, moon, stars and planets above the visible horizon but it can also be used for measuring the vertical and horizontal angles subtended by charted objects. Vertical sextant angles of objects of a known height, with the appropriate corrections applied, can be looked up in tables listing angles against range. There are several books which include tables of this nature, but one of the most compact and useful, containing tables specifically for this purpose, is Lecky's "The Danger Angle and Offshore Distance Table'.

The procedure is quite simple. For instance, a lighthouse is shown on the chart to have an elevation of 130 feet. The sextant, held vertically, is used to measure the angle between the centre of the lantern of the lighthouse (the charted elevation) and the sea surface beneath it. Suppose the angle measured was 0° 19.8′. The first correction that must be applied is the Index Error of the instrument itself (see Chapter 8).

| | |
|---|---|
| Sextant Altitude of lighthouse | 0° 19.8′a |
| Index Error | +01.2 |
| Observed Altitude | = 0° 21.0′ |

Consulting Part I of Lecky's tables—Heights of objects from 50 to 1100 feet and distances from 0.1 to 7·5 miles—the range of the lighthouse can be read straight out as being 3.5 miles. The only correction to be applied when using this part of the tables is that for Index Error, as shown. A range is therefore obtained very simply.

Part II of these tables covers heights of objects from 200 to 18,000 feet and distances from 5 to 110 miles. This section is normally used when trying to find the range of the land whilst you are still a considerable distance away from it. The peak of the island of Fogo in the Cape Verde Islands has an elevation of 9281 feet and may be seen at great distances. Approaching from the South, from your D.R. position you estimate that you are 48 miles off it but wish to find the range accurately for a fix. Using Part II of Lecky's tables, there are two additional corrections which must be applied to the Sextant Altitude before entering the tables to obtain the range. These are the Angle of Dip (for the observer's height of eye above the water) and the other (1/12th of the estimated distance) for refraction—the bending of the light rays in the atmosphere. The Sextant Angle is therefore corrected for:

a. Index Error.
b. Dip (from another small table in Lecky's)
c. Refraction—1/12th of estimated distance (correction in minutes of arc).

FOGO, elevation 9281 feet. Your height of eye in the boat is 10 feet, estimated distance off, 48 miles. Having taken the Sextant Altitude between the peak and the horizon between you and the island, this turns out to be 1° 20.0′.

| | | |
|---|---|---|
| Sextant Altitude | 1° 20.0′ | |
| Index Error | +01.2 | (same as before) |
| Observed Altitude | 1° 21.2 | |
| Dip | —03.4 | (always minus—this for height of eye of 10 feet) |
| | 1° 17.8 | |
| Refraction (1/12th of estimated distance) | —04.0 | (always minus—this for $\frac{48}{12}$=4 minutes of arc) |
| Angle = | 1° 13.8′ | |

Looking up 1° 13.8′ in Part II of the tables against an altitude of 9200 feet gives a range of 52.1 miles for Fogo's peak. Note that Part II of Lecky's tables are intended for use with objects lying *beyond* the sea horizon.

The plotting of a visual range and bearing fix on the chart is very straightforward. The bearing must, of course, first be corrected for compass errors to obtain the true bearing before it is plotted. Theoretically also, when using bearings of objects at very long range there is a discrepancy between observed bearings and those taken from, or plotted on, the chart, due to distortions involved in the construction of the Mercator projection. However, for practical purposes, this is of no significance except in high latitudes. It is important, though, to plot 'long range' bearings very carefully.

Therefore, plot your true bearing first, draw the position line from the object, and then with a pair of compasses or dividers set to the range obtained, draw the range circle from the object on the chart. The point of intersection of the bearing position line and the range circle is the fix position. Then write the time against it. This fix is shown in Figure 18M.

## IV. FIXING BY TWO OR MORE RANGES

If the ranges of two or more objects are taken almost simultangeously, using vertical sextant angles or any other means,the position circles from these objects intersecting on the chart will give a fix. This is a good method of fixing if suitably defined objects of known height are available. (Figure 18N.)

## V. FIXING BY USING HORIZONTAL SEXTANT ANGLES

This is a highly accurate method of fixing which can be used when precision is vital, e.g. for laying a buoy or mooring in the correct spot or in charting the position of a navigational mark. It is a method that has been used by surveyors for many years in the normal course of their business. Its main disadvantage is that it is a comparatively slow method of fixing and the plotting of horizontal sextant angles is a somewhat pernickety job.

Horizontal sextant angles are taken by holding the sextant on its side, handle underneath; one object is viewed directly through the clear glass section of the horizon glass, and the reflected image of the other object is superimposed on it by moving the scale micrometer or vernier. The angle between the two can then be read off.

A minimum of three charted objects must be used to obtain a fix, as the angles on each side of the centre object are needed. If there is a large height difference between any of the objects as viewed from the observer's position, it may be difficult to superimpose the objects correctly; therefore objects of more or less equal height are preferred. The objects chosen should also be more than 30° apart to give a good cut. Ambiguity can again exist if the objects and the boat all lie on the circumference of the same circle (as commented upon earlier when discussing 'cocked hats'). It is therefore safer to choose a centre object nearer to you than the other two—or objects in a straight line—so that such a situation cannot arise.

Once the angles have been observed, they can be plotted onto the chart with the aid of a 'Station Pointer' or 'Douglas Protractor'. The Station Pointer is the more accurate of the two.

To use the Station Pointer, set the two angles on the outer arms. Supposing that the central object was a church, the left-hand object a lighthouse, and the right-hand object the right-hand edge of a cliff (Figure 19A).

The Station Pointer, with its arms set to the appropriate angles, is laid on the chart and moved around until the *bevelled* edges of each leg pass exactly through the charted position of the objects observed. The boat's position is then the centre of the instrument, normally marked by a small 'V' notch through which a pencil mark can be made on the chart.

The result is a good, accurate fix achieved independently of any compass bearing.

The Douglas Protractor is designed for use on exactly the same principles, but it merely consists of a transparent plastic square, with an engraved, matt upper surface graduated round the edge in degrees, and with a small hole at its centre.

The left and right angles are marked with a pencil on the matt surface, and lines drawn from the edge to the centre of the protractor. It is then placed on the chart and moved until the

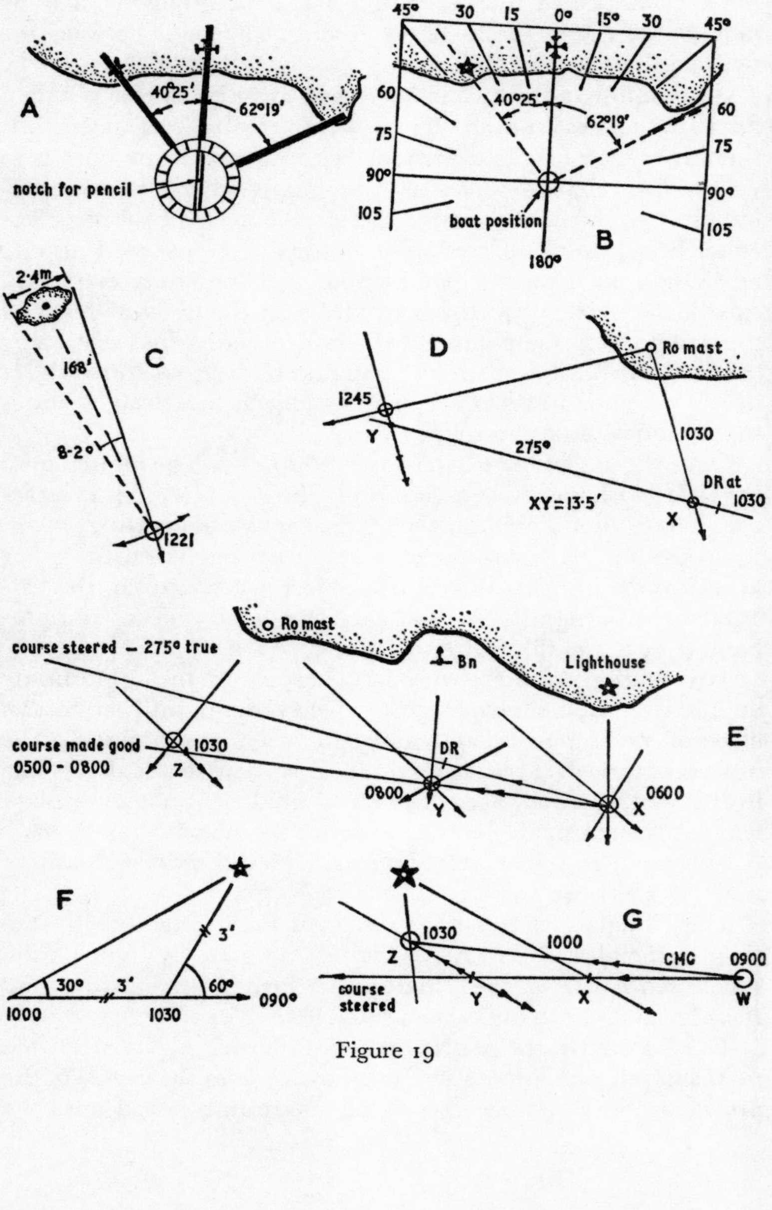
A
40°25'
62°19'
notch for pencil
45°
30
15
0°
15°
30
45°
60
75
90°
105
40°25'
62°19'
boat position
180°
B
2·4m
168'
C
8·2°
1221
D
Ro mast
1245
Y
1030
275°
DR at
XY=13·5'
1030
X
Ro mast
course steered – 275° true
Bn
Lighthouse
course made good
0500 - 0800
1030
Z
DR
0800
Y
0600
X
E
F
3'
30°
3'
60°
1000
1030
090°
G
1030
1000
Z
CMG
0900
course
steered
Y
X
W

Figure 19

lines all pass through the charted positions of the objects. The centre then marks the boat's position on the chart. (Figure 19B.)

## VI. FIXING BY BEARING AND A HORIZONTAL SEXTANT ANGLE

With an accurate instrument, such as a sextant, for measuring angles, it is possible to get the range of an island (or other object) by measuring the angle subtended by its edges. Clearly the greater the angle subtended, the better the range accuracy is likely to be. The formula for calculating the range is:

$$\text{Range} = \frac{360 \times A}{2\,\pi \times a}$$

where A = Distance between objects (edges) in miles

$a$ = Measured angle in degrees

$\pi = 3.14 \text{ or } \frac{22}{7}$

Therefore, with an island 2.4 miles across and a subtended angle of 8° 12′, the range will be:

$$\text{Range} = \frac{360 \times 2.4}{2\,\pi \times 8.2} = \frac{360 \times 2.4 \times 7}{2 \times 22 \times 8.2} = 16.8 \text{ miles}$$

If a simultaneous compass bearing is obtained of one edge, a fix is obtained (Figure 19C). Note in this case that the range should strictly be measured from the centre of the base-line between the two objects.

This is not, perhaps, a type of fix that is very commonly used, perhaps because the maths. involved are a little laborious? However, it provides a way of getting a fix if the angle of cut between two bearings is too shallow to provide a reasonable answer.

## VII. THE RUNNING FIX

Quite often, particularly in the more out of the way places with a featureless coastline, it is difficult to find more than one identifiable charted object that you can use for fixing. Beggars cannot choose, so a less accurate method of fixing has to be used. The Running Fix is that most commonly used in these circumstances. A bearing is taken of the only object visible, the time noted and the position line from this bearing plotted on the chart. Later when the bearing of the object has altered sufficiently—due to the boat's movement—a second bearing is taken and plotted. The first position line is then 'transferred' up to the second line, allowing for the boat's course and speed over the period between the two bearings. This should be the boat's course and speed 'over the ground' if it can be estimated from past history. If not, the boat's course and speed through the water must be used. It is this 'distance run'—or the distance travelled in the time between the two bearings—which is the weakness of this type of fix. Any inaccuracy in the estimation of the distance run will show up as fix inaccuracy, without, of course, the navigator being aware of it.

An example again. Variation 8° E, Deviation 3° W for a course of 270° by compass, speed 6 knots. A wireless mast bears 340° at 1030 and 050° at 1245. What is the boat's position at this time? (See Figure 19D.)

| | | | |
|---|---|---|---|
| Compass bearing | 340° | Compass bearing | 050° |
| Total correction | +5 | | +5 |
| True bearing | 345 | True bearing | 055 |

True Course 275°

As you do not know precisely where you are at 1030, it is normal to take the point on the position line nearest the D.R. position—thus Point X. From Point X, allow for a run of 2¼ hours at 6 knots = 13.5 miles. Plot this, Point Y, on your presumed track since 1030. Then draw a parallel to the 1030 position line through this point. The intersection of the transferred position line (conventionally shown with two arrows) and the 055° (1245) bearing is the fix position at 1245.

The more accurate the run between fixes, the better the fix itself. If you had been moving along a coastline prior to obtaining this running fix and had been able to take two three-point fixes in the previous few hours, you might estimate your distance run for the running fix differently (Figure 19E).

Between 0600 and 0800, although steering a course of 275°

True, the course made good over the ground (measured with a parallel ruler between Points X and Y) is found to be 271°. Also, by measuring XY with a pair of dividers, the distance made good in the 2 hours was 14 miles = a Speed Made Good of 7 knots. From Fix Y, now lay off a course of 271° and allow $2\frac{1}{2}$ hours at 7 knots = 17.5 miles to bring you to a 1030 Estimated Position—Point Z.

The beacon has now disappeared, but the transferred position line from the Radio Mast at 0800, combined with the 1030 bearing of the same object, gives you a fix at 1030 to the North of your Estimated position but South of the course steered.

*Doubling the Angle on the Bow*

An old dodge, but this in fact is really a form of running fix. If the time is taken when an object is at a certain angle on the bow—say 30° on the port bow—and it is taken again when the object is twice this angle on the bow (60°), the distance run between the two times will be equal to the boat's distance off the object at the later time. For example, in Figure 19F, a lighthouse was 30° on the Port bow at 1000, course 090° speed 6 knots, and was found to be 60° on the bow at 1030. The boat should then be 3 miles off the object at 1030.

This check is not much use if any tidal stream or current is present, i.e. the course and speed made good is different from that steered. If a set is expected, it is better to use a Running Fix allowing for an unknown tidal stream. Finally, in the case of 45° (doubling to 90°) on the bow—called a 'four point bearing'—the distance run will be equal to the distance off the object when it is abeam.

*Running Fix with unknown Tidal Stream* (Figure 19G)

Suppose that a fix was taken previously at Point W at 0900. The lighthouse bore 308° at 1000 and 346° at 1030. Presuming that you were at Point X at 1000, the speed made good is $\frac{WX}{\text{Time}} = \frac{4}{1} = 4$ knots. Then insert Point Y allowing for half an hour's run at the speed over WX, i.e. $\frac{4}{2} = 2$ miles for the distance XY. Your fix position at 1030 is then point Z, the course made good WZ, and the current YZ, setting towards Z.

*Other Dodges*

Some less positive methods of obtaining position are given below. Strictly speaking, these are not 'fixing' methods in the accurate sense, but may nevertheless be very useful in giving a good indication of the boat's probable position.

*Using Soundings*

A sharply defined fathom line can, on many occasions, be a good navigational aid, particularly when approaching land. Perhaps you might be approaching Cape St. Vincent from the West in thick weather and be uncertain of your exact position. The bottom shelves steeply in this area and the 100 fathom line is sharply defined about 5 miles off the Cape. If your boat is luxurious enough to be fitted with an echo sounder that will reach that far down, well and good. Providing you keep outside this depth, the boat cannot come to any harm from grounding—the hazard is more likely to come from large ships rounding the Cape. But not all boats have echo sounders nor a lead line much more than 20 fathoms in length, so it is a question of getting the best out of the equipment held on board. A sharply defined 20 fathom line is an obvious target in these circumstances for the gaining of a position line at the least risk, and a glance at the chart will soon show whether this is a practical proposition.

In many places, the seabed slopes too gently or is too undulating for the crossing of a particular fathom line to immediately ring up three red cherries. However, unless the bottom is absolutely flat, there is a technique which can assist in the use of soundings to establish a position. This is to take soundings regularly, noting the time and depth, and adjusting the latter for the likely height of tide. At the same time, keep a careful record of the course and speed. After a long enough time to allow a reasonable picture to be built up, place a piece of tracing paper on the chart, marking on it a line to indicate North. Then estimate the boat's probable course and speed made good over the period, laying off this course on the tracing paper. Finally, plot and mark the depth of each sounding along the track, using the chart scale to measure the distances between soundings. The resulting sounding pattern on the tracing paper is then moved around on the chart, keeping the North/South line roughly parallel with the chart meridians, until the pattern fits the sounding marks on the chart in the vicinity of the probable position.

*Horizon Range*

At night, the approximate range of a lighthouse can be estimated from the moment when it just appears above, or dips below, the horizon. Visibility must be good, of course, and the light of sufficient power or intensity to reach the horizon and some distance beyond it. The intensity of all lights is given in the Admiralty Light Lists, but a glance at the chart usually gives sufficient clue on this point as the major lights will show long visibility ranges. It is the actual raising of the light itself above the horizon or disappearance below it that must be taken. The Loom—the beam only visible sweeping around the horizon—may be seen long before or after the actual light is seen.

A table in Lecky's, mentioned previously under vertical sextant angles, can be used for calculating the range; there are similar tables in each List of Lights, and in other marine navigational tables, such as Inman's *Nautical Tables* and Reed's *Almanack*.

Taking Cape St. Vincent as an example; the light there is described as Fl. 5 sec 264 ft. 23 M—i.e. it flashes white once every 5 seconds, has an elevation of 264 feet, and is visible to an observer, whose height of eye is 15 feet, at 23 miles. If your height of eye is 10 feet, the light should appear above the horizon at the following range:—

| | |
|---|---|
| Range of light for its 264 feet elevation | 18.7 miles |
| Range of light for your 10 feet elevation | 3.63 miles |
| Light appears or dips at | 22.33 miles |

*Range from Sound Reflection*

Only of practical use really when one is confronted with the situation of being stuck in the middle of a Scottish Loch or Norwegian Fjord with fog coming down, depth too great for anchoring, and the beer running out.

The echo of any noise you care to make (which might be considerable) will return to you, so that:

$$\text{Distance in feet of boat from cliff} = \frac{\text{Time in seconds} \times 1130}{2}$$

or if this sum is too difficult to work out in your head, the rough distance is:

$$\text{Distance in Cables} = \text{Time interval in seconds} \times 0.9$$

## SOME PILOTAGE HINTS

*'Shooting Up' Objects*

In an unfamiliar area, it is important to be able to identify objects quickly and accurately; otherwise a lot of time will be wasted trying to plot position lines that do not meet and close cocked hats that refuse to shut. Or, there may be a very conspicuous object in the area which is not charted but would otherwise be useful for fixing.

There are several ways of 'shooting up' an object. The quickest method is to get the unknown object in transit with a charted object, and at the same time take a bearing of the two. The bearing, laid off with a parallel ruler on the chart from the known object, will often identify the doubtful one.

This system can also be used in another way. For example, going up a narrow channel, it may be necessary to identify buoys quickly and there is not much time for plotting fixes (see Figure 20A):

The target is the mooring buoy at X, passing between the Port hand buoy at Y, and the Starboard hand buoy at Z. From the chart, it is obvious that there are a lot of buoys in this channel and it may be difficult, on rounding the corner, to identify one from the other. To save the embarrassment of either going aground or pinching someone else's mooring, before reaching this point find out from the chart the true bearings of these three buoys when they are in transit with the pier-head on the South shore. Convert these bearings to compass bearings by applying the variation, and the deviation for a heading of 065°. Then, quick bearings of 080°, 110° and 115° on the pier-head on the approach will help identify these buoys.

Getting an uncharted object marked on the chart; if the

uncharted object is taken successively in transit with two or more charted objects, its position can then be fixed (Figure 20B).

A bearing was taken (305° True) of the uncharted beacon in transit with a church at 0915; at 0930, the beacon, bearing 060°

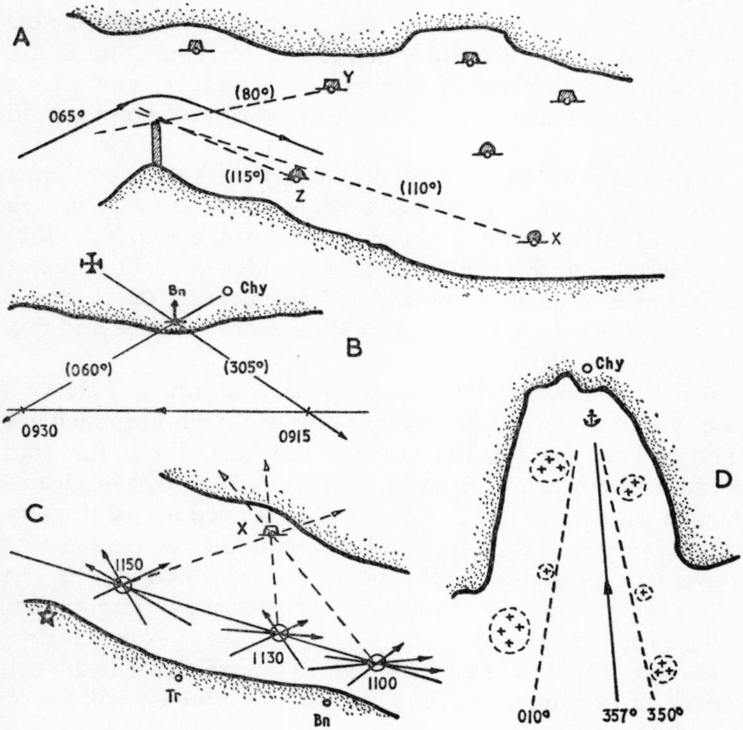


Figure 20

True, was in transit with a chimney. The intersection of the two position lines fixes the position of the beacon.

If an object cannot be got into transit with other charted marks, the alternative is to take a series of three-point fixes and on each occasion take a bearing of the uncharted object. Again the intersection point of these bearings from the individual fixes will give the object's position.

For example, in Figure 20C, to establish the position of the buoy at X. Take a bearing of the buoy on each occasion of taking the fix bearings at 1100, 1130 and 1150. Convert the compass bearings to True, plot the fixes and the position line through the buoy from each fix. Eureka!

*Lines of Bearing and Clearing Bearings*

During pilotage in confined waters, leading bearings and clearing bearings are extremely useful in making the correct approach and in clearing dangers. For instance, you wish to enter a harbour whose mouth is cluttered by rocks on each side (Figure 20D).

From the chart you can see that if you can keep the chimney bearing 357° True, this line of bearing will lead safely into the anchorage. Also that if the bearing increases, e.g. to 358° True, you must be to Port of track and an alteration of course to starboard is needed. Conversely, if the bearing decreases—to 356° True or less—you must be to Starboard of track and must alter course to Port.

It is also as well to have clearing bearings up your sleeve in case you are forced to leave your planned approach—by another boat leaving harbour for example. From the chart establish the bearings from an object which will lead clear of dangers on either side. The object chosen need not be the same as that being used for a leading bearing, but in this example it is convenient to do so. If the bearing of the chimney is kept between 350° and 010° True, you cannot run over the rocks on either side.

For use, the True bearings will, of course, have to be converted into compass bearings as it is those which will be observed during the approach. Assuming that the Variation is 8° W and that the Deviation (on a compass course of North) is 3° E. The total correction is therefore 5° W. The leading bearing will then be 002° Compass, and the clearing bearings 015° to 355°. These should be noted down before you make the approach:

| | |
|---|---|
| Leading Bearing | 002° Compass |
| Clearing Bearings | Keep chimney between 015° (Port) to 355° (Star)<br>or <015° (P)> 355° (S) |

The closer the mark, the more sensitive will your leading or clearing bearings be. So do not choose objects that are too far away. One degree subtends a distance of one mile at 60 miles. Therefore a degree error at 2 miles is the equivalent of about 70 yards off track, and at $\frac{1}{2}$ a mile—about 17 yards.

# 7

# Tidal Stream and Current Problems

Tidal streams and currents vary greatly in strength not only from one ocean or sea to another but also between adjacent local waters. The coasts of the North Atlantic are uncomfortably placed in this respect, the tidal range being generally large, and therefore strong tidal streams abound. Tidal streams of over 5 knots are not unknown round the British Isles.

Due respect must therefore be paid to these streams and currents, particularly in small craft where the boat's speed in poor sailing conditions may well be less than the rate of the stream.

With respect to chartwork, there are several individual problems—where tidal streams are concerned—that need to be solved:

1. To shape a particular course to a given point allowing for a known tidal stream; and as a refinement to this, to arrive at a particular place at a particular time allowing for the set.
2. To estimate the boat's position after a given time allowing for a known tidal stream.
3. To find the rate and direction in which an unknown tidal stream is setting.
4. To use this information to arrive at a destination at a given time.

1. *To shape a course to reach a given point allowing for a known tidal stream* (Figure 21A)

In this case, you know your starting and end positions, the direction and rate of the tidal stream, and the boat's speed through the water. Wanted—the course.

You are at Point X and wish to reach Point Z, i.e. the line XZ

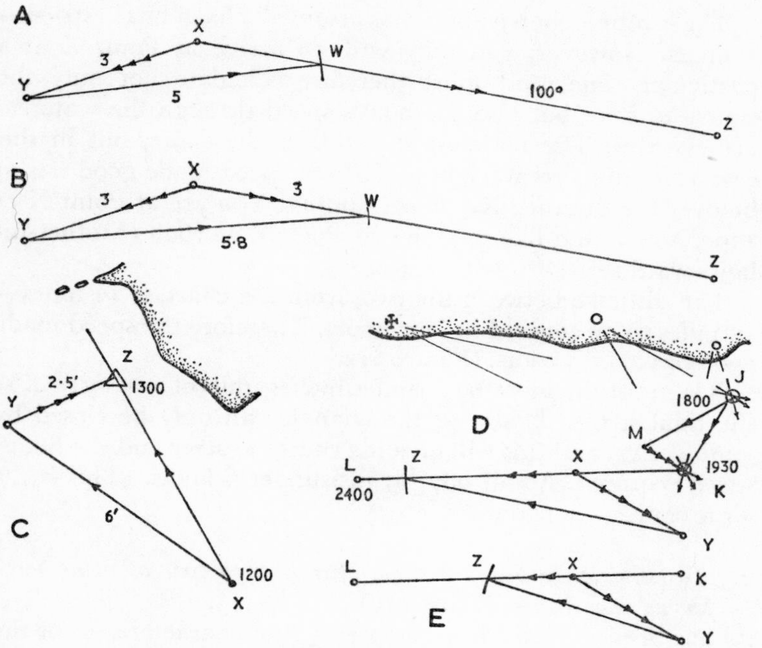


Figure 21

must be the course made good (C.M.G.). The tidal stream is setting (towards) 250° at 3 knots, and your boat's speed is 5 knots. The construction is quite simple. Draw XZ and XY, the set and rate of the tidal stream—250° length 3 miles (if using a 1 hour triangle on the chart). Set your boat's speed (5 miles) on a pair of dividers or compasses and with one point on Y, cut off along the line XZ. Then:

YW = The course to steer
XZ = The course made good
XW = The speed made good along XW/WZ

Measuring XW off the same scale as that used previously for XY and YW, you can find its value in knots (about 2.3 knots).

Then $\frac{\text{XZ (distance to go)}}{\text{XW (speed made good)}}$

allows the calculation of the time of arrival at Point Z.

The example shown above has assumed a fixed boat's speed—5 knots. However, you may wish to arrive at Point Z at a particular time, and must therefore calculate not only the course to steer but also the boat's speed through the water to achieve this. The basic construction is the same, but in this case you must first work out what the speed made good has to be over the distance XZ. For example: You are at Point X at 1200, and would like to arrive at Point Z at 1600 (Mother-in-law's waiting).

The distance between the two, from the chart, is 12 miles—12 miles to be covered in four hours. Therefore the speed made good has to be 3 knots. (Figure 21B.)

XW must therefore be 3 (miles/inches/units of any sort). XY, the tidal stream is also 3; the triangle can only be closed by joining YW, and this will give the course to steer and the boat's speed required. About 087° at just under 6 knots. (This is the True course—not compass.)

2. *To estimate the boat's position after a given time allowing for a known tidal stream* (Figure 21C)

As has been explained in Chapter 5, the characteristics of the streams around the British Isles are now fairly well known and information about them for particular times and places is given in the Tidal Diamonds on the charts. Let us say that the Diamond off the Isle of Wight at 1200 on October 2nd states that the stream is setting 065° at 2.5 knots. Your position is south of St. Catherine's Point Light and making for the Needles, steering 305° True at 6 knots. Wanted—your E.P. at 1300.

Lay off XY—your course and speed through the water—305° 6 knots. Then lay off YZ, the direction and rate of the tidal stream, which was 065° 2.5 knots. Your E.P. at 1300 is then Point Z, your course made good the direction XZ, and the speed made good is the distance XZ.

With any of these problems, you can use any units you like in the speed triangle, i.e. 1 knot = 1 inch, mile, centimetre or what have you. But as you are working on the chart, it is more convenient to use the chart scale and make, when you can, 1 knot = 1 mile. On some charts of small scale, this would not be satisfactory and then you could use 1 knot = 5/10 miles. Nor is it necessary solely to use one hour's run for each vector of the speed triangle. Providing the same time is allowed for each vector, any period can be used—$\frac{1}{2}$ hour, 2 hours' worth, etc. These comments only apply to the *speed triangle*. Distances must, of course, be measured on the scale of the chart and it is easy to forget this if using different units for speed.

In the example above, it is obviously simpler to use an hour triangle at 1 knot = 1 mile. Note also the conventional arrow symbols on the sides of the triangle:

| | |
|---|---|
| Course and speed through the water | One arrow |
| Course and speed made good | Two arrows |
| Tidal stream direction and rate | Three arrows |

3. *To find the set and rate of an unknown Tidal Stream, and*
4. *to use this information to reach a destination at a given time* (Figure 21D).

In this example, a fix is obtained at Position J at 1800. The second fix is obtained at Position K at 1930, the course and speed steered in the interval being 240°—3 knots. You wish to reach Position L by midnight if possible, and the distance between K and L on the chart is 15 miles. What course and speed should be steered assuming that the tidal stream or current remains the same?

JK must be the course made good between 1800 and 1930. The course and speed steered during this time are then laid off from J—240° at 3 knots=240° for 4.5 miles over $1\frac{1}{2}$ hours. This gives Point M. MK must be the direction and distance by which the boat has been set to the South-East over the $1\frac{1}{2}$ hours, i.e. 120° for 2 miles. Therefore the current must be setting towards 120° at a rate of $\frac{2 \times 2}{3} = 1.3$ knots.

You wish to make good a course between K and L and have $4\frac{1}{2}$ hours in which to cover the 15 miles. That means a speed made good of $\frac{15 \times 2}{9} = 3.3$ knots.

Anywhere along the line KL, lay off the tidal stream vector XY—120° at 1.3 knots. Then from X along XL, lay off the speed to be made good—XZ—which is 3.3 knots. Then YZ is the course to steer, and the length YZ the boat's speed required—about 4.5 knots.

Alternatively, if insufficient wind or power is available to make this speed, you might want to find out at what time you will reach Position L at your present speed through the water.

Lay off the current vector as before—then with the dividers set to your speed, cut off along XL at Z. XZ is then the speed made good (Figure 21E).

XZ turns out to be about 1.8 knots. Therefore it will take $\frac{15}{1.8} = 8$ hours 20 minutes to reach Position L at your present speed, and you will not be there until 0350. Bad luck!

# 8

# Navigational Instruments and Aids

Magnetic Compasses and chart table instruments have been described in earlier chapters, but there are other instruments and navigational aids in common use which are of interest to the small boat owner.

Perhaps one ought to start with the Gyro Compass. This is not generally used in the smaller boats, mainly no doubt because of its cost, but also because it needs a continuous power supply of one sort or another and a certain amount of maintenance. Small North-seeking gyro compasses are fitted in some fairly small craft, and it is fair to say that they are generally more accurate than a magnetic compass—and, of course, no irritating corrections for variation and deviation to worry about. They do suffer from errors, but unless something is radically wrong, these are normally of a fixed variety. Incidentally, gyro error is normally classified as 'High' or 'Low'. A gyro error of 1° High means that the compass is over-reading by 1°, and therefore one degree must be subtracted from the compass reading to obtain True course or bearing. They have another advantage in that a number of repeaters can normally be driven off the master compass, and therefore a good all-round view of the horizon can be gained for the taking of bearings, and the steering repeater sited in an ideal position.

However, all these advantages must be set against the price. Before leaving the subject, though, it might be worth mentioning an interesting (and cheap) adaptation of a gyro compass which I came across a few years ago. The owner of this boat had on board a small portable air-blown free gyroscope, which had probably been originally fitted in an aircraft. This was connected by a rubber pipe, through a reducing valve, to a compressed air bottle. The procedure was then quite simple—when a bearing was required, the gyro was fished out of its stowage (it only weighed about a pound) the card aligned to the True boat's heading, and the air switched on. It ran up to full speed in a few seconds and could then be used anywhere in the cockpit area for the taking of bearings. Its advantages over a hand bearing magnetic compass were that the card was absolutely steady, however much the boat was moving about, and it was not, of course, affected by any local magnetism in the boat.

A free gyro, as was this instrument, is not North-seeking and suffers from the normal apparent drift due to the Earth's rotation. This is 15° per hour, so four minutes would have to elapse before the gyro drifted one degree. This is more than enough time to take the bearings for a three-point fix. If anybody wishes to copy this idea, then I should refer them to the owner, Wing Commander Crammond, but I regret that I do not know his whereabouts at the present time. Perhaps, also, similar instruments are now available commercially.

*Sextants*

There are two main types; those that use the horizon as a reference (marine) and those that use a bubble (aircraft). The marine type is the more accurate—*naturally*!—but to dispose of the opposition first. Bubble sextants are difficult to use at sea—bubble acceleration caused by the boat's or ship's movement makes it very difficult to hold the bubble steady. This is vital, of course, to the accuracy of the readings obtained as the bubble *is* the horizon. It is unfortunate that they cannot be used satisfactorily at sea, for they would give the facility for taking astronomical sights throughout the night, which is not possible with the normal marine sextant.

Marine sextants have remained largely unchanged in design for many years, although the method of taking readings from the sextant scale has been altered, the 'micrometer' drum replacing the now obsolescent 'vernier' scale. A standard marine sextant, capable of being read to an accuracy of about 0.1 of a minute of arc, now costs about £75 new. A second-hand vernier sextant can probably be picked up for about £50. Such an accuracy is desirable for astronomical sights, but is not really needed for horizontal and vertical sextant angles. There are a few smaller marine sextants about, which may be picked up second-hand for about £10 to £20, that only read to the nearest minute of arc but are good enough for sextant angle fixing.

A marine sextant has two mirrors, the top one (called the Index glass) being fully silvered and the bottom one (the

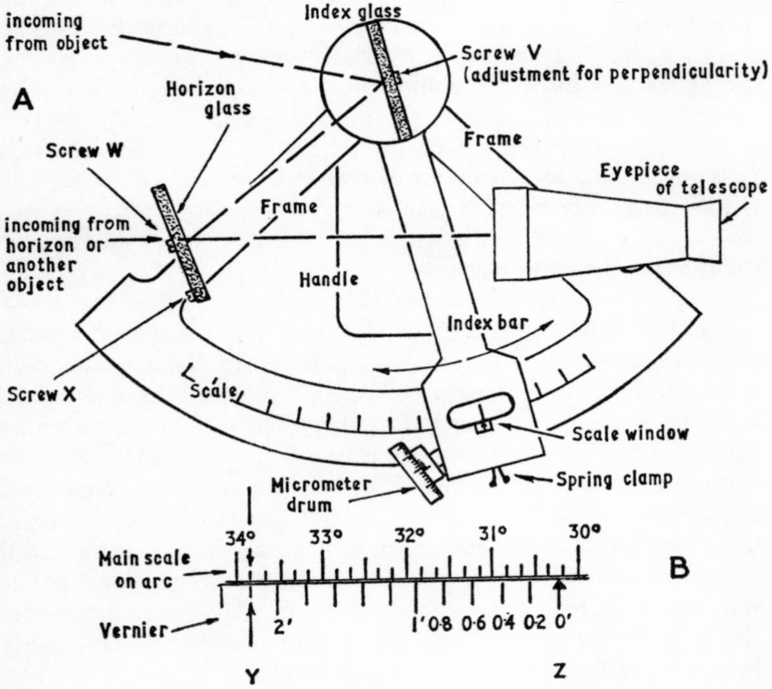


Figure 22

Horizon glass) half silvered and half clear glass. An observer can thus see the horizon (or any object) directly through the clear part of the Horizon glass. At the same time, he can see the image of another object reflected from the top mirror to the silvered portion of the bottom mirror. If the two mirrors are exactly parallel, the reflected and direct images of a single object should be coincident and the scale reading 0° 00.0′.

Figure 22A gives a rough idea of what a sextant looks like when held vertically (as it would be when taking sights) and viewed from the side. This is *only* for those who are not familiar with the instrument, and it does not show various details, such as frame supports, shades, etc.

The Index Bar is movable over the whole arc of the scale, the spring clamp grips holding it locked at roughly the desired angle once they are released by the thumb and forefinger. Fine adjustments, also affecting both the reflected image and the scale reading, are made by turning the micrometer drum. The sextant can be set, therefore, to about the right angle by moving the Index Bar until the pointer in the scale window shows the correct number of degrees and the micrometer drum will click into place near this reading once the spring clamps are released. It is important to refrain from trying to move the Index Bar without, or only partially, releasing the spring clamps as this may cause damage to the worm on the micrometer drum or to the rack on the sextant scale.

Suppose that the instrument is to be used for the taking of a vertical sextant angle. First make sure that the smaller (star) telescope is fitted. (The sextant box normally contains two telescopes; the longer (sun) telescope gives greater magnification but produces an inverted image.) Then focus the telescope using the horizon or any other convenient object. Holding the sextant by its handle in the right hand, point it at the object keeping the horizon in view through the clear glass. Release the clamps with your left hand, and move the index bar until the top of the object heaves into view in the silvered part of the horizon mirror. Let the clamps go, and move the micrometer drum until the horizon and the top of the object are exactly level with each other. Now read off—and on!

Reading off an angle is much easier with a micrometer than a vernier sextant, particularly when the light is not very good. Micrometer first; the degrees in the angle are read off opposite the pointer in the scale window—simple! The minutes are read off a scale inscribed around the drum opposite a pointer attached to the frame—also simple. The *points* of a minute are dealt with on the vernier principle; beside the frame pointer near the micrometer drum, you will find marked divisions—0.2, 0.4, 0.6, 0.8 of a minute. Whichever of these divisions is most nearly opposite one of the divisions on the micrometer drum, that division gives the points of a minute.

The vernier sextant is used in the same way, except that fine adjustments are made by a tangent screw attached to the bottom of the index bar, and the method of scale reading is different. In the first place, the main scale on the arc is more finely graduated than on a micrometer sextant, so a movable magnifying glass is attached to the frame to assist. Secondly, the degrees, minutes, and points of a minute have to be deduced from the relative positions of the main scale on the arc and the vernier scale outside it.

The vernier scales used on sextants can be read to 0.2 minutes of arc. In Figure 22B above, it is shown as a straight scale for ease of illustration although, in practice, it would be curved.

Each of the small divisions on the main scale is the equivalent of 10 minutes of arc. Therefore the arrowhead indicating the zero of the Vernier scale is showing a main scale reading of between 30° 10′ and 30° 20′ (Point Z). The starting point is, then, 30° 10′. To find the additional minutes and points of a minute, look along both scales until a mark on each *exactly* coincides. This occurs at the point marked Y. There, the reading on the vernier scale is 2.2′. Thus the whole angle is 30° 10′ plus 02.2 = 30° 12.2′.

When taking horizontal sextant angles, the sextant is laid on its side, handle underneath, and the left-hand object viewed through the horizon glass. The actual image of one object is then brought into coincidence with the reflected image of the other on the dividing line between the clear/silvered parts of the horizon glass. This is easier to do than it sounds and should cause no problems.

*Sextant Errors*

Some errors, to which a sextant is susceptible, can be removed by its owner, and these are listed below in the *order of treatment* (potential doctors please note):

a. *Perpendicularity.* The index glass must be perpendicular to the plane (frame) of the instrument.

*Remedy*: Put the index bar roughly in the centre of the arc. Hold the sextant horizontally and look into the index glass from outside—towards the arc of the scale. The reflected image of the arc should join (be in line with) the direct view of the arc. If it is not, take the small cover off Screw V (see diagram) and turn this screw until all is as it should be.

b. *Side Error*. The horizon glass must be perpendicular to the plane of the instrument.

*Remedy*: Set the scale reading to about zero. Hold the sextant vertical and point it at a star. When the index bar is moved by the micrometer, the reflected image of the star should pass directly over the direct image. If it does not, but passes to one side or other, take the cover off Screw W (this is the one on top of the glass away from the sextant frame) and adjust it until the correct answer is achieved.

c. *Index Error*. Mirrors not exactly parallel when the scale reading is zero.

*Remedy*: This is the most important error, but the one which must be dealt with last as adjustments of the other two errors may well affect the index error.

By day, choose a section of the horizon which is clear and sharply defined, hold the sextant vertical, and adjust the micrometer until the reflected and direct images of the horizon are exactly in line horizontally. Then read off the scale. At night, select a bright star, point the sextant directly at it with the scale reading set near zero, and superimpose the two images. Read off the scale as before.

The Index Error readings thus obtained may either be 'on' or 'off' the scale. Supposing that when the images or the horizon, or a star, were brought exactly into coincidence, the scale showed a reading of 0° 02.3′. This is 'on' the scale and indicates that the sextant was over-reading by that amount. The Index Error was then—02.3′ of arc. If, on the other hand, when image coincidence was obtained, the scale reading was 58.8′, the sextant would be under-reading and the error 'off the scale' by 01.2′. The Index Error was then +1.2 minutes.

Either way, this Index Error can be corrected by adjusting Screw X—at the side of the frame of the horizon glass. Watching the horizon or star through the sextant, and with the scale set precisely to 0° 00.0′, turn the screw until the images are coincident. Finally, re-check the Side Error, and have a last Index Error check.

*Note*: Particularly with vernier sextants, it may not be worth trying to remove Index Error if it is less than about 3 minutes. It can always be allowed for by adding or subtracting its value to/from observed altitudes or bearings.

*Care of a Sextant*

Sextants are fairly expensive instruments and should therefore be looked after properly. One or two points:

a. Always pick a sextant up by its frame, and carry it either by the frame or handle. Never carry it by the mirrors, or put it down on a surface upside down so that the mirrors are taking the instrument's weight. Obviously it should not be knocked about, and its replacement in its box after use is a good safety precaution.

b. Do not stow the sextant away for long periods encrusted with salt—wipe the mirrors and frame over carefully with a soft cloth and fresh water. Then oil the scale and drum.

## INSTRUMENTS FOR MEASURING DEPTH

*The Echo Sounder*

Quite a few boats are now fitted with echo sounders, so perhaps a few remarks about these instruments would not come amiss. The basic principle on which they work is very simple—A sound pulse is transmitted from an oscillator attached to the boat's hull below the waterline, reaches the seabed, and is reflected back to a receiver in the boat. As the speed of sound in water is known (although it varies between waters of different densities—sea and fresh for example), if the time is measured between transmission and reception of the pulse, a depth can be obtained.

In shallow water, the time interval between transmission/reception is very short and can only be measured by electronic means. Most echo sounders, too, work on a supersonic frequency so that one cannot actually hear the transmissions. The main difference between the various echo sounders on the market is in the way the information received is displayed.

Some of the more sophisticated echo sounders use a recorder-type display where the incoming bottom echoes are marked on a moving trace of impregnated paper by an electric stylus. This has the advantage of enabling a permanent record to be kept of the seabed over which the boat has passed. This can be very useful in constructing a sounding trace—as explained in Chapter 6—and it is more or less essential for surveying work. However, most yachtsmen are only interested in knowing what their depth is at the time, and there are other types of display which show this more simply. Some take the form of a circular gas-filled tube, marked with an appropriate depth scale. When a transmission is made, a spot of light starts rotating around the tube, and this brightens when any echo is received from the seabed. This is quite a good, practical display. The choice of an echo sounder really depends, once again, on your pocket. There should not be a great deal that can go wrong with an echo sounder, as it is a comparatively simple piece of electronic equipment, but on the whole the more expensive—the more reliable.

One point before leaving this subject—be careful to check whether your echo sounder is set to read from the *waterline* or the *keel*. In certain circumstances, this might make quite a difference!

*The Lead Line*

For those working on a limited budget, the luxury of being able to take a depth at will merely by pressing a switch is probably unattainable. So, one must resort to the bracing life and time-honoured handraulic methods; the lead line, or possibly in very shallow water, a suitably marked stave or boathook.

A lead line should be made out of very pliant rope that does not kink or retain any loops or bights once it is uncurled. Ordinary stranded rope is no good for this purpose but special plaited rope is sold. The rope is spliced through a ring on the top of the lead weight. The lead itself has a hollowed-out section at the bottom which can be 'armed' with tallow. When the lead is recovered, it should be therefore possible to identify the nature of the bottom from whatever is sticking to the tallow. If it comes up with a bit of shell attached—shells. If it comes up with a fly-button . . . ?

There is a standard way of marking a boat's lead line, although this will presumably be changed when the charts all become metric. There is no necessity to use this method if you do not wish to do so—any method will do providing it is clear to the user—but it has the advantage of being familiar to a lot of people. The fathoms—starting from the lead end—are marked as follows:

| | | |
|---|---|---|
| One fathom | One strip of leather | (One Fathom |
| Two fathoms | Two strips of leather | =6 Feet) |
| Three fathoms | Three strips of leather | |
| Five fathoms | Piece of White bunting (cloth) | |
| Seven fathoms | Piece of Red bunting | |
| Ten fathoms | Piece of leather with a hole in it. | |
| Thirteen fathoms | Piece of Blue bunting | |
| Fifteen fathoms | Piece of White bunting | |
| Seventeen fathoms | Piece of Red bunting | |
| Twenty fathoms | Two strips of leather with holes in them | |

These are known as 'Marks'. So if the lead line came out of the water with a piece of blue bunting resting on the surface, the leadsman should, in his best Metro-Goldwyn-Mayer voice, call out 'By the Mark—Thirteen'. The intervening fathoms, between the marks, are estimated and called 'Deeps'. 'Deep Four', 'Deep Six' and so on.

When taking a sounding with the lead line, the rope is coiled up in the leadsman's inboard hand, leaving a small bight between it and where the other hand is grasping the lead line a few feet above the lead. The lead is then swung and released to land in the water a sufficient distance ahead to allow for the boat's speed—the object being to take the sounding when the lead line is up and down in the water right underneath the leadsman. In most boats, with a fairly low freeboard, this can be achieved with an under-arm swing, but in ships the man has to be more ambitious and whirl the lead overhead with consequent hazard to the rest of the community. However, if when first practising you instruct your friends to go below, there should be no need for the first-aid box.

*Logs*

These either measure speed directly or distance directly; no matter, for one can easily be deduced from the other. There are basically two types:

1

a. A bottom log that works on pressure differences between static pressure orifice and an impact pressure orifice—The Pitometer Log.

b. A bottom or towed log that is driven by a small propeller, or impeller, whose rotation is measured by electrical or mechanical means—The Chernikeef Log, Walker's Patent Log and others.

*The Pitometer Log*

This measures speed directly, and is generally used in large ships. It works on very much the same principle as an aircraft's pitot tube. A rodmeter projects below the hull, one shielded pipe in it measuring the static pressure at the depth of the rodmeter and the other pipe facing forward to sense the full pressure due to the ship's speed through the water. These two pressures are compared in a mercury (or other) differential, which also acts as a damping device, and a speed produced. This can be converted into distance for display on another dial.

*Advantages.* No moving parts outside the hull which can get clogged with sea-weed, etc. Very accurate, particularly at low/medium speeds.

*Disadvantages.* Comparatively expensive and rather bulky for small craft. Needs an opening in the hull for the rodmeter.

*The Chernikeef Log*

This measures distance directly, and has been used in all sizes of craft. A small propeller is sited at the end of a shaft projecting through the hull (this can normally be raised and lowered through a gland—as can the rodmeter of a Pitometer Log). As the vessel moves through the water, the impeller rotates and this rotation is transmitted electrically to a distance recorder. This is converted to speed for display on another dial.

*Advantages.* Possible to measure very slow speeds as the impeller is quite sensitive. The log can also be adjusted (calibrated) to a certain degree by bending the impeller blades.

*Disadvantages.* Expense, and the impeller can become quite easily fouled, or perhaps damaged, by debris, seaweed, etc. Needs an opening in the hull for the impeller shaft.

*The Walker's Patent Log*

This measures distance directly, and is the log (or variety of log) most used in small craft—and in big ships too, sometimes as an emergency log. It is a towed log, where a small screw propeller, called the rotator, is towed at the end of a long piece of log line. The boat's passage through the water causes the rotator to turn, the rotation being transmitted via the line to a mechanical register fitted on a bracket in the stern of the boat. A metal wheel, called a governor, is sometimes used in the line just astern of the boat, clear of the water, to smooth out the rotation. The distance travelled can either be read off the mechanical register or, with electric logs, transmitted to a receiver sited elsewhere in the boat.

Streaming the log is quite simple. The rotator is paid out over the stern and the inboard end attached by its hook to the ring of the mechanical register. If the governor wheel is going to be used, this should be attached by its short length of line to the register—before the rotator is paid out—and kept inboard. The rotator is then paid out and the inner end of the log line clipped to the governor. This is then eased over the stern.

Recovering the log is a bit more difficult, particularly if the boat is making sternway—so haul in the log in good time! First recover the governor, unclip it, and if the boat is still making headway, pay out the inner end of the log line over the other side of the stern whilst hauling in the rotator. When the latter is inboard, the rest of the line can be brought in free of turns.

The log line is made specially for this purpose. Generally speaking, the longer the length used the better for the purposes of accuracy, but do not forget that if the boat's speed is very low, the rotator may trail across the bottom and possibly get snarled up in something down there. The recommended lengths* are:

| | | |
|---|---|---|
| Maximum speed | 10 knots | — 40 fathoms |
| | 15 knots | — 50 fathoms |
| | 18 + knots | — 65 fathoms |

*Advantages.* Cheap and simple. No holes in the hull. Spares easily carried.

*Disadvantages.* The main disadvantage is that the log has to be

* A lot of boats only use about 75–100 feet of line with fairly good results.

hauled in whenever the boat goes very slowly, stops, or goes astern (unless in deep water—or in powered boats you want a rope round the screw). The rotator is liable to be fouled with seaweed; in foreign parts, also, it seems to have a strong fascination for sharks—with somewhat violent results. It is also inaccurate in a heavy following sea.

Other methods of estimating speeds, without the use of a log, are mentioned in the next chapter.

### *Radio Aids*

There are quite a number of Radio Aids in constant use by both ships and aircraft, but not all of these, mainly for reasons of equipment expense or complexity, are available to the average yachtsman.

Those aids which require sophisticated equipment on board include the following:

a. *Radar*. Very widely used at sea by ships and an invaluable aid to navigation. Some large yachts and power boats have been fitted with radar, and there are, in fact, fairly small and portable radar sets now being produced commercially. However, the vast majority of small craft are without it.

b. *Loran*. This is an ocean aid, and Loran transmitting stations are sited in the North Atlantic, Pacific, and other areas. There are two systems, one of which is very accurate, but an expensive receiver is required.

c. *Decca*. A very well known, and widely used, coastal aid of high accuracy. Decca chains surround the coasts of the British Isles, the North Sea, the Channel and a growing number of localities abroad. The chains are generally usable within about 250 miles of the master station. However, again a special receiver is required.

The Admiralty publishes special charts, overlaid with the Decca and Loran lattices. These can be obtained from any chart agent, together with small booklets showing the world-wide coverage of these aids.

### *Coastal Radio Direction Finding Stations*

This is an aid which falls into an intermediate category as far as yachtsmen are concerned. RDF stations are spread around the coasts and will, on request and on the booking of a small fee (free, I believe, in an emergency), take a bearing of the vessel asking for it and then transmit the bearing back to the boat. This, of course, means that the boat must be fitted with both a transmitter and receiver which can be tuned to the correct frequencies. Details of the services provided by these coastal RDF stations can be found in the Admiralty List of Radio Signals. These stations should not be confused with Radio Beacons (q.v. below). The station takes the bearing, not the boat. With morse (continuous wave) transmissions from the boat, bearings can be taken over very long ranges and this really is probably most useful as an ocean aid.

*Consol*

This is an ocean radio aid, freely usable by all who care to read the signals on an ordinary radio receiver. The range of Consol

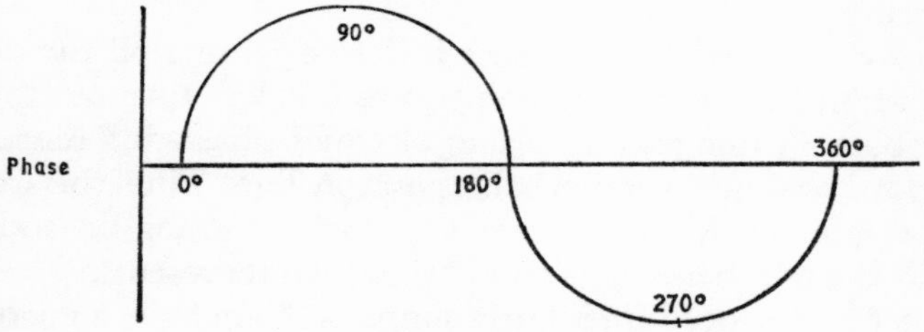


Figure 23

is about 1500 miles and there are stations at Bushmills (Northern Ireland), Stavanger (Norway), Ploneis (France) and Seville (Spain). Consol charts are published, and details of frequencies, together with Consol position tables, are given each year in the Admiralty List of Radio Signals, and in other almanacks and publications.

The system works on the principle of phase-differences, and was developed from the radio aid the Luftwaffe used to find targets in Britain during the night bombing raids of the war. A very simple explanation of phase-difference; drawn diagrammatically a radio wave can be considered, not as a straight line, but as a Sine wave (see Figure 23).

The wave starts at phase 0°, rises to a maximum value in one direction 90°, sinks again to zero 180°, and rises to a maximum in the other direction 270° before returning to zero again. During this time, depending on the wave length (frequency) of the transmitter, the wave will have covered a certain physical distance over the land or the water. If a boat was precisely half-way between two transmitters of the same wave length, the two signals would reach it at exactly the same phase, e.g. both at the 90° point. If, on the other hand, the boat was further away from one transmitter than the other, the signals would arrive out of phase, each with its own phase value, e.g. one at 90° the other at 270°.

The Decca receiver uses this principle by actually measuring the phase differences with considerable accuracy. In Consol, the transmissions are modulated in such a way that a different sound is produced in an ordinary radio receiver as the phase-difference changes. For example, a continuous note at zero phase-difference, 30 dots and 30 dashes at maximum phase-difference, and so on.

If on a chart or a map a line is drawn joining all the points equidistant from two fixed points, this line will turn out to be a hyperbola. A line joining places of equal phase-difference will therefore produce a hyperbolic position line. Thus the Consol lattice on a chart is a lattice of hyperbolic position lines identified by the number of dots and dashes each represents.

The fixing procedure is fairly simple—if you have a good ear! (I haven't, and have never yet achieved a really satisfactory Consol fix.) Tune into a Consol station. The frequency is given on the chart as well as in the books. Listen until you hear the call-sign and then start counting the dots and dashes. The total received until the start of the next cycle should add up to 60. It is not always easy to discern when the dashes start or the dots end as they merge into a continuous note (called the equi-signal). If, for example, only 4 dots and 50 dashes are heard, i.e. 6 symbols are missing, the numbers should be adjusted to read 7 dots and 53 dashes. On the chart, pencil in the position line equivalent to this Consol reading nearest the D.R., and then tune to the next station and repeat the procedure. This should give a Consol fix. Remember that this is an ocean aid and that therefore great accuracy cannot be expected in the fix—as a rough estimate 5 to 10 miles might be taken as a guide, but a lot depends on the radio conditions, distance from the transmitters, and the efficiency of the count.

*Radio Direction Finding (D/F)*

Quite a number of boats these days are fitted with a Radio Direction Finder—or D/F set as it is commonly called. These sets usually have some form of rotatable circular 'loop' aerial. By tuning the receiver to a particular shore station and moving the aerial around, a maximum or minimum received signal is found. The signals are then amplified, and compared aurally or on a meter. When the minimum signal is reached, the plane of the boat's aerial must be at right-angles to the bearing of the transmitter. The aerial bearing at this time will, of course, be relative to the boat's heading, but if the aerial bearing is related to the compass, the compass bearing of the transmitter can be established.

D/F sets of this nature can give quite good results over comparatively short ranges—in the order of 100 miles. The accuracy of the bearing obtainable depends on a number of factors. The character of the set being used is one, range from the transmitter another. A principal cause of trouble is the interference between radio waves (from the same transmitter) that follow a ground-wave path and those following a sky-wave track. The latter pass out through the Earth's atmosphere but are reflected back by an ionospheric layer, to return to Earth at varying distances from the transmitter. A pure ground-wave signal will give the best bearings, but particularly at dusk and dawn, and also at night, sky-wave signals may intrude. With loop aerials, in sky-wave interference conditions, serious errors may result at ranges over 25 miles from the transmitter. As far as reception is concerned, there is no indication whether a sky- or ground-wave is being received.

Another possible cause of trouble in some ship or boat-borne D/F sets is that a receiprocal bearing may be obtained (i.e. a bearing that is 180° out). A simple loop aerial cannot differentiate between the signal strengths received when it is turned through exactly 180°. Some sets are fitted with a sensing device which removes this difficulty, and the prospective purchaser of a D/F set would be wise to check this point.

Another cautionary note: depending on the position of the

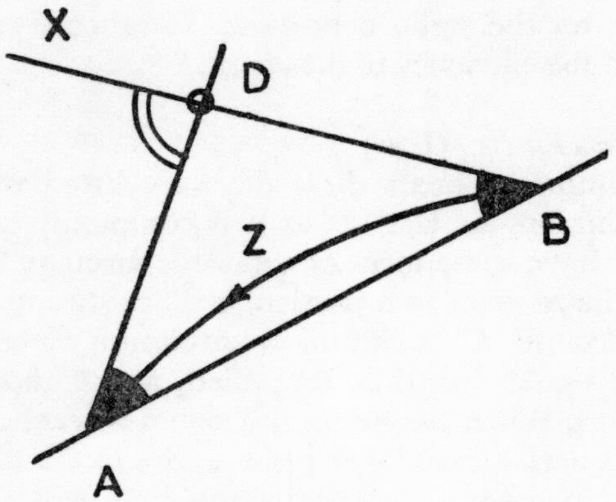


Figure 24

aerial in the boat, other fittings such as masts and shrouds may deflect the incoming signals on certain relative bearings. A D/F set should, therefore, be calibrated 'in situ' to discover whether any of these local errors are important enough to be added to or deducted from the received bearings.

Over ranges of less than 100 miles, the fact that radio waves travel on Great Circle paths can be ignored and little error will result if the bearing of the transmitter is plotted directly on a Mercator chart. But over long ranges—100–2000 miles or so—the difference between the Great Circle and Rhumb Line bearings must be taken into account.

In Figure 24, the boat at Point A takes the D/F bearing of a transmitter at Point B. The radio wave from the transmitter follows the Great Circle course BZA; the tangents to this curve from A and B meet at D. The bearing of B from A on the chart is the Rhumb Line bearing AB. A correction has therefore to be applied to the observed bearing equal to the angle DAB. This angle is called 'Half convergency' as it is equal to half the convergency angle ADX.

The correction for this angle is always applied *towards the Equator* as the Great Circle curves towards the Pole. Its value can be found from the following formula:

$$\text{Half-Convergency (in minutes)} = \begin{matrix}\text{D' Long}\\ \text{(minutes)}\end{matrix} \times \frac{\text{Sine Mean Lat.}}{2}$$

(The D′ Long. and Mean Lat. between boat and transmitter)

An easier way to find the value of Half-Convergency is to use the Traverse Table (the use of this table is described in Chapter 9).

Enter the D' Long. as the Distance and Mean Latitude as the Course.

$$\frac{\text{Departure}}{2} = \text{Half-Convergency in minutes.}$$

There are now many Radio Beacons, a lot of them fitted in lighthouses, sited around the coasts. They transmit at fixed intervals, normally sending their identification signal in Morse Code first, followed by a long dash to facilitate D/F bearings. The Radio Beacons (Ro. Bn. on the chart) and Coast Radio stations transmitting for D/F purposes are ringed with magenta circles. However, the charts give no details on callsigns, frequencies and times of transmission. Complete details of these are given in the Admiralty List of Radio Signals, from diagrams available at chart agents or from the makers of the D/F sets. (Coast Radio Stations may only transmit on request.)

A D/F set is therefore very useful to the yachtsman. Two plotted D/F bearings will give a fix, and many an owner must have been grateful for his D/F set on a foggy night in mid Channel. However, the limitations and possible causes of error in D/F bearings must always be remembered. It is an Aid—not God!

# 9

# Practical Navigation in a Boat

The preceding chapters have described some of the tools and techniques used in the art of navigation; the time now approaches when theory has to be put into practice. The real work should start well before you slip from your moorings or leave that nice, comfortable jetty where the boat has been lying whilst you have been busy with other things.

*Planning the Passage*

The first thing to ensure is that all the charts, tables, books, instruments, are on board to cover the whole route and any possible diversions it may be necessary to make should the weather turn foul or one of the crew go sick; precautions first, fun later. Having done this check, start to plan your passage on the charts. Even though you may not be able to stick to them rigidly because of weather conditions, it is always a good plan to draw in the intended tracks on the chart; whilst doing this, mark back from your destination the distance to go so that, on whichever chart happens to be in use at the time, you know how far away the target lies. On long passages, it will probably be necessary to draw out the general track initially on a fairly small scale chart, and then transfer it to larger scale charts, amending it as necessary to skirt any dangers that may show up more clearly on the latter. If in any doubt as to how wide a berth to give a particular point or headland, always consult the Pilot for the advice it may offer. This will also give pretty detailed information on the tidal streams and currents to be expected even if the chart is a bit vague on the subject. Look out for any unhealthy spots which it would be unwise to approach too closely at unfavourable moments, e.g. unlit rocks or sandbanks in the middle of a pitch-dark night. There are a host of things to look for, and they cannot all be listed here, but it is largely a matter of common sense.

When transferring positions or courses from one chart to another, it is usually easiest to do this as a range and bearing from a particular point common to both charts. It can also be done by Lat. and Long., but this is usually slower and in a few areas still the Lats. and Longs. vary slightly between charts.

When you have got the general track marked on the chart, perused any doubtful places or corners, got a general idea of the tidal streams or currents likely to be met, the final considerations should be planning for leaving harbour and the weather.

To take the weather first. In U.K. waters, it is very simple to get a weather forecast. They are given frequently on the ordinary radio programmes, and the Met Office will give a forecast on request by telephone. Abroad it is not quite so easy, as the local radio (damn it) may be incomprehensible. However, in harbour it is usually possible to get a local forecast from the yacht club, port authorities, or even a chap stroking a piece of seaweed on the beach; it is a question of where English is spoke, if you do not know the country's mother tongue, rather than a lack of available forecasters—which are legion everywhere. At sea, details of area forecasts in English are given once again in the Admiralty List of Radio Signals. General weather information is also contained in the Pilots. Your final decision on when to leave harbour must rest, obviously, on the local and area weather forecast and the state of the tide. Even if the wind is fair, it may be prudent to wait until the worst of the flood tide has passed before making out an estuary. On the other hand, it may be worth beating out against the tide for a while if greater advantage can be gained later. It is impossible to generalise on this.

Finally, the plan for leaving harbour. If it is a familiar place, there should be no problems in getting away to sea from it. A strange place, though, may require a little more careful study, and perhaps a written-down plan for leaving harbour with courses on leading marks, turning points, and clearing bearings taken off the chart and noted down for quick reference as you proceed to sea. It is a good thing to keep one section of your 'navigator's notebook' (see next paragraph) reserved for leaving harbour plans of this nature. Once the problem has been studied and the plan made, it can be used again at any time in the future with adjustments for different states of the tide, wind, etc. In power driven boats, the same plan can be used time and again without any alterations.

*The Navigator's Notebook*

It is a good practice when making a passage to keep a small notebook in which all important navigational data can be entered. It serves as the place in which alterations of course and speed, bearings taken for fixes, compass checks, etc., can be written. It is better to write such data down in a notebook rather than scribbling it on odd scraps of paper or on the chart where it stands in grave danger of being erased or lost or confused with other detail. One good reason for recording these facts on a systematic basis is because of the necessity sometimes to go back and check previous fixes, courses, etc. Suppose a mark is not sighted at the expected time; the wise sailor then goes quickly back over his previous work to make sure that no glaring errors have been made. If the facts have been recorded, an error may be found and corrected in time. If no obvious error has been made, then there is at least some comfort in that knowledge.

The notebook need not be an elaborate affair. A small pocketbook with a pencilled margin down the left-hand side of each page is all that is required . . . time being inserted in the margin and relevant details across the rest of the page opposite the appropriate times.

An example of a typical page of a navigator's notebook is given opposite.

*Taking Bearings for Fixes*

Whether taking bearings by using the Azimuth circle on a compass or by means of a hand-bearing compass, it is necessary to observe them accurately—and preferably quickly. The best routine in this respect is:

a. Identify the objects you are going to use both visually and on the chart.

b. Write the names of the objects into your notebook.

c. Take the bearings of the objects in quick succession. If the boat is moving fast and the objects are close, take the bearing of

| | |
|---|---|
| 0900 | No. 1 buoy ⊢— 1 cable<br>Co. 270 – Spd. 4 Kts. |
| 0915 | A/Co. to 250 C. |
| 0920 | Compass check<br>Leading Lights in transit - 295 C.<br>(Makes deviation 4 W) |
| 0925 | Spd. 6 kts. |
| 0925 | Chimney 285 C.<br>Tower (Barry Pt) 000 C.<br>Sinter Bn. 060 C. |
| 0925 | T.S. since 0900 setting 270, 1 kt. |
| 0940 | A/Co. to 090 C (078 T) |
| 0945 | →⊣ Challenger Pt. 180 C. |
| 1000 | Wind gusty. Reefed Sp. 3 kts. |

⊢— means abeam to Port · —⊣ means abeam to Starboard
→⊣ means right-hand edge

the object nearest the beam *last*; this is the bearing that will be changing most rapidly.

d. Do *not* write the bearing of each object down before taking the bearing of the next. This wastes time. Instead remember the last two digits of each bearing in your head, e.g. 09, 24, 67 for bearings of 309, 024 and 067. Then go to the notebook and write the two figures against the appropriate objects. The first digit in each case can easily be added by inspection of the chart. It is considerably easier to retain six figures in your head than nine whilst taking the bearings.

e. If you are in a boat that is only fitted with a steering compass with no horizon view from it, and there is no hand bearing compass available, the boat must be pointed at each object in succession and the ship's head read off the compass when the bow is pointed directly at each object.

*Speed through the Water*

If you do not possess a log, judgement of speed or distance run through the water is largely a matter of experience with that particular craft. The amount of wake produced at varying boat speeds is normally a good clue. Sailing craft, too, have a fairly set maximum speed which is a function of their waterline length. In powered craft, the maximum speed is also well defined in calm water but more difficult to assess when heading into a lop. Intermediate speeds can also be pretty accurately estimated in power boats from the engine revolutions and a speed/revolution table devised (a measured mile in your locality?).

Particularly when the boat is moving fairly slowly, quite an accurate judgement of speed can be made by using the 'Dutchman's Log'. If the time taken for a piece of wood or other marker to travel from the bow to the stern is measured, the boat's speed can be calculated. If, therefore, you can persuade your wife to throw her handbag in the water (latest hat will not do—too much windage) abreast the bow and clear of the hull and, with a stop-watch, time the period elapsed until you fish it out of the water exactly abreast the stern, you will obtain an estimate of the boat's speed. The shorter the boat, of course, the more critical the timing becomes from all points of view.

In a boat of 30 foot length, suppose the marker thrown over parallel with the bow takes 5 seconds to reach the stern. Then:

$$\text{Speed (knots)} = \frac{\text{Distance (miles)}}{\text{Time (hours)}}$$

$$\text{or in this case Speed} = \frac{30}{6000} \times \frac{60 \times 60}{5} = 3.6 \text{ knots.}$$

*Surface Drift*

When the wind has been blowing steadily over the sea for some time, the water surface responds to the frictional drag of the air by moving in the general direction towards which the wind is blowing. This drift current, or surface drift as it is called, is deflected by the Earth's rotation to the right of the wind direction in the Northern Hemisphere and to the left in the Southern. The amount of deflection can be as much as 45°. The strength of these currents varies greatly depending on how long the wind has been blowing, its strength, and the amount of fetch (distance over the sea) it has had. An average figure for the current's rate is usually taken to be somewhere about 1/50th of the wind speed. Therefore a 20 knot wind blowing in a steady direction for a few hours could be expected to produce about ½ a knot of surface drift.

The major currents in the open oceans are, of course, caused by winds such as the Trade Winds blowing consistently over vast areas of ocean for many months of the year. The surface drift mentioned here is a much more local effect.

*Leeway*

In any boat, but particularly sailing craft, leeway (movement bodily downwind) has to be taken into account when estimating the course made good. This is over and above any current or tidal stream that may be present. The amount of leeway a boat makes is governed by its hull form, the relative wind direction and the boat's speed. Boats with deep keels or large centreboards will make less leeway than shallower, less stiff craft. Again, judging the amount of leeway a boat will make under given conditions is largely a matter of experience with that boat —and the helmsman! The angle the wake makes with the boat's heading is a good indicator; generally, the faster the boat is moving in a forward direction, the less the leeway made.

*Navigating out of Sight of Land*

It is is not the intention in this book to include anything about the taking and working of astronomical sights, frequently the only method of obtaining a reliable fix out of sight of land. A knowledge of astro-navigation is essential if you are contemplating any really long voyages in the open sea.

A point that might be made here is that the accuracy of your D.R. or Estimated position is bound to degrade with time—unless you are very lucky and all the drift and current errors conveniently cancel each other out; sadly, this does not often happen! The staleness of the D.R. or E.P. (since the last fix) is thus a big factor in any estimation of how closely you know your true position. The longer time away from a fix, the bigger the guess.

When the shore and known fixing marks are visible, it is not always necessary to keep a record of every tack or alteration of course as a fix can be taken when necessary to re-establish the boat's position. When out of sight of land, however, there is a real need to keep as accurate a D.R. or E.P. as possible. This can be done by plotting on the chart every course and every distance run on that course. If, when sailing, frequent tacking is necessary to make ground to windward, the plotting of each tack may be a tedious business. The alternative is to use a Traverse Table.

These are contained in most sets of nautical tables. Basically, all they do is to give a tabular solution for a right-angled triangle, the hypoteneuse being the Distance Run, the other sides being the distance resolved North–South (D′ Lat.) and the distance resolved East–West (Departure). The course steered is the angle (or 90° minus the angle) between these two sides (Figure 25).

Suppose a boat is steering 060° and runs on this course for 3 miles. Solving the triangle by plane trigonometry:

$$\text{Sine } 30^\circ = \frac{\text{D}'\text{ Lat.}}{3} \quad \text{Therefore D}'\text{ Lat.} = 3 \text{ Sine } 30$$

$$= 3 \times 0.5$$

$$= 1.5 \text{ miles or minutes of latitude.}$$

$$\text{Cos } 30° = \frac{\text{Departure}}{3} \quad \text{Therefore Dep.} = 3 \text{ Cos } 30$$
$$= 3 \times 0.866$$
$$= 2.598 \text{ miles East (or West).}$$

*Note.* The Departure can be plotted directly on the chart—using the Latitude scale for its measurement—or converted into a difference of Longitude (D′ Long.) by the formula:

D′ Long. = Departure × Secant Middle Latitude.

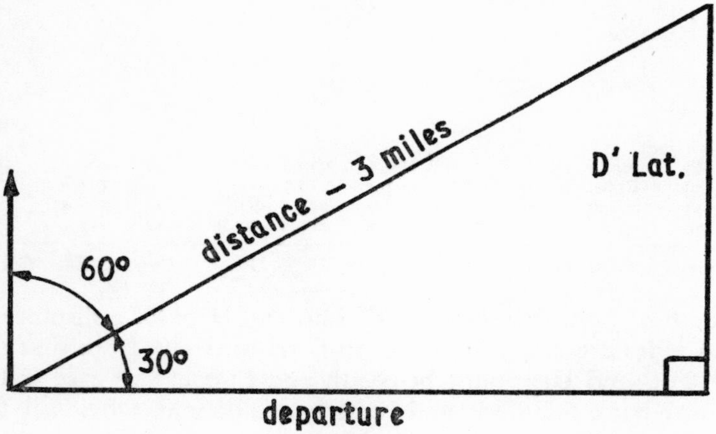


Figure 25

A Traverse Table, such as that given in Inman's Nautical Tables, may be arranged as on page 142.

The arrangement of this table is such that if the course angle is less than 45° the table is entered from the left-hand edge and the column headings at the top used for D′ Lat. and Dep. If the course is greater than 45°, the table is entered from the right-hand edge and the column headings at the bottom used. For example:

| | D′ Lat. | Dep. |
|---|---|---|
| Course 031°—Distance 2.98 miles | 2.55 N | 1.53 E |
| Course 060°—Distance 3.00 miles | 1.5 N | 2.598 E |

For courses between 090°–180°, the table is used as if the course was South . . . ° East; a course of 176° is the equivalent of

DISTANCE

| Course | 297 | | 298 | | 299 | | 300 | | Course |
|---|---|---|---|---|---|---|---|---|---|
| | D'Lat. | Dep. | D'Lat. | Dep. | D'Lat. | Dep. | D'Lat. | Dep. | |
| 1 | 297.0 | 5.2 | 298.0 | 5.2 | 299.0 | 5.2 | 300.0 | 5.2 | 89 |
| 2 | 296.8 | 10.4 | 297.8 | 10.4 | 298.8 | 10.4 | 299.8 | 10.5 | 88 |
| 3 | 296.6 | 15.5 | 297.6 | 15.6 | 298.6 | 15.6 | 299.6 | 15.7 | 87 |
| 4 | 296.3 | 20.7 | 297.3 | 20.8 | 298.3 | 20.9 | 299.3 | 20.9 | 86 |
| , | | | | | | | | | |
| , | | | | | | | | | |
| , | | | | | | | | | |
| 30 | 257.2 | 148.5 | 258.1 | 149.0 | 258.9 | 149.5 | 259.8 | 150.0 | 60 |
| 31 | 254.6 | 153.0 | 255.4 | 153.5 | 256.3 | 154.0 | 257.1 | 154.5 | 59 |
| 32 | 251.9 | 157.4 | 252.7 | 157.9 | 253.6 | 158.4 | 254.4 | 159.0 | 58 |
| , | | | | | | | | | |
| , | | | | | | | | | |
| , | | | | | | | | | |
| , | | | | | | | | | |
| , | | | | | | | | | |
| , | | | | | | | | | |
| , | | | | | | | | | |
| , | | | | | | | | | |
| 42 | 220.7 | 198.7 | 221.5 | 199.4 | 222.2 | 200.1 | 222.9 | 200.7 | 48 |
| 43 | 217.2 | 202.6 | 217.9 | 203.2 | 218.7 | 203.9 | 219.4 | 204.6 | 47 |
| 44 | 213.6 | 206.3 | 214.4 | 207.0 | 215.1 | 207.7 | 215.8 | 208.4 | 46 |
| 45 | 210.0 | 210.0 | 210.7 | 210.7 | 211.4 | 211.4 | 212.1 | 212.1 | 45 |
| Course | Dep. | D'Lat. | Dep. | D'Lat. | Dep. | D'Lat. | Dep. | D'Lat. | Course |

South 4° East, therefore the D' Lat. and Dep. for a distance of 3.0 miles are 2.933 S and 0.209 E respectively. Note that the D' Lat. and Dep. must be South and East in this case as the course has a Southerly and Easterly component. Obviously the D'Lat. is going to be much the larger as a course of 176° is within a few degrees of due South. Similarly, for courses between 180°–270°, the course is considered as being South . . .° West, and between courses 270°–360° North . . .° West.

The following example illustrates the use of the Traverse Table in practice. A boat starts in position 50° 10.0′ N 4° 06.0′ W. It then steers these courses:

| | D' Lat. | | Dep. | |
|---|---|---|---|---|
| | N | S | E | W |
| Course 004° for 30 miles | 29.93 | — | 2.09 | — |
| Course 148° for 30 miles | — | 25.44 | 15.90 | — |
| Course 182° for 30 miles | — | 29.98 | — | 1.05 |
| Course 312° for 30 miles | 20.07 | — | — | 22.29 |
| | 50.00 | 55.42 | 17.99 | 23.34 |

As a result of these manœuvres the boat has travelled:

| | |
|---|---|
| 50.00 N | 17.99 E |
| 55.42 S | 23.34 W |
| 5.42 S | 5.35 W |

In other words, it finishes up 5.42′ or minutes of Latitude or miles South of the starting position and 5.35 miles West. The simplest way of re-plotting the boat's position on the chart is to set these 'Northings' and 'Eastings' successively on the dividers—Departure is in *miles* East or West and therefore the Latitude scale can be used for both quantities—and plot the appropriate distances South and West from the initial position. Alternatively, the Longitude of the final position can be found by converting Departure into D′ Long.—by a table or:

$$\begin{aligned} \text{D}' \text{ Long.} &= \text{Departure Sec. Lat.} \\ \text{Therefore D}' \text{ Long.} &= 5.35 \text{ Sec. } 50^\circ\ 05' \\ &= 5.35 \times 1.55 \\ &= 8.3' \text{ West.} \end{aligned}$$

| | | |
|---|---|---|
| Starting Position | 50° 10.0′ N | 4° 06.0′ W |
| Difference | 0° 05.4′ S | 0° 08.3′ W |
| Final Position | 50° 04.6′ N | 4° 14.3′ W |

In practice, there would be no point in working to two places of decimals of a minute; that was done for illustration. The final position can then be plotted on the chart quite normally using the Latitude and Longitude scales. To simplify matters, a table for converting Departure into D′ Long. is usually placed next to the Traverse Table in the book.

Having worked the courses and distances run in one of these ways to give the D.R. position, allowance must then be made for leeway, tidal stream and current to arrive at the Estimated Position. As for leeway, it is probably easier to allow for it by judging the difference it will have made on each of the courses steered and then working the adjusted courses in the Traverse Table instead of the straight courses steered. The effect of tidal stream or current over the *whole* period can be estimated and applied as one correction on the chart.

*Making a Landfall*

The acid test! But there is nothing more satisfying than to arrive at the right place at the right time in spite of all the quirks of wind, tide and weather.

The careful work required in keeping an accurate E.P. in the open sea, aided by soundings and radio aids if you have any to use, will now begin to pay off. On the other hand, a bit of careless work out in the middle may result in a good deal of nail-biting and the aspect of a selection of objects which bear no resemblance to the chart around your destination.

To take the worst case first; suppose that you have been at sea for a couple of days or more without a proper fix—it has been blowing hard and the visibility is not good. Your destination is sited on a low-lying, pretty featureless coast with no conspicuous off-shore marks; there are rocks well to seaward and the current is strong. Under these conditions, the wisest course of action would be to wait or anchor until the visibility improved. However, there may be urgent reasons for an early arrival in harbour. As the D.R. position may be well out, it is too optimistic to expect a neat arrival straight off the harbour entrance. The first objective must be to get a positive identification of some land or sea mark as early as possible. Consequently, although the coastline may appear to be absolutely featureless from the chart, an exhaustive study of the Sailing Directions (Admiralty Pilot) may have revealed some clues which would help, i.e. the colour and appearance of the coastline along particular stretches, any small hills or dunes that are slightly more easily seen than than the rest. Having then amassed as much local information as possible, aim to close the coast in an area where the offshore dangers are least and where there is a good chance of identifying some feature on the shore from a good range. An alternative course of action is to deliberately aim to miss your destination on one side or other by an amount at least equal to your probable reckoning errors. Then, on finally making the coast, there can be no doubt in which direction to turn parallel with the coastline in order to reach harbour.

At other times, you may be making a landfall on a coastline which has plenty of artificial or natural features, but with tricky approach channels through shallows before the harbour can be reached. A bold headland is easy to identify, but a coastline backed by broken country may prove more difficult to sort out. If there is a major lighthouse in the area, it may be preferable to approach the coast in the dark to gain a positive identification of the light by its characteristics. It is often more difficult to pick out a lighthouse by day, even if the visibility is good, if it does not stand out well against the background. Conversely, off some major rivers and ports, the host of navigational marks and shore lights may tend to confuse by their sheer weight of numbers. At night, it may not be easy to pick up the smaller navigational lights against a background of street lighting, traffic moving on shore, etc. Such considerations as these should be taken into account when approaching a strange harbour from seaward. Glean as much information as you can beforehand from the chart and Sailing Directions and then use common sense in planning the approach taking into account the possible inaccuracies in your Estimated Position. A lot of landfalls are quite straightforward but there may be the odd one which may cause some anxiety unless the appropriate precautions have been taken.

*Approaching the Coast in bad visibility*

As has already been said, the golden rule if visibility is bad and you are uncertain of your position is to *anchor* as soon as the water becomes sufficiently shallow—preferably clear of shipping lanes. It is most dangerous to press on in the blind faith that the mark which should have appeared some time back will sooner or later appear. Another 'must', if it is possible, is to get a good fix before the visibility completely closes down; this applies when making a passage along the coast as well.

*Anchoring in a chosen spot*

There are many factors to be taken into account when choosing an anchor berth. The principal ones are:

(i) What is the direction of the prevailing wind? (Avoid anchoring if you can possibly do so on a Lee shore—if the anchor drags you will end up on the rocks, or at best have a most uncomfortable time.)

(ii) Is there sufficient room to swing safely round the anchor without danger to yourself or other craft?

(iii) Is it reasonably easy to get out of your selected berth if the wind or current changes direction or strengthens?

(iv) Is the holding ground good?

(v) Is the water too deep for comfort?

There are a lot of supplementary considerations (or perhaps not so supplementary?) such as nearness to landing places, bright lights, etc., but the first requirement for an easy mind is to ensure the boat's safety. A swinging time ashore is slightly spoilt if on your return you find that the boat has had ideas of its own.

No anchor, however good its design, will hold properly if an insufficient length of cable or anchor warp is veered. The holding power of an anchor depends to a large extent on the angle the chain makes with the stock of the anchor whilst it is lying on the bottom. If the length of chain near the stock is lying flat, as the strain comes on, the initial pull on the anchor is applied to it almost horizontally. The anchor then provides the maximum resistance to being dragged along the bottom, and the additional cable acts as a spring (due to its weight) preventing any sudden jerk on the anchor. On the other hand, if only just enough cable is let out to allow the anchor to reach the bottom, the pull coming onto it is applied almost vertically and as a sudden jerk; thus the anchor is easily pulled out of its ground and has little holding power.

The amount of cable that should be veered depends on a number of factors such as the depth of water, strength of wind, duration of stay in that berth, etc. As a good general rule, if you intend to remain at anchor for an appreciable length of time, it is advisable to let out a length of cable equal to *Eight Times the Depth of Water*, i.e. in a depth of 5 fathoms veer 40 fathoms of anchor chain. This is a generously safe allowance under normal conditions. Some people might say that this is far too much, but if there is otherwise plenty of swinging room, it is much better to err on the safe side. Should a gale blow up and you intend to remain at anchor, it would be advisable to let out all the cable you have got—the more out, the better the chance of the anchor holding.

The quality of the various holding grounds has been discussed in a previous chapter. Sand and mud are the best, rock and weed the worst. Apart from the danger of an anchor slipping and dragging quickly across a rocky bottom, there is also the risk of it jamming in a crevice when you least want it to do so. If you have to anchor on a rocky bottom, it is a wise precaution to buoy the anchor before letting it go—this means attaching a rope (of greater length than the depth of water!) to its crown. Before the anchor is let go, the anchor buoy is thrown clear over the side. If the anchor subsequently jams on being weighed, the buoy can be recovered and a pull put on the anchor from a different direction to try and pull it out. If this in turn fails, the next recourse is to try and sail it out, i.e. heave in until the cable is taut, secure it and then gather headway until the cable draws aft. A strain from a different direction is thus applied, and this can be tried successively on different headings. If this also fails, and you are in a hurry to get clear, the final solution is to slip the cable, buoying it from the inboard end before you do so. This end of the cable can then be recovered on return to the berth.

The amount of swinging room to be allowed when coming to a single anchor depends on the length of cable out, the depth of water, and the length of the boat. The proximity of other craft and dangers must be taken into account. For example, a 27 foot boat anchoring in 4 fathoms might veer 32 fathoms of cable. The theoretical radius to be allowed for the swinging circle is therefore:

$$27 \text{ feet} + 6\ (32 - 4) \text{ feet} = 27 + 6 \times 28 = 195 \text{ feet or } 65 \text{ yds.}$$

However a safety margin should be allowed (over and above this) of at least two boat's length from the nearest underwater hazard. When anchoring near other boats, it may not be possible to allow this comfortable combined margin, but it is as well to remember that although normally they will all swing to the current or wind at about the same time, sometimes things do not work out so neatly. It is not unknown for two boats anchored close to each other to swing to a shift of wind in the opposite directions. Complications can result!

Tidal stream or current has a greater effect than wind on the direction at which a boat will lie at anchor, particularly when the boat is deep in the water. A $\frac{1}{2}$ knot of current may have a greater effect than a 15–20 knot wind if they are in opposition.

*The approach to the Anchor Berth*

There is sometimes a bit more to the actual business of anchoring than just dropping the anchor over the bow, securing the cable and then reaching for the brandy and ginger ale. Such methods may suffice in the summer in the Mediterranean, but are not liable to lead to security of tenure under English weather conditions. The aim on anchoring must be to lay the cable out across the bottom in a straight line, then any strain applied to the anchor and cable at a later time is taken up steadily. If, on the other hand, the cable is piled up in an untidy heap, coil upon coil, any subsequent boat movement will rapidly unwind the coil until a sudden jerk is imparted to the anchor which may break it out. The same disadvantage lies, to a lesser extent, in laying the cable out in a bight (loop) across the bottom.

There are basically two ways of laying the cable out in a straight line on anchoring—by means of a dropping or running moor. The former is principally used when it does not matter a great deal whether the boat is precisely in its correct berth; the latter when it is important to anchor bang in the right spot. Whichever method is used, it is always preferable to anchor heading into the wind or tidal stream. In a dropping moor, the anchor is let go as soon as the chosen berth is reached; the boat is then immediately allowed to drop back downwind or current laying out the cable as it goes. The cable will then be laid out straight and the boat drops gently back, its speed being controlled by the rate of veering the anchor cable. No jerk need be imparted to the cable at any stage.

With a running moor, the boat must have a certain amount of headway as the berth is reached so that control of the boat's heading can be maintained right up to the last moment. This allows a more precise positioning of the anchor on the bottom. For example the intention is to anchor precisely in the position shown in Figure 26.

The wind is from the NNE and there is no tidal stream. The approach course is from the SW on a heading of 030° True. First, a look at the chart may show a conveniently placed object which can be used as a headmark during the final approach; here the church is well placed to give a good lead into the anchorage and one which will clear the rocks on both sides of the entrance. (Very rarely does it happen that absolutely no

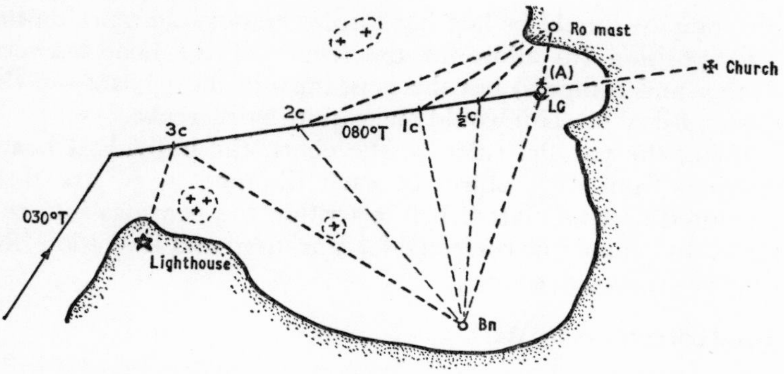


Figure 26

(The anchoring position is marked by the symbol under (A))

suitable mark is available—in this case, if the church had not been there, the point just to the South of it on the shore would have served just as well on a slightly different bearing; though it might have been more difficult to identify.)

With a parallel ruler set between the church and anchor berth, lay off the bearing on the chart and measure it from the compass rose (080° T). If the church is kept bearing 080° True during the run in, the boat must be on the right line of approach to the anchoring position.

However, the boat has to get onto this line from its approach heading of 030° and this leads to a word about 'turning bearings'. When using a bearing to decide the moment to turn onto a new course, always use the bearing of an object as near as possible *parallel to the new course.* Doing this much reduces the chance of ending up off the intended track due to compass error or some other reason. In the above diagram, it would be much better to use a bearing of the church for the final turn rather than a bearing of the lighthouse.

Next, distance to go, or distance along the track is required. This is obtained by selecting charted objects on the beam during the final approach and measuring the bearings these should subtend at given distances from the berth. In the sketch, the 3 cable, 2 cable, 1 cable, and ½ cable (100 yards) distances to go have been marked, with finally the L.G. ('Let go') position shown. To be clinically accurate, the distance from the bow to the compass should be laid back from the berth in marking the L.G. position (to allow for the length of the boat between anchor and compass) but this is peanuts in short boats—in the Queen Elizabeth II it would make quite a difference.

Using the parallel ruler on the chart, the individual beam bearings from each object at each distance to go are then measured and tabulated. Before writing the bearings into the table, they should be corrected for Variation and Deviation (for the final approach course):

Total correction 10°W.

Run 080° True—090° Compass on Puddlecome Church

| Distance | Lighthouse | Ro. Mast | Beacon |
|---|---|---|---|
| 3 c. | 195° Compass | — | 125° C. |
| 2 c. | — | 070° C. | 140° C. |
| 1 c. | — | 055° C. | 165° C. |
| ½ c. | — | 038° C. | 182° C. |
| L.G. | — | 007° C. | 196° C. |

This preparatory work does not take long and with the facts written into a notebook, a perfectly safe and accurate anchoring can be achieved without any need for fixing.

If the church is kept bearing 090° by compass by adjustment of the course to allow for leeway, the boat can be brought in at the right speed for anchoring by checking the beam bearings. It is wise to have more than one set of beam bearings up your sleeve, as one can then be checked against another; also it may be found that one mark is obscured, at an important time, by other craft or be difficult to identify against the background.

With a running moor, if there is no wind or current, headway can be maintained until the right amount of cable is out. If there is a strong headwind or current, however, it is better to take the way off the boat as soon as the anchor is down and then drop slowly back on the chain. Whichever way the approach is made, if it is impossible to anchor into wind or tide, the anchor should be dropped on the weather side (i.e. up wind or current). When the way comes off the boat, the anchor chain will then lead out clear as the boat drops to leeward. If the anchor is let go on the lee side, the chain may foul the keel or scrape paint from the bow.

*Entering Harbour*

Shiphandling is a subject in its own right, and it is not the intention to discourse on it here. However, one or two small navigational hints may be of some help. On entering a harbour, always keep an eye on other craft at anchor and buoys to see in which direction the current is setting. Then identify the berth for which you are making as early as possible and approach it, if you can, against the current. The same applies when aiming to pick up a mooring. In sailing boats, the direction of your final approach to a berth is inevitably determined by the wind direction. In power boats, if the wind is blowing off a jetty, the approach should be made at a wider angle than normal and the boat turned to parallel up with the jetty at the last moment. If the wind is blowing onto a jetty, a shallower approach aiming a short distance off the jetty should be made.

In any type of boat bigger than dinghy size, always have an anchor ready for letting go when entering (or leaving) harbour. You never know when this may come in useful and it is always a good precaution even when you intend to berth alongside or at buoys. In a strange place, you may not know what the berthing arrangements are in that port on first entering, and therefore keep a good look out for groups of other yachts, customs posts and harbour offices where you may be able to obtain quick enlightenment (or a rude answer).

Finally to emphasise once again the importance of looking up all the information you can on the tides, currents, depths, conspicuous objects *before* entering a strange place. A frantic thumbing-through the Pilot with one hand, whilst steering with the other, and holding the sheets with your big toe may not breed confidence amongst your fellow men and girls.

# 10

# The Rule of the Road at Sea

Any boat owner must have a working knowledge of the International Regulations for the Prevention of Collision at Sea. Otherwise, he may become something of a menace to others on the sea.

The Rule of the Road at Sea is published, in this country, under an Order in Council entitled 'The Collision Regulations (Ships and Seaplanes on the Water) and Signals of Distress (Ships) Order 1965'. This is obtainable from H.M. Stationery Office for the princely sum of $7\frac{1}{2}$ new pence.

By international agreement, the rules were slightly changed in 1960, principally to take into account the fitting of radar into a large number of ships since the war. There are altogether some 31 Rules covering—the lights and shapes to be carried by day and night—the sound signals to be made in poor visibility—and the Steering and Sailing rules. Some of them have been re-grouped for convenience and paraphrased here, without altering their sense; and some of the more important have been quoted verbatim—where this is the case, they are shown within inverted commas. Some commentary is also given on specific rules. These alterations and additions are not 'official' but merely aimed to help interpretation; needless to say, I disclaim any responsibility for accidents that might be deemed to have arisen as a result of reading these pages!

The Rules should not be confused with the 'Racing Rules' which are only in force when boats are actually competing, or about to compete, in a race. The Racing Rules are themselves based on the Rule of the Road but have major additions and modifications to the basic rules.

*Rules 1 and 30—Preliminary Statements and Definitions*

(i) The International Regulations do not interfere with Local

Authority rules made for particular harbours, rivers, and inland waterways. (Rule 30.)

(ii) The rules concerning lights must be complied with from sunset to sunrise in all weather conditions. Additionally, these lights may be shown in daylight hours in restricted visibility, or in any other circumstances when it is considered prudent to do so. No other lights should be exhibited which can be confused with the prescribed lights. (Rule 1.)

(iii) A power-driven vessel is a vessel that is *using* an engine at the time, whether or not this is her principal means of motive power, whether or not she is also sailing, and whether or not it is an inboard or outboard motor. (Rule 1.)

(iv) If a boat is using an engine with sails also hoisted, it should, by day, carry a black cone, point downwards, in the fore part of the boat where it can best be seen. The cone should be at least two feet in diameter at its base! (Rule 14.) (This, you may imagine, is not always complied with, but in the event of an accident . . .)

(v) A vessel is considered to be underway as soon as it is no longer attached by any means to the shore or bottom; i.e. it is underway as soon as the last line is slipped from a jetty or a buoy, and as soon as the anchor is clear of the bottom when weighing. A boat aground is obviously not underway! (Rule 1.)

(vi) Vessels are considered to be in sight of one another only when one can be observed visually from the other. (This is put in principally to disqualify radar contact as a 'sighting'.) (Rule 1.)

(vii) Regarding lights, 'Visible' means visible on a dark night in a clear atmosphere. (Rule 1.)

(viii) The term 'engaged in fishing' means fishing with nets, lines or trawls but does *not* include fishing with trolling lines (streamed when underway over the stern or from outriggers). (Rule 1.)

### *Rules 2 to 14—Lights and Shapes*

#### POWER DRIVEN VESSELS UNDERWAY

1. *A power driven vessel underway and over 150 feet in length must carry:*

(a) *Two White Masthead Lights.* These must be sited on different mastheads or parts of the ship's superstructure and the forward one must be at least 15 feet lower than the after. Both must be visible from 5 miles over an arc of visibility stretching from two points ($22\frac{1}{2}°$) abaft the beam on one side, through the bow, to two points abaft the other beam, i.e. from the relative bearings of Red $112\frac{1}{2}°$ to Green $112\frac{1}{2}°$. (Rule 2.)

(b) *Bow Lights.* On the port bow, a RED light showing from right ahead (but not across the bow) to two points abaft the port beam. On the starboard bow, a Green light showing from right ahead to two points abaft the starboard beam. The visibility distance of both lights must be 2 miles. (Rule 2.)

(c) *Overtaking Light.* This is a white light, carried on the stern, visible over 12 points (135°) through the stern, i.e. from $67\frac{1}{2}°$ on one quarter—through the stern—to $67\frac{1}{2}°$ on the other quarter. Visibility distance 2 miles. (Rule 10.)

*Summary*

A ship underway over 150 feet in length seen from dead ahead will show both masthead lights, one above the other, and both bow lights. As you cross its bow from port to starboard (at a goodly range, I trust), the red light will go out, the green light remain visible, and the two white lights separate. At long range, the bow lights may not indeed be visible and the masthead lights, by their aspect, will give the first indication of the ship's course.

If your relative position is more than $22\frac{1}{2}°$ abaft the beam of the other ship, its white overtaking light *only* can be seen.

2. *A Power Driven Vessel less than 150 feet in length must carry*:

One White Masthead Light, Bow Lights, and an overtaking light. It can carry a second white masthead light (as above) but is not obliged to do so. All these lights have the same characteristics as those defined in paragraph (1) a, b, and c, for ships over 150 feet long. (Rule 2.)

3. *A Power Driven Vessel less than 65 feet long must carry*:

(a) *A White Masthead Light* not less than 9 feet above the gunwale and visible 3 miles. (Rule 7.)

(b) *Bow Lights* showing RED and GREEN over the same arcs as those of a ship but need only be visible at 1 mile. (Rule 7.)

or (c) The bow lights instead of being individually sited one on each side may be combined into the same lampholder (Combined Lantern) but they must show over the same arcs as before (Red—Port: Green—Starboard) and the combined lantern should be not less than 3 feet beneath the white masthead light. (Rule 7.)

(d) *Overtaking Light.* In small vessels, it may not be possible on account of bad weather or other sufficient cause for the overtaking light to be a permanent fixture. However, if a permanent overtaking light is not carried, a white portable light or electric torch should be kept at hand ready for use and be shown in sufficient time to prevent collision. (Rule 10.)

4. *Power Driven Vessels less than 40 feet long may carry*:

(a) *The White Masthead Light* at less than 9 feet above the gunwale but not less than 3 feet above the sidelights or combined lantern. (Rule 7.)

(b) *Separate Sidelights or a Combined Lantern.* (Rule 7).

(c) *A fixed or temporary overtaking light* under the same conditions as for a boat of 65 foot length. (Rule 10.)

SAILING VESSELS UNDERWAY

These follow the same lights, depending on their size, as the equivalent length large power driven vessels with the important exception that they *never* carry the white masthead lights of the power vessel. They *may* carry instead on their foremast a RED light over a GREEN light *both* visible from Red $112\frac{1}{2}°$ through the bow to Green $112\frac{1}{2}°$.

These foremast lights should be visible at 2 miles. (Rule 5.)

A large sailing vessel underway may therefore be showing a Red over a green at the foremast, its bow lights, and an overtaking light. The latter lights are obligatory, in accordance with the rules for craft of their size, even though the foremast lights are not.

*Sailing Vessels and Boats under oars less than 40 feet long must carry*:

(a) *Bow lights.* If they are not fitted with the normal sidelights, they must carry, where it can best be seen, a combined lantern Red to port and Green to starboard, visible 1 mile. If it is not possible to fix this light permanently, it should be kept ready for immediate use and exhibited in sufficient time to prevent collision. (You will get no marks under these rules if Red is allowed to show to starboard or Green to port.) (Rule 6.)

(b) *Overtaking Light.* The normal overtaking light. Again, if it is not possible for this light to be fixed, a torch or lantern showing a white light should be kept at hand ready for immediate use. (Rule 10.)

*Small Rowing Boats under Oars or Sail*:
The full concessions are allowed here. They need only have an electric torch or lantern showing a white light available which must be exhibited in time to prevent collision. (Rule 7.)

*Notes on Sailing vessels and Rowing Boats*
As can be seen, there is no stipulated maximum length, under the 40 foot limit, above which the use of the single white light is forbidden. This is left to the discretion of the owner, noting that it is in his own interest to carry the proper lights if it is practicable so to do.

VESSELS TOWING OR BEING TOWED

(a) *The Towing Vessel. A Power-driven vessel if towing (or pushing) other craft, and if the length of tow exceeds 600 feet, must carry*:

*Three White Masthead Lights* in a vertical line on the same mast and visible from two points abaft the beam, through the bow, to two points abaft the other beam. These, therefore, show over the same arc as the normal masthead lights but are carried *instead* of them. (Rule 3.) (By day, the towing vessel shows a black diamond shape.)

*Bow Lights*. As normal.

*Overtaking Light* as normal, or a small white light abaft the funnel or superstructure for the tow to steer by, but this light must not show forward of the beam. (Rule 3.)

(b) *Power Vessel Towing—with a length of tow of less than 600 feet must carry*:

*Two White Masthead Lights* again in a vertical line and having the same characteristics as the 3 Masthead Lights mentioned above. (Rule 3.)

*Bow Lights*.

*Overtaking Light* or a small light for the tow to steer by.

(c) *Power Driven Vessels of less than 65 feet in length Towing. When towing or pushing other vessels, they must carry*:

*The Two White Masthead Lights* not less than 4 feet apart, and one of them must be in the same position as the ordinary white masthead light, i.e. an extra masthead light is needed. (Rule 7.)

*Bow Lights* or a Combined Lantern.

*Overtaking Light* or a small white light for the tow to steer by. (Rules 7 and 10.)

*The Vessels being Towed*

A vessel being towed must carry bow lights (or a Combined Lantern) and if it is the last vessel in the tow, an Overtaking Light. Craft in the middle of the tow must carry Bow Lights, but need not carry a proper overtaking light, but may show a small white light aft. (Rules 5 and 10.)

*Vessels being Pushed by a Tug or another Boat*

(a) A single vessel being pushed carries the normal sidelights or a Combined Lantern.

(b) A group of boats being pushed must be lit as one vessel, i.e. the port bow light on the port outer craft and the starboard bow light on the starboard outer craft. (Rule 5.)

## LIGHTS AND SHAPES FOR SPECIAL CIRCUMSTANCES

*A Vessel Not Under Command*

A vessel is said to be 'Not under Command' when she has had some sort of failure such as a steering or engine breakdown which precludes her from manœuvring; she cannot therefore keep out of the way of other ships or craft. At night she shows, where they can best be seen, TWO RED LIGHTS vertically one above the other, visible all round the horizon at a distance of at least 2 miles. A power vessel will switch off her masthead lights, but the sidelights should be kept on in both power and sailing craft, under these circumstances, if the vessel is making way through the water. By day, a vessel not under command hoists two black balls in lieu of the red lights (colloquially known as N.U.C. Balls). (Rule 4.)

*Vessels Carrying out Special Operations*

Vessels engaged in special tasks such as laying Telegraph Cables or navigational marks, surveying, carrying out underwater operations, warships replenishing at sea or operating aircraft, cannot easily get out of the way of other vessels approaching them. Instead of the normal masthead lights, at night they carry 3 lights in a vertical line—RED WHITE RED—visible all round the horizon for at least 2 miles.

By day, these lights are replaced by shapes in a vertical line—RED BALL WHITE DIAMOND RED BALL. (Rule 4.)

*Minesweepers*

When engaged in minesweeping operations or exercises, a minesweeper carries at night a GREEN all round light at her fore truck, and a GREEN light at the yard on the side on which danger exists. If the minesweeper has sweeps out on both sides, therefore, a TRIANGLE OF GREEN LIGHTS will be visible; these lights show all round the horizon; and are in addition to her ordinary lights.

By day, black balls are shown in lieu of the lights. (Rule 4.)

The showing of these lights or shapes by sweepers indicates that it is dangerous for other craft to approach closer than 3000 feet astern or 1500 feet on the beam on the side (or sides) from which the sweeps are streamed. (Rule 4.)

*Note*: When any of these vessels (Not under Command, conducting special operations or minesweeping) are actually making way through the water they continue to show their bow lights and overtaking light.

*Pilot Vessels*

*Steam Pilot Vessel*. When on her station and underway at night will show her bow lights, overtaking light, and a special signal of an all round WHITE LIGHT over an all round RED LIGHT at the masthead. She should also show an intermittent all round white flare or light every ten minutes or so. When not under way, she switches off her bow and overtaking lights. If she is not engaged on pilotage duty at the time, she merely shows the ordinary lights for a similar vessel of her length. (Rule 8.)

*Sailing Pilot Vessel.* When on her station on pilotage duty under way, she carries her bow and overtaking lights plus an all round WHITE LIGHT at the masthead, and burns one or more white flares or lights at intervals not exceeding 10 minutes. If not engaged on pilotage duty, she shows the lights appropriate to a vessel of her size only. (Rule 8.)

*Fishing Vessels.* Rule 9

When they are not actually *engaged* in fishing, they should show the normal lights appropriate to a vessel of their length. When they *are* fishing, they carry the following lights:

*Trawlers.* These craft drag a dredge net or other gear along or near the sea bed; when trawling they carry:

*GREEN LIGHT over a WHITE LIGHT* visible all round the horizon.

*Masthead Light.* The ordinary masthead light may also be carried, but it must be lower and abaft the green and white lights.

*Bow Lights and Overtaking Light.* These are shown, as normally, if the trawler is making way through the water; if she is stopped, they are switched off.

*Fishing Vessels Other than Trawlers.* They carry:

*RED LIGHT over a WHITE LIGHT* visible all round the horizon.

*Bow Lights and Overtaking Light* if making way through the water.

*Fishing Vessels with gear extending more than 500 feet horizontally into the Seaway carry*:

*RED LIGHT over a WHITE LIGHT.*

*An additional all round WHITE LIGHT* in the direction in which the gear is extended. This extra light should show above the sidelights but not at a greater height than the other all round white.

*Bow and Overtaking Lights* if they are making way through the water.

A drift net vessel may well be showing lights of this nature.

*Note.* Fishing vessels may use a flare or searchlight to attract the attention of approaching vessels endangered by or endangering their gear. They also sometimes use clusters of bright working lights.

By day fishing vessels over 65 feet long when engaged in fishing carry two black cones point to point where they can best be seen. If they are less than 65 feet long, they may carry a basket instead. If their outlying gear extends more than 500 feet horizontally, they should carry an additional cone—point up—in the direction of the outlying gear.

*Vessels at Anchor.* Rule 11

*A Vessel more than 150 feet long* when at anchor must show, at night, an all round white light forward and an all round white light aft; the latter being lower than the forward light. Both should be visible at 3 miles.

*A Vessel less than 150 feet long* need not carry the after light, and the required visibility range of the forward anchor light is 2 miles.

By day, the anchor lights of vessels of any size are replaced by a black ball carried in the forepart of the craft.

*A Vessel Aground.* Rule 11

Must show at night:

*Anchor Lights* appropriate to her size.

*TWO RED LIGHTS* one above the other, visible all round the horizon.

By day, she should show three black balls in a vertical line.

*Rules 15 and 16—Sound Signals and Conduct in Restricted Visibility*

*A power-driven vessel 40 or more feet in length* must have a whistle or siren, a mechanical foghorn and a bell.

*A Sailing vessel 40 or more feet in length* must have a mechanical foghorn and a bell.

*A vessel of less than 40 feet in length*, a rowing boat (or a sea-plane!) shall not be obliged to make any of the following signals in poor visibility, but must make some other efficient sound signals at intervals of less than one minute.

This is the gist of the slightly ambiguous statement in Rule 15 (ix) but it means that, in bad visibility, as in good, the onus for avoiding collision is on all seafarers and not only on those in large ships. In any sort of cruising boat, 40 feet long or not, there should be some form of foghorn or claxon. The investment of a few shillings in one of these will seem justified if you find yourself in the middle of a busy shipping lane in dense fog.

*Means*

The signals described below for vessels underway should be given by:

(i) power driven vessels on a whistle or siren.
(ii) sailing vessels on the foghorn.
(iii) vessels being towed on a siren, whistle or foghorn.

*Duration of Blasts.* Rule 1

A 'short blast' means a blast of about 1 second's duration. A 'prolonged blast' (or 'long' blast) means a blast of between 4–6 seconds' duration.

*Sound signals in Bad Visibility*

In fog, mist, falling snow, heavy rainstorms or any other condition similarly restricting visibility, whether by day or night, the following sound signals must be made:

(The signals are listed here in tabular form for ease of reference, although in the Rules they are written out in full.)

| Made by | Made on | Interval | Signal |
|---|---|---|---|
| Vessel more than 350 feet in length AT ANCHOR | Bell forward Gong aft | Every minute | Rapid ringing of bell for 5 seconds, followed by gong aft |
| Vessel less than 350 feet long AT ANCHOR | Bell | Every minute | Rapid ringing of bell for 5 seconds |
| Any vessel AT ANCHOR sensing danger from an approaching vessel | Siren, whistle or foghorn | As requisite | One short—one prolonged—one short blast ('R' in Morse) |
| A vessel AGROUND | Bell forward (Gong aft) | Every minute | 3 distinct strokes on the bell, followed by 5 seconds' rapid ringing, followed by 3 strokes. (Gong as before for large ship) |

| Made by | Made on | Interval | Signal |
|---|---|---|---|
| Power vessel MAKING WAY through the water | Siren or whistle | Every two minutes | One prolonged blast |
| Power vessel under way but STOPPED | Siren or whistle | Every two minutes | Two prolonged blasts with an interval of one second between |
| Sailing vessel on the STARBOARD TACK | Foghorn | Every minute | One blast |
| Sailing vessel on the PORT TACK | Foghorn | Every minute | Two blasts |
| Sailing vessel with the wind ABAFT the beam | Foghorn | Every minute | Three blasts |
| A vessel TOWING, picking up CABLES, OR NOT UNDER COMMAND | Siren, whistle or foghorn | Every minute | One prolonged blast followed by two short blasts ('D' in Morse) |
| A vessel being TOWED, or if more than one vessel is being towed, the last vessel in the tow | Siren, whistle or foghorn | Every minute | One prolonged blast followed by three short blasts ('B' in Morse) |
| A vessel ENGAGED IN FISHING, when under way or at anchor. (Does not apply to vessels trolling, which give normal signals) | Siren, whistle or foghorn | Every minute | One prolonged blast followed by two short blasts ('D' in Morse) |
| A power driven PILOT VESSEL when engaged in PILOTAGE | Siren or whistle | As requisite | The normal fog signal followed by four short blasts ('H' in Morse) |

*Conduct of Vessels in Fog or Poor Visibility.* Rule 16
A great number of the collisions at sea occur in fog or in conditions of poor visibility, and Rule 16 is one of the most important of the 'Rules of the Road'. It is therefore quoted here in full:

(a) 'Every vessel, or seaplane when taxi-ing on the water, shall, in fog, mist, falling snow, heavy rainstorms or any other condition similarly restricting visibility, go at a moderate speed, having careful regard to the existing circumstances and conditions.

(b) A power-driven vessel hearing, apparently forward of her beam, the fog signal of a vessel the position of which is not ascertained, shall, so far as the circumstances of the case admit, stop her engines, and then navigate with caution until the danger of collision is over.

(c) A power driven vessel which detects the presence of another vessel forward of her beam before hearing her fog signal or sighting her visually may take early and substantial action to avoid a close quarter situation but, if this cannot be avoided, she shall, so far as the circumstances of the case admit, stop her engines in proper time to avoid collision and then navigate with caution until danger of collision is over.'

Many court cases have been won and lost on the careful (or otherwise) observance of this rule. Paragraph (c) was inserted in the 1960 version of the Regulations to take into account the detection of other vessels by radar, but it does not absolve the radar fitted vessel from stopping engines if a close quarter situation develops.

From the yachtsman's point of view, it must be borne in mind that the very conditions which produce bad visibility, heavy rain for instance, can also degrade radar performance due to the attenuation of radar signals by the water droplets in the atmosphere. The degree of attenuation suffered depends in part on the frequency of the individual radar set. At worst, small craft may escape detection at all by a ship's radar, particularly if the sea is rough and the small echo from a yacht or boat becomes lost amongst the signals reflected by the waves (this is usually called 'Sea Clutter'). At best, a boat which only presents a small reflecting area to the radar pulse cannot be detected at long range. It is, therefore, foolish to assume that a small boat or yacht is bound to be detected by a ship's radar and to take any unnecessary risks based on this assumption; for example trying to get across the bows of a ship in fog.

*Sound Signals made by Vessels in Sight of one another*. Rule 28
Certain other sound signals are restricted for use when vessels are *in sight of one another* by day or by night. It is important that,

when taking avoiding action in close quarter situations, one boat's intentions should be made clear to the other. This can and must be done by the use of sound signals which are tabulated below:

| Made by | Made on | Signal |
|---|---|---|
| A power driven vessel about to ALTER COURSE TO STARBOARD | Siren or whistle | One short blast |
| A power driven vessel about to ALTER COURSE TO PORT | Siren or whistle | Two short blasts |
| A power driven vessel about to PUT ITS ENGINE(S) ASTERN | Siren or whistle | Three short blasts |
| A power driven vessel having the right of way over another vessel, but in doubt whether the other vessel is taking sufficient action to avert a collision. (This signal is optional) | Siren or whistle | At least 5 short and rapid blasts |

*Note.* Strictly speaking the last signal only applies to one power vessel having the right of way over another. However, there are other Rules (20 and 25) which state that a power driven vessel, of less than 65 feet in length, or a sailing vessel must not hamper, in a narrow channel, the safe passage of a ship which can only navigate inside the channel. A liner or heavy tanker entering or leaving harbour may well be unable to alter course to avoid small craft in the channel and may use this signal to indicate that she is concerned, i.e. 'Get out of my way—I cannot get out of Yours'.

## SOUND SIGNALS—INTERNATIONAL CODE

These are not taken from the Regulations for the prevention of Collisions at sea, but from the International Code of Signals. They are included here for easy reference.

| Signal | Morse Letter | Meaning |
|---|---|---|
| S.O.S. | ···———··· | Distress Signal. |
| F | ··—· | I am disabled. Communicate with me. |
| K | —·— | You should stop your ship instantly. |
| L | ·—·· | You should stop. I have something important to communicate. |

| | | |
|---|---|---|
| O | – – – | Man overboard. |
| P | ·– –· | Your lights are out or burning badly. |
| R | ·–· | The way is off my ship. You may feel your way past me. (Could be used in fog.) |
| U | ··– | You are standing into danger. |
| V | ···– | I require assistance. |
| W | ·– – | I require medical assistance. |

These signals may be made by flags, morse, semaphore, or on a siren, whistle or foghorn in any weather.

## THE STEERING AND SAILING RULES (17–30)

Green to Green and Red to Red
Perfect safety, go Ahead,
If to starboard Red appear
'Tis your duty to keep clear.

A well known doggerel verse which summarises some of the main rules regarding power vessels. This section of the Regulations is, of course, the vital section. Those rules already covered about Lights, Shapes, and Sound Signals are designed to assist the mariner to comply with the main Steering and Sailing Rules which follow here.

They cover the actions to be taken by individual vessels approaching one another where a risk of collision may exist. They tell the seaman which vessel must give way under various circumstances and when the other vessel may be expected to take the appropriate avoiding action.

Some of the basic principles are:

(i) A vessel must establish first that, if she holds her present course and speed, a risk of collision with another vessel exists.

(ii) Power-Driven (Steam) vessels generally give way to Sail or boats under oars.

(iii) When two Sailing Vessels approach each other, the vessel which, owing to the wind direction, would lose more ground by an alteration of course has the right of way.

(iv) In all circumstances where a risk of collision exists, the vessel that has the right of way (sometimes called the 'privileged' vessel) should hold her course and speed whilst the other vessel (sometimes called the 'Giving Way' or 'Burdened' vessel) must alter course and/or speed. However, if at the last moment collision appears to be inevitable, both vessels must do their best, by altering course or speed, to avoid it if at all possible or at least lessen the impact.

(v) When obeying these rules, the action taken by the burdened vessel must be positive and taken in ample time.

(vi) An Overtaking vessel, whether power or sail driven, must always avoid the vessel being overtaken. This overrides all the other rules.

(vii) All vessels not engaged in fishing (except those 'Not under Command', engaged in special operations, etc.) must keep out of the way of vessels actually engaged in fishing. This does not, however, give fishing vessels the right to obstruct a fairway.

*Sailing Vessels*. Rule 17
This Rule is quoted here *in toto*:

'(a) When two sailing vessels are approaching one another, so as to involve risk of collision, one of them shall keep out of the way of the other as follows:

(i) When each has the wind on a different side, the vessel which has the wind on the Port side shall keep out of the way of the other.

(ii) When both have the wind on the same side, the vessel which is to Windward shall keep out of the way of the vessel which is to Leeward.

(b) For the purposes of this Rule, the windward side shall be deemed to be the side opposite to that on which the mainsail is carried or, in the case of a square-rigged vessel, the side opposite to that on which the largest fore-and-aft sail is carried.'

*Interpretation*
This rule was reworded in the 1960 Revision, and is now stated in much simpler form than it used to be. It is quite straightforward—Port Tack gives way to Starboard Tack when the boats have the wind on the opposite sides. The Windward boat gives way to the Leeward boat when they are both on the same tack.

Some examples of these rules, applied to particular situations,

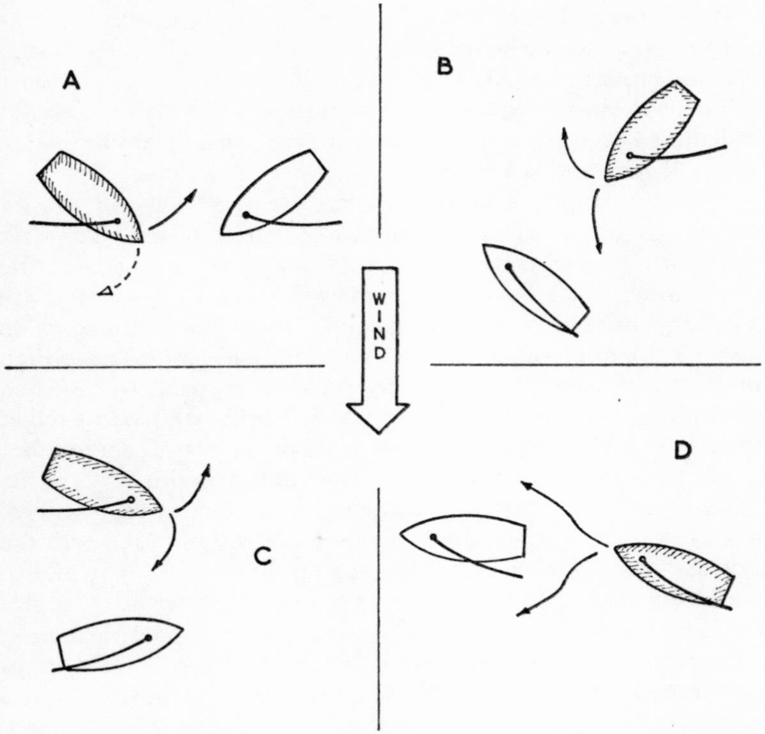


Figure 27

are given in Figures 27A. In all the diagrams, the 'burdened' or 'giving way' vessel is shown shaded.

The Overtaking Rule (Rule 24) is over-riding; in Example (d), the Leeward boat is overtaking the other from a relative position more than two points abaft the Windward boat's beam, and in this situation the Leeward boat has to give way. (The obligations of vessels overtaking others are further expanded in the section under Rule 24.)

*Power-Driven Vessels.* Rules 18 and 19
The basic rule is quite straightforward and is quoted here in full: 'When two power-driven vessels are crossing, so as to involve risk of collision, the vessel which has the other on her own starboard side shall keep out of the way of the other.'

In the normal course of events, the 'burdened' vessel, which has the other to starboard of her, will alter course to starboard to avoid crossing ahead of the other ship.

The end-on situation is usually more difficult; this is covered in Rule 18 and this is quoted in full here (apart from the paragraph covering seaplanes):

'When two power-driven vessels are meeting end on, or nearly end on, so as to involve risk of collision, *each* shall alter her course to starboard, so that each may pass on the port side of the other. This Rule only applies to cases where vessels are meeting end on, or nearly end on, in such a manner as to involve risk of collision, and does *not* apply to two vessels which must, if both keep on their respective courses, pass clear of each other. The only cases to which it does apply are when each of the two vessels is end on, or nearly end on, to the other; in other words to cases in which, by day, each vessel sees the masts of the other in a line, or nearly in a line, with her own; and by night, to cases in which each vessel is in such a position to see both the sidelights of the other. It does not apply to cases in which a vessel sees another ahead crossing her own course; or by night, to cases where the red light of one vessel is opposed to the red light of the other or where the green light of one vessel is opposed to the green light of the other or where a red light without a green light or a green light without a red light is seen ahead, or where both green and red lights are seen anywhere but ahead.'

*Interpretation*

The italics in the above text are mine, but see to what great (and almost exhausting lengths) the Rule goes in defining the 'end on' situation. This is because it is one of the trickiest for assessment of the correct action to be taken. There is also always the doubt as to what action the other ship is going to take in these circumstances.

Suppose, for instance, two ships or boats sight each other at comparatively long range very fine on each other's starboard bow. Each knows that if a true end on situation exists, each should alter course to starboard, viz. Figure 28A.

But if the other is not exactly dead ahead, the two vessels may pass clear starboard to starboard, viz. Figure 28B.

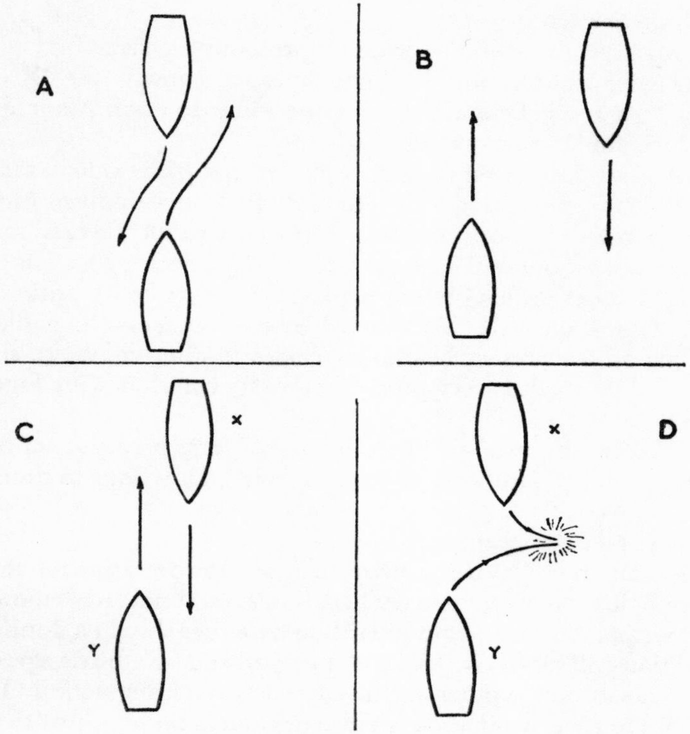


Figure 28

Dangerous doubts can arise if the situation is allowed to develop into a close range situation when neither is sure whether the other will hold her course with the hope of passing clear to starboard or alternatively is about to alter course thinking it to be too risky to hold her present heading (Figure 28c).

In ship 'X', the skipper is doubtful whether by continuing on his present course he will risk a head-on collision with 'Y'. The same thoughts may well be in 'Y's mind. For both of them at this late stage there are four possible alternatives:

(a) To hold their present course and speed in the hope that they will pass clear.

(b) To alter course to starboard.

(c) To alter course to port.

(d) To take the way off the ship by going astern.

Of these alternatives:

(a) May well be a vain hope and is probably too risky.

(b) Is a viable action if there is room enough for 'X' to cross 'Y's track before they get too close to each other and similarly for 'Y' to cross 'X's track.

(c) An *early* alteration to port on both ships' parts at long range would have prevented this situation from developing, but a late alteration to port on either ship's part would be extremely dangerous. Suppose 'X' at the last moment decided he was not going to clear on his present course and made an alteration to port. Simultaneously 'Y', worried by the closeness 'X' is likely to pass on his present heading, decides to alter course to starboard. The result is very probably a nasty collision. (See Figure 28D.)

(d) Taking the way off by going astern may be a way of preventing collision (or minimising its results) if it is done in time.

*Avoiding the Close Quarter Situation*

The main point to remember in any interpretation of these rules is that the close quarter situation should never be allowed to develop. An early, bold alteration by either ship in a doubtful situation will clarify it from the start and remove the dangerous elements of doubt inherent in late, hasty avoiding action. This applies to all combinations of circumstances and not just to the particular case quoted above.

A good indication whether a close quarter situation is likely to develop later is given by carefully watching the compass bearing of every approaching vessel. If the compass bearing does not change, and both vessels maintain their present courses and speeds, the two will collide. If the bearing is only changing slowly, the two craft may approach each other later at too close a range for comfort (the rate of change of bearing is, of course, a function of the range between the two ships). In the long range nearly head-on situation, the watching of the other ship's bearing can be the only reliable way of judging the relative motion of the two ships, particularly if the other ship is yawing in a seaway and it is difficult to determine its mean course.

At long range, the compass bearing is normally taken of the other ship's bridge or mast. At short ranges, it may be necessary to take the bearing of the right section of the other vessel; for example, if you are trying to cut under a super-tanker's stern, the bearing of its stern, and not its bridge, must be watched (cutting across such a ship's bows is in no way recommended—bearings or no).

*Power Driven Vessel's Meeting a Sailing Vessel.* Rule 20
'When a power driven vessel and a Sailing vessel are proceeding in such directions as to involve risk of collision, except as provided for in Rules 24 (the Overtaking Rule) and 26 (Vessels engaged in fishing) the power driven vessel shall keep out of the way of the Sailing Vessel.

This rule shall not give the Sailing vessel the right to hamper, in a narrow channel, the safe passage of a power driven vessel which can navigate only inside such channel.'

*Interpretation*
This rule is quite simple. Normally, sailing vessels have the right of way over power-driven vessels—except when they are overtaking, meet power driven vessels engaged in fishing, in a narrow channel, not under command, etc.

*Holding Course and Speed.* Rule 21
'Where by any of these rules one of two vessels is to keep out of the way, the other shall keep her course and speed. When, from any cause, the latter vessel finds herself so close that collision cannot be avoided by the action of the giving-way vessel alone, she also shall take such action as will best aid to avert collision (see Rules 27 and 29).'

*Interpretation*
It is important that the 'privileged' vessel holds her course and speed, otherwise early avoiding action by the 'burdened' ship may be nullified and the situation confused.

*Avoiding Crossing Ahead.* Rule 22
'Every vessel which is directed by these rules to keep out of the way of another vessel shall, so far as possible, take positive early action to comply with this obligation, and shall, if the circumstances of the case admit, avoid crossing ahead of the other.'

*Slackening Speed.* Rule 23

'Every power driven vessel which is directed by these rules to keep out of the way of another vessel shall, on approaching her, if necessary, slacken her speed, or stop, or reverse.'

*The Overtaking Rule.* Rule 24

'NOTWITHSTANDING ANYTHING CONTAINED IN THESE RULES, every vessel OVERTAKING any other shall keep out of the way of the overtaken vessel.

Every vessel coming up with another vessel from any direction more than 22½° (2 points) abaft her beam, i.e. in such a position with reference to the vessel she is overtaking that at night she would be unable to see either of that vessel's sidelights, shall be deemed to be an overtaking vessel; and no subsequent alteration of the bearing between the two vessels shall make the overtaking vessel a crossing vessel within the meaning of these Rules, or relieve her of the duty of keeping clear of the overtaken vessel until she is finally past and clear.

If the overtaking vessel cannot determine with certainty whether she is forward or abaft this direction from the other vessel, she shall assume that she is an overtaking vessel and keep out of the way.'

*Interpretation*

The capitals in this extract from the rules have been added for emphasis, as this is one of the most important regulations. Its application hinges on the relative direction of approach between the two vessels.

In Figure 29A, two power driven vessels are on converging courses but 'X' has just sighted 'Y', and 'Y's' bearing appears to be steady. A risk of collision must therefore exist.

'X' must keep clear of 'Y' in this case, as 'X' has the other on, or just before her beam on the starboard side. 'X's' best action would be to alter course to starboard to pass under 'Y's' stern, whilst 'Y' must hold her course and speed. Alternatively 'X' could reduce speed to allow 'Y' to pass unhindered ahead.

Suppose, however, a slightly different situation exists (Figure 29B) when 'X' is first aware of 'Y's' presence. The bearing is' again steady, but 'Y' is now approaching 'X' from well abaft

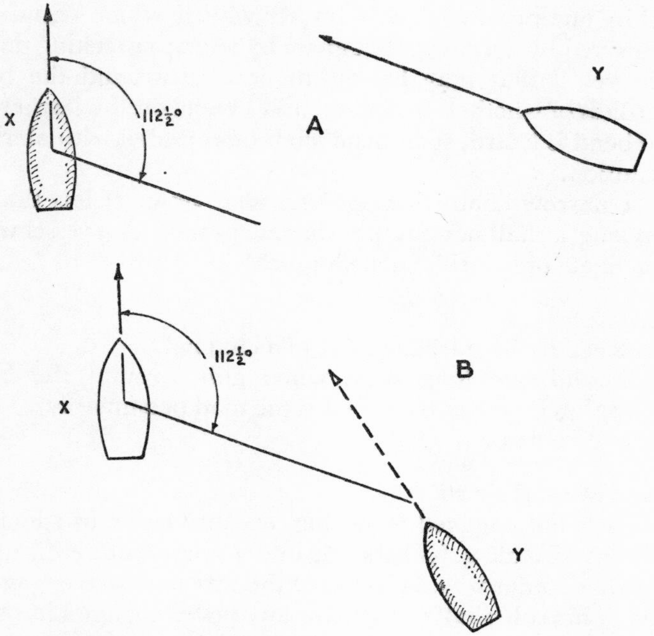


Figure 29

her beam; she must also be going faster than 'X', for otherwise her bearing as seen from 'X' would be drawing aft.

'Y' is now an overtaking vessel, being more than 2 points abaft 'X's' beam, and is obliged to alter course to avoid 'X' and keep out of 'X's' way until she is completely clear ahead. If 'Y' is doubtful when first seeing 'X' whether or not she is more than two points abaft 'X's' beam, she must assume that she is an overtaking vessel.

*Narrow Channels*. Rule 25

'In a narrow channel every power-driven vessel when proceeding along the course of the channel shall, when it is safe and practicable, keep to that side of the fairway or mid-channel which lies on the starboard side of such vessel.

Whenever a power-driven vessel is nearing a bend in a channel where a vessel approaching from the other direction cannot be seen, such power-driven vessel, when she shall have arrived within one half mile (½ mile) of the bend, shall give a

M

signal by one prolonged blast on her whistle which signal shall be answered by a similar blast given by any approaching power-driven vessel that may be within hearing around the bend. Regardless of whether an approaching vessel on the farther side of the bend is heard, such bend shall be rounded with alertness and caution.

In a narrow channel a power-driven vessel of less than 65 feet in length shall not hamper the safe passage of a vessel which can navigate only inside such channel.'

*Interpretation*

(a) Keep to the starboard side of a channel.

(b) Sound one long blast when going round the bend! (only applies to power boats—and the mad presumably).

(c) Self evident.

*Fishing Vessels*. Rule 26

'All vessels not engaged in fishing, except vessels to which the provisions of Rule 4 (vessels not under command, etc.) apply, shall, when under way, keep out of the way of vessels engaged in fishing. This rule shall not give to any vessel engaged in fishing the right of obstructing a fairway used by vessels other than fishing vessels.'

*Limitations of Craft*. Rule 27

'In obeying and construing these rules due regard shall be had to all dangers of navigation and collision, and to any special circumstances, including the limitations of the craft involved, which may render a departure from the above rules necessary in order to avoid immediate danger.'

*Neglect of Seamanlike Precautions*. Rule 29

'Noting in these Rules shall exonerate any vessel, or the owner, master or crew thereof, from the consequences of any neglect to carry lights or signals, or of any neglect to keep a proper look-out, or of the neglect of any precaution which may be required by the ordinary practice of seamen, or by the special circumstances of the case.'

*Interpretation*

So Watch Out!

*Local Harbour Regulations*. Rule 30

'Nothing in these rules shall interfere with the operation of a special rule duly made by local authority relative to the navigation of any harbour, river, lake, or inland water, including a reserved seaplane area.'

*Distress Signals*. Rule 31

'When any vessel . . . is in distress and requires assistance from other vessels or from the shore, the following shall be the signals to be used or displayed by her, either together or separately, namely:

(i) A gun or other explosive signal fired at intervals of about a minute. (Not normally a thing to hand!)

(ii) A continuous sounding with any fog-signalling apparatus.

(iii) Rockets or shells, throwing red stars fired one at a time at short intervals.

(iv) A signal made by radiotelegraphy or by any other signalling method consisting of the group ···–––··· (S.O.S.) in the Morse Code.

(v) The International Code signal of distress indicated by N.C.

(vi) A signal consisting of a square flag having above or below it a ball or anything resembling a ball.

(vii) A signal sent by radiotelephony consisting of the spoken word "Mayday".

(viii) Flames on the vessel (as from a burning tar barrel, oil barrel, etc.

(ix) A rocket parachute flare or a hand flare showing a red light.

(x) A smoke signal giving off a volume of orange-coloured smoke.

(xi) Slowly and repeatedly raising and lowering arms outstretched to each side.

The use of any of the foregoing signals, except for the purpose of indicating that a vessel or seaplane is in distress, and the use of any signals which may be confused with any of the above signals, is prohibited.'

*Comment*

Most yachts are unlikely platforms for the production of rockets or shells, and 'flames on the vessel' sounds as if that might be even more distressing, so perhaps most will be reduced to alternative (xi). Anyway, make like a seagull, and do your best.

*Summary of the Main Points of the Steering and Sailing Rules*

(1) Where power-driven vessels are concerned, give way to vessels on the starboard hand if a risk of collision exists. In most circumstances, the 'burdened' vessel should alter to starboard. The 'end on' situation; both vessels should alter to starboard if a risk of collision exists.

(2) Early action should be taken to prevent a close quarter situation developing. The vessel that has the right of way should maintain her course and speed unless collision becomes imminent, when she should also take any action that will avoid a collision.

(3) Power gives way to sail—and oars.

(4) With two sailing craft, the one that is on the Port Tack (wind on opposite sides) or to Windward (wind on same side) is obliged to give way.

(5) Overtaking vessels, whether under power or sail, must keep clear of the vessel being overtaken. This rule overrides all others.

(6) Vessels actually engaged in fishing must be avoided by other craft.

(7) Speed must be moderate in fog. Make correct sound signals.

(8) Small craft must not hamper large craft in a channel.

(9) Vessels not under command, or engaged in special operations, cannot be expected to keep clear of other craft and should themselves be avoided. These craft carry special signals by day and night.

(10) Keep a good look-out. Perhaps the most important of all.

# Appendix I: The Morse Code

This has been included as an 'Aide Memoire' in the Appendix as it is really the most important and widely used means of communication at sea—apart from the human voice. If not regularly used, I find that one's Morse is inclined to get a bit rusty and that a little 'revision' is required. So here it is:

| *Letters* | *Morse* | *Letter* | *Morse* |
|---|---|---|---|
| A | ·– | N | –· |
| B | –··· | O | ––– |
| C | –·–· | P | ·––· |
| D | –·· | Q | ––·– |
| E | · | R | ·–· |
| F | ··–· | S | ··· |
| G | ––· | T | – |
| H | ···· | U | ··– |
| I | ·· | V | ···– |
| J | ·––– | W | ·–– |
| K | –·– | X | –··– |
| L | ·–·· | Y | –·–– |
| M | –– | Z | ––·· |

| *Numbers* | *Morse* | *Special Signals* | *Morse* |
|---|---|---|---|
| 0 | ––––– | Call—A's | ·–·–·– |
| 1 | ·–––– | Answer—T | – |
| 2 | ··––– | Break—BT | –···– |
| 3 | ···–– | From—DE | –·· · |
| 4 | ····– | Repeat—IMI | ··––·· |
| 5 | ····· | Erase—E's | ······ |
| 6 | –···· | End—AR | ·–·–· |
| 7 | ––··· | Received and Understood—R | ·–· |
| 8 | –––·· | | |
| 9 | ––––· | | |

# Appendix 2: Great Circle Sailing

This is really only of interest to the sailor who plans to cover very long distances, but the subject was mentioned earlier in the book and a brief expansion of it may not be out of place here.

Over very long distances—across the Atlantic for example—there is a significant advantage in distance to be gained by following a Great Circle track, as opposed to a Rhumb Line track. The amount of distance saved depends on the D′ Lat. and D′ Long. of the departure point and destination. If one place is directly North or South of the other, there is no saving—nor is there any in an East/West direction near the Equator. With other combinations, particularly in high latitudes, the disparity between the two tracks may be quite large.

By far the simplest way of transferring the Great Circle track onto an ordinary navigational (Mercator) chart is to use a Great Circle Sailing chart or diagram. The latter consists of a blank Latitude/Longitude grid drawn on the Gnomonic projection (Admiralty Chart No. 5029 is an example of this type of diagram which can be bought at any chart agent). To use it, plot the departure point and destination on the diagram by Latitude and Longitude and draw a straight line between them. Then every 200 miles or so, prick off a point and write down its Latitude and Longitude. All that then needs to be done is to plot these intermediate points on the Mercator chart. The result is a series of chords (courses to steer) to the Great Circle curve on the navigational chart (Figure 30).

By all appearances, the Rhumb line AB looks the shortest distance between A and B, but this is not so. Due to the expanding Latitude scale, the curve A X Y Z B is in fact the shortest distance. Your initial Great Circle course would therefore be the direction AV and the final course ZB. There is only one final rule to remember. Great circles always curve TOWARDS the

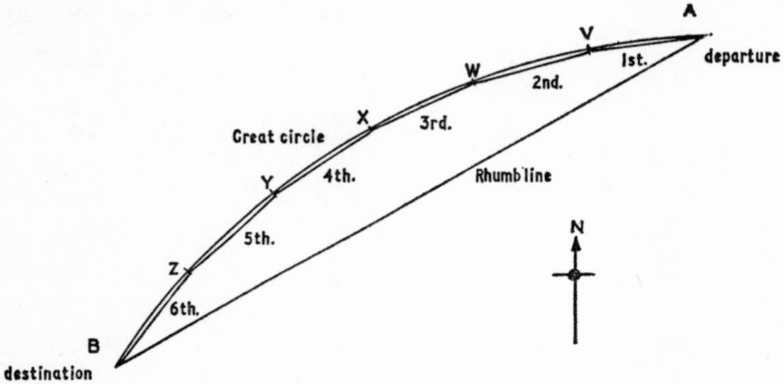


Figure 30

the Pole with their apex, and therefore in the Southern Hemisphere the circle will curve the opposite way to the one shown in the diagram above.

There are other methods of finding the initial and subsequent courses between two places and the Great Circle distance

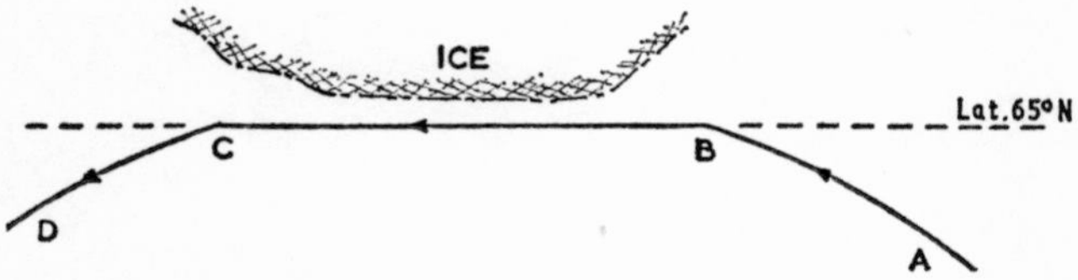


Figure 31

between them. Great Circle tables have been published in this country and elsewhere; enter these tables with the Latitude and Longitude of the departure point and destination, and the initial Great Circle course and distance (and subsequent courses) are tabulated directly.

Sometimes it is not possible to steer a complete Great Circle course—for example a Great Circle track between the North of

Scotland and Halifax would approach the ice pack in the North Atlantic at certain seasons of the year. A limiting Latitude is normally chosen to allow for such natural hazards and the Great Circle track modified to reach this Latitude, run along it, and then curve down again to the final destination (Figure 31).

# Index

Compiled by Gordon Robinson